In most
genomic,
inevitable
tion sequen
have becom
and the prob
multi-omics d

This book pr
long-lasting imp
the field, introduc
experimental data,

George Tseng com
genomics from the F
fessor of biostatistics,
at the University of Pit
computational method d

Debashis Ghosh is profess
Informatics at the Colorado
of Colorado Anschutz Medi
opment of the statistical me
platform used in cancer resear
methodology has been funded
He has published more than 16
book chapters in statistical and sc

Xianghong Jasmine Zhou complet
of Technology (ETH Zurich) and co
University. She is currently Director a
ogy and Bioinformatics program at
Dr. Zhou is the PI of the NIH center for
within the MAPGen consortium. She h
integrative genomics, addressing the "B
enormous amount of extremely diverse g
She was a recipient of several awards, inc.
and a NSF Career award.

INTEGRATING OMICS DATA

Edited by

GEORGE TSENG
University of Pittsburgh

DEBASHIS GHOSH
University of Colorado

XIANGHONG JASMINE ZHOU
University of Southern California

32 Avenue of the Americas, New York, NY 10013-2473, USA

Cambridge University Press is part of the University of Cambridge.

It furthers the University's mission by disseminating knowledge in the pursuit of education, learning, and research at the highest international levels of excellence.

www.cambridge.org
Information on this title: www.cambridge.org/9781107069114

© Cambridge University Press 2015

First published 2015

Printed in the United States of America

A catalog record for this publication is available from the British Library.

Library of Congress Cataloging in Publication Data
Tseng, George.
Integrating omics data / George Tseng, University of Pittsburgh, Debashis Ghosh, University of
Colorado, Xianghong Jasmine Zhou, University of Southern California.
pages   cm
ISBN 978-1-107-06911-4 (hardback)
1. Genomics – Statistical methods.   2. Meta-analysis.   I. Title.
QH438.4.S73T74   2015
572.8′6–dc23   2014048684

ISBN 978-1-107-06911-4 Hardback

Contents

Contributors   *page* vii

Introduction   1

Part A: Horizontal Meta-Analysis
  1. Meta-Analysis of Genome-Wide Association Studies:
      A Practical Guide   *Wei Chen*   9
  2. MetaOmics: Transcriptomic Meta-Analysis Methods for
      Biomarker Detection, Pathway Analysis and Other Exploratory
      Purposes   *SungHwan Kim, Zhiguang Huo, Yongseok Park, and
      George C. Tseng*   39
  3. Integrative Analysis of Many Biological Networks to Study
      Gene Regulation   *Wenyuan Li, Chao Dai, and Xianghong
      Jasmine Zhou*   68
  4. Network Integration of Genetically Regulated Gene Expression
      to Study Complex Diseases   *Zhidong Tu, Bin Zhang,
      and Jun Zhu*   88
  5. Integrative Analysis of Multiple ChIP-X Data Sets Using
      Correlation Motifs   *Hongkai Ji and Yingying Wei*   110

Part B: Vertical Integrative Analysis (General Methods)
  6. Identify Multi-Dimensional Modules from Diverse Cancer
      Genomics Data   *Shihua Zhang, Wenyuan Li, and Xianghong
      Jasmine Zhou*   135
  7. A Latent Variable Approach for Integrative Clustering of
      Multiple Genomic Data Types   *Ronglai Shen*   155
  8. Penalized Integrative Analysis of High-Dimensional Omics
      Data   *Jin Liu, Xingjie Shi, Jian Huang, and Shuangge Ma*   174

9. A Bayesian Graphical Model for Integrative Analysis of TCGA Data: BayesGraph for TCGA Integration *Yanxun Xu, Yitan Zhu, and Yuan Ji* 205

10. Bayesian Models for Flexible Integrative Analysis of Multi-Platform Genomics Data *Elizabeth J. McGuffey, Jeffrey S. Morris, Ganiraju C. Manyam, Raymond J. Carroll, and Veerabhadran Baladandayuthapani* 221

11. Exploratory Methods to Integrate Multisource Data *Eric F. Lock and Andrew B. Nobel* 242

Part C: Vertical Integrative Analysis (Methods Specialized to Particular Data Types)

12. eQTL and Directed Graphical Model *Wei Sun and Min Jin Ha* 271

13. MicroRNAs: Target Prediction and Involvement in Gene Regulatory Networks *Panayiotis V. Benos* 291

14. Integration of Cancer Omics Data into a Whole-Cell Pathway Model for Patient-Specific Interpretation *Charles Vaske, Sam Ng, Evan Paull, and Joshua Stuart* 310

15. Analyzing Combinations of Somatic Mutations in Cancer Genomes *Mark D. M. Leiserson and Benjamin J. Raphael* 337

16. A Mass-Action-Based Model for Gene Expression Regulation in Dynamic Systems *Guoshou Teo, Christine Vogel, Debashis Ghosh, Sinae Kim, and Hyungwon Choi* 362

17. From Transcription Factor Binding and Histone Modification to Gene Expression: Integrative Quantitative Models *Chao Cheng* 380

18. Data Integration on Noncoding RNA Studies *Zhou Du, Teng Fei, Myles Brown, X. Shirley Liu, and Yiwen Chen* 403

19. Drug-Pathway Association Analysis: Integration of High-Dimensional Transcriptional and Drug Sensitivity Profile *Cong Li, Can Yang, Greg Hather, Ray Liu, and Hongyu Zhao* 425

Index 445

Color plates follow page 134

Contributors

Veerabhadran Baladandayuthapani, Department of Biostatistics, UT MD Anderson Cancer Center, Houston, TX

Panayiotis V. Benos, Department of Computational and Systems Biology, University of Pittsburgh, Pittsburgh, PA

Myles Brown, Center for Functional Cancer Epigeneitcs, Dana-Farber Cancer Institute, Boston, MA

Raymond J. Carroll, Department of Statistics, Texas A&M University, College Station, TX

Wei Chen, Department of Pediatrics, University of Pittsburgh, Pittsburgh, PA

Yiwen Chen, Department of Bioinformatics and Computational Biology, Division of Quantitative Sciences, UT MD Anderson Cancer Center, Houston, TX

Chao Cheng, Department of Genetics, Geisel School of Medicine at Dartmouth, Hanover, NH; Institute for Quantitative Biomedical Sciences, Geisel School of Medicine at Dartmouth, Lebanon, NH; Norris Cotton Cancer Center, Geisel School of Medicine at Dartmouth, Lebanon, NH

Hyungwon Choi, Saw Swee Hock School of Public Health, National University of Singapore

Chao Dai, Molecular and Computational Biology, University of Southern California, Los Angeles, CA

Zhou Du, Howard Hughes Medical Institute, Program in Cellular and Molecular Medicine, Boston Children's Hospital, and Department of Genetics, Harvard Medical School, Boston, MA

Teng Fei, Center for Functional Cancer Epigeneitcs, Dana-Farber Cancer Institute, Boston, MA

Debashis Ghosh, Department of Biostatistics and Informatics, Colorado School of Public Health, University of Colorado Denver

Min Jin Ha, Department of Biostatistics, MD Anderson Cancer Center, Houston, TX

Greg Hather, Takeda Pharmaceuticals International Co., Cambridge, MA

Jian Huang, Department of Statistics and Actuarial Science, University of Iowa

Zhiguang Huo, Department of Biostatistics, University of Pittsburgh, Pittsburgh, PA

Hongkai Ji, Department of Biostatistics, Johns Hopkins Bloomberg School of Public Health, Baltimore, MD

Yuan Ji, Program of Computational Genomics and Medicine, NorthShore University HealthSystem; Department of Public Health Sciences, University of Chicago

Sinae Kim, Department of Biostatistics, School of Public Health, Rutgers University

SungHwan Kim, Department of Biostatistics, University of Pittsburgh, Pittsburgh, PA

Mark D. M. Leiserson, Department of Computer Science and Center for Computational Molecular Biology, Brown University, Providence, RI

Cong Li, Program in Computational Biology and Bioinformatics, Yale University, New Haven, CT

Wenyuan Li, Molecular and Computational Biology, University of Southern California, Los Angeles, CA

Jin Liu, Centre for Quantitative Medicine, Duke-NUS Graduate Medical School

Ray Liu, Takeda Pharmaceuticals International Co., Cambridge, MA

X. Shirley Liu, Center for Functional Cancer Epigeneitcs, Dana-Farber Cancer Institute, Boston, MA

Eric F. Lock, Division of Biostatistics, University of Minnesota, Minneapolis, MN

Shuangge Ma, Department of Biostatistics, Yale University School of Statistics; Capital University of Economics and Business, China

Ganiraju C. Manyam, Department of Bioinformatics and Computational Biology, UT MD Anderson Cancer Center, Houston, TX

Elizabeth J. McGuffey, Mathematics Department, United States Naval Academy, Annapolis, MD

Jeffrey S. Morris, Department of Biostatistics, UT MD Anderson Cancer Center, Houston, TX

Sam Ng, Department of Biomolecular Engineering, Center for Biomolecular Science and Engineering, University of California at Santa Cruz, Santa Cruz, CA

Andrew B. Nobel, Department of Statistics and Operations Research, University of North Carolina, Chapel Hill, NC

Yongseok Park, Department of Biostatistics, University of Pittsburgh, Pittsburgh, PA

Benjamin J. Raphael, Department of Computer Science and Center for Computational Molecular Biology, Brown University, Providence, RI

Ronglai Shen, Department of Epidemiology and Biostatistics, Memorial Sloan Kettering Cancer Center, New York, NY

Xingjie Shi, Department of Statistics, Nanjing University of Finance and Economics, China School of Statistics and Management, Shanghai University of Finance and Economics, China

Joshua Stuart, Department of Biomolecular Engineering, Center for Biomolecular Science and Engineering, University of California at Santa Cruz, Santa Cruz, CA

Wei Sun, Department of Biostatistics, Department of Genetics, University of North Carolina, Chapel Hill, Chapel Hill, NC

Guoshou Teo, Saw Swee Hock School of Public Health, National University of Singapore

George C. Tseng, Department of Biostatistics, University of Pittsburgh, Pittsburgh, PA

Zhidong Tu, Icahn Institute of Genomics and Multiscale Biology, Icahn School of Medicine at Mount Sinai, New York, NY; Department of Genetics and Genomic Sciences, Icahn School of Medicine at Mount Sinai, New York, NY

Charles Vaske, NantOmics, Culver City, CA

Christine Vogel, Center for Systems Biology, Department of Biology, New York University

Yingying Wei, Department of Statistics, The Chinese University of Hong Kong, Hong Kong

Yanxun Xu, Department of Applied Mathematics and Statistics, Johns Hopkins University

Can Yang, Department of Mathematics, Hong Kong Baptist University, Kowloon Tong, Hong Kong, China

Bin Zhang, Icahn Institute of Genomics and Multiscale Biology, Icahn School of Medicine at Mount Sinai, New York, NY; Department of Genetics and Genomic Sciences, Icahn School of Medicine at Mount Sinai, New York, NY

Shihua Zhang, National Center for Mathematics and Interdisciplinary Sciences, Academy of Mathematics and Systems Science, Chinese Academy of Sciences, Beijing

Hongyu Zhao, Department of Biostatistics, Yale School of Public Health, New Haven, CT; Program in Computational Biology and Bioinformatics, Yale University, New Haven, CT

Xianghong Jasmine Zhou, Molecular and Computational Biology, University of Southern California, Los Angeles, CA

Jun Zhu, Icahn Institute of Genomics and Multiscale Biology, Icahn School of Medicine at Mount Sinai, New York, NY; Department of Genetics and Genomic Sciences, Icahn School of Medicine at Mount Sinai, New York, NY

Yitan Zhu, Program of Computational Genomics and Medicine, NorthShore University HealthSystem

Introduction

In the past two decades, high-throughput experimental techniques such as mass spectrometry, microarrays, and next-generation sequencing have revolutionized biomedical research with abundant genome-scale data. The fruits of this research are paving the way toward precision medicine. "Ultra-big" data sets are routinely generated now that the cost of these experiments has greatly decreased. As a result, more and more of these data sets are made available in the public domain. This abundance of data permits us to study biological processes and disease mechanisms in a multifaceted manner, drawing insights from DNA variations (e.g., genotyping and mutation), RNA transcription (e.g., gene or isoform expression and fusion transcripts), gene regulation by epigenetic changes (e.g., methylation, protein–DNA interaction, and miRNA expression), and protein expression/modification.

The enormous scope of high-throughput results creates many statistical and computational obstacles to storing, analyzing, integrating, and interpreting the data. Generally speaking, the research community is pursuing two kinds of integrative studies: horizontal meta-analysis (data from different cohorts, often from different labs) and vertical multi-omics analysis (multiple experiments performed on the same cohort). Either of these may also integrate results from the growing pathway and pharmacogenetics databases. The vast range of available data and new biomedical questions that can be answered calls for research teams with multidisciplinary quantitative expertise, including in computer science, statistics, applied math, and machine learning. This edited book collects state-of-the-art computational and statistical methods recently developed in the booming field of omics data integration. Its purpose is to showcase a wide range of cutting-edge methods and tools for our readers, in hopes of inspiring new biological and methodological research techniques to advance the field.

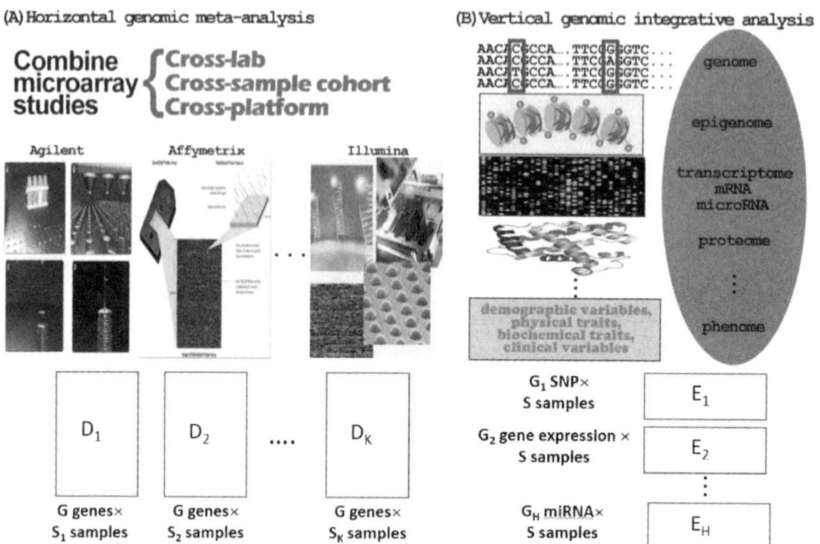

Figure I.1 (A) Horizontal omics meta-analysis. (B) Vertical multi-omics integrative analysis.

The microarray boom of the late 1990s introduced a now common convention for raw omics data: samples are arranged on the columns of the matrix, while gene features are on the rows (the main reason being that Microsoft Excel could only manage 256 columns at the time). This is in contrast to the traditional statistical convention to place samples on the rows, but in this book we keep the popular bioinformatic convention. Therefore, when multiple omics data sets from different labs are combined, the studies are integrated horizontally (Figure I.1A). In this context, many of the data integration problems now being published are analogous to traditional meta-analysis. This is why we name cross-cohort data integration "horizontal omics meta-analysis" in the preceding paragraphs and in Chapters 1–5. Alternatively, when multiple types of omics experiments are performed on the same cohort, the data sets are vertically aligned (Figure I.1B). The next set of chapters describes various types of "vertical multi-omics integrative analysis." Chapters 6–11 cover methods applicable to any type of omics data (e.g., clustering or dimension reduction methods not specific to the biological property or structure of the omics data). Chapters 12–19 cover methods that are specific to certain omics data types.

In the following we give an overview of the book's contents, based on the different biological purposes and quantitative techniques described herein.

Dimension reduction. Multi-omics data sets have naturally drawn attention to many dimension reduction methods. Chapter 2 introduces a variant of principal component analysis (MetaPCA), whereas Chapter 11 describes a method called *joint and individual variation explained* (JIVE). The first is designed for horizontal analysis, and the second is for vertical analysis. Chapter 6 proposes variations of the partial least squares (PLS) and nonnegative matrix factorization (NMF) methods, named *sparse multi-block partial least squares* (sMBPLS) regression and joint NMF. These methods reduce dimensionality and identify coherent modules in vertical multi-omics cancer data.

In addition, many published methods that have applied latent variable models and/or matrix factorization can also be considered dimension reduction techniques. The iCluster method in Chapter 7 and the iFad and iPad methods in Chapter 19 are examples of these.

Unsupervised analysis. An incrasingly popular type of analysis in omics data is to identify novel disease subtypes of clinical importance in a complex disease via unsupervised machine learning (also known as cluster analysis). Chapter 2 introduces the MetaSparseKmeans method for horizontal meta-clustering analysis. Chapter 7 proposes an iCluster method developed for multi-omics analysis. Chapter 11 develops a Bayesian consensus clustering (BCC) method that tracks both consensus and source-specific clustering in multi-omics data.

Integration with biological pathway information. Integration of omics data with public pathway databases sheds light on the key functional pathways associated with an underlying disease mechanism or other experimental perturbations. The methods described in Chapters 2, 15, and 19 include such pathway-based analyses. Chapter 2 uses the MetaPath algorithm to combine multiple transcriptomic studies for pathway analysis. Chapter 15 surveys different approaches to identifying significantly mutated pathways in cancer patients. Chapter 19 integrates transcriptome profiles, drug response profiles, and a pathway database to form drug-pathway associations.

Meta-analysis methods and the homogeneity/heterogeneity issue. Many horizontal omics meta-analysis problems have settings similar to traditional meta-analysis, but the new data structures and biological questions are inspiring novel developments. Chapters 1–5 cover this area. Chapter 1 describes new methods and practical guidelines for meta-analysis of genome-wide association studies (GWAS). An increasingly relevant problem in data integration is managing homogeneity and heterogeneity in the analysis (see Chapters 2, 5, 8, and 11). The adaptively weighted meta-analysis approach in Chapter 2 directly searches a feature-dependent subset of studies with concordant signals for horizontal meta-analysis. Chapter 5 applies the concept of motifs to handle exponentially increasing homogeneity/heterogeneity patterns in ChIP-chip and ChIP-seq meta-analysis. Chapter 8 provides homogeneity and heterogeneity regularization models for outcome association analysis. Chapter 11 develops a joint principal component analysis (PCA) framework that can separate the homogeneous and heterogeneous signals during dimension reduction and a Bayesian consensus clustering (BCC) method that tracks consensus and source-specific information in clustering formation.

Graphical model and network analysis. Graphical and network methods are powerful tools to model and elucidate associations, message flows, and gene regulation in biological systems. Many methods in this book make use of graphical and network models (e.g., Chapters 2, 3, 4, 9, 12, 13, 14, and 15). Chapter 2 describes the MetaDiffNetwork method for identifying recurrent network modules that are highly connected in one condition but altered in another condition across multiple transcriptomic studies. Chapter 3 reviews several novel graph mining algorithms to identify frequent and heavy subgraphs across a series of large weighted graphs and discover frequent coupled subgraphs in a series of two-layered graphs. Chapter 4 describes computational methods for modeling genetic information flow in networks and studying differential connectivity in co-expression networks. Chapter 9 proposes a graphical model using a Bayesian approach to study regulatory relationships of multi-omics data. Chapter 12 extends the expression quantitative trait loci (eQTL) analysis to directed graphical models. Chapter 13 discusses integrative methods for inferring miRNA regulatory networks. Chapter 14 presents a probabilistic graphical model that integrates diverse omics data to infer cancer patient-specific pathway activities, as well as a mathematical model to isolate important subnetworks. Chapter 15 contains network-based approaches to identify recurrent combinations of mutated genes in cancer genomes.

Bayesian modeling and inference. Hierarchical Bayesian models provide a natural solution to interpreting many multi-omics data structures and answering biological questions. The potential downsides of Bayesian analysis include arguable prior distribution specifications and the high computing cost of Monte Carlo simulations. Chapters 5, 9, 10, 11, and 19 contain examples of Bayesian approaches to data integration. Chapter 6 applies an EM algorithm to derive the posterior probabilities in the outcome association analysis. Chapter 9 adopts a Bayesian inference for a Markov random field model that can investigate multi-omics regulatory relationships. Chapter 10 proposes a multilayer Bayesian hierarchical model to integrate miRNA, copy number variation, methylation, mRNA expression, and clinical phenotype. Chapter 11 develops a Bayesian consensus clustering model using conjugate priors and Gibbs sampling for inference. Chapter 19 applies an advanced collapsed Gibbs sampling technique to speed up the posterior probability approximation.

Regularization and penalization methods. The techniques of feature regularization and penalization have gradually gained popularity in genomic research. This trend arises naturally because the high dimensionality of the models often diminishes their stability and obscures interpretation. Regularization methods "shrink" the effect sizes of the majority of features so that they provide zero contribution to the model, thereby achieving a model with limited dimensionality and good theoretical properties. The penalization and regularization methods are seen in Chapters 2, 6, 7, 8, 10, and 19. Chapter 2 applies regularization in the meta-analysis framework of sparse K-means when combining multiple transcriptomic studies to identify disease subtypes (the MetaSparseKmeans method). Chapter 6 applies network regularization in the joint NMF method. The iCluster method in Chapter 7 uses regularization in the latent variable model before it performs clustering analysis. Chapter 8 performs penalization and feature selection in the high-dimensional association models. The Bayesian hierarchical model in Chapter 10 incorporates ideas from the statistical regularization literature for combining multiple levels of omics data. Chapter 19 adopts regularization in the drug-pathway association analysis.

Data integration to study gene regulation. Gene regulation is a complex process, subject to multilevel controls. Much of the data integration effort has been devoted to deciphering the mechanisms and implications of gene regulation. Chapters 3, 4, 5, 9, 12, 13, 16, 17, and 18 contain computational and statistical methods to study various aspects of gene regulation. Chapter 3 presents methods that integrate many microarray or RNA-seq data sets to reconstruct transcriptional regulatory networks and splicing regulatory networks and explore how transcription and splicing simultaneously take place. Chapter 4 reviews several computational approaches that model the flow of genetic information to gene expression in biological networks. Chapter 5 develops a novel statistical framework for integrative analyses of ChIP-X data to improve peak calling and study allele-specific binding. Chapter 9 proposes a Bayesian graphical model to study regulatory relationships involving copy number variation, DNA methylation, and mRNA expression. Chapter 12 reviews methods to estimate directed graphical models with eQTL data. Chapter 13 focuses on predicting microRNA targets and microRNA regulatory networks. Chapter 16 discusses a model-based approach to quantitatively dissect the contributions of RNA-level and protein-level regulation in the variation in gene expression. Chapter 17 discusses statistical models to quantify the relationship between TF binding, histone modification, and gene expression. Chapter 18 presents some integrative analysis approaches to identify lncRNAs that are specific to cancer subtypes and predict those that are potential drivers of cancer progression.

PART A
HORIZONTAL META-ANALYSIS

1

Meta-Analysis of Genome-Wide Association Studies: A Practical Guide

WEI CHEN

Abstract

Meta-analysis is an effective approach to combining summary statistics across multiple studies. This approach has been widely used in recent genome-wide association studies (GWAS) and next-generation sequencing (NGS) studies. As a result, numerous disease-susceptibility loci, which cannot be found in a singe GWAS, have been identified through the meta-analysis of multiple studies. In this chapter, we give an overview how meta-analysis techniques can be used in consortium projects and provide guidance for future studies. Sections 1.1.1 and 1.1.2 cover background information on GWAS and imputation techniques, which play a key role in the meta-analysis of multiple studies. Section 1.2.1 discusses the methods of meta-analysis for single variant tests and provides a basic workflow of meta-analysis in a typical consortium project. Section 1.2.2 presents an application of Section 1.2.1 from a meta-analysis of age-related macular degeneration (AMD). Next, Section 1.2.3 discusses a method for meta-analysis for a gene-level test. Section 1.2.4 presents an application of Section 1.2.3 from a meta-analysis of plasma lipid levels. Section 1.2.5 provides a discussion of popular software for meta-analysis of genetic studies. Finally, Section 1.3 closes the chapter and discusses future directions.

1.1 Introduction

1.1.1 Meta-Analysis of Genome-Wide Association Studies

In the past decade, new technologies have enabled researchers to examine genetic and genomic data on a whole-genome scale. A genome-wide association study (GWAS) is known as a popular design for assessing thousands to millions of common and rare genetic variants associated with a disease or a trait. Thousands of disease-susceptible variants have been discovered through the GWAS of hundreds or thousands of individuals [1, 2]. To summarize the findings of these studies, the National Human Genome Research Institute has

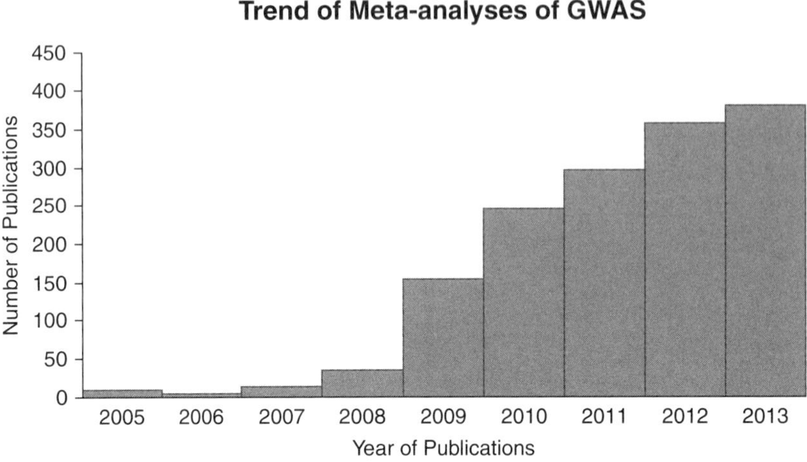

Figure 1.1 Number of publications by year from 2005 to 2013.

organized a catalog of published genome-wide association studies with frequent updates (http://www.genome.gov/gwastudies/). However, single-center GWAS typically has a limited number of samples, thus the power to detect those variants is small, especially for variants with small to modest effect sizes, as observed in many complex diseases. Although it is ideal to combine genetic data for as many individuals as possible, local institutional review board policy makes sharing of individual-level data from each study site difficult or impossible. In such situations, meta-analysis becomes a popular and powerful approach for combining summary statistics from multiple GWAS by increasing the sample size without sharing individual-level data. Multiple consortia have been founded to exchange and merge genetic data sets from multiple sites, with the central goal of identifying more disease-susceptibility loci [3]. Figure 1.1 illustrates the rapidly increase in the number of publications in PubMed using search terms "meta analysis" and "GWAS" from 2005 to 2013. If we focus on one disease or trait, we see a greatly increased number of participating studies and total sample sizes.

1.1.2 Imputation

One technical difficulty in combining different studies comprises the various genotyping platforms, which differ in density, position, and genotyping accuracy. Consequently, complete summary statistics of only a small set of

Single Nucleotide Polymorphisms (SNPs) would be available from each contributing study, which might reduce the power to detect causal variants. Statistical imputation techniques have been successful in inferring genotypes [4–8]. The rationale is to infer genotypes that were not directly genotyped in the GWAS (target) but can be "imputed" using external reference panels (e.g., HapMap or 1000 Genomes project data sets) under the assumption that the linkage disequilibrium (LD) pattern of the reference panel is similar to the target study subjects. Imputation not only increases the genomic coverage of the target study but also provides an effective way to harmonize marker sets and their association results across multiple studies [6]. Multiple computationally efficient programs (e.g., MACH[9], IMPUTE[7], and BEAGLE[10]) have been made available in the community. Imputed quality is usually assessed by program-specific metrics. For example, the correlation r^2 between imputed and expected genotypes across all samples is often used as an empirical filter. SNPs with r^2 less than a certain threshold value (e.g., 0.5) will be excluded before the data are merged to avoid a severe impact of imputation uncertainty. Comprehensive surveys can be found elsewhere [4, 6, 11].

1.1.3 Outline

There are several excellent review papers from experts in this area [12–15]. We will reemphasize the key steps in a meta-analysis using real examples in a tutorial manner and add newly developed methods for gene-based rare variant tests. We have organized the chapter as follows. In the first part, we describe a standard workflow for the meta-analysis of GWAS for a single-marker test. In the second part, we present a statistical method recently developed to perform gene-based association tests and conditional analyses for rare variants. We illustrate the methods using two real examples from large-scale consortium projects.

1.2 Methods and Applications

1.2.1 Methods for Meta-Analysis of GWAS for Single Marker Test

Previous review papers on the meta-analysis of GWAS provide comprehensive discussions on multiple aspects [12–15]. In this chapter, we give more practical and detailed guidance to readers based on our experiences and lessons from completed meta-analysis projects to complement previous reviews. We hope the readers of this chapter will learn basic concepts and techniques of meta-analysis of GWAS to perform their own analyses.

Table 1.1 *Workflow of meta-analysis of GWAS*

Stage 1: Project preparation
a. Form a consortium
b. Set up committees (e.g., steering, data management, and analysis)
c. Design an analysis plan and circulate to all participants

Stage 2: Data cleaning and freeze
a. Collect summary statistics from each study (upload to a central data repository)
b. Quality check (imputation quality, strand issue, frequency comparison, etc.)
c. Re-collect summary statistics from the studies that have corrections
d. Generate a cleaned data set for each study and freeze all data sets

Stage 3: Primary meta-analysis
a. Combine summary statistics (e.g., fixed effects model)
b. Assess heterogeneity (e.g., Q and I^2 statistics)
c. Cross-validate among independent analysis groups

Stage 4: Replication study and additional analysis
a. Pick top signals using a predefined threshold
b. Send out for replication in independent cohorts
c. Combine all data

Stage 5: Manuscript writing and plans for future work
a. Form a writing team and summarize author contributions
b. Draft the main consortium paper and other companion papers
c. Propose plans for future analysis

Table 1.1 presents a typical workflow for a GWAS meta-analysis from a common consortium including multiple studies. We list several key steps and elaborate a few critical points. Some stages can be simplified for a relatively small meta-analysis (e.g., two or three studies in total).

A detailed description of each step is beyond the scope of this chapter. Here we focus on a few important items that are critical in practice according to our experience. First, we provide an example statistical analysis plan, which should be distributed to study partners prior to the meta-analysis. Readers can use it as a reference or modify it to fit their own purposes. A tip here is that a detailed and clear analysis plan will save significant effort in correcting human mistakes in data preparation and will increase collaboration efficiency.

Sample Statistical Analysis Plan

1. Fill out one copy of the "Study_Descriptives_Template" for each study and submit the completed file to XX, XX (XX@XX.edu).

2. Please have the member of your group who is uploading the data contact XX (XX@XX.edu) for instructions and your username/password.

3. Perform imputation and GWAS analyses as described in the agreed "Meta_Analysis_Plan" document.

3.1. Perform the analysis for each SNP under an additive model and keep *at least* four digits after the decimal place for all statistics (the use of more precision is encouraged). For *p*-values, keep at least four significant digits (e.g., *p*-values of 0.0000 are not useful but *p*-values of 1.234e-20 are OK).

3.2. Follow our definitions for Effect_allele (allele corresponding to change in betas) and Other_allele (the noneffect allele). Thus, for an additive model, a positive effect size estimate will indicate that Effect_allele is more common in cases than controls.

(Note: We do not specify which allele should be the effect allele – you may choose the effect allele for each SNP in whatever way best suits you (or your analysis software), but please do report your choice of Effect_allele and Other_allele as requested below.)

3.3. Provide the EAF (effect allele frequency) instead of the minor allele frequency for each SNP.

4. Provide the results of each analysis in a separate file, in the format listed under "Results file format," and named according to the scheme described in "File naming scheme" (see below for both). Following the requested format and naming scheme for your results will greatly assist us in collecting and processing the data from many different groups while minimizing errors.

5. Provide information describing your method of imputation and the quality control metrics you used. This needs to be provided for each of the data sets you imputed separately (e.g., particularly if cases and controls were imputed separately).

6. Please provide association results in tab-delimited plain text files, including a single header line with the following columns in the order exemplified in Table 1.2.

Note: All numeric data can be specified in either scientific or decimal notation and should be specified with at least four decimal places. Integer data should be supplied as a single integer number with no decimal point. Please code missing values in any column as a single period character ("."). No quotes should be used for any data cells or headers. No row indices column or any other extra columns should be provided.

Quality Control

Because multiple groups are participating in a meta-analysis, quality checking is a critical step to the downstream analysis. According to our experience, errors can occur in multiple steps, and a solid plan for quality control is required to minimize the impact of multiple error sources. We describe our procedure step

Table 1.2 *Explanation of column names in result file*

Column name	Description	Data format
MarkerName	dbSNP ID of the marker	Character string
Strand	strand on which the alleles are reported. Typically should be "+" for every SNP.	Single character "+" or "−"
N	number of subjects analyzed	positive integer[a]
N_Cases	number of cases analyzed	positive integer[a]
N_Controls	number of controls analyzed	positive integer[a]
Effect_allele	the allele associated with phenotypic traits (corresponding to change in betas, not necessary to be the risk allele)	a single uppercase character "A" "C" "G" or "T"
Other_allele	indicating the other (non-effect) allele	a single uppercase character "A" "C" "G" or "T"
EAF	effect allele frequency (range 0–1)	Numeric data[b]
EAF_cases	estimated frequency of the effect allele in cases	Numeric data[b]
EAF_controls	estimated frequency of the effect allele in controls	Numeric data[b]
Information_type	a code indicating the type of data in the "Information" column: 0 = if the SNP was not tested using imputation/genotyping uncertainty, in which case the following column should be missing (e.g., for directly genotyped SNPs), 1 = if the following column contains "r2_Hat" from MACH, 2 = if the following column contains "proper_info" from IMPUTE	0, 1, or 2
Information	a value (range 0–1) corresponding to the information content output from the association testing (corresponding to the data type specified in the "Information_type" column above)	Numeric data[b]
BETA	the regression coefficient indicating change per effect allele. If no regression coefficient is available, please provide a numeric value that indicates whether the allele was associated with increased or decreased trait values.	Numeric data[b]
SE	the standard error of "BETA" above	Numeric data[b]
P	the two-sided *p*-value for the association (not adjusted for genomic control)	Numeric data[b]

Note: Please code missing values in any column as a single period character ("."). No quotes should be used around any data cells or headers. No row indices column or any other extra columns should be provided.

[a] All numeric data can be specified in either scientific or decimal notation and should be specified with at least four decimal places.

[b] Integer data should be supplied as a single integer number with no decimal point.

Table 1.3 *An example of result file submitted to the AMD consortium*

Marker Name	Strand N	N_ case	N_ ctrls	Effect_ Al	Other_ Al	EAF	Info_ Type	Info	BETA	SE	P
rs11111	+1900	1000	900	A	C	0.3255	2	0.8901	0.0302	0.0036	0.001234
rs22222	+2000	1000	1000	T	G	0.5891	2	0.9301	0.0302	0.0036	0.1234
rs33333	+2000	1500	500	G	T	0.1000	2	0.3000	0.0000	0.0590	0.5000

by step. This is typically done after all results have been uploaded to a central database from participant studies.

1. *Check file format/integrity*. First, we check the completeness of data and analyses. We need to check if each column is present, if each file has a correct column header and is in accordance with the requested result file format (see Table 1.3), and if marker names overlap with the imputation panel and all chromosomes are available.
2. *Compare reference allele frequency of controls with allele frequencies of imputation reference panel*. The most common mistake in the meta-analysis of GWAS is the mislabeling of reference alleles. In practice, the reference allele can be any type among minor alleles, risk alleles, and random alleles. We need to check the consistency between allele labels and reported frequencies. A very effective approach is to check pairwise allele frequencies. Figure 1.2 shows the pairwise comparison of allele frequencies after matching allele frequencies by allele labels. The upper panel shows the expected patterns from the comparisons between different populations (data generated from the 1000 Genomes Project). Populations that have less genetic similarity tend to have worse concordance of allele frequencies. The bottom panel shows three common mistakes: (1) reported allele frequencies for alternative alleles or the columns of reference and alternative alleles are swapped; (2) a subset of SNPs has swapped allele labels or frequencies, and (3) one study only reports minor allele frequencies in the result file. The best solution is to confirm your suspicions with the study group and ask for corrected files.
3. *Perform SNP-wise quality control*. Once we rule out any systematic errors in the reported files, we apply a series of filters for quality checking (Table 1.4). Any SNP that fails the quality check will be removed or masked prior to final meta-analysis.
4. *Report per study summary statistics of SNPs that passed the central quality control (see Table 1.4). If available, check positive controls (known association signals) as an optional step*. For certain diseases (e.g., macular

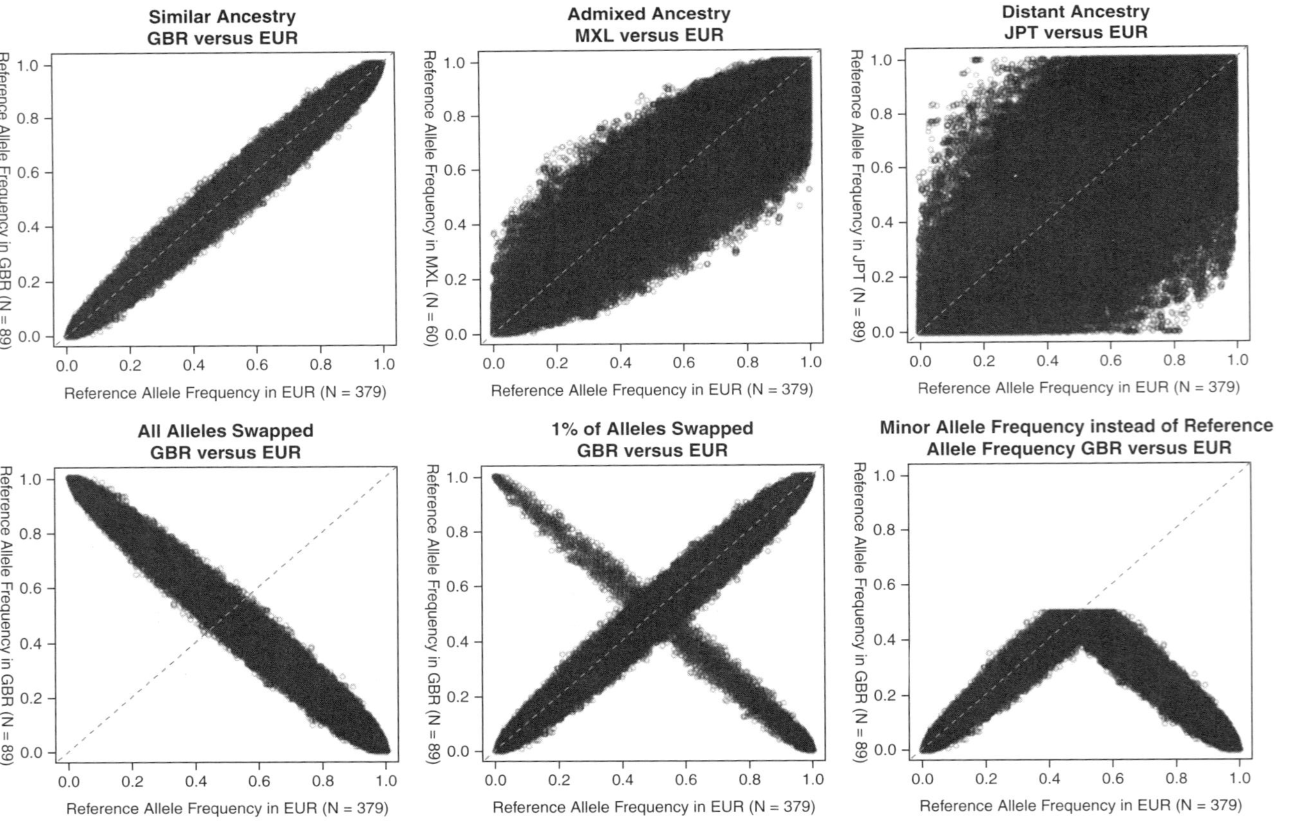

Figure 1.2 Comparison of allele frequencies between different populations.

Table 1.4 *Checklist of SNP-wise quality control*

Check	Thresholds
Total number of SNPs	should be comparable between studies because of imputation reference panel
Strand orientation	will have to be flipped to "+" for meta-analysis
Low imputation quality	filter settings depending on software
Duplicated SNPs	remove
Monomorphic variants	remove
Extreme effect sizes	$\mid BETA \mid > 5$
Extreme standard errors	$(> 5, <= 0)$
SNPs with extreme p-values	$p = 0$
SNPs with missing values	
SNPs that passed QC	small samples will have fewer SNPs

degeneration, type I diabetes) that have known markers with strong effect sizes, we can check those markers and flag the study if, for example, all or most markers are insignificant.

5. *Summarize quality checking and available studies and sample sizes and report back to the consortium.* After extensive data cleaning, a central data repository will store all cleaned data. Analysis groups (sometimes different from the data cleaning groups) will have access to the repository and perform meta-analysis according to the statistical plan.

Overview of Meta-Analysis Methods for Common Variants

After the quality checking step, a complete data freeze will be available to analysis groups to perform the core meta-analysis. User-friendly software packages have made meta-analysis relatively straightforward for analysts. However, each program provides different method options, and users are often confused about which method to use or do not know how to interpret the output. Understanding some basic concepts will greatly help the analysis and will aid in avoiding potential mistakes. We give a brief review of basic concepts and popular methods. Meta-analysis has been popular in various topics of medical research. The rationale is to combine and contrast summary results from multiple studies to characterize their similarities and differences. The null hypothesis (underlying truth) varies among different applications. In summary, we would like to answer if nonzero effect sizes exist in (1) all studies, (2) at least one study, and (3) at least r studies. More details can be found in Chapter 2. For GWAS, we typically focus on the first assumption that there is a common nonzero effect size with or without variation in all studies. We focus on the methods for this application in this chapter.

We briefly describe the theory and math for meta-analysis. A comprehensive description can be found in classic books about meta-analysis [16–18]. Assume that there are m studies in the final meta-analysis. Let N_i, β_i, and σ_i be the sample size, the effect size (e.g., log odds ratio (OR) for a binary phenotype or regression coefficient for a continuous phenotype), and its standard error of the ith study for a single SNP, where $i = 1, 2, \ldots m$.

We describe several conventional methods in the following. The simplest type of method based on p-values from each study is called *Fisher's method:* $T_{fisher} = -2 \times \sum_{i=1}^{k} \log(p_i) \sim \chi^2_{2k}$, or the closely related approach *Stouffer's Z-score method*, $T_z = \sum_{i=1}^{K} (\Phi^{-1}(P_i)/\sqrt{K}) \sim N(0, 1)$, where p_i is the association p-value from the ith study and $\Phi^{-1}(x)$ is the inverse of normal cumulative distribution function.

The second type of method is called the *fixed effects model (FE)*. This model assumes that each study has the same true effect size and combines the effect size under a weighted scheme. The most popular weight is based on the standard error of effect size:

$$Z = \sum_i w_i \times Z_i \Big/ \sqrt{\sum_i w_i^2} \quad w_i = \sqrt{N_i} \, p = 2 \times \Phi(-|Z|)$$

The combined effect size can be estimated by the weighted mean of an individual estimate, $\hat{\beta} = \sum_i w_i \times \beta_i / \sum_i w_i$, $var(\hat{\beta}) = 1 / \sum_i w_i$, $w_i = 1/\sigma_i^2$, where σ_i^2 is the standard error of effect size in the ith study. If the inverse variance is unavailable, weight is often assigned as the square root of effect sample size $\sqrt{N_{eff}}$, which is the sample size for a continuous trait, or is calculated by $N_{eff} = c \times N_{case} \times N_{control}/(N_{case} + N_{control})$ for a binary trait, where the scale parameter c has no impact on the calculation. If some studies are family based, the effective sample size cannot be theoretically determined and is often assigned with an empirical number (e.g., the number of families). The fixed effects model is the most popular method in practice.

Heterogeneity

If the true effect size varies among studies, we need to take into account such variation in our models. We can decompose the observed variation among the estimated effect sizes into true variation and random error. The most common method is called the *random effects model*, which models between-study variations τ^2 in the modified weight: $w_i^R = 1/(\sigma_i^2 + \tau^2)$. To assess the heterogeneity, two statistics are popular in practice: Cochran's Q statistic, $Q = \sum_{i=1}^{k} w_i \times (\beta_i - \hat{\beta})^2$, follows a χ^2 distribution with the degree of freedom $k - 1$, and the I^2 statistic, $I^2 = (Q - df)/Q \times 100\% = (\tau^2/(\tau^2 + var(\hat{\beta}))) \times 100\%$, which is the proportion of the observed variance that is a real variation of effect size.

We present a fictive example in Table 1.5. There are two studies, one with 500 cases and 500 controls and the other with 250 cases and 250 controls. The same five SNPs are reported in both studies. Given the reported effect size (BETA) and standard error (SE), we combined the two studies using a fixed effects model with the inverse variance scheme described earlier. The first two SNPs have opposite directions, while the other three have the same direction. The SNP3 has a combined p-value 3×10^{-8}, which implies that this SNP is interesting and merits further validation.

Although the standard methods described earlier have been widely used in practice, their limitations in the context of GWAS have also been discussed and novel methods have been proposed. We briefly describe a few models that are useful in special settings.

Modified Random Effects Model (RE)

Han et al. [19] observed that the random effects model is typically less powerful than the fixed effects model in practice. They presented a modified random effects approach that uses an alternative null hypothesis. They showed that this method is more powerful than FE when heterogeneity presents in the meta-analysis of GWAS.

Methods for Overlapping Subjects

In reality, when multiple studies are combined, it is unavoidable to have overlapping or related subjects in two or more studies. The reasons, for example, include duplicate visits in different clinical sites, use of pubic controls, and use of common cases in different contributed studies. Ignoring the issue of overlapping subjects will greatly inflate type I error and reduce the power of detecting causal variants [20–22]. Several approaches have been adapted in practice. For example, overlapping subjects are often split among studies or are removed in one study but kept in the other, with the result that each subject is only calculated once. Lin et al. proposed a method to explicitly model the correlation between studies with application to both individual-level and summary data [20]. They showed that the method is superior to the splitting approach in simulated and real data sets.

Table 1.5 *Fictive example for a meta-analysis of two studies (Study 1: 500 cases, 500 controls; Study 2: 250 cases, 250 controls)*

| | Study 1 | | | | | Study 2 | | | | | Studies 1 and 2 combined | | | |
| | EAF | | | | | EAF | | | | | | | | |
Marker	Cases	Controls	BETA	SE	P value	Cases	Controls	BETA	SE	P value	BETA	SE	P value	Direction
SNP1	0.201	0.201	0.000	0.112	1.00	0.204	0.202	0.012	0.157	0.94	0.004	0.091	0.96	−+
SNP2	0.100	0.101	−0.011	0.149	0.94	0.102	0.102	0.000	0.209	1.00	−0.007	0.121	0.95	−+
SNP3	0.248	0.164	0.520	0.114	4.8E-06	0.254	0.172	0.497	0.159	1.7E-03	0.512	0.092	3.0E-08	++
SNP4	0.449	0.447	0.008	0.090	0.93	0.456	0.442	0.057	0.127	0.66	0.024	0.073	0.74	++
SNP5	0.300	0.298	0.010	0.098	0.92	0.298	0.296	0.010	0.139	0.94	0.010	0.080	0.90	++

Note: SNP3 reached genome-wide significance after meta-analyzing Study 1 and Study 2. EAF, effect allele frequency; BETA, effect estimate; SE, standard error.

Method:

– Per study: summary statistics from logistic regression models (performed in R).

– Metal/classical approach, uses effect size estimates and standard errors.

Summary and Individual Data

If all raw data are available, the intuitive way to analyze the data is to pull all data together and perform a single analysis (mega-analysis), which can apply a uniform criterion and plan for quality checking and analysis. However, Lin et al. showed that meta-analysis is as efficient as mega-analysis if the analysis in each study is performed properly and follows similar guidance on quality checking and analysis [21]. These results are very encouraging and useful to the consortium where individual study data are difficult to obtain but a consensus statistical plan is enforced in all studies.

Interpretation of Meta-Analysis Results

The results from a meta-analysis are often difficult to interpret because of potential heterogeneity. Han et al. interpreted meta-analyses of genome-wide association studies [23]. They proposed a new approach to differentiate different types of studies.

1.2.2 Application 1: Meta-Analysis of AMD GWAS

So far, we have illustrated the basic concept and workflow of meta-analyses. Here, we present a real example from the AMDGene Consortium [3]. AMD is a leading cause of blindness in elderly people. As one of the most successful examples of genome-wide association studies of complex diseases, numerous loci have been identified to be associated with AMD susceptibly. In recent years, much progress has been made in identifying genetic contributors to AMD susceptibility. Linkage and association studies have identified a series of susceptibility loci, from a complement factor pathway and other notable genes [24–33]. Two large independent GWAS further expanded the gene list to *TIMP3, LIPC, and CETP*; the last two are known to be associated with high-density lipoprotein cholesterol (HDL) [34, 35]. In 2010, the AMD Gene Consortium of 15 international research groups was formed to accelerate the genetic research in AMD. A meta-analysis of multiple AMD GWAS with more than 7,600 cases and more than 50,000 controls identified 19 significant loci, including 7 novel ones [3] (Figure 1.3). This study was to that date the largest genetic study of AMD ever conducted. We participated in this study and went through all steps that we described earlier in this chapter. We applied a fixed effects model to this analysis using an inverse variance weighting scheme. Table 1.6 is a summary of our meta-analysis. There are two stages of analysis. In the first stage, the discovery study, we performed a meta-analysis on 15 available AMD GWAS. In the second stage, the follow-up study, we picked

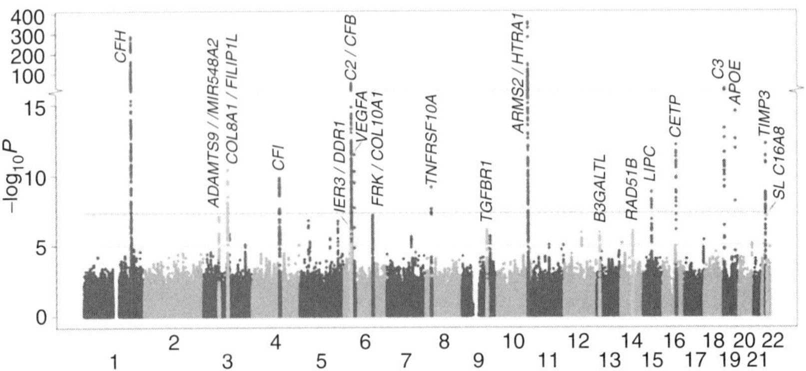

Figure 1.3 Meta-analysis of age-related macular degeneration.

the top 40 independent SNPs ($P < 10^{-5}$) and genotyped them into multiple independent cohorts that did not overlap with the 15 GWAS. Then, we applied a fixed effects model again to combine the results from discovery and follow-up studies. We presented our conclusions based on the joint results. For example, a hit in gene *B3GALTL* that reached p-value $= 2 \times 10^{-6}$ in the discovery study was observed with $p = 0.002$ in the follow-up study and, in the joint analysis, surpassed the threshold for genome-wide significance ($P < 5 \times 10^{-8}$).

1.2.3 Meta-Analysis of Gene-Level Tests of Rare Variant Associations

We have focused on the meta-analysis of GWAS by combining single markers. Current GWAS provide a powerful platform for assessing the association between common variants and complex traits. However, because of the low linkage disequilibrium between common and rare variants, they can be still underpowered for detecting rare variant associations [36]. Compared to common variants, rare variants are more likely to be functional, and therefore it can be easier to interpret the identified associations. Recent advances in exome sequencing and the development of exome genotyping arrays are enabling explorations of the very large reservoir of rare coding variants in humans and are expected to accelerate the pace of discovery in human genetics [37].

Testing rare variants individually can be underpowered, particularly when each rare variant only has moderate effect sizes. Alternatively, rare variants can be examined using association tests that group alleles in a gene or other functional unit. Compared to tests of individual alleles, this grouping can increase power, especially when applied to large samples where several rare variants are observed in the same functional unit [36]. The simplest rare variant

Table 1.6 *Summary of loci with associations reaching genome-wide significance*

SNP/risk allele	Chr.	Position	Nearby genes	EAF	Discovery		Follow-up		Joint		
					P	OR	P	OR	P	OR	95% CI
Loci previously reported with $p < 5 \times 10^{-8}$											
rs10490924/T	10	124.2 Mb	*ARMS2*	0.30	4×10^{-353}	2.71	2.8×10^{-190}	2.88	4×10^{-540}	2.76	[2.72–2.80]
rs10737680/A	1	196.7 Mb	*CFH*	0.64	1×10^{-283}	2.40	2.7×10^{-152}	2.50	1×10^{-434}	2.43	[2.39–2.47]
rs429608/G	6	31.9 Mb	*C2–CFB*	0.86	2×10^{-54}	1.67	2.4×10^{-37}	1.89	4×10^{-89}	1.74	[1.68–1.79]
rs2230199/C	19	6.7 Mb	*C3*	0.20	2×10^{-26}	1.46	3.4×10^{-17}	1.37	1×10^{-41}	1.42	[1.37–1.47]
rs5749482/G	22	33.1 Mb	*TIMP3*	0.74	6×10^{-13}	1.25	9.7×10^{-17}	1.45	2×10^{-26}	1.31	[1.26–1.36]
rs4420638/A	19	45.4 Mb	*APOE*	0.83	3×10^{-15}	1.34	4.2×10^{-7}	1.25	2×10^{-20}	1.30	[1.24–1.36]
rs1864163/G	16	57 Mb	*CETP*	0.76	8×10^{-13}	1.25	8.7×10^{-5}	1.17	7×10^{-16}	1.22	[1.17–1.27]
rs943080/T	6	43.8 Mb	*VEGFA*	0.51	4×10^{-12}	1.18	1.6×10^{-5}	1.12	9×10^{-16}	1.15	[1.12–1.18]
rs13278062/T	8	23.1 Mb	*TNFRSF10A*	0.48	7×10^{-10}	1.17	6.4×10^{-7}	1.14	3×10^{-15}	1.15	[1.12–1.19]
rs920915/C	15	58.7 Mb	*LIPC*	0.48	2×10^{-9}	1.14	0.004	1.10	3×10^{-11}	1.13	[1.09–1.17]
rs4698775/G	4	110.6 Mb	*CFI*	0.31	2×10^{-10}	1.16	0.025	1.08	7×10^{-11}	1.14	[1.10–1.17]
rs3812111/T	6	116.4 Mb	*COL10A1*	0.64	7×10^{-8}	1.13	0.022	1.06	2×10^{-8}	1.10	[1.07–1.14]
Loci reaching $p < 5 \times 10^{-8}$ for the first time											
rs13081855/T	3	99.5 Mb	*COL8A1-FILIP1L*	0.10	4×10^{-11}	1.28	6.0×10^{-4}	1.17	4×10^{-13}	1.23	[1.17–1.29]
rs3130783/A	6	30.8 Mb	*IER3–DDR1*	0.79	1×10^{-6}	1.15	3.5×10^{-6}	1.16	2×10^{-11}	1.16	[1.11–1.20]
rs8135665/T	22	38.5 Mb	*SLC16A8*	0.21	8×10^{-8}	1.16	5.6×10^{-5}	1.13	2×10^{-11}	1.15	[1.11–1.19]
rs334353/T	9	101.9 Mb	*TGFBR1*	0.73	9×10^{-7}	1.13	6.7×10^{-6}	1.13	3×10^{-11}	1.13	[1.10–1.17]
rs8017304/A	14	68.8 Mb	*RAD51B*	0.61	9×10^{-7}	1.11	2.1×10^{-5}	1.11	9×10^{-11}	1.11	[1.08–1.14]
rs6795735/T	3	64.7 Mb	*ADAMTS9*	0.46	9×10^{-8}	1.13	0.0066	1.07	5×10^{-9}	1.10	[1.07–1.14]
rs9542236/C	13	31.8 Mb	*B3GALTL*	0.44	2×10^{-6}	1.12	0.0018	1.08	2×10^{-8}	1.10	[1.07–1.14]

tests consider the number of potentially functional alleles in each individual, but the tests can be refined to weigh variants according to their likely functional impact [38] to allow for imputed or uncertain genotypes [39] or to allow variants that increase and decrease risk to reside in the same gene [40, 41] (a feature that is important when the same gene harbors hypermorph and hypomorph alleles). The optimal strategy for testing rare variant associations depends on the underlying genetic architecture of traits, which are often unknown a priori. Therefore, it is often necessary to perform a combination of rare variant tests where each one favors a slightly different alternative model.

In the second part of this chapter, we describe a set of methods we developed that are suitable for rare variant association [42]. A few statistical methods have been developed for performing meta-analysis of rare variant association. They all start with simple statistics that can be calculated in an individual study (single-site score statistics and their covariance matrix, which summarizes the linkage disequilibrium information and relatedness among sampled individuals). We then show that, when these statistics are shared, a wide variety of gene-level association tests can be executed centrally – including both weighted or unweighted burden tests with fixed [43] or variable frequency thresholds [38] and sequence kernel association tests (SKAT) that accommodate alleles with opposite effects within a gene [40, 41].

In addition to performing meta-analysis of marginal associations, we also describe methods for performing conditional meta-analysis, which is important for dissecting independent association signals from shadows of known loci, as well as a Monte Carlo simulation–based approach for empirical assessment of significance when only summary level data is available. These meta-analysis approaches generate comparable results to sharing individual-level data (and, in fact, have identical results when allowing for between-study heterogeneity in nuisance parameters, such as trait means, variances, and covariate effects [42]). As an illustration of the approach, we analyze blood lipid levels in more than 18,500 individuals genotyped with exome genotyping arrays. Our analysis of blood lipid levels provides examples of loci where the signal for gene-level association tests exceeds the signal for single-variant tests and shows that our approach can recover signals driven by very rare variants (frequency $<0.05\%$). Given that very large sample sizes are required for successful rare variant association studies, we expect that our methods (and refined versions thereof) will be widely useful.

Our approach is based on the insight that analogues of most gene-level association tests can be constructed using single-variant test statistics and knowledge of their correlation structures. As shown in Liu et al. [42], simple and weighted burden tests, variable threshold tests, and tests allowing for variants with opposite effects can be constructed in this manner. We meta-analyze single-variant statistics using the Cochran-Mantel-Haenszel method, calculate variance-covariance matrices for these statistics, and construct gene-level association tests by combining the two. Importantly, rare variant statistics calculated in this way are less vulnerable to artifacts due to population stratification than statistics generated by naively pooling individual-level data. As in other meta-analysis settings, sharing summary statistics accelerates the overall analysis process, mitigates concerns about participant confidentiality, and reduces the risk that data will be used for unapproved analyses (as always, to avoid violating the trust of research subjects, we strongly recommend that investigators sharing summary statistics agree that these will not be used to identify research subjects).

Meta-Analysis Approaches

This section starts with a summary of notation and proceeds to describe the statistics to be shared between studies and methods for single-variant meta-analysis. We then show that the statistics for different gene-level tests can be calculated using summary-level data, enabling efficient meta-analysis.

For simplicity, we describe our strategy for analysis of a single gene. Let J be number of variant nucleotide sites genotyped in at least one study. For study k, let n_k denote the number of samples phenotyped and genotyped, and let the vector $\mathbf{y_k} = (Y_{1,k}, \ldots, Y_{N_k,k})^T$ denote the quantitative trait residuals (after adjustment for any covariates), with variance σ_k^2. Within each study k, we encode genotype information in matrix X_k where each entry $X_{i,j,k}$ represents the genotype for individual i at site j, coded as the number of alternative alleles. We encode missing genotypes in the data set as the average number of minor alleles in individuals, genotyped for that marker. The multisite genotype for individual i is denoted by the row vector $\mathbf{x}_{i,\bullet,k}$, and the genotypes for all N_k individuals at site j are given by column vector $\mathbf{x}_{\bullet,j,k}$. For the ease of presentation, we define the mean genotype matrix $\bar{X}_k$, where the (i, j)th element is $\left(\sum_i X_{i,j,k} \right) / N_k$.

Summary Statistics to Be Shared

For each study, we first calculate and share a vector of score statistics $\mathbf{u_k} = (X_k - \bar{X}_k)^T \mathbf{y_k}$, a corresponding variance-covariance matrix $V_k = \hat{\sigma}_k^2 N_k \operatorname{cov}(X_k) = \hat{\sigma}_k^2 (X_k - \bar{X}_k)^T (X_k - \bar{X}_k)$, and allele frequencies for each marker $p_{j,k} = \sum_i X_{i,j,k} / 2N_k$. Note that V_k effectively describes linkage disequilibrium relationships between the variants being examined. To perform quality control, we also share mean and variance for the quantitative trait residuals, genotype call rate, and Hardy-Weinberg equilibrium p-values at each variant site.

Meta-Analysis of Single-Variant Association Test Statistics

We first combine single-variant association test statistics across studies using the Cochran-Mantel-Haenszel method. Specifically, we calculate a score statistic at each site as

$$t_{j,\bullet} = U_{j,\bullet}/\sqrt{V_{j,j,\bullet}}$$

where $U_{j,\bullet} = \sum_k U_{j,k}$ and $V_{j,j,\bullet} = \sum_k V_{j,j,k}$. For ease of presentation, we denote the vector of single-variant association tests after meta-analysis as $\mathbf{u} = \sum_k \mathbf{u_k}$. Under the null, this vector is distributed as multivariate normal with mean vector $\mathbf{0}$ and covariance matrix $\sum_k \boldsymbol{V}_k$.

Burden Tests That Assume Variants Have Similar Effect Sizes

For a simple burden test in study k, the impact of multiple rare variants in a region can be modeled using a shared regression coefficient in a model that takes the form

$$Y_{i,k} = \beta_{0,k} + \beta_{\text{BURDEN}}\, C_{\text{BURDEN}}(\mathbf{x_{i,\bullet,k}}) + \varepsilon_{i,k}$$

where $\varepsilon_{i,k} \sim \text{N}(0, \sigma_k^2)$ and $C_{\text{BURDEN}}(\mathbf{x_{i,\bullet,k}})$ is a function that takes genotypes for a single individual as input and returns the count of rare alleles (the "rare variant burden") in the gene being examined. When individual-level data are available and nuisance parameters $\beta_{0,k}$ and σ_k^2 are allowed to vary between studies, the score statistic for a rare variant burden test becomes

$$U_{\text{BURDEN}} = \sum_k U_{\text{BURDEN},k} = \sum_k w^{\mathbf{T}}\mathbf{u_k} = w^{\mathbf{T}}\mathbf{u}$$

which is equal to a linear sum of (weighted) single variant score statistics.

 Under the null, this statistic is approximately normally distributed with mean 0 and variance $V_{\text{BURDEN}} = w^T \left(\sum_k \boldsymbol{V}_k\right) w$, enabling significance tests. Here, $\mathbf{w}$ is the vector of weights, which is $w = (w_1, \ldots, w_J)$, with each element w_j representing the weight assigned to variant j according to its allele frequency or its computationally predicted functional impact. The preceding formula makes it clear that, when nuisance parameters are allowed to vary between studies, the same burden score statistics that could be calculated by sharing individual data can be equivalently calculated using shared summary statistics.

Variable Threshold Tests with an Adaptive Frequency Threshold

In variable threshold test, rare variant burden statistics are calculated for each observed variant minor allele frequency threshold, and significance is evaluated for the maximum of these statistics. Given a specific variant frequency threshold F, we define the resulting burden score statistic as

$$U_{\mathrm{BURDEN}(F)} = \mathbf{v}_{\mathbf{F}}^{\mathbf{T}}\vec{U}$$

Here, $\mathbf{v_F}$ is a vector of indicators where the jth element equals 1 if the pooled minor allele frequency at variant site j is less than F and zero otherwise. For convenience, we also define a matrix of indicators for minor allele frequency thresholds $\mathbf{\Phi} = (\mathbf{v_{F_1}}, \mathbf{v_{F_2}}, \ldots, \mathbf{v_{F_J}})$. After a burden statistic is calculated for each potential frequency threshold, these are standardized, dividing each statistic by its corresponding variance, and the maximum statistic is identified:

$$T_{VT} = \max_{F}\left\{T_{\mathrm{BURDEN}(F)}\right\}, \text{ where } T_{\mathrm{BURDEN}(F)} = U_{\mathrm{BURDEN}(F)}\Big/ \sqrt{\mathbf{v}_F^T \sum_k V_k \mathbf{v}_F}$$

Significance for this statistic can be evaluated using the cumulative distribution function for the multivariate normal distribution [38]. Specifically, given the definition of the covariance between burden statistics calculated using different allele frequency thresholds, we have

$$\left(T_{\mathrm{BURDEN}(F_1)}, \ldots, T_{\mathrm{BURDEN}(F_M)}\right) \sim \mathrm{MVN}\left(\mathbf{0}, \Phi\left(\sum\nolimits_k V_k\right)\Phi^T\right)$$

The p-value for the VT test statistic is given by

$$\begin{aligned}
p &= 1 - \Pr(T_{\mathrm{VT}} \le t_{\mathrm{VT}}) \\
&= 1 - \Pr(T_{\mathrm{BURDEN}(F_1)} \le t, \ldots, T_{\mathrm{BURDEN}(F_M)} \le t) \\
&= 1 - F_{\mathrm{MVN}}(t, \ldots, t)
\end{aligned}$$

where F_{MVN} is the distribution function for the multivariate normal distribution $\mathrm{MVN}\left(\mathbf{0}, \Phi\left(\sum_k V_k\right)\Phi^T\right)$.

Burden Tests That Assume a Distribution of Variant Effect Sizes (e.g., SKAT Tests)

The simple burden test and variable threshold test can be underpowered when variants with opposite phenotypic effects reside in the same gene and are grouped together, because the shared regression coefficient can average close to zero in that situation. To accommodate this setting, we consider an underlying distribution of rare variance effect sizes with mean zero and test whether the variance of this distribution τ is greater than zero.

When individual-level data are available, association analysis in study k is performed using the following model:

$$Y_{i,k} = \beta_{0,k} + \sum_j \beta_j X_{i,j,k} + \varepsilon_{i,k}, \quad \text{where} \quad \varepsilon_{i,k} \sim \mathrm{N}\left(0, \sigma_k^2\right)$$

We make inferences about rare variant effect sizes $\beta = (\beta_1, \beta_2, \ldots, \beta_J)$ by assuming these follow a common distribution with mean zero and variance τ. Under the null, $\tau = 0$. Following Wu et al. [44], we derive the score statistic for this model and show that it can be calculated on the basis of per-study summary statistics:

$$Q = \left(\sum_k \mathbf{u_k}\right)^T K \left(\sum_k \mathbf{u_k}\right)$$

Here, K is the kernel matrix that compares multisite genotypes. A default choice [44] is a diagonal matrix $K = \mathrm{diag}(w_1, w_2, \ldots, w_J)$, with w_j being the weight assigned to variant site j. The statistic Q follows a mixture chi-square distribution [45], which means that Q is equivalent in distribution to a weighted sum of independent chi-square random variables. The weights (or mixture proportions) are given by the eigenvalues for the matrix $(\sum_k V_k)^{1/2} K (\sum_k V_k)^{1/2}$.

Monte Carlo Method for Empirical Assessment of Significance

The previous sections describe how a series of gene-level test statistics can be calculated and, for each one, propose a strategy for evaluating significance using asymptotic distributions. In practice, evaluating the required numerical integrals can be challenging because variance-covariance matrices are sometimes singular or nearly singular.

Note that single-variant test statistics are distributed as

$$\sum_k \mathbf{u_k} = \sum_k \mathbf{y_k^T}(X_k - \bar{X}_k) \sim \mathrm{MVN}\left(\mathbf{0}, \sum_k V_k\right)$$

Then, to evaluate significance empirically, one can sample random vectors from the distribution $\mathrm{MVN}\left(\mathbf{0}, \sum_k V_k\right)$ and calculate gene-level rare variant test statistics for each of these sampled random vectors, resulting in an empirical distribution for any gene-level statistic [46]. As usual, p-values can then be evaluated by comparing the test statistics for the original data with those in this empirical distribution. For computational efficiency, we use an adaptive algorithm where a larger number of vectors is sampled when assessing small p-values and fewer vectors are sampled when assessing larger p-values [47].

Conditional Analyses

It is well known that, because of linkage disequilibrium, one or more common causal variants can result in shadow association signals at other nearby common variants. For common variants, Yang et al. [48] have shown that linkage disequilibrium relationships between variants, estimated from external reference panels, can be used to enable conditional analysis in meta-analysis settings. For rare variants and gene-level tests, accurately describing relationships between variants is crucial, and we recommend against using external reference panels.

For simplicity, we only discuss how to obtain single-variant test statistics conditional on known loci. Subsequent gene-level association tests can be similarly constructed using the same idea as marginal analysis. To describe our strategy for conditional analyses, we first decompose the genotype matrix into two components: a matrix of genotypes X_k for variants to be tested for independent association and a matrix of genotypes Z_k for variants that should be included as covariates in the null regression model (and, thus, controlled for). To facilitate presentations, we denote $W_k = (X_k, Z_k)$.

Conditional Analysis for Gene-Level Tests That Use a Shared Regression Coefficient for Rare Variants

When individual-level data are available, conditional analysis considers a model similar to

$$Y_{i,k} = \beta_{0,k} + \beta_{\text{BURDEN}}\, C_{\text{BURDEN}}(\vec{X}_{i,\bullet,k}) + \vec{\alpha}_k^T \vec{Z}_{i,k} + \varepsilon_{i,k}$$

This analysis could be readily carried out by repeating analysis and recalculating score statistics for each study, but this is not required. Instead, the score statistics that result from the conditional analysis can be readily estimated using summary information.

Let $\tilde{T}_{\beta_{\text{BURDEN}},k}$, $\tilde{U}_{\beta_{\text{BURDEN}},k}$, and $\tilde{V}_{\beta_{\text{BURDEN}},\beta_{\text{BURDEN}},k}$ denote test statistics, score statistics, and their variances from conditional analysis, analogous to statistics previously defined for unconditional analysis. As usual, $\tilde{T}_{\beta_{\text{BURDEN}},k} = \tilde{U}_{\beta_{\text{BURDEN}},k}/\sqrt{\tilde{V}_{\beta_{\text{BURDEN}},\beta_{\text{BURDEN}},k}}$. To derive the component statistics, we use the approach of Lin and Tang [49], to show that

$$\tilde{U}_{\beta_{\text{BURDEN}},k} = ((\mathbf{y_k} - \bar{Y}_k) - (Z_k - \bar{Z}_k)\hat{\alpha}_k)^T X_k$$

$$\text{with } \hat{\boldsymbol{\alpha}}_{\mathbf{k}} = \left((Z_k - \bar{Z}_k)^{\mathrm{T}}(Z_k - \bar{Z}_k)\right)^{-1}(Z_k - \bar{Z}_k)^{\mathrm{T}}\mathbf{y_k}$$

and that

$$\tilde{V}_{\beta_{\text{BURDEN}},\beta_{\text{BURDEN}},k} = \hat{\phi}^2 w^T \left((X_k - \bar{X}_k)^T(X_k - \bar{X}_k)\right) w$$

$$-\hat{\phi}^2 w^T \left((X_k - \bar{X}_k)^T(Z_k - \bar{Z}_k)\right)\left((Z_k - \bar{Z}_k)^T(Z_k - \bar{Z}_k)\right)^{-1}$$

$$\left((Z_k - \bar{Z}_k)^T(X_k - \bar{X}_k)\right) w$$

$$\text{with } \hat{\phi}^2 = 1/N_k \times ((\mathbf{y_k} - \bar{Y}_k) - (Z_k - \bar{Z}_k)\hat{\boldsymbol{\alpha}}_{\mathbf{k}})^T((\mathbf{y_k} - \bar{Y}_k) - (Z_k - \bar{Z}_k)\hat{\boldsymbol{\alpha}}_{\mathbf{k}})$$

Now, we can verify that $\tilde{U}_{\beta_{\text{BURDEN}},k}$ and $\tilde{V}_{\beta_{\text{BURDEN}},\beta_{\text{BURDEN}},k}$ can be calculated using shared summary level statistics, because all key terms in the preceding equations can be extracted from the list of single-variant score statistics and from the variance-covariance matrix of single-marker association test statistics (which we have shared), because

$$(W_k - \bar{W}_k)^{\mathrm{T}}(W_k - \bar{W}_k) = \begin{pmatrix} (X_k - \bar{X}_k)^T (X_k - \bar{X}_k) & (X_k - \bar{X}_k)^T (Z_k - \bar{Z}_k) \\ (Z_k - \bar{Z}_k)^T (X_k - \bar{X}_k) & (Z_k - \bar{Z}_k)^T (Z_k - \bar{Z}_k) \end{pmatrix}$$

Finally, meta-analysis burden score statistics can be calculated as

$$\tilde{T}_{\beta_{\text{BURDEN}},\bullet} = \sum_k \tilde{U}_{\beta_{\text{BURDEN}},k} \Big/ \sqrt{\sum_k \tilde{V}_{\beta_{\text{BURDEN}},\beta_{\text{BURDEN}},k}}$$

Thus, conditional meta-analysis statistics can be calculated from shared single-variant statistics and their variance-covariance matrix, as desired.

1.2.4 Application 2: Meta-Analysis of Plasma Lipid Levels

We applied the developed meta-analysis approach to a meta-analysis of blood lipid levels in 18,699 individuals of European ancestry genotyped with Illumina Exome arrays and drawn from seven studies: the Women's Health Initiative [50], the Ottawa Heart Study, the Malmö Diet and Cancer Study – Cardiovascular Cohort (MDC) [51], the PROCARDIS Precocious Coronary Artery Disease Case Series [52], the PROCARDIS Control series, and the Nord-Trøndelag Health Study (HUNT) myocardial infraction cases and matched controls [53]. Overall, 171,193 variants were polymorphic in at least one individual. Among these variants, 125,702 – the vast majority – have frequency $<1\%$.

We started by meta-analyzing single variant association test results. The resulting test statistics appear well calibrated, with genomic control value <1.05 for all three traits, both for common and for rare variants. At a significance threshold of $p < 3 \times 10^{-7}$ (corresponding to 0.05 / 171,193), we found significantly associated variants (with MAF $< 5\%$) at *LPL, ANGPTL4, LIPG, CD300LG, LIPC, APOB, HNF4A* for HDL [54]; *PCSK9, BCAM-CBLC-PVR* (neighboring APOE), and *APOB* for LDL [54]; and *ANGPTL4, LPL,* and *APOB* for TG [54]. Except for the variants in *LIPC* and *APOB*, all other significantly associated variants have frequency of $>1\%$, reflecting the limited power of single-variant association tests for rare alleles.

We next carried out gene-level tests. Again, test statistics appear well calibrated, with genomic control value <1.05. At a significance threshold of $p < 3.1 \times 10^{-6}$ (corresponding to 0.05/16,153 and thus allowing for the number of genes tested), we observed association at *LIPC*, *LPL*, *ANGPTL4*, *LIPG*, *HNF4A*, and *CD300LG* for HDL; at the *PCSK9*, *APOE*-locus (as well as nearby genes *PVR*, *BCAM*, and *CBLC*), and *LDLR* for LDL; and at *ANGPTL4* and *LPL* for triglycerides (Table 1.7). Reassuringly, these signals point to loci identified in previous genome-wide association studies and/or resequencing studies. Importantly, note that our approach was able to appropriately identify the signal in *LDLR*, which is driven by several very rare variants (each with frequency $< .00052$) that nearly always increase blood LDL cholesterol levels, and that, at several other loci, gene-level p-values exceeded the best single-variant p-value in the gene (Table 1.8).

An added convenience of sharing single-variant statistics together with their covariance matrices, as we propose, is that it facilitates conditional analyses, extending an idea used by Yang et al. [48] for analysis of common variants in GWAS meta-analysis. In this real data application, we reexamined two of the LDL-associated loci in detail, *LDLR* and *APOE-BCAM-CBLC-PVR*. For *LDLR*, we examined the relationship between rare variant signals and three nearby common variants [54]. Specifically, we conditioned on genotypes for three common variants (rs6511720, rs2228671, and rs72658855) exhibiting significant association in the region and found that *LDLR* rare variant association remains significant (p-value 4.6×10^{-7}) (Table 1.8). For the *APOE-BCAM-CBLC-PVR* locus, after conditioning on the common variant showing strongest association in the region (rs7412), gene-level associations at *BCAM*, *CLBC*, and *PVR* become nonsignificant, suggesting that these rare variant signals are the result of regional linkage disequilibrium with more common and well-described variants in *APOE*.

1.2.5 GWAS-Tailored Software

To realize the benefits of developed methodologies and apply them to high-throughput data analysis, providing an efficiently implemented, well-maintained, and carefully documented software package can be critical. Existing packages for single-marker tests include METAL [55], MebABEL [56], GWAMA [57], PLINK [58], and multiple R packages (e.g., GWAtoolbox [59]). They support different input file formats, which are flexible and easy to generate. All packages have implemented the fixed effects model, but only GWAMA also implemented the random effects model. All packages output heterogeneity statistics Q and I^2. These tools have been compared thoroughly in previous reviews.

The meta-analysis software can be particularly essential for the analysis of rare variant associations. To implement a simple rare variant association test,

Table 1.7 *Results for meta-analysis of gene-level rare variant association test*

Gene	Gene position[a]	Burden-1	Burden-5	SKAT-1	SKAT-5	VT	MAF cutoff	Direction of single-variant association statistics[b]	Estimates of genetic average effect (s.d units) for rare variants under different MAF thresholds		
									0.01	0.05	VT
HDL											
LIPC[c]	chr15:58.7Mb	$\mathbf{1.4\times10^{-12}}$	$\mathbf{3.5\times10^{-7}}$	$\mathbf{1.8\times10^{-9}}$	1.4×10^{-2}	$\mathbf{4.5\times10^{-12}}$	3.7×10^{-3}	$-+++++--+-$	0.5	0.1	0.5
LPL	chr8:19.8Mb	9.7×10^{-1}	$\mathbf{2.5\times10^{-24}}$	3.5×10^{-1}	$\mathbf{5.0\times10^{-13}}$	$\mathbf{1.5\times10^{-23}}$	2.5×10^{-2}	$(-)-(-)+-++$	–	−0.3	−0.3
ANGPTL4[c]	chr19:8.4Mb	2.2×10^{-2}	$\mathbf{2.9\times10^{-19}}$	2.2×10^{-2}	$\mathbf{3.0\times10^{-19}}$	$\mathbf{1.8\times10^{-18}}$	2.6×10^{-2}	$(+)--++-+++$	–	0.3	0.3
LIPG[c]	chr18:47.1Mb	2.2×10^{-5}	$\mathbf{6.4\times10^{-19}}$	2.1×10^{-5}	$\mathbf{2.9\times10^{-9}}$	$\mathbf{4.4\times10^{-18}}$	1.3×10^{-2}	$-++----(+)+$	–	0.4	0.4
HNF4A	chr20:43.0Mb	7.5×10^{-1}	$\mathbf{2.8\times10^{-7}}$	6.8×10^{-1}	$\mathbf{2.5\times10^{-7}}$	$\mathbf{1.5\times10^{-6}}$	4.1×10^{-2}	$(-)---+-+$	–	−0.1	−0.1
CD300LG	chr17:41.9Mb	4.9×10^{-1}	$\mathbf{8.5\times10^{-7}}$	5.2×10^{-1}	1.0×10^{-5}	$\mathbf{3.1\times10^{-6}}$	3.3×10^{-2}	$(-)+-(+)$	–	−0.1	–
LDL											
PCSK9[c]	chr1:55.5Mb	1.8×10^{-2}	$\mathbf{7.4\times10^{-19}}$	8.1×10^{-2}	$\mathbf{5.5\times10^{-17}}$	$\mathbf{2.0\times10^{-28}}$	1.3×10^{-2}	$(-)--(-)--+-++-$	–	−0.3	−0.5
BCAM	chr19:45.3Mb	1.7×10^{-1}	$\mathbf{1.6\times10^{-18}}$	1.5×10^{-1}	3.0×10^{-5}	$\mathbf{2.6\times10^{-17}}$	3.6×10^{-2}	$(-)+++(-)+-+++--$ $-+(-)+--+--++$	–	−0.1	−0.1
CBLC	chr19:45.3Mb	9.4×10^{-1}	$\mathbf{2.0\times10^{-15}}$	4.4×10^{-1}	1.5×10^{-4}	$\mathbf{1.0\times10^{-14}}$	4.4×10^{-2}	$-(-)--+-(-)(+)$	–	−0.1	−0.1
PVR	chr19:45.2Mb	6.1×10^{-2}	$\mathbf{3.0\times10^{-10}}$	4.8×10^{-2}	6.3×10^{-2}	$\mathbf{1.1\times10^{-9}}$	4.9×10^{-2}	$(-)++--+$	–	−0.1	−0.1
LDLR[c]	chr19:11.2Mb	1.8×10^{-3}	4.7×10^{-5}	3.8×10^{-2}	2.5×10^{-1}	$\mathbf{2.4\times10^{-7}}$	5.2×10^{-4}	$+++++++++-+++$ $+--+$	–	–	0.8
TG											
ANGPTL4[c]	chr19:8.4Mb	2.6×10^{-2}	$\mathbf{1.2\times10^{-24}}$	3.7×10^{-2}	$\mathbf{3.9\times10^{-25}}$	$\mathbf{7.1\times10^{-24}}$	2.6×10^{-2}	$(-)+---+---$	–	−0.3	−0.2
LPL[c]	chr8:19.8Mb	6.8×10^{-1}	$\mathbf{7.7\times10^{-20}}$	2.6×10^{-1}	$\mathbf{1.8\times10^{-11}}$	$\mathbf{4.6\times10^{-19}}$	2.5×10^{-2}	$(+)+(+)--+-$	–	0.2	0.2

Note: Associations that attain exome-wide significance ($p < 3.1\times10^{-6}$) are displayed. Five gene-level association tests were used to analyze the data: simple burden tests with 1% or 5% cutoff (Burden-1 and Burden-5), SKAT tests with 1% or 5% cutoff (SKAT-1 and SKAT-5), and variable-threshold (VT) tests that analyze variants with MAF < 5%. Significant p-values for each test are displayed in boldface. For the associations that are significant, estimates of average genetic effect are also shown.

[a] Gene position is defined based upon hg19, GRCh37 Genome Reference Consortium Human Reference 37.

[b] Direction of single site statistics for variants with MAF < 5%. Variants within parenthesis have frequency >1%.

[c] The loci with one or more gene-level association signals exceeding the top single-variant signal.

Table 1.8 *Results of conditional association analysis for LDL and variants in LDLR*

				Single-variant association analysis				
RS	Ref	Alt	MAF	Original p-value	Conditional p-value	Original effect[a] estimate[a]	Conditional effect estimate[a]	Annotation
rs6511720	G	T	0.11	2×10^{-38}	–	-0.22	–	Intron
rs2228671	C	T	0.11	4×10^{-22}	–	-0.16	–	Synonymous
rs2738459	A	C	0.49	4×10^{-8}	–	-0.06	–	Intron
rs1166957	G	A	0.046	8×10^{-4}	0.201	0.08	0.03	Nonsynonymous
rs1396241	G	A	0.0001	8×10^{-4}	0.001	1.68	1.61	Nonsynonymous
rs1397913	G	A	0.0005	8×10^{-4}	0.002	0.77	0.7	Nonsynonymous
rs1997741	C	A	3×10^{-5}	0.004	0.002	2.88	3.14	Stop_Gain
rs1441727	G	A	3×10^{-5}	0.024	0.037	2.26	2.11	Nonsynonymous
rs1416739	G	A	3×10^{-5}	0.048	0.056	1.98	1.92	Nonsynonymous
rs1506739	C	T	6×10^{-5}	0.056	0.031	1.35	1.54	Nonsynonymous
rs2894208	C	T	6×10^{-5}	0.151	0.158	1.02	1.01	Nonsynonymous
rs1390431	T	A	0.0001	0.21	0.241	0.63	0.59	Nonsynonymous
rs1393616	G	A	3×10^{-5}	0.266	0.2	1.11	1.29	Nonsynonymous
rs1439929	G	A	8×10^{-5}	0.358	0.233	0.53	0.7	Nonsynonymous
rs1378539	G	A	0.0018	0.391	0.212	-0.11	-0.16	Nonsynonymous
rs1330650	C	T	6×10^{-5}	0.511	0.47	0.47	0.51	Nonsynonymous
rs1486986	G	A	0.0001	0.539	0.585	0.27	0.25	Nonsynonymous
rs2007276	G	A	3×10^{-5}	0.603	0.667	0.52	0.43	Nonsynonymous
rs5928	G	A	3×10^{-5}	0.892	0.851	-0.14	-0.19	Nonsynonymous
rs1462001	C	G	0.0002	0.997	0.99	0	0	Nonsynonymous

			Gene-level test			
Gene	p-value before conditioning	p-value after conditioning	MAF cutoff before conditioning	MAF cutoff after conditioning	Estimate of genetic effect before conditioning[a]	Estimate of genetic effect after conditioning[a]
LDLR	2.4×10^{-7}	4.6×10^{-7}	5.2×10^{-4}	5.2×10^{-4}	0.75	0.73

Note: We performed conditional association analysis for variants in *LDLR*, conditioning on three common variants (rs6511720, rs2228671, and rs72658855) that are strongly associated in single-variant analyses (i.e., with p-value $< 3\times10^{-7}$). The rs number, reference and alternate alleles, minor allele frequencies, p-values before and after conditioning, estimates of effect size per copy of the alternative alleles, and annotation information are displayed for nonsynonymous and loss-of-function variants. Gene-level association test results are at the bottom of the table.

[a] In standard deviation units.

multiple sources of information will need to be integrated. For instance, it is often desirable to aggregate functionally important rare variants in a gene, which will require annotation information in the implementation of rare variant tests. Multiple software packages were available for performing gene-level association tests. Existing packages include MASS [49], MetaSKAT [41], and RAREMETAL [42].

Although most of the software packages implement methods that are similar in theory, their functionality and supported file formats vary greatly. In the following, we summarize a few key differences in their implementations.

1. *Supported file format.* All three methods implement functions for generating summary-level data: MASS takes input from Score-Seq or Score-Seq-TDS, MetaSKAT supports bed or ped files as input, whereas RAREMETAL supports the VCF file format as well as genotype of phenotype files in Ped files.
2. *The analysis of related samples.* RAREMETAL supports the analysis of both related and unrelated samples, whereas MASS and MetaSKAT only support the analysis of unrelated samples.
3. *Random effects meta-analysis.* RAREMETAL only implements fixed effects meta-analysis, given that heterogeneity of genetic effects in different cohorts are rarely observed. Conversely, both MetaSKAT and MASS implement features for performing random effect meta-analysis, which is more powerful when there are large genetic effect heterogeneities between different studies.
4. *Conditional association analysis.* Among these methods, RAREMETAL implements methods for conditional meta-analysis, which is useful for dissecting genuine signals from shadows of known loci.

1.3 Summary and Future Perspective

In this chapter, we illustrated the basic principles of meta-analysis for genetic association studies, provided practice guidance, and walked through two real examples from large-scale consortium projects. Meta-analysis of multiple GWAS provides an effective and powerful way to identify more disease-associated loci and to improve our understanding of complex diseases. The GWAS-specific methodology for meta-analysis described in this chapter has gained success in many consortium projects. Despite early successes of sequencing studies, there is a growing recognition that most rare variants involved in complex diseases have moderate effect sizes. Powerful detection of rare variant association would benefit from the analysis of large data sets. Given the difficulty of pooling individual-level data across different studies, meta-analysis will continue to play an important role for the next wave of gene-mapping efforts.

As sequencing technologies continue to develop, 1000 genomes has become a practical goal in the next couple years. Meta-analysis will be focused on the analysis of sequence data. One clear challenge is how to make existing methods scalable for the analysis of sequence data and how to accommodate heterogeneities in the data set. In addition, it is also of great interest to develop methods that facilitate meta-analysis of epistasis effects and methods that will advance our understanding of the genetic architecture of rare variants using aggregated summary statistics.

To further understand the pathology and etiology of complex diseases, researchers are integrating different levels of high-throughput data, including DNA, RNA, epigenetics, and proteomics. Meta-analysis will continuously be an effective approach for these types of "vertical" data. This, however, is beyond the scope of this chapter and is discussed in other chapters.

References

1 Hindorff, L. A., et al., *Potential etiologic and functional implications of genome-wide association loci for human diseases and traits.* Proc Natl Acad Sci U S A, 2009. **106**(23): pp. 9362–7.

2 McCarthy, M. I., et al., *Genome-wide association studies for complex traits: consensus, uncertainty and challenges.* Nat Rev Genet, 2008. **9**(5): pp. 356–69.

3 Fritsche, L. G., et al., *Seven new loci associated with age-related macular degeneration.* Nat Genet, 2013. **45**(4): pp. 433–9, 439e1–2.

4 Marchini, J., and B. Howie, *Genotype imputation for genome-wide association studies.* Nat Rev Genet, 2010. **11**(7): pp. 499–511.

5 Nothnagel, M., et al., *A comprehensive evaluation of SNP genotype imputation.* Hum Genet, 2009. **125**(2): pp. 163–71.

6 Li, Y., et al., *Genotype imputation.* Annu Rev Genomics Hum Genet, 2009. **10**: pp. 387–406.

7 Howie, B. N., P. Donnelly, and J. Marchini, *A flexible and accurate genotype imputation method for the next generation of genome-wide association studies.* PLoS Genet, 2009. **5**(6): e1000529.

8 Marchini, J., et al., *A new multipoint method for genome-wide association studies by imputation of genotypes.* Nat Genet, 2007. **39**(7): pp. 906–13.

9 Liu, E. Y., et al., *MaCH-Admix: Genotype Imputation for Admixed Populations.* Genet Epidemiol, 2013. **37**(1): pp. 25–37.

10 Browning, B. L. and S. R. Browning, *A unified approach to genotype imputation and haplotype-phase inference for large data sets of trios and unrelated individuals.* Am J Hum Genet, 2009. **84**(2): pp. 210–23.

11 Howie, B., et al., *Fast and accurate genotype imputation in genome-wide association studies through pre-phasing.* Nat Genet, 2012. **44**(8): pp. 955–9.

12 Evangelou, E., and J. P. Ioannidis, *Meta-analysis methods for genome-wide association studies and beyond.* Nat Rev Genet, 2013. **14**(6): pp. 379–89.

13 Begum, F., et al., *Comprehensive literature review and statistical considerations for GWAS meta-analysis.* Nucleic Acids Res, 2012. **40**(9): pp. 3777–84.

14 Zeggini, E., and J. P. Ioannidis, *Meta-analysis in genome-wide association studies.* Pharmacogenomics, 2009. **10**(2): pp. 191–201.

15 de Bakker, P. I., et al., *Practical aspects of imputation-driven meta-analysis of genome-wide association studies.* Hum Mol Genet, 2008. **17**(R2): pp. R122–8.

16 Borenstein, M., *Introduction to meta-analysis.* 2009, Chichester, UK: John Wiley, 421 pp.

17 Leandro, G., *Meta-analysis in medical research: the handbook for the understanding and practice of meta-analysis.* 2005, Malden, Mass.: BMJ Books/Blackwell, 98 pp.

18 Sutton, A. J., *Methods for meta-analysis in medical research.* 2000, Chichester, UK: John Wiley, 317 pp.

19 Han, B., and E. Eskin, *Random-effects model aimed at discovering associations in meta-analysis of genome-wide association studies.* Am J Hum Genet, 2011. **88**(5): pp. 586–98.

20 Lin, D. Y., and P. F. Sullivan, *Meta-analysis of genome-wide association studies with overlapping subjects.* Am J Hum Genet, 2009. **85**(6): pp. 862–72.

21 Lin, D. Y., and D. Zeng, *Meta-analysis of genome-wide association studies: no efficiency gain in using individual participant data.* Genet Epidemiol, 2010. **34**(1): pp. 60–6.

22 Lin, D. Y., and D. Zeng, *On the relative efficiency of using summary statistics versus individual-level data in meta-analysis.* Biometrika, 2010. **97**(2): pp. 321–32.

23 Han, B., and E. Eskin, *Interpreting meta-analyses of genome-wide association studies.* PLoS Genet, 2012. **8**(3): e1002555.

24 Fisher, S. A., et al., *Meta-analysis of genome scans of age-related macular degeneration.* Hum Mol Genet, 2005. **14**(15): pp. 2257–64.

25 Klein, R. J., et al., *Complement factor H polymorphism in age-related macular degeneration.* Science, 2005. **308**(5720): pp. 385–9.

26 Dewan, A., et al., *HTRA1 promoter polymorphism in wet age-related macular degeneration.* Science, 2006. **314**(5801): pp. 989–92.

27 Rivera, A., et al., *Hypothetical LOC387715 is a second major susceptibility gene for age-related macular degeneration, contributing independently of complement factor H to disease risk.* Hum Mol Genet, 2005. **14**(21): pp. 3227–36.

28 Jakobsdottir, J., et al., *Susceptibility genes for age-related maculopathy on chromosome 10q26.* Am J Hum Genet, 2005. **77**(3): pp. 389–407.

29 Edwards, A. O., et al., *Complement factor H polymorphism and age-related macular degeneration.* Science, 2005. **308**(5720): pp. 421–4.

30 Fagerness, J. A., et al., *Variation near complement factor I is associated with risk of advanced AMD.* Eur J Hum Genet, 2009. **17**(1): pp. 100–4.

31 Gold, B., et al., *Variation in factor B (BF) and complement component 2 (C2) genes is associated with age-related macular degeneration.* Nat Genet, 2006. **38**(4): pp. 458–62.

32 Maller, J. B., et al., *Variation in complement factor 3 is associated with risk of age-related macular degeneration.* Nat Genet, 2007. **39**(10): pp. 1200–1.

33 Yates, J. R., et al., *Complement C3 variant and the risk of age-related macular degeneration.* N Engl J Med, 2007. **357**(6): pp. 553–61.

34 Cho, Y. M., et al., *Multifactor-dimensionality reduction shows a two-locus interaction associated with Type 2 diabetes mellitus.* Diabetologia, 2004. **47**(3): pp. 549–54.

35 Chen, W., et al., *Genetic variants near TIMP3 and high-density lipoprotein-associated loci influence susceptibility to age-related macular degeneration.* Proc Natl Acad Sci U S A, 2010. **107**(16): pp. 7401–6.

36 Li, B., and S. M. Leal, *Methods for detecting associations with rare variants for common diseases: application to analysis of sequence data.* Am J Hum Genet, 2008. **83**(3): pp. 311–21.

37 1000 Genomes Project, et al., *An integrated map of genetic variation from 1,092 human genomes.* Nature, 2012. **491**(7422): pp. 56–65.

38 Price, A. L., et al., *Pooled association tests for rare variants in exon-resequencing studies.* Am J Hum Genet, 2010. **86**(6): pp. 832–38.

39 Zawistowski, M., et al., *Extending rare-variant testing strategies: analysis of noncoding sequence and imputed genotypes.* Am J Hum Genet, 2010. **87**(5): pp. 604–17.

40 Lee, S., et al., *Optimal unified approach for rare-variant association testing with application to small-sample case-control whole-exome sequencing studies.* Am J Hum Genet, 2012. **91**(2): pp. 224–37.

41 Lee, S., et al., *General framework for meta-analysis of rare variants in sequencing association studies.* Am J Hum Genet, 2013. **93**(1): pp. 42–53.

42 Liu, D.J., et al., *Meta-analysis of gene-level tests for rare variant association.* Nat Genet, 2014. **46**(2): pp. 200–4.

43 Madsen, B. E., and S. R. Browning, *A groupwise association test for rare mutations using a weighted sum statistic.* PLoS Genet, 2009. **5**(2): e1000384.

44 Wu, M. C., et al., *Rare-variant association testing for sequencing data with the sequence kernel association test.* Am J Hum Genet, 2011. **89**(1): pp. 82–93.

45 Lin, X., *Variance component testing in generalised linear models with random effects.* Biometrika, 1997. **84**(2): pp. 309–26.

46 Lin, D. Y., and F. Zou, *Assessing genomewide statistical significance in linkage studies.* Genet Epidemiol, 2004. **27**(3): pp. 202–14.

47 Besag, J., and P. Clifford, *Sequential Monte Carlo p-values.* Biometrika, 1991. **78**(2): pp. 301–4.

48 Yang, J., et al., *Conditional and joint multiple-SNP analysis of GWAS summary statistics identifies additional variants influencing complex traits.* Nat Genet, 2012. **44**(4): pp. 369–75, S1–3.

49 Lin, D. Y., and Z. Z. Tang, *A general framework for detecting disease associations with rare variants in sequencing studies.* Am J Hum Genet, 2011. **89**(3): pp. 354–67.

50 Anderson, G. L., et al., *Implementation of the Women's Health Initiative study design.* Ann Epidemiol, 2003. **13**(9 Suppl): pp. S5–17.

51 Kathiresan, S., et al., *Polymorphisms associated with cholesterol and risk of cardiovascular events.* N Engl J Med, 2008. **358**(12): pp. 1240–9.

52 Barlera, S., et al., *PROCARDIS: A current approach to the study of the genetics of myocardial infarct.* Ital Heart J Suppl, 2001. **2**(9): pp. 997–1004.

53 Krokstad, S., et al., *Cohort profile: the HUNT study, Norway.* Int J Epidemiol, 2013. **42**(4): pp. 968–77.

54 Teslovich, T. M., et al., *Biological, clinical and population relevance of 95 loci for blood lipids.* Nature, 2010. **466**(7307): pp. 707–13.

55 Willer, C. J., Y. Li, and G. R. Abecasis, *METAL: fast and efficient meta-analysis of genomewide association scans.* Bioinformatics, 2010. **26**(17): pp. 2190–91.

56 Aulchenko, Y. S., et al., *GenABEL: an R library for genome-wide association analysis.* Bioinformatics, 2007. **23**(10): pp. 1294–96.

57 Magi, R., and A. P. Morris, *GWAMA: software for genome-wide association meta-analysis.* BMC Bioinformatics, 2010. **11**: pp. 288.
58 Purcell, S., et al., *PLINK: a tool set for whole-genome association and population-based linkage analyses.* Am J Hum Genet, 2007. **81**(3): pp. 559–75.
59 Fuchsberger, C., et al., *GWAtoolbox: an R package for fast quality control and handling of genome-wide association studies meta-analysis data.* Bioinformatics, 2012. **28**(3): pp. 444–45.

2

MetaOmics: Transcriptomic Meta-Analysis Methods for Biomarker Detection, Pathway Analysis and Other Exploratory Purposes

SUNGHWAN KIM, ZHIGUANG HUO, YONGSEOK PARK, AND
GEORGE C. TSENG

Abstract

In this chapter, we present a MetaOmics software suite to combine multiple transcriptomic studies for meta-analysis. MetaOmics contains more than a dozen in-house-developed methods and consists of seven subpackages for different data analysis and biological objectives: MetaQC for quality control assessment, MetaDE for differentially expressed gene detection, MetaPath for pathway enrichment analysis, MetaPCA for dimension reduction, MetaClust for clustering analysis, MetaNetwork for network analysis, and MetaPredict for prediction analysis. With the increasing number of experimental data accumulated in the public domain, application of related omics meta-analysis methods provides increased statistical power and validated conclusions to improve disease treatment and mechanism understanding.

2.1 Introduction

With the advances in high-throughput experimental technology in the past decades, the production of genomic data has become affordable and large genomic data are prevalent in recent biomedical research. Effective data management and analysis tools are essential to fully decipher the biological information inside the tremendous amount of experimental data. In the past decade, enormous bodies of transcriptomic data have been accumulated from microarray experiments, which resulted in several large public data depositories, such as Gene Expression Omnibus (GEO) and ArrayExpress. Recent development of next generation sequencing (NGS) technology accelerated the data accumulation in databases like Sequence Read Archive (SRA). In general, each individual study often has small or moderate sample size. As a result, the statistical power of candidate marker or pathway detection in each study is often limited, the reproducibility of the conclusions is relatively low, and the generalizability of the inferred information has been frequently criticized.

Combining multiple studies has emerged as an appealing practice because of improved statistical power and estimation accuracy, while it may also provide validation about the final conclusion. Many "transcriptomic meta-analysis" methods have been developed and widely applied in the real data analysis. In the literature, however, most of the methods were proposed to identify candidate marker genes differentially expressed between two or multiple conditions. Similar "meta-analysis" ideas can be extended for enriched pathway detection, clustering analysis, dimension reduction, and network and disease classification analysis (see Ramasamy et al. (2008) and Tseng et al. (2012) for more details). In this chapter, we first introduce statistical methods in the "MetaOmics" software suite, including those still under development in our lab. Section 2.2 discusses issues in data preprocessing and a "MetaQC" package for quality control and decision of inclusion/exclusion criteria. Section 2.3 covers meta-analysis methods for candidate marker gene detection in the "MetaDE" package. Section 2.4 presents methods and issues for pathway enrichment analysis when applying meta-analysis (the MetaPath package). Section 2.5 introduces some ongoing development of meta-analysis exploratory tools for dimension reduction (MetaPCA), disease subtype discovery (MetaSparseKmeans), gene module clustering (MetaGeneClust), differential network module identification (MetaDiffNet), and prediction analysis (MetaPredict). Section 2.6 provides a prostate cancer example to illustrate the use of MetaQC, MetaDE, and MetaPath. Finally, Section 2.7 provides closing remarks and discussions. Table 2.1 is a list of purposes and contents of all subpackages in the MetaOmics software suite. All the packages and source codes are available at http://www.pitt.edu/~tsengweb/MetaOmicsHome.htm.

2.2 Data Preprocessing and MetaQC

Before performing a meta-analysis, data must be retrieved and preprocessed adequately. Ramasamy et al. (2008) outlined a seven-step practical guideline to conducting microarray meta-analysis for detecting differentially expressed genes: (1) identify suitable microarray studies; (2) extract the data from studies; (3) prepare the individual data sets; (4) annotate the individual datasets; (5) resolve the many-to-many relationship between probes and genes; (6) combine the study-specific estimates; and (7) analyze, present, and interpret results. In step 1–4, the research objectives and inclusion/exclusion criteria are identified and a comprehensive literature search is performed. Each individual study is retrieved and preprocessed (e.g., normalization, filtering, and missing value imputation) using conventional methods. In step 5, genes or probes across different studies are matched to resolve the many-to-many relationship.

Table 2.1 *Summary of subpackages in the MetaOmics software suite*

Section	Subpackages	Purpose	Methods available in the package	
			Methods from the literature	Methods developed in our lab
1.1	MetaQC	Quality control and inclusion/exclusion criteria	–	MetaQC (Kang et al., 2012)
1.2	MetaDE	Detection of differentially expressed genes	Fisher's method, Stouffer's method, minP, maxP, FEM, REM, RP, RS, PR, SR	AW-Fisher's method (Li and Tseng, 2011), rOP (Song and Tseng, 2014), Meta-MCC (Lu et al., 2010), MetaACV (Wang et al., 2012), MetaImpute (Tang et al., 2014)
1.3	MetaPath	Detection of enriched pathways	–	MetaPath (Shen et al., 2010)
1.4	MetaPCA	Dimension reduction	JIVE (Lock et al., 2013)	MetaPCA (unpublished)
	MetaClust	Cluster analysis for disease subtype discovery or gene module identification	–	MetaSparseKmeans (unpublished), MetaGeneClust (unpublished)
	MetaNetwork	Detection of conserved or differential network	–	MetaDiffNet (unpublished), MetaLA (unpublished)
	MetaPredict	Prediction analysis	–	MetaTSP (unpublished)

A conventional practice is to match all gene features or probes in each study to a unique gene symbol or UniGene ID. The many-to-one relationship is then simply solved by taking an average or representative probe (e.g., the one with the largest interquartile range). Then meta-analysis methods introduced in Section 2.3 can be used in step 6.

In step 1, the inclusion/exclusion of transcriptomic studies may significantly impact the accuracy and performance of downstream meta-analysis. Including as many informative studies as possible increases statistical power, while including poor quality or problematic studies will dilute information, weaken statistical power, or even distort final biological conclusions. In the literature, researchers often used ad hoc expert opinion or naive threshold decided by sample size or platform. In the MetaQC package (Kang et al., 2012), we provided an objective and quantitative tool to help determine the inclusion/exclusion of studies for meta-analysis. Specifically, MetaQC calculates the following six quantitative quality control (QC) measures: (1) internal homogeneity of coexpression structure among studies (IQC), (2) external consistency of coexpression pattern with pathway database (EQC), and (3) accuracy and (4) consistency of differentially expressed gene detection (AQCg and CQCg) or enriched pathway identification (AQCp and CQCp). Each QC index is defined using $-\log(p\text{-value})$, where p-values are from formal hypothesis testing in each QC criterion. A summary of the QC criteria is provided in the following.

1. *Internal quality control (IQC)* is the first QC criterion, in which the internal homogeneity of the coexpression structure of each study is evaluated using the IQC index. IQC compares pairwise differences of the studies in an unsupervised manner (without any prior or external information other than the expression profile data). The aim is to identify potentially inconsistent or outlier studies from quantified coexpression dissimilarity. It applies the correlation of correlations measure that was previously used in reproducibility analysis of gene coexpression patterns across different studies, which is called integrative correlation coefficients (Parmigiani et al., 2012). If the coexpression structure of a problematic study is significantly different from the other studies, its IQC index (calculated using a p-value from the Wilcoxon rank-sum test) will produce a small score and a warning message is generated.

2. *External quality control (EQC)* is supervised by external pathway information from relevant databases. Pathway information such as functional or coregulated gene sets is obtained from established databases such as KEGG, GO, Biocarta, and Reactome. This information is applied to evaluate the consistency of the coexpression structure for a given study and subsequently to

determine the quality of a study. EQC uses a similar concept as integrative correlation that is used in IQC, and a problematic study with inconsistent coexpression structure generates a small EQC score. EQC relies on the selected pathway set, but the evaluation of a study is independent.

3. *Accuracy quality control (AQCg and AQCp) and consistency quality control (CQCg and CQCp)* aim at quantifying the reproducibility (accuracy or consistency) of DE genes (subscript "g") or pathways (subscript "p") identified by each individual study compared to those identified using meta-analysis of all the other studies. In AQC, the DE genes or pathways detected by the meta-analysis of all the other studies are considered the truth, and the accuracy of DE genes or pathways detected by a given study is calculated to derive the QC index. For CQC, the similarity (consistency) of the DE genes or pathways detected by a given study and the meta-analysis of all other studies are calculated.

After generating six QC indexes, visualization plots and summarization tables are generated using principal component analysis (PCA) biplots and standardized mean ranks (SMR) to assist in visualization and decision. PCA biplot is a popular technique to simultaneously show observations (studies) and relative positions of variables (six QC indexes) in a two-dimensional space so that the performance of each observation (study) can be interpreted by each variable intuitively. In MetaQC, each microarray study is projected from six-dimensional QC measures to a two-dimensional PC subspace. The direction of each QC measure is juxtaposed on top of the two-dimensional subspace using arrows. Specifically, the coordinates of each quality criterion are determined by its correlation to the two driving PCs. The origin of the biplot is taken as the statistical threshold with Bonferroni correction (i.e., projected from $\log(0.05/\#studies)$ in each of the QC measure dimensions), suggesting that studies located in the opposite (negative) area of arrows are candidate outlier studies. The scale of each QC measure is standardized before PCA to avoid dominance of a particular QC measure from scaling issues. In addition to biplot visualization, we also define a quantitative summary score by calculating the ranks of each QC measure among all studies and then compute a standardized mean rank (SMR) of each study: $\frac{\text{mean rank of all QC measures}}{\#\text{ of studies}}$. By definition, $0 < SMR \leq 1$, and large SMR represents a likely problematic study. Note that our visualization and summarization tools are not meant for an automated "proof" of problematic studies but rather are suggested for further inspection to detect potential technical (small sample size or flawed protocol) or biological (irrelevant biological objective or potential tissue contamination) causes of their low quality and to determine their exclusion from meta-analysis. In the prostate cancer example in Section 2.6, MetaQC will be applied to determine problematic studies to be removed from meta-analysis.

2.3 MetaDE for Marker Gene Detection

2.3.1 Meta-Analysis Methods Implemented in MetaDE

Microarray meta-analysis can be categorized into three types: combining *p*-values, combining effect sizes, and combining ranks. *P*-value combination methods can be further divided into "evidence aggregation methods" and "order statistics methods." Note that many methods covered in this subsection are also applicable to GWAS meta-analysis in Chapter 1. Although there are many common features and considerations in GWAS and transcriptomic meta-analysis, they differ in at least two aspects. (1) In GWAS, the measurements from the assays are concluded to assess genotyping in the DNA level (e.g., 0, 1, or 2 representing dosage of minor frequency allele at a particular SNP). The genotyping is not expected to change in different tissues under different conditions, and the genotyping accuracy is very high (usually $>99\%$). Conversely, gene expression levels are in continuous measurements and can be variable in different tissues and conditions. (2) In GWAS meta-analysis, it is common to aggregate information across studies to assess the effect size of association between genotyping and case-control phenotype. For transcriptomic meta-analysis, the design across studies can be quite variable, and different array platforms are often used. Different studies may not provide comparable effect size measures. As a result, effect size combination methods are more popular in GWAS meta-analysis, whereas *p*-value and rank combination methods are more robust and suited to transcriptomic meta-analysis.

Combine p-values: Evidence Aggregation Methods

In this type of method, *p*-values are transformed into evidence scores, and all evidence scores are summed to form a total score. In general, the transformation function uses an inverse cumulative density function (CDF) $F^{-1}(\cdot)$, and so the total score $T = \sum_{k=1}^{K} F^{-1}(P_k)$. The Fisher's and Stouffer's methods are special cases where $F(\cdot)$ are CDFs of χ^2 distribution with 2 degrees of freedom and standard normal distribution, respectively. One less common method uses logit transformation, where $F(x) = 1/(1 + e^{-x})$ and $T = \sum_{k=1}^{K} \log(P_k/(1 - P_k))$. In addition, Li and Tseng (2011) developed the AW-Fisher's method to improve the statistical power when studies are heterogeneous. Here we discuss Fisher's, Stouffer's, and AW-Fisher's methods in detail.

Fisher's method (Fisher, 1925) uses log-transformation of *p*-values from individual studies. Fisher's statistic or total evidence score is $T_{\text{Fisher}} = -2 \cdot$

$\sum_{k=1}^{K} \log(P_k)$, where P_k is the p-value from the kth study. Here T_{Fisher} follows a chi-square distribution with $2K$ degrees of freedom under the null hypothesis of all studies. The method is admissible and asymptotically Bahadur optimal under some simplified Gaussian scenarios (Birnbaum, 1954). Note that there is an option to perform permutation in the MetaDE package instead of using a chi-square distribution to obtain p-values. This permutation option is also available for the Stuffer's and AW-Fisher's methods. The permutation analysis demands larger computation but avoids distributional assumptions and also better accommodates gene dependence structure.

Stouffer's method uses inverse normal transformation of p-values. Stouffer's statistic is $T_{\text{Stouffer}} = \sum_{k=1}^{K} z_k / \sqrt{K}$ ($z_k = \Phi^{-1}(1 - P_k)$), where Φ is the CDF of a standard normal distribution. It can be shown that $T_{\text{Stouffer}} \sim N(0, 1)$ under null hypothesis. Similar to Fisher's method, smaller p-values contribute a larger total score, but the contribution of a single small p-value is not as large as it is in Fisher's method.

The adaptively weighted Fisher's method was first introduced by Li and Tseng (2011) to be used in omics data analysis where the heterogeneity of studies is prevalent. The classical Fisher's method has a major drawback in biological applications in that it does not distinguish homogeneous and heterogeneous signals when combining multiple studies. For example, the Fisher's method gives the same Fisher's statistics when combining five p-values 0.00001, 1, 1, 1, 1 or five p-values 0.1, 0.1, 0.1, 0.1, 0.1, while the biological interpretations of these two biomarkers are very different (see Table 2.2 for more details). AW-Fisher's method is targeting to distinguish different p-value combinations by assigning a binary weight (0 or 1) to each study when combining evidence scores. Let $\vec{w} = (w_1, \ldots, w_K)$, then $T(\vec{w}) = -2 \sum_{k=1}^{K} w_k \cdot \log(P_k)$, where $w_k = 0$ or 1. For a given weight $\vec{w}$, the p-value can be calculated as treating studies with weight 0 as unobserved, that is, $p(T(\vec{w})) = F_{\vec{w}}^{-1}(T(\vec{w}))$, where $F_{\vec{w}}^{-1}$ is the inverse CDF of a chi-square distribution with $2 \cdot \sum_{k=1}^{K} w_k$ degrees of freedom. The AW-Fisher's statistic is the smallest p-value for all $\vec{w}$, that is, $T_{AW} = \min_{\vec{w}} p(T(\vec{w}))$. One important feature of this method is the information about which studies contribute to the evidence aggregation. Here a study with weight 1 is considered from a potential alternative hypothesis (nonzero effective size). For example, an adaptive weight of $(1,1,1,1,1)$ indicates that all five studies contributed to differential analysis, whereas $(1,0,0,0,0)$ means only the first one contributed to the final inference on differential analysis. Besides microarray meta-analysis, the concept of adaptive weights has been applied in GWAS meta-analysis (Bhattacharjee et al., 2012; Han et al., 2012), eQTL meta-analysis (Flutre et al., 2013), and a meta-analysis of multiple mutants to detect differential phosphorylation (Zhang et al., 2013). Fisher's, AW-Fisher's, and minimum p-value methods (to be introduced later) were shown to be admissible under simplified Gaussian scenarios, which indicates that each method has its own advantage under different alternative hypothesis settings. We have further shown that AW-Fisher's method always provides near-optimal statistical power when it is not theoretically optimal. In an on-going project, the consistency of the adaptive weight estimate ($\vec{w}^*$) and asymptotic Bahadur optimality of AW-Fisher's method were proved. A fast and accurate p-value calculation algorithm is also discussed, and a function to calculate p-value based on importance sampling and spline smoothing will be provided in the MetaDE package, which is essential for omics applications.

Order Statistics Methods

Minimum p-value (minP) uses the minimum p-value from the K studies as the test statistic (Tippett, 1931). Here $T \sim \text{Beta}(1, K)$ under the null hypothesis. This method detects a DE gene when at least one very small p-value exists.

Maximum p-value (maxP) uses the maximum p-value from K studies as the test statistic (Wilkinson, 1951). Here $T \sim \text{Beta}(K, 1)$ under the null hypothesis. This method targets DE genes that have relatively small p-values in all studies.

The rth ordered p-value (rOP) uses the rth ordered p-value from the K studies. Under the null hypothesis, $T \sim \text{Beta}(r, K - r + 1)$. The minP and maxP methods are special cases of rOP. The rOP method is considered a robust form of maxP (where r is set as greater than $0.5 \cdot K$) to identify candidate markers differentially expressed in a "majority" of studies (Song et al., 2014). A data-driven algorithm has also been developed to determine r and been implemented in MetaDE.

Combining Effect Sizes

Fixed effects model (FEM) applies a simple linear model to combine the effect sizes from K studies.

Random effects model (REM) (Choi et al., 2003) is a method that is extended from FEM by allowing random effects for each study to address the heterogeneity of studies. A homogeneity test (Cochran, 1954) can be used to determine the existence of random effects and before selecting FEM or REM.

Combining Ranks of Statistics

RankProd (RP) and RankSum (RS) are based on the biological belief that if a gene repeatedly ranks high in terms of fold change in multiple experiments, the gene is more likely a DE gene (Breitling et al., 2004). These two methods are implemented in the RankProd package (Hong et al., 2013) and are also available in the MetaDE package.

Product of ranks (PR) and Sum of ranks (SR) are two methods using the product or sum of the DE evidence ranks inside each study (Dreyfuss et al., 2009). Let R_{gk} be the rank of p-value of gene g among all genes in study k; then the test statistics of PR and SR methods are $PR_g = \prod_{k=1}^{K} R_{gk}$ and $SR_g = \sum_{k=1}^{K} R_{gk}$, respectively. P-values based on these test statistics can be calculated analytically or from a permutation test. Note that the ranks ordered from the smallest to largest (the default) are more sensitive than ranks ordered largest to smallest in the PR method, whereas rank order makes no difference in the SR method.

2.3.2 Hypothesis Settings Behind the Methods

Different meta-analysis methods target different underlying alternative hypothesis settings. Consider a situation to combine K transcriptomic studies using a meta-analysis method where each study contains G genes for information integration. Let θ_{gk} be the underlying true effect size for gene g in study k ($1 \le g \le G$, $1 \le k \le K$). For gene g, two complementary hypothesis settings have been used:

$$HS_A : \left\{ H_0 : \bigcap_k \{\theta_{gk} = 0\} \text{ versus } H_a^{(A)} : \bigcap_k \{\theta_{gk} \ne 0\} \right\}$$

$$HS_B : \left\{ H_0 : \bigcap_k \{\theta_{gk} = 0\} \text{ versus } H_a^{(B)} : \bigcup_k \{\theta_{gk} \ne 0\} \right\}$$

Similar hypothesis settings considered by Birnbaum (1954) and other types were also discussed, for example, the union-intersection test (UIT) (Roy, 1953) and intersection-union test (Berger, 1982), also known as conjunction and disjunction tests. In HS_A, the targeted biomarkers are those differentially expressed in all studies, whereas HS_B pursues biomarkers differentially expressed in at least one of all studies. In terms of biological meanings, HS_A is stringent and desirable to identify consistent and homogeneous biomarkers across all studies. HS_B, however, is useful when heterogeneity is expected. For example, when studies are analyzing different tissues to be combined (e.g., study 1 uses epithelial tissues and study 2 uses blood samples), it is reasonable to identify tissue-specific biomarkers detected under HS_B. Note that HS_B is identical to the classical UIT (Roy, 1953). Most popular transcriptomic meta-analysis methods target HS_B, including classical Fisher's method, Stouffer's method, minP, and AW-Fisher's method. The maximum p-value method (maxP) is probably the only method designed to target HS_A. By taking the maximum of p-values from combined studies as the test statistic, the method requires that all p-values be small for a gene to be detected.

The HS_A hypothesis setting and maxP method are obviously too stringent in light of the generally noisy nature of microarray experiments. When K is

Table 2.2 *Five hypothetical genes to compare different meta-analysis methods and hypothesis settings*

	Gene A	Gene B	Gene C	Gene D	Gene E
Study 1	0.1	1E-3*	1E-5*	0.25	0.15
Study 2	0.1	1	1	0.25	0.15
Study 3	0.1	1	1	0.25	0.15
Study 4	0.1	1	1	0.25	0.15
Study 5	0.1	1	1	0.25	0.9
Fisher's method (HS_B)	0.011*	0.182	0.011*	0.179	0.119
AW-Fisher's method (HS_B)	0.075	0.009*	1E-04*	0.637	0.287
	(1,1,1,1,1)	(1,0,0,0,0)	(1,0,0,0,0)		
Stouffer's method (HS_B)	0.002*	1	1	0.066	0.100
minP (HS_B)	0.410	0.005*	5e-05*	0.763	0.556
maxP (HS_A)	1E-5*	1	1	0.001*	0.590
rOP ($r = 4$) (HS_r)	5E-4*	1	1	0.016*	0.002*

*$p < 0.05$.

large, HS_A is not robust and inevitably detects very few genes. Instead of requiring differential expression in all studies, biologists are more interested in, for example, "biomarkers that are differentially expressed in more than 70% of the combined studies." This leads to a robust hypothesis setting HS_r (Song et al., 2014). Let $\Theta_{gh} = \left\{ \sum_{k=1}^{K} I(\theta_{gk} \neq 0) = h \right\}$ be the situation that exactly h out of K studies are differentially expressed. The new robust hypothesis setting becomes

$$HS_r : \left\{ H_0 : \bigcap_k \{\theta_{gk} = 0\} \text{ versus } H_a^{(r)} : \bigcup_{h=r}^{K} \Theta_{gh} \right\}$$

where $r = \lceil p \cdot K \rceil$, $\lceil x \rceil$ is the smallest integer no less than x and p $(0 < p \leq 1)$ is the minimal percentage of studies required to call differential expression (e.g., $p = 70\%$).

Table 2.2 shows the results when we consider five genes in each study to illustrate different meta-analysis methods under different underlying hypothesis settings. Here gene A has marginally significant p-values ($p = 0.1$) in all five studies; genes B and C have a strong p-value in study 1 ($p = 1E - 3$ and $p = 1E - 5$) but no signal $p = 1$ in the other four studies; gene D is similar to gene A but has much weaker statistical significance ($p = 0.25$ in all five studies); gene E differs from gene D in that the first four studies have small p-values ($p = 0.15$) but study 5 has a large p-value ($p = 0.9$). P-values for six meta-analysis methods derived from classical parametric inference (independent and uniform null p-values) are shown in the table. Comparing Fisher's method and minP for HS_B, minP is sensitive to a study that has a very small p-value (e.g., genes B and C), whereas Fisher's method, as an evidence aggregation method, is more sensitive when all or most studies are marginally statistically significant (e.g., gene A). Comparing genes A and C, Fisher's method gives the same p-values while these two genes have very different biological interpretations. AW-Fisher's method provides a better solution with the estimated adaptive weights (1,1,1,1,1) for gene A and (1,0,0,0,0) for gene C, which indicate the homogeneity and heterogeneity natures of studies in the meta-analysis. We also find that AW-Fisher's method is statistically powerful in both extremes in gene A and gene B, whereas Fisher's method and minP are sensitive in only one extreme. Note that gene D and gene E cannot be detected using Fisher's method, AW-Fisher's method, Stouffer's method, and minP methods that are designed for HS_B. However, p-values to detect gene D from maxP and rOP methods are 0.001 and 0.016, respectively, indicating differential expression. Gene E cannot be identified by maxP ($p = 0.59$) but can be detected by rOP at $r = 4$ ($p = 0.002$). Gene E gives a good motivation for rOP in that maxP may be too stringent when many studies are combined, whereas rOP provides additional robustness when one or a small portion of studies do not have differential expression. In omics meta-analysis, genes similar to gene E are common due to the noisy nature of high-throughput genomic experiments or when a low-quality study is accidentally included in the meta-analysis. In summary, the choice of the best method for a given application depends on the biological objectives and data structure, which is illustrated in the next section with a comparative study.

2.3.3 Discussion of Several Related Issues

Selection of the Best Method

As was discussed in the previous section, the best meta-analysis method for a given application depends on the biological objectives and underlying hypothesis settings. A comprehensive comparative study on the 12 meta-analysis methods based on simulations and four evaluation criteria using six real applications has been published to provide guidance in real practice (Chang et al., 2013). In the simulations, combining five transcriptomic studies with 2000 genes was considered. Let K_0 be the number of differentially expressed studies, that is, $\sum_{k=1}^{K} I(\theta_k \neq 0) = K_0$. In each simulation, we simulated 50 samples in each group and the 2000 genes consisting of 1000 non-DE genes ($K_0 = 0$) and 200 DE genes of the types $K_0 = 1, 2, \ldots, 5$. Figure 2.1A shows the number of detected DE genes of each K_0 category when using FDR $= 5\%$. The result elucidates the types of hypothesis settings behind the methods: maxP, PR, and SR

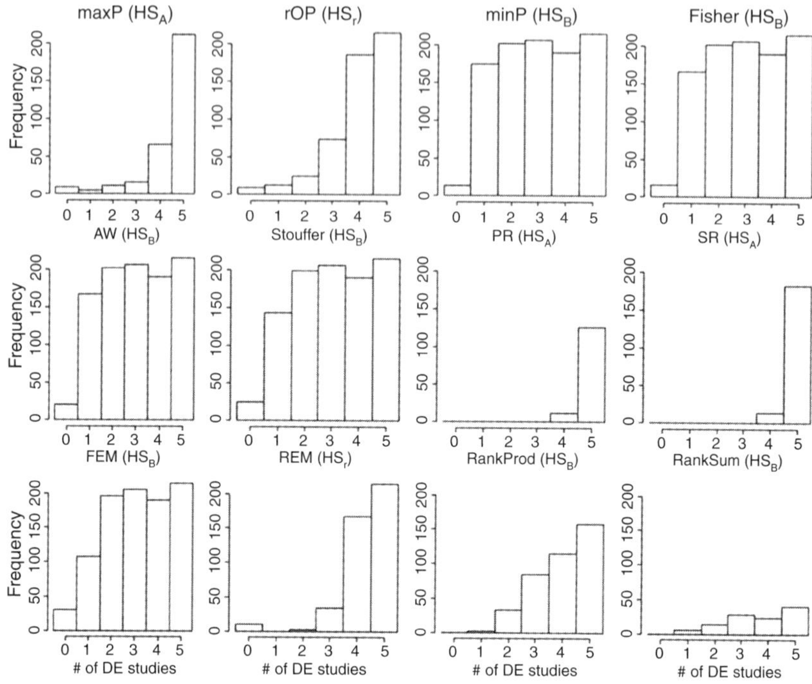

Figure 2.1 The histograms of the true number of DE studies among detected DE genes under FDR = 5% in each method. X-axes: number of DE studies K_0; Y-axes: number of true detected DE genes. From Chang et al. (2013).

for HS_A; minP, Fisher's method, AW, Stouffer's method, FEM, RankProd, and RankSum for HS_B; rOP and REM for HS_r. Six transcriptomic meta-analysis applications (prostate cancer with normal vs. primary cancer, prostate cancer with primary cancer vs. metastasis, brain cancer, major depressive disorder, lung disease, and breast cancer survival) were used to evaluate these 12 methods using four evaluation criteria: detection capability, biological association, stability, and robustness. Table 2.3 shows the final conclusion of method performance using rank sum over six applications. Among the HS_A methods, PR is recommended. Among HS_B methods, both Fisher's method and AW-Fisher's method performed well. But AW-Fisher's method generates adaptive weights that provide better biological interpretation and is preferred in general. In HS_r methods, rOP is recommended. Our paper also applied a multidimensional scaling plot to visualize similarity of detected DE genes across the 12 methods (Figure 2.2A) and used an entropy measure to quantify whether the marker genes in the meta-analysis data are homogeneous (Figure 2.2B). In the entropy box plots of Figure 2.2B, the data sets of "prostate cancer (Pri vs. Meta)"

Table 2.3 *Ranks of method performance in the four evaluation criteria*

	Targeted HS	Detection capability	Biological association	Stability	Robustness	Rank sum	MDS
PR	HS_A	12	4	4	6	26	1
SR	HS_A	11	6	9	7	33	1
maxP	HS_A	9	10	12	11	42	2
rOP	HS_r	7	5	10	10	32	2
REM	HS_r	10	11	5	8	34	3
Fisher's method	HS_B	1	2	3	3	9	1
AW	HS_B	2	3	6	2	13	1
Stouffer's method	HS_B	3	1	8	4	16	1
minP	HS_B	4	7	7	1	19	1
RankProd	HS_B	8	8	1	5	22	3
RankSum	HS_B	6	12	2	12	32	3
FEM	HS_B	5	9	11	9	34	3

generate a smaller entropy, indicating heterogeneous signals of many candidate markers, and so applying AW-Fisher's method for HS_B is more adequate. On the contrary, data sets for MDD obtain larger entropy. Most DE genes are consistent and homogeneous across studies, and HS_A is more appropriate.

Homogeneity and Heterogeneity

In traditional meta-analysis, a random effects model is designed to characterize the homogeneity and heterogeneity of combined effect sizes. The REM model and its associated Cochran's homogeneity test, however, rely on a strong model (Gaussian distribution) assumption and are less applicable in omics applications. Among the aforementioned methods, only the AW-Fisher's method can help characterize gene-specific and genome-wide homogeneity and heterogeneity in the meta-analysis. We expect more active research of homogeneity and heterogeneity analysis in different types of omics data integration in the future.

Concordance of Expression Change Direction

Methods of combining p-values have a potential drawback compared to those combining effect sizes in that they may detect genes with discordant directions of expression changes across studies (e.g., up-regulated in one study but down-regulated in another study). These candidate genes tend to be spurious and should be avoided unless any biological reason exists (e.g., different studies apply different tissues such that discordant expression changes are expected). Owen (2009) applied a one-sided p-value correction

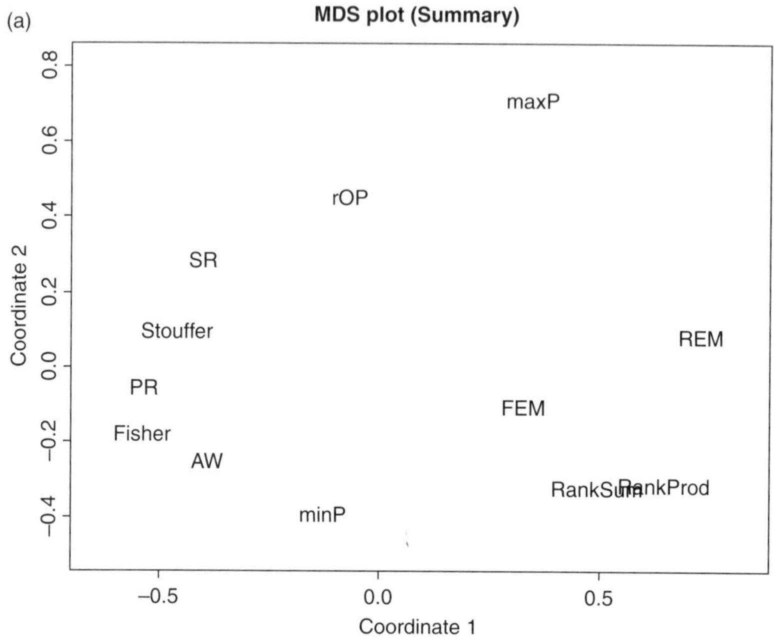

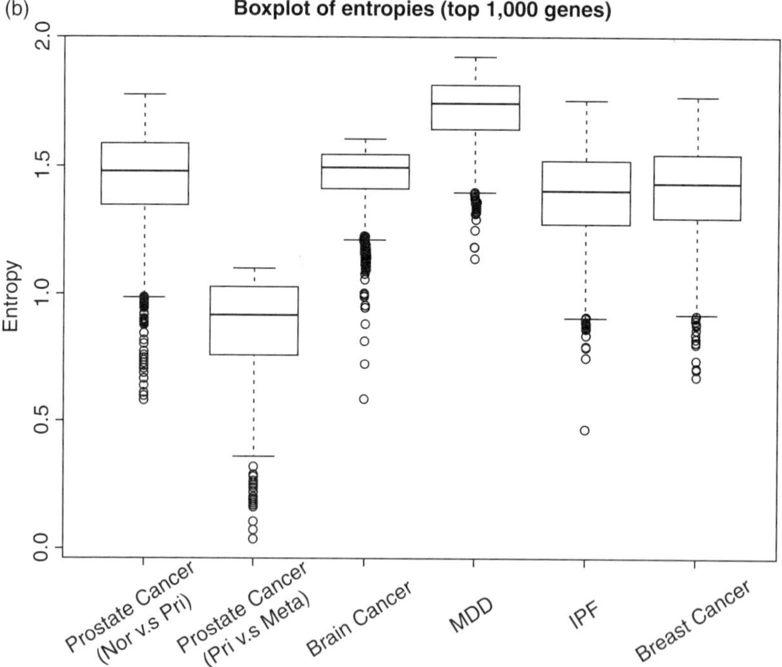

Figure 2.2 Characterization of methods and data sets. From Chang et al. (2013).

for Fisher's method. Specifically, he considered the modified test statistic $T^C_{\text{Fisher}} = \max(T^L_{\text{Fisher}}; T^R_{\text{Fisher}})$, where $T^L_{\text{Fisher}} = -2\sum_{k=1}^{K}\log(\tilde{p}_k)$, $T^R_{\text{Fisher}} = -2\sum_{k=1}^{K}\log(1-\tilde{p}_k)$ and $\tilde{p}_k$ is the left one-sided p-value of study k. The method attempts to avoid detection of discordant DE genes across studies by offsetting discordance signals. However, this method has an immediate drawback in that it does not have a closed-form solution for fast p-value evaluation. In our experience, this type of method cannot guarantee to exclude discordant DE genes (Li and Tseng, 2011). In an ongoing project, we are developing a thresholding version of AW-Fisher's method to guarantee concordance of DE gene detection, and the method will be included in MetaDE in the future. When combining studies with multiple class labels (e.g., multiple disease subtypes or treatments), the issue of discordant expression patterns is also common. We have developed a multiclass correlation (MCC) method for meta-analysis of these types of data (Lu et al., 2009) to avoid the problem.

Inference and Computing

Meta-analysis for DE gene detection involves large-scale computing. For methods applying parametric models and inferences (e.g., Fisher's method, Stouffer's method, minP, maxP, rOP, FEM, and REM), computing can be much simplified. A p-value for each gene can be sequentially derived, and the Benjamini-Hochberg (BH) procedure is often used to control a false discovery rate. For methods without a closed-form solution and requiring permutation analysis (e.g., the RankProd and RankSum), computing is several-hundred-fold heavier. For AW-Fisher's method, we have recently applied a fast importance sampling technique for p-value evaluation (unpublished). Note that because the parametric inference followed by the BH procedure does not account for gene dependence structure, permutation analysis (by randomly permuting samples) is sometimes preferred for more accurate inference, and thus we provide it as an option in MetaDE. But in most applications, if the goal is to prioritize and identify the top candidate markers for pathway analysis and hypothesis generation, parametric inference followed by BH correction will save computing time and can serve the biological purpose.

Other Special Situations

The MetaDE package contains methods for the following two special situations commonly encountered in transcriptomic meta-analysis.

First, raw data and thus p-values are often not available for some data sets, and only a list of statistically significant DE genes (p-value less than a threshold) is provided in the publication (Griffith et al., 2006). Although many journals and funding agencies have encouraged or enforced data sharing policies, the situation has only improved moderately. In the motivating colorectal cancer example in Tang et al. (2014), three of the seven data sets have raw data available from GEO. The remaining four studies only have lists of 1950, 130, 448, and 245 DE genes under p-value thresholds 1E-7, 0.006, 0.001, and 0.009. As a result, we have developed mean and multiple imputation methods for Fisher's and Stouffer's methods for meta-analysis. The result shows significant improvement in statistical power and biological findings by pathway enrichment analysis. These imputation methods for transcriptomic meta-analysis are included in the MetaDE package.

Second, we combined eight transcrimptomic studies of postmortem brain tissues from major depressive disorder patients and matched controls. These studies have small sample sizes (due to difficulty in obtaining postmortem brain tissues), weak signal, paired design, and up to 7–10 confounding variables (e.g., age, pH value of brain, alcohol taking, drug history). We developed a random intercept model with variable selection using Bayesian information criteria (BIC) and meta-analysis (MetaACV) to account for the three difficulties (Wang et al., 2012). The result showed improved statistical power in disease biomarker detection and identified clinical variables that frequently impacted expression patterns in disease biomarker detection. Since such an experimental design with a small sample size and a large number of confounding variables is quite common for complex diseases with difficult-to-obtain tissues, we have included MetaACV in the MetaDE package.

2.4 MetaPath for Pathway Analysis

It is common that DE gene lists obtained from different studies associated with the same disease status or phenotype often have little overlap (Ein-Dor et al., 2005; Tan et al., 2003), while pathway analysis (aka gene set enrichment analysis) often generates more consistent results (Manoli et al., 2006). As a result, a meta-analysis framework to improve the statistical power of pathway enrichment analysis is appealing to provide a more robust and powerful tool than a single study of pathway analysis. Figure 2.3 presents the general framework of pathway enrichment analysis for a single study (Figure 2.3A) and three variations of meta-analysis pathway enrichment (MAPE) methods in the Meta-Path package (Figure 2.3B–2.3D) (Shen et al., 2010). Suppose a data matrix $\{x_{gs}\}$ ($1 \leq g \leq G, 1 \leq s \leq S$) represents the gene expression intensity for gene g and sample s. A binary phenotype label is available for each sample: $\{y_s\}$ ($1 \leq s \leq S$) and $y_s \in \{0, 1\}$ (e.g., 0 represents normal patients and 1 represents tumor patients). A pathway database matrix $\{z_{gp}\}$ ($1 \leq g \leq G, 1 \leq p \leq P$) represents the pathway information of P pathways where $z_{gp} = 1$ when gene

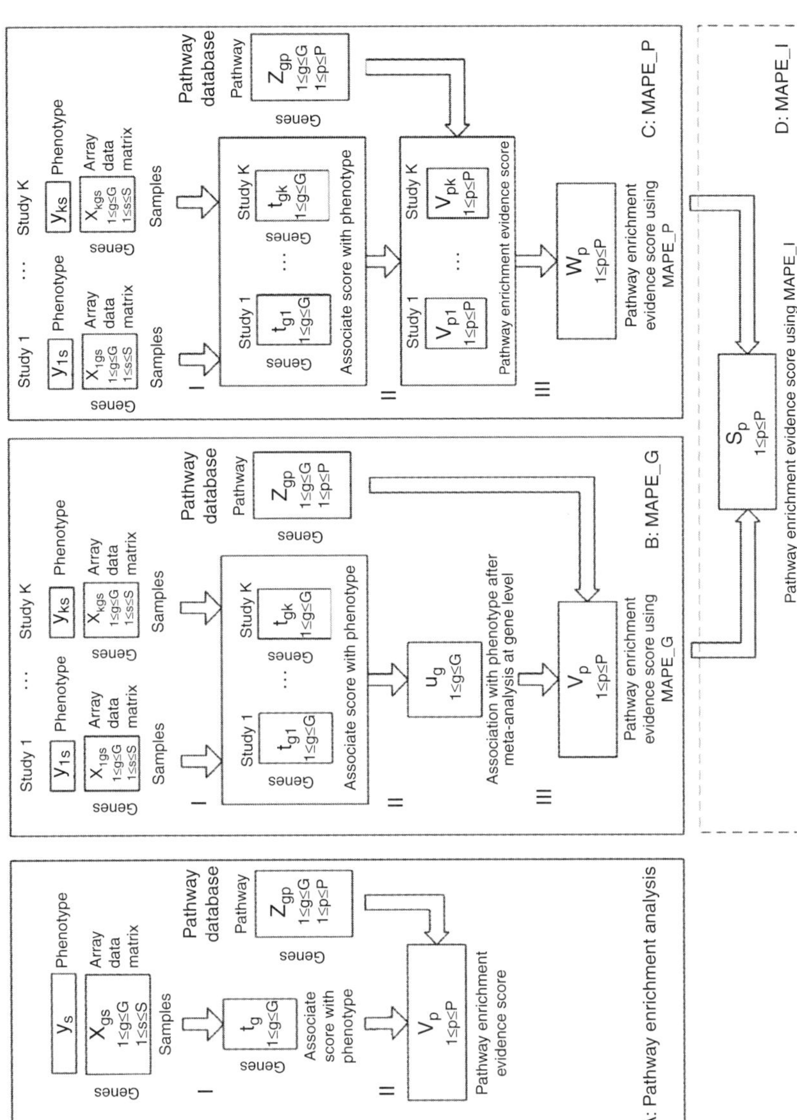

Figure 2.3 Diagram for the three MAPE methods in the MetaPath package. (a) Pathway enrichment analysis for an individual study. (b) MAPE_G, C. MAPE_P, D. MAPE_I. From Shen et al. (2010).

g belongs to pathway p and $z_{gp} = 0$ otherwise. In step I of Figure 2.3A, the association score of phenotype in each gene g is first calculated as t_g (usually by t-statistics or a correlation measure). In step II, a pathway enrichment evidence score v_p is calculated for each pathway p (e.g., Kolmogorov-Smirnov (KS) statistics or mean of t-statistics). A genewise and/or samplewise permutation test is then performed to assess the statistical significance of v_p. Q-value of each pathway is then assessed, and the false discovery rate is controlled at 5% or more loosely at 10% or 20%.

To implement meta-analysis for pathway enrichment, denote by $\{x_{kgs}\}$ ($1 \leq k \leq K$, $1 \leq g \leq G$, $1 \leq s \leq S_k$) the expression intensity of gene g and sample s in study k in Figure 2.3B and 2.3C. $\{y_{ks}\}$ ($1 \leq k \leq K$, $1 \leq s \leq S_k$) and $y_{ks} \in \{0, 1\}$ represents the phenotype labels for sample s in study k. Figure 2.3B shows the procedure for the MAPE_G method. In step I, the association scores with phenotype are calculated in each study (i.e., $\{t_{gk}\}$, $1 \leq g \leq G$, $1 \leq k \leq K$). In step II, meta-analysis is performed for biomarker detection and produces a new association score after meta-analysis at the gene level (i.e., $\{u_g\}$, $1 \leq g \leq G$). In step III, the pathway enrichment analysis is performed as in step II of Figure 2.3A. The evidence scores $\{v_p\}$, corresponding Q-values $\{q(v_p)\}$, and a list of enriched pathways are then generated. MAPE_G can be viewed as a natural combination of a meta-analysis for biomarker detection (steps I and II) followed by a pathway enrichment analysis (step III) in a sequential manner. In Figure 2.3C, an alternative MAPE_P framework is shown. The step I of association scores for each study is identical to that in MAPE_G. In step II, instead of performing meta-analysis at the gene level, we perform pathway enrichment analysis in each individual study to obtain the studywise pathway enrichment evidence scores: $\{v_{pk}\}$ ($1 \leq k \leq K$, $1 \leq p \leq P$). The meta-analysis on the pathway level is then performed in step III to derive the combined evidence score w_p. Finally, the corresponding Q-values $\{q(w_p)\}$ are assessed using permutation analysis. Shen et al. (2010) showed that MAPE_G and MAPE_P have complementary statistical power to detect different types of biologically interesting pathways. As a result, an integrative method, named MAPE_I, was developed to incorporate the complementary advantages of both methods (Figure 2.3D). Specifically, a minP statistic is used by taking the minimum p-values from MAPE_G and MAPE_P for each pathway (i.e., $s_p = \min(p(v_p), p(w_p))$). The statistical inference and control of FDR are similarly performed using permutation analysis. Three key elements in MAPE can have multiple options: the association statistics of gene expression and the phenotype, pathway enrichment statistics, and meta-analysis statistics. In the current MetaPath package, we included t-statistics for association; the

KS-test for pathway enrichment; and Fisher's method, maxP, minP, and roP for meta-analysis. We demonstrate a prostate cancer example in Section 2.6 using the MetaQC, MetaDE, and MetaPath packages.

2.5 Other Exploratory Tools

In addition to MetaQC, MetaDE, and MetaPath packages, four additional packages are currently in production or in publication and are briefly introduced.

2.5.1 MetaPCA for Dimension Reduction

Dimension reduction tools such as principal component analysis (PCA), multidimension scaling (MDS), and nonnegative matrix factorization (NMF) have been popularly used in high-dimensional omics data investigation. When multiple transcriptomic studies are available, PCA can be applied to each study and identify its PC subspace for dimension reduction and data mining. Each study, however, generates distinct PC subspace in this case and cannot be generalized or compared. As a result, we have developed a meta-analysis PCA (MetaPCA) framework to perform simultaneous dimension reduction and identify a common PC subspace across multiple transcriptomic studies. We explored two MetaPCA approaches by maximizing sum of variance (SV) decomposition and maximizing the sum of squared cosines (SSC). Through simulations and real data applications, we demonstrated that MetaPCA borrows information across studies to find more robust and accurate PC subspace. Figure 2.4 shows a MetaPCA application in a classical yeast cell cycle example (Spellman et al., 1998). The data contain four studies of cell cycle time series data experimentally synchronized by four chemical procedures (alpha, elutriation, CDC15, and CDC28). In individual PCA analyses, alpha, CDC28, and ELU exhibited a good cyclic pattern while CDC15 shows a mysterious oscillating pattern that has been previously identified (Li et al., 2002). By MetaPCA using SV and SSC approaches, it finds common PC subspaces that show clearer cyclic patterns in all four studies. The MetaPCA R package is publicly available in CRAN and will be incorporated into the MetaOmics software suite. The MetaPCA package also includes implementation of a recent JIVE (Lock et al., 2013) method for simultaneous PC decomposition of multiple omics studies (see Chapter 11). Although JIVE was designed for vertical integration of multiple omics data, it can be applied to horizontal meta-analysis as well. Our analyses have shown that the MetaPCA approaches perform better than JIVE in finding common PC subspaces.

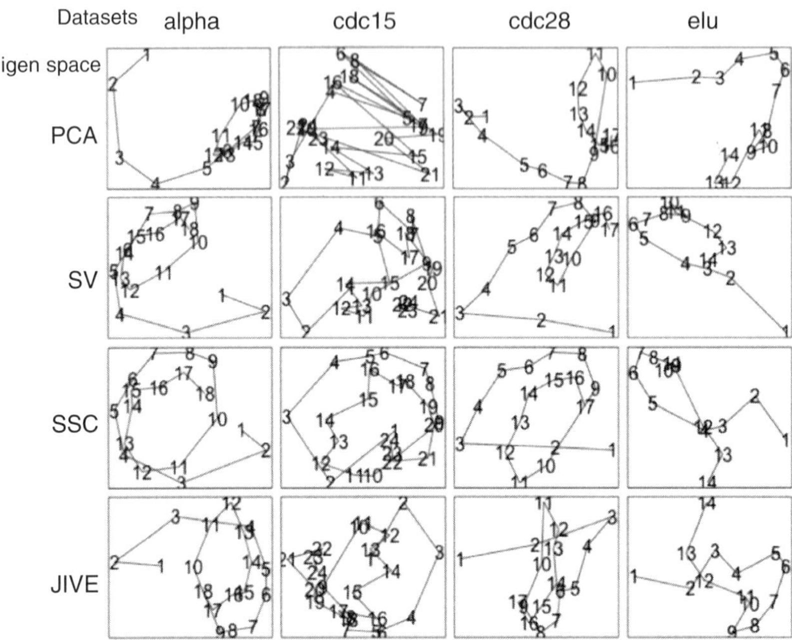

Figure 2.4 Four data sets (18 time series samples in alpha, 24 samples in cdc15, 17 samples in cdc28, and 14 samples in elu) in Spellman yeast cell cycle data were projected to PC subspaces. First row: separated single study PCA analyses; second and third rows: two MetaPCA approaches by SV and SSC optimization; fourth row: JIVE method.

2.5.2 MetaClust for Disease Subtype Discovery and Gene Module Clustering

Disease Subtype Discovery (MetaSparseKmeans)

Disease phenotyping via omics data sets provides new capabilities to identify novel disease subtypes that can potentially benefit patients by personalized medicine. The first step toward this goal is usually an exploratory cluster analysis on the samples to identify disease subtypes. In breast cancer, for example, many studies (Lehn et al., 2014) have demonstrated five clear subtypes (Basel, Luminal-A, Luminal-B, Normal-like, and Her2) that have distinct clinical, survival, and drug responses. One major difficulty in the field is that each transcriptomic study claims its own "intrinsic genes" and phenotyping model, making its translational application difficult. In the MetaClust package, we have developed a MetaSparseKmeans method to combine multiple transcriptomic studies for a unified robust and accurate classification model that is applicable

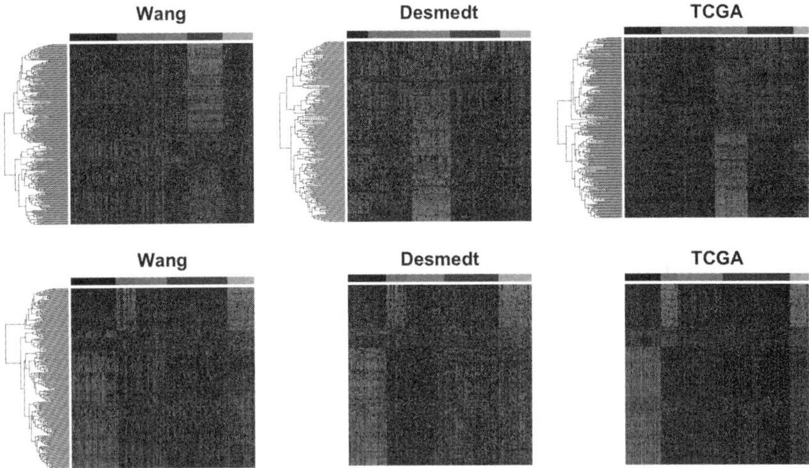

Figure 2.5 MetaSparseKmeans result for breast cancer data of three studies from different platforms. Each row represents a gene that is concordant across studies. In each study, patients are totally divided into five clusters, represented by five unique colors in the color bar at the top of the data matrix of each study. The top three figures are individual clustering results. The bottom three figures are meta-clustering results.

to future patients. The method selects the common intrinsic gene set, performs simultaneous sample clustering, and guarantees concordant matching of subtype expression patterns across studies. We demonstrated success of the method by simulations and applications to breast cancer and leukemia. Figure 2.5 shows the breast cancer example of clustering by each individual study (top) or by combining three studies for simultaneous clustering of a common classification model (bottom).

Gene Module Identification (MetaGeneClust)

In addition to clustering samples, one may be interested in clustering genes to find tightly coexpressed gene modules in the transcriptomic experiments. These coexpressed gene modules may reflect common pathways, regulatory mechanisms, or molecular functions. Because the gene clustering results are usually unstable, a meta-analysis framework by combining multiple studies is appealing to reduce false positives. In the MetaClust package, we are developing two MetaGeneModule approaches by combining gene pair distances or by combining gene clustering results in the meta-analysis. A preliminary application by combining gene pair distances on eight major depressive disorder transcriptomic studies has identified a conserved BDNF, glutamate, and GABA-enriched module that relates to human depression (Chang et al., 2014).

2.5.3 MetaNetwork for Differential Network Detection

Investigating the structure of two coexpression networks might provide additional insight into underlying coexpression patterns of disease relevant regulatory changes. The inference of differential coexpression modules in a single transcriptomic study is, however, usually inaccurate and unstable due to complex biological systems, measurment errors, and large biological variations. Combining multiple transcriptomic studies to identify repeated differential coexpression modules conceptually will provide increased statistical power and reduce false positives. In an ongoing project, we have developed a MetaDiffNet method to detect differentially coexpressed modules between different disease conditions across multiple studies. MetaDiffNet outputs module information with consistent edge connection density differences between two conditions by optimizing a weighted target function. It contains several steps to control the false discovery rate using permutation analysis. The optimization task is extremely difficult to detect large modules so we limited the size of modules to be less than 30 genes and used pathway-guided module assembly to identify biologically meaningful modules. We have demonstrated this method using simulated data sets and two real examples of combining four breast cancer data sets (ER+ vs. ER-) and eight major depressive disorder data sets (disease vs. control). The identified modules were validated and explored by pathway enrichment analysis. Figure 2.6 shows an identified differential coexpressed module in breast cancer ER+ vs. ER− comparison. The network densities (i.e., the percentage of connected edges among all possible edges) decreased from 0.68–0.83 in ER+ patients in the four studies to 0.21–0.28 in ER−. The 16 genes were found highly enriched (5 out of 16) in the KEGG immune response pathway. In a separate effort, we have also developed a meta-analytic framework for the liquid association method (MetaLA) to identify altered gene coexpression structure in different cellular status. We plan to include both MetaDiffNet and MetaLA methods in the MetaNetwork packages.

2.5.4 MetaPredict for Prediction Analysis by Combining Multiple Studies

Geman et al. (2004) developed a robust and accurate "top scoring pair" (TSP) algorithm for prediction analysis in expression profile. The rank-based prediction rule is nonparametric and transparent in interpretation. Although it has been shown to be accurate in a single study using cross-validation, the accuracy

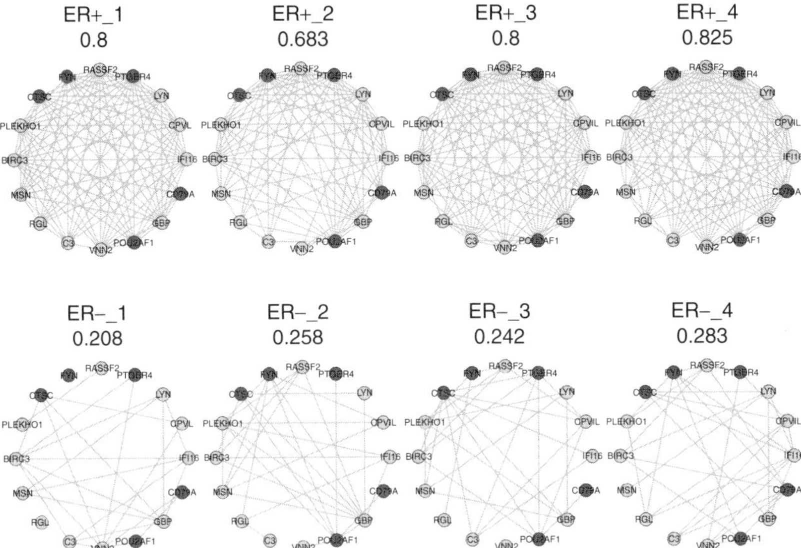

Figure 2.6 A differentially coexpressed module was identified when combining four transcriptomic studies. The four columns in the figure represent coexpression networks (ER+ and ER− separated) from four different transcriptomic studies. First row: coexpression network of ER+ patients. The coexpression network densities were between 0.68 and 0.83. Second row: coexpression network of ER− patients. The network densities dropped to 0.21–0.28. The gene module is enriched in the KEGG immune response pathway ($p < $ 1E-8).

dropped dramatically when we attempted to perform interstudy prediction (i.e., construct the prediction model in one study and bring the model to predict in another study), which is an essential task for transcriptomic data analysis for reproducible and translational research. In an ongoing project, we have developed a meta-analysis version of the TSP algorithm that combines multiple transcriptomic studies to construct a prediction model. In our simulation and real data analysis using lung diseases, breast cancer, and various cancers in TCGA data sets, the proposed MetaTSP method outperforms the single-study TSP method in interstudy prediction.

2.6 Prostate Cancer Example to Demonstrate MetaQC, MetaDE, and MetaPath

In the first publication for MetaOmics software suite (Wang et al., 2012), we included the MetaQC, MetaDE, and MetaPath packages and demonstrated the functionalities using a prostate cancer example. The three R packages allow

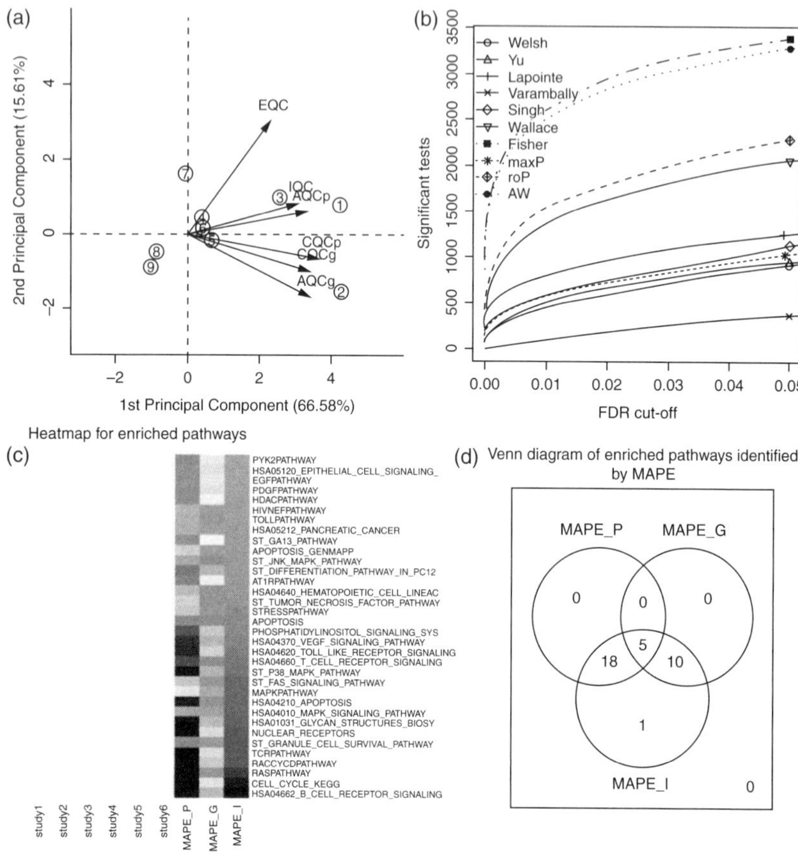

Figure 2.7 (A) PCA biplot from MetaQC. (B) Number of detected DE genes under different q-value thresholds. (C) Heatmap showing $-\log(q - values)$ of detected pathways in single studies, MAPE_P, MAPE_G, and MAPE_I. (D) Venn diagram of detected pathways by the three MAPE methods. From Wang et al. (2012).

flexible input format of experimental data and four different types of outcome variables (case/control, multi-class, continuous and survival). These packages also allow missing values in the individual experimental study or missing values caused by mismatched genes across studies (i.e., genes covered in one study but not covered in another study). For several computationally intensive routines, the packages have options to use multicore parallel computing for timely implementation. In a prostate cancer example, there were nine studies collected containing adjacent normal and primary cancer samples: Welsh et al. (2001), Yu et al. (2004), Lapointe et al. (2004), Varambally et al. (2005),

Singh et al. (2002), Wallace et al. (2008), Nanni et al. (2006), Tomlins et al. (2007), and Dhanasekaran et al. (2001). After matching genes using official gene symbols, preprocessing, and filtering, 4441 genes were used for meta-analysis. Figure 2.7A shows results of PCA biplot by MetaQC. Three of the nine studies (studies 7–9; Nanni, Tomlins, and Dhanasekaran) were identified as low quality and were removed from meta-analysis. Figure 2.7B shows the number of detected DE genes under different FDR thresholds in the remaining six single-study analyses and meta-analyses by Fisher's method, maxP, rOP ($r = 4$), and AW methods. It is clear that meta-analysis usually detects more candidate markers, except for maxP. Finally, Figures 2.7C and 2.7D show a heatmap of detected pathways (q-value < 0.2 in any method) and a Venn diagram of pathways detected by MAPE_P, MAPE_G, and MAPE_I using Meta-Path. The majority of the detected pathways appeared to be cancer related. Single-study analyses showed very weak pathway enrichment; MAPE_P and MAPE_G appeared to have complementary detection power (identified 23 and 15 pathways with only 5 in common). MAPE_I adopted advantages from both MAPE_P and MAPE_G and detected the largest number of pathways (34 pathways).

2.7 Conclusions

Using meta-analysis to combine multiple transcriptomic studies brings new statistical and computational challenges. In the MetaOmics software suite, we included seven R packages for quality control (MetaQC), detecting differentially expressed genes (MetaDE), identifying associated biological pathways (MetaPath), dimension reduction (MetaPCA), cluster analysis (MetaClust), differential network detection (MetaNetwork), and classification analysis (MetaPredict). This suite includes a wide variety of bioinformatics tools to combine multiple transcriptomic studies, which will significantly enhance the understanding of disease mechanisms and help generate novel biological hypotheses. The result will ultimately benefit patients through translating novel findings into more effective treatments or disease management programs.

References

Aizenman, M., and Barsky, D. J. 1987. Sharpness of the phase transition in percolation models. *Comm. Math. Phys.*, **108**, 489–526.

Berger, R. L. 1982. Multiparameter hypothesis testing and acceptance sampling. *Technometrics*, **24**, 295–300.

Bhattacharjee, S., Rajaraman, P., Jacobs, K. B., Wheeler, W. A., Melin, B. S., Hartge, P., GliomaScan Consortium, Yeager, M., Chung, C. C., Chanock, S. J., and

Chatterjee, N. 2012. A subset-based approach improves power and interpretation for the combined analysis of genetic association studies of heterogeneous traits. *Am. J. Hum. Genet.*, **90**, 821–835.

Birnbaum, A. 1954. Combining independent tests of significance. *J. Am. Stat. Assoc.*, **49**, 559–574.

Breitling, R., Armengaud, P., Amtmann, A., and Herzyk, P. 2004. Rank products: A simple, yet powerful, new method to detect differentially regulated genes in replicated microarray experiments. *FEBS Lett.*, **573**, 83–92.

Chang, L. C., Jamain, S., Lin, C. W., Rujescu, D., Tseng, G. C., and Sibille, E. 2014. A conserved BDNF, glutamate- and GABA-enriched gene module related to human depression identified by coexpression meta-analysis and DNA variant genome-wide association studies. *PLoS One*, **9**, 3.

Chang, L., Lin, H., Sibille, E., and Tseng, G. C. 2013. Meta-analysis methods for combining multiple expression profiles: Comparisons, statistical characterization and an application guideline important diagnostic biomarkers. *BMC Bioinformatics*, **14**, 368.

Choi, J. K., Yu, U., Kim, S., and Yoo, O. J. (2003). Combining multiple microarray studies and modeling interstudy variation. *Bioinformatics*, **19**, i84–90.

Cochran, W. G. 1954. The combination of estimates from different experiments. *Biometrics*, **1**, 101–129.

Dhanasekaran, S. M., Barrette, T. R., Ghosh, D., Shah, R., Varambally, S., Kurachi, K., Pienta, K. J., Rubin, M. A., and Chinnaiyan, A. M. 2001. Delineation of prognostic biomarkers in prostate cancer. *Nature*, **412**(6849), 822–826.

Dreyfuss, J. M., Johnson, M. D., and Park, P. J. 2009. Meta-analysis of glioblastoma multiforme versus anaplastic astrocytoma identifies robust gene markers. *Mol. Cancer*, **4**, 71.

Ein-Dor, L., Kela, I., Getz, G., Givol, D., and Domany, E. 2005. Outcome signature genes in breast cancer: Is there a unique set? *Bioinformatics*, **21**, 171–178.

Fisher, R. A. 1925. *Statistical Methods for Research Workers*. Oliver and Boyd.

Flutre, T., Wen, X., Pritchard, J., and Stephens, M. 2013. A statistical framework for joint eQTL analysis in multiple tissues. *PLoS Genet.*, **9**, e1003486.

Geman, D., d'Avignon, C., Naiman, D., Winslow, R., and Zeboulon, A. 2004. Gene expression comparisons for class prediction in cancer studies. *Proceedings 36'th Symposium on the Interface: Computing Science and Statistics*.

Griffith, O. L., Melck, A., Jones, S. J., and Wiseman, S. M. 2006. Meta-analysis and meta-review of thyroid cancer gene expression profiling studies identifies important diagnostic biomarkers. *J. Clin. Oncol.*, **24**, 5043–5051.

Hammersley, J. M. 1957. Percolation processes: Lower bounds for the critical probability. *Ann. Math. Statist.*, **28**, 790–795.

Hammersley, J. M. 1961. Comparison of atom and bond percolation processes. *J. Math. Phys.*, **2**, 728–733.

Hammersley, J. M., and Mazzarino, G. 1994. Properties of large Eden clusters in the plane. *Combin. Probab. Comput.*, **3**, 471–505.

Han, B., and Eskin, E. 2012. Interpreting meta-analyses of genome-wide association studies. *PLoS Genet.*, **8**, 3.

Hong, F., Breitling, R., McEntee, C. W., Wittner B. S., Nemhauser, J. L., and Chory, J. 2006. RankProd: A bioconductor package for detecting differentially expressed genes in meta-analysis. *Bioinformatics*, **22**, 2825–2827.

Kang, D. D., Sibille, E., Kaminski, N., and Tseng, G C. 2012. MetaQC: Objective quality control and inclusion/exclusion criteria for genomic meta-analysis. *Nucleic Acids Res.*, **40**, e15.

Kesten, H. 1990. Asymptotics in high dimensions for percolation. Pages 219–240 of: Grimmett, G. R., and Welsh, D. J. A. (eds.), *Disorder in Physical Systems: A Volume in Honour of John Hammersley*. Oxford University Press.

Lapointe, J., Li, C., Higgins, J. P., van de Rijn, M., Bair, E., Montgomery, K., Ferrari, M., Egevad, L., Rayford, W., Bergerheim, U., et al. 2004. Gene expression profiling identifies clinically relevant subtypes of prostate cancer. *Proc. Nat. Acad. Sci. USA*, **101**(3), 811–816.

Lehn, S., Tobin, N. P., Sims, A. H., Stl, O., Jirstrm, K., Axelson, H., and Landberg, G. 2014. Decreased expression of Yes-associated protein is associated with outcome in the luminal A breast cancer subgroup and with an impaired tamoxifen response. *BMC Cancer*, **14**, 119.

Li, J., and Tseng, C. G. 2011. An adaptively weighted statistic for detecting differential gene expression when combining multiple transcriptomic studies. *Ann. Appl. Stat.*, **5**, 994–1019.

Li, K., Yan, M., and Yuan, S. 2002. A simple statistical model for depicting the cdc 15-synchronized yeast cell-cycle regulated gene expression data. *Stat. Sin.*, **12**, 141–158.

Lock, E. F., Hoadley, K. A., Marron, J. S., and Nobel, A. B. 2013. Joint and individual variation explained (JIVE) for integrated analysis of multiple data types. *Ann. Appl. Stat.*, **7**, 523–542.

Lu, S., Li, J., Song, C., Shen, K., and Tseng, C. G. 2009. Biomarker detection in the integration of multiple multi-class genomic studies. *Bioinformatics*, **26**, 333–340.

Manoli, T., Gretz, N., Grne, H. J., Kenzelmann, M., Eils, R., and Brors, B. 2006. Group testing for pathway analysis improves comparability of different microarray datasets. *Bioinformatics*, **22**, 2500–2506.

Menshikov, M. V. 1985. Estimates for percolation thresholds for lattices in $\mathbf{R}^n$. *Dokl. Akad. Nauk SSSR*, **284**, 36–39.

Menshikov, M. V., Molchanov, S. A., and Sidorenko, A. F. 1986. Percolation theory and some applications. Pages 53–110 of: *Probability Theory. Mathematical Statistics. Theoretical Cybernetics, Vol. 24 (Russian)*. Akad. Nauk SSSR Vsesoyuz. Inst. Nauchn. i Tekhn. Inform. Translated in *J. Soviet Math.* **42**, 1766–1810 (1988).

Nanni, S., Priolo, C., Grasselli, A., D'Eletto, M., Merola, R., Moretti, F., Gallucci, M., De Carli, P., Sentinelli, S., Cianciulli, M., et al. 2006. Epithelial-restricted gene profile of primary cultures from human prostate tumors: A molecular approach to predict clinical behavior of prostate cancer. *Mol. Cancer Res.*, **4**(2), 79–92.

Owen, A. B., 2009. Karl Pearson's meta analysis revisited. *Ann. Stat.*, **37**, 3867–3892.

Parmigiani, G., Garrett-Mayer, E. S., Anbazhagan, R., and Gabrielson, E. 2004. A cross-study comparison of gene expression studies for the molecular classification of lung cancer. *Clin. Cancer Res.*, 10, 2922.

Ramasamy, A., Mondry, A., Holmes, C. C., and Altman, G. D. 2008. Key issues in conducting a meta-analysis of gene expression microarray datasets. *PLoS Med.*, **5**, e15.

Roy, S. 1953. On a heuristic method of test construction and its use in multivariate analysis. *Ann. Math. Stat.*, **24**, 167–318.

Shen, K., and Tseng, G. C. 2010. Meta-analysis for pathway enrichment analysis when combining multiple genomic studies. *Bioinformatics*, **10**, 1316–1323.

Singh, D., Febbo, P. G., Ross, K., Jackson, D. G., Manola, J., Ladd, C., Tamayo, P., Renshaw, A. A., D'Amico, A. V., Richie, J. P., et al. 2002. Gene expression correlates of clinical prostate cancer behavior. *Cancer Cell*, **1**(2), 203–209.

Song, C., and Tseng, C. G. 2014. Hypothesis setting and order statistic for robust genomic meta-analysis. *Ann. Appl. Stat.,* 777–800.

Spellman, P. T., Sherlock, G., Zhang, M. Q., Iyer, V. R., Anders, K., Eisen, M. B., Brown, P. O., Botstein, D., and Futcher, B. 1998. Comprehensive identification of cell cycle–regulated genes of the yeast *Saccharomyces cerevisiae* by microarray hybridization. *Mol. Biol. Cell.,* **12**, 3273–3297.

Stouffer, S. A., Suchman, E. A., DeVinney, L. C., Star, S. A., Jr., and Williams, R. M. 1949. *The American Soldier, Vol. 1: Adjustment during Army Life.* Princeton University Press.

Tan, P. K., Downey, T. J., Spitznagel, E. L., Jr., Xu, P., Fu, D., Dimitrov, D. S., Lempicki, R. A., Raaka, B. M., and Cam, M. C. 2003. Evaluation of gene expression measurements from commercial microarray platforms. *Nucleic Acids Res.,* **19**, 5676–5684.

Tang, S., Ding, Y., Sibille, E., Mogil, J. S., Lariviere, W. R., and Tseng, G. C. 2014. Imputation of truncated *p*-values for meta-analysis methods and its genomic application. *Ann. Appl. Stat.,* **8**, 2150–2174.

Tippett, L. H. C. 1931. *The Methods of Statistics.* Williams and Norgate.

Tomlins, S. A., Mehra, R., Rhodes, D. R., Cao, X., Wang, L., Dhanasekaran, S. M., Kalyana-Sundaram, S., Wei, J. T., Rubin, M. A., Pienta, K. J., et al. 2007. Integrative molecular concept modeling of prostate cancer progression. *Nat. Genet.,* **39**(1), 41–51.

Tseng, G. C., Ghosh, D., and Feingold, E. 2012. Comprehensive literature review and statistical considerations for microarray meta-analysis. *Nucleic Acids Res.,* **40**, 3785–3799.

Varambally, S., Yu, J., Laxman, B., Rhodes, D. R., Mehra, R., Tomlins, S. A., Shah, R. B., Chandran, U., Monzon, F. A., Becich, M. J., et al. 2005. Integrative genomic and proteomic analysis of prostate cancer reveals signatures of metastatic progression. *Cancer Cell,* **8**(5), 393–406.

Vyssotsky, V. A., Gordon, S. B., Frisch, H. L., and Hammersley, J. M. 1961. Critical percolation probabilities (bond problem). *Phys. Rev.,* **123**, 1566–1567.

Wallace, T. A., Prueitt, R. L., Yi, M., Howe, T. M., Gillespie, J. W., Yfantis, H. G., Stephens, R. M., Caporaso, N. E., Loffredo, C. A., and Ambs, S. 2008. Tumor immunobiological differences in prostate cancer between African-American and European-American men. *Cancer Res.,* **68**(3), 927–936.

Wang, X., Kang, D. D., Shen, K., Song, C., Lu, S., Chang, L. C., Liao, S. G., Huo, Z., Tang, S., Ding, Y., Kaminski, N., Sibille, E., Lin, Y., Li, J., and Tseng, G. C. 2012. An R package suite for microarray meta-analysis in quality control, differentially expressed gene analysis and pathway enrichment detection. *Bioinformatics,* **19**, 2534–2536.

Wang, X., Lin, Y., Song, C., Sibille, E., and Tseng, G. C. 2012. Detecting disease-associated genes with confounding variable adjustment and the impact on genomic meta-analysis: With application to major depressive disorder. *Bioinformatics,* **13**, 52.

Welsh, J. B., Sapinoso, L. M., Su, A. I., Kern, S. G., Wang-Rodriguez, J., Moskaluk, C. A., Frierson, H. F., and Hampton, G. M. 2001. Analysis of gene expression identifies candidate markers and pharmacological targets in prostate cancer. *Cancer Res.,* **61**(16), 5974–5978.

Wilkinson, B. 1951. A statistical consideration in psychological research. *Psychol. Bull.,* **48**, 156–158.

Yu, Y. P., Landsittel, D., Jing, L., Nelson, J., Ren, B., Liu, L., McDonald, C., Thomas, R., Dhir, R., Finkelstein, S., et al. 2004. Gene expression alterations in prostate cancer predicting tumor aggression and preceding development of malignancy. *J. Clin. Oncol.*, **22**(14), 2790–2799.

Zhang, Y., Kweon, H. K., Shively, C., Kumar, A., and Andrews, P. C. 2013. Towards systematic discovery of signaling networks in budding yeast filamentous growth stress response using interventional phosphorylation data. *PLoS Comput. Biol.*, **9**, e1003077.

3

Integrative Analysis of Many Biological Networks to Study Gene Regulation

WENYUAN LI, CHAO DAI, AND
XIANGHONG JASMINE ZHOU

Abstract

Gene expression is a complex process regulated and coordinated through the interactions of cellular components. In recent years, there has been an increasing accumulation of high-throughput biological network data characterizing these interactions at multiple levels and under different conditions and perturbations. Integrative analysis of many such networks is an important means of understanding the complexity of gene regulation. Therefore, there is an urgent need for powerful integrative methods capable of analyzing many biological networks simultaneously. In this chapter, we review three state-of-the-art studies that develop computational methods for identifying three types of patterns in multiple networks: transciptional modules, splicing modules, and coupled transcription-splicing modules. The atlas of these modules facilitates understanding of gene regulation processes, through reconstructing transcriptional regulatory networks, splicing regulatory networks, and exploring how transcription and splicing simultaneously take place.

3.1 Introduction

Gene regulation is an intricate system involving precise interactions between tens of thousands of cellular components, and sophisticated systems coordinating diverse activities across multiple layers. Biological networks are undoubtedly the most expressive data source for describing this interconnectivity. Understanding the complexities of these networks has become a mainstay of systems biology, and is rapidly growing into a new subfield known as network biology. High-throughput genomic technologies are providing unprecedented opportunities to systematically characterize diverse types of biological networks, including protein-protein interactions, metabolic reactions, signaling and transcription-regulatory networks, and coexpression relationships. These networks characterize the variety of the interactions and activities of cellular components and have great potential to reveal underlying gene regulation mechanisms.

Extensive research indicates that gene regulation is governed by multiple, complex, and extensively coupled networks, which can be generated under different conditions and perturbations. Therefore, a single network is insufficient to discover network patterns with multiple facets, subtle signals, and dynamic features, which are often embedded in multiple networks with diverse conditions. A paradigmatic example is the high-order cooperativity analysis of gene regulation presented in this chapter. In addition, single networks often include many false positive signals (edges) incurred by experimental biases and errors, whereas integrative analysis of many networks can produce more reliable results by focusing only on those patterns that frequently occur across multiple networks. Therefore, there is an urgent need for novel computational methods supporting the integrative analysis of many biological networks.

In this chapter, we review three state-of-the-art integrative analyses that can identify gene-regulation modules from many massive biological networks. First, we describe a method to identify heavily connected subgraphs that frequently occur in many edge-weighted biological networks (Li et al., 2011). We used gene coexpression networks derived from 130 human microarray data sets as a testing system for this method. A substantially large fraction of the identified subgraphs are transcriptional regulatory modules, which can be used to reconstruct transcriptional networks. Second, we apply the same method to find splicing regulatory modules in many exon co-splicing networks derived from 38 human RNA-seq data sets (Dai et al., 2012). This research defines a splicing module as a set of cassette exons co-regulated by the same splicing factors. Splicing modules can grant new insights into functions associated with post-transcriptional regulation, and reveal that the same exons can dynamically participate in different pathways depending on different conditions and which other exons are co-spliced. Third, we identify coupled transcription-splicing modules in a series of paired gene coexpression and exon co-splicing networks, each pair being derived from a single RNA-seq data set (Li et al., 2012). We hypothesize that when the genes of a transcriptional module have exons that also form a splicing module, it is likely to implicate that transcription and splicing simultaneously take place. The modules therefore reveal which transcription and splicing factors participate in coupling mechanisms, and how they are mediated through protein-protein interactions.

In addition to presenting the biological discoveries, this chapter introduces state-of-the-art network mining algorithms. All the methods introduced in this chapter are advanced pattern mining techniques based on the tensor that models a large set of massive, edge-weighted networks and enables the efficient module discovery. Specifically, a set of networks is modeled as a third-order tensor (i.e., a three-dimensional array), granting access to a wealth of established numerical tools for efficient pattern discovery. Although heuristic graph-based data mining algorithms have been developed for similar goals, they face two major limitations compared with tensor-based methods. (1) The general strategy of their searching heuristics is a stepwise reduction of the large search space, but each step involves one or more arbitrary cutoffs that may be hard for users to select and control. (2) Most graph-based approaches to analyzing multiple networks are restricted to unweighted networks, which are less informative than the original weighted networks. Weighted networks are often perceived as harder to analyze (Newman, 2004) than unweighted networks, but advanced numerical optimization techniques are equally applicable to either unweighted or weighted networks when expressed in tensor form. In addition, tensor-based methods can easily incorporate additional constraints representing prior knowledge.

3.2 Reconstructing Transcriptional Regulatory Modules

In this section, we propose a tensor-based computational method to efficiently and systematically identify *recurrent heavy subgraphs* (RHSs), which are coexpression gene clusters that frequently occur in many gene coexpression networks. We show that the likelihood of a gene cluster in RHS being a transcription module increases significantly with its recurrence in the networks, highlighting the importance of the integrative approach. Moreover, our high-order analysis reveals dynamic cooperativeness in transcription regulatory networks.

3.2.1 Methods and Materials

Given m networks with the same n vertices but different topologies, we represent the whole system as a third-order tensor $\mathcal{A} = (a_{ijk})_{n \times n \times m}$ (see an example in Figure 3.1). Each element a_{ijk} is the nonnegative weight of the undirected edge between vertices i and j in the kth network. An RHS is described by two membership vectors: (1) the *gene membership vector* $\mathbf{x} = (x_1, \ldots, x_n)^T$, where $x_i = 1$ if gene i belongs to the RHS and $x_i = 0$ otherwise; and (2) the *network membership vector* $\mathbf{y} = (y_1, \ldots, y_m)^T$, where $y_j = 1$ if the RHS appears in network j and $y_j = 0$ otherwise. The "heaviness" of a RHS is quantified as the summed weight of all its edges: $H_{\mathcal{A}}(\mathbf{x}, \mathbf{y}) = \frac{1}{2} \sum_{i=1}^{n} \sum_{j=1}^{n} \sum_{k=1}^{m} a_{ijk} x_i x_j y_k$. Discovering recurrent heavy subgraphs can be formulated as a discrete combinatorial optimization problem: *among all RHSs of fixed size (K_1 member genes and K_2 member networks), look for the heaviest*. This is a NP-hard integer programming problem and therefore not solvable in reasonable time. Instead, we solve a continuous optimization problem with the same objective by

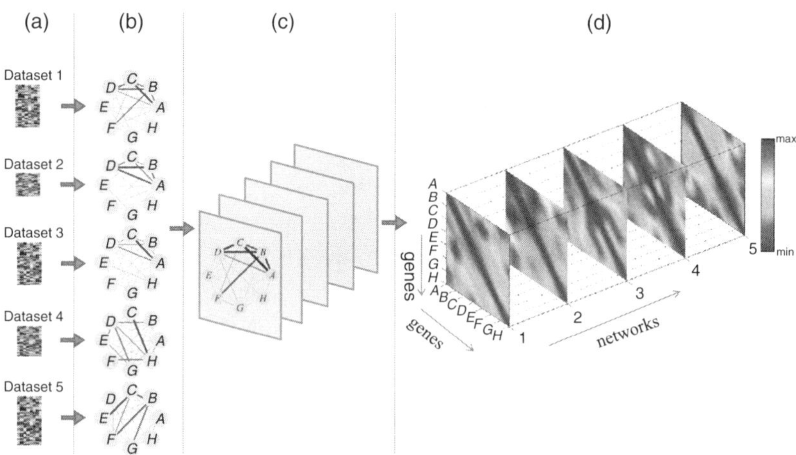

Figure 3.1 Illustration of the tensor representation for multiple networks and a recurrent heavy subgraph. (A) Microarray data sets are modeled as (B) a collection of coexpression networks. (C) These coexpression networks can be "stacked" together into (D) a third-order tensor such that each slice represents the adjacency matrix of one network. The weights of edges in the coexpression networks and their corresponding tensor elements are indicated by the color scale to the right of the figure. In (D), after reordering the tensor using the gene and network membership vectors, it becomes clear that the subtensor in the top-left corner of the tensor (formed by genes A, B, C, D in networks $1, 2, 3$) corresponds to a recurrent heavy subgraph.

relaxing the integer constraints to continuous constraints. This optimization problem is formally expressed as follows:

$$\max_{\mathbf{x} \in \mathbb{R}^n_+, \mathbf{y} \in \mathbb{R}^m_+} \quad H_{\mathcal{A}}(\mathbf{x}, \mathbf{y})$$

$$\text{subject to} \begin{cases} f(\mathbf{x}) = 1 \\ g(\mathbf{y}) = 1 \end{cases} \tag{3.1}$$

where $f(\mathbf{x})$ and $g(\mathbf{y})$ are vector norms that can determine what types of modules to be discovered. We discuss them later. These equations define a tensor-based framework for the RHS identification problem. By solving Equation (3.1), users can easily identify the top-ranking networks (after sorting the tensor by $\mathbf{y}$) and top-ranking genes (after sorting each network by $\mathbf{x}$) contributing to the objective function. After rearranging the networks in this manner, the heaviest RHS occupies a corner of the 3D tensor. We then mask this RHS with zeros and search for the next heaviest RHS.

A RHS is a subset of genes that are heavily connected to each other in as many networks as possible. We can obtain solution membership vectors that correspond to this intuition by carefully choosing the two norms $f(\mathbf{x})$ and $g(\mathbf{y})$.

In the first case, we adopt the mixed norm $L_{p,2}(\mathbf{x}) = \alpha\|\mathbf{x}\|_p + (1-\alpha)\|\mathbf{x}\|_2$ ($p < 1$ and $0 < \alpha < 1$) for $f(\mathbf{x})$. Since L_p favononzerors sparse vectors and L_2 favors uniform vectors, a suitable choice of α should yield vectors with a few similar, nonzero elements and many elements that are close to zero. In the second case, we use $L_q(\mathbf{y})$ ($q > 1$) for $g(\mathbf{y})$. This norm favors uniform vectors, because we want the RHS to appear in as many networks as possible. After performing a simulation study, we found that the parameters $p = 0.8$, $\alpha = 0.2$, and $q = 10$ lead to the discovery of the most RHSs. Because the vector norm $f(\mathbf{x})$ is non-convex, we used the multi-stage convex relaxation method to perform the optimization. This algorithm is known to have good numerical properties for non-convex optimization problems (Zhang, 2010). For more details regarding our vector norm choices and the tensor-based optimization method, refer to Li et al. (2011). The code is available at `http://zhoulab.usc.edu/CMTensor/`.

The RHSs can be intuitively obtained by including those genes and networks with large membership values. In practice, we first rank the genes and networks in decreasing order of their membership values in $\hat{\mathbf{x}}$ and $\hat{\mathbf{y}}$. Then we extract two representative RHSs that are as large as possible while satisfying the heaviness threshold: the pattern that occurs in the most networks while having at least a certain number of top-ranking genes, and one with the largest number of top-ranking genes that appears in at least a certain number of top-ranking networks. Both patterns are included in our results for further analysis. Therefore, each solution to the optimization problem yields two RHSs.

To test the method, we selected 130 human microarray data sets, each with at least 20 samples, from the NCBI's Gene Expression Omnibus. Each microarray data set is modeled as a coexpression network using jackknife Pearson correlations as the edge weights. However, we do not use raw correlations: to make the coefficients comparable across data sets, we first "normalize" them with a z-transform following the procedure introduced in Xu et al. (2008). The absolute value of this normalized correlation is used as the edge weight in the coexpression network.

3.2.2 Results

Transcriptional Module Analysis

After applying our method to 130 microarray data sets generated under various experimental conditions, we identified 4,327 RHSs. Each RHS contains ≥ 5 member genes, appears in ≥ 5 networks, and has a "heaviness" (defined as the average weight of its edges in networks where the RHS appears) ≥ 0.4. The average size of these patterns is 8.5 genes, and the average recurrence is

10.1 networks. To assess the biological significance of the identified RHSs, we evaluate the extent to which these RHSs represent functional modules, transcriptional regulatory modules, and protein complexes.

Because the genes in an RHS are strongly co-expressed in multiple data sets generated under different conditions, they are likely to represent a transcription module. To evaluate this possibility, we used the 191 ChIP-seq profiles generated by the Encyclopedia of DNA Elements (ENCODE) consortium (Thomas et al., 2007). If the member genes of an RHS are highly enriched in the targets for any regulatory factor, then that factor is likely to actively regulate the RHS under the experimental conditions of our data sets. In this case, we consider the RHS module to be transcriptional homogenous. For example, if we require an enrichment q-value < 0.05, then 56.4% of the 4,327 RHSs are transcription homogenous (compared to only 1.4% of randomly produced RHSs). The percentage of modules that are transcription homogenous increases rapidly with heaviness and recurrence. This highlights the importance of integrating multiple data sets for analysis. The five most frequently enriched regulators are *c-Myc*, *Pol2*, *DNase*, *TAF II*, and *E2F4*. Remarkably, 2,108 (48.7%) of the modules are enriched in at least two factors, 1,926 (44.5%) in at least three factors, and 1,807 (41.8%) in at least four factors. These remarkable statistics highlight the combinatorial nature of transcriptional regulation.

High-Order Cooperativity and Regulation in Transcription Regulatory Networks

The discovery of many RHS modules spanning a variety of experimental or disease conditions enables us to investigate high-order coordination. We applied our previously proposed *second-order analysis* to uncover the cooperativity among the transcription modules (Zhou et al., 2005), thereby reconstructing transcriptional networks. Intuitively, if two transcription modules always form two coexpression clusters under the same set of conditions (i.e., in the same data sets), it in fact suggests that their respective transcription factors (TFs) are active simultaneously. Conversely, if the transcription modules never form coexpression clusters in the same data sets, their factors are likely to be inactive simultaneously. This cooperativity between two sets of transcription factors can be quantified using the second-order expression correlation (Li et al., 2011). We focus on the 57 TFs with enriched targets in our modules. Among these, we identified 25 TF pairs that regulate two different transcription modules with high second-order correlations.

We traced the potential sources of cooperativity in these pairs using genome-wide TF binding data and protein-protein interaction data (Breitkreutz et al., 2008). Given two modules, controlled respectively by transcription factors TF1

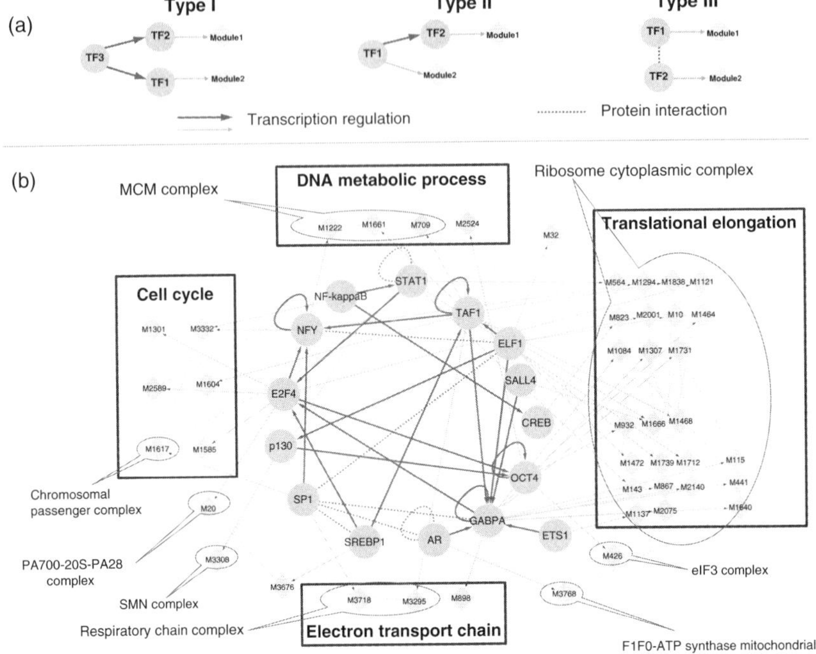

Figure 3.2 Reconstruction of transcriptional regulatory networks. (A) Three types of transcription networks that could explain a second-order correlation between transcriptional modules. Given two modules controlled by two different transcription factors, TF1 and TF2, respectively, the coactivation of the two modules implies cooperativity between TF1 and TF2. In a type I network, the activities of TF1 and TF2 are controlled by a common transcription factor(s) TF3; in a type II network, the activity of TF2 is controlled by TF1 or vice versa; in a type III network, TF1 and TF2 interact at the protein level. (B) A regulatory network reconstructed from the derived transcription networks. Green circles denote transcription factors, yellow boxes are transcription modules defined by RHSs, blue ovals denote protein complexes represented by the RHSs, and blue boxes highlight distinct biological processes.

and TF2 (for simplicity, we assume these are individual TFs rather than sets of factors), there are at least three possible direct causes of cooperativity between TF1 and TF2 (Figure 3.2A): the expressions of TF1 and TF2 are activated by a common transcription factor TF3 (a type I transcription network), or TF1 activates the expression of TF2 (a type II transcription network), or TF1 and TF2 interact at the protein level (a type III transcription network). In the special case where a module pair shares the majority of their genes, the cooperativity between TF1 and TF2 is known to be combinatorial control.

We identified 33 transcription networks, among which 10 are Type I, 19 are Type II, and 4 are Type III. These transcription networks interconnect to form a partial cellular regulatory network (Figure 3.2). Four networks are involved in the cell cycle: the type I network involving *SREBP1* and *TAF1/E2F4*, the two type II networks involving *STAT1* and *E2F4* as well as *SP1* and *NFYA*, and the type III network involving *ELF1* and *SP1*. The roles of these networks are supported by independent evidence of cooperative roles of those transcription factors reported in the literature (Takahashi et al., 2008; Li et al., 2011). Other transcription networks participate in translational elongation, rRNA processing, RNA splicing, DNA replication, DNA packaging, and electron transport, among others. Notably, our reconstructed transcriptional regulatory network includes 35 modules that represent protein complexes, which provide a mechanistic explanation for the correlated activities of those protein complexes, as shown in Figure 3.2B.

3.3 Reconstructing Splicing Regulatory Modules

Alternative splicing is a ubiquitous gene regulatory mechanism that dramatically increases the complexity of the proteome. However, the mechanisms for regulating alternative splicing are poorly understood, and study of coordinated splicing regulation has been limited to individual cases. A central concept in transcription regulation is the *transcription module*, defined as a set of genes that are co-regulated by the same transcription factor(s). Analogously, such coordinated regulation also occurs at the splicing level (Nagoshi et al., 1988; Hedley and Maniatis, 1991; Jernej Ule et al., 2005). For example, the splicing factor *Nova* regulates exon splicing for a set of genes that shape the synapse (Jernej Ule et al., 2005). However, the study of such coordinated splicing regulation has thus far been limited to individual cases (Hedley and Maniatis, 1991; Jernej Ule et al., 2005; Hentze and Kuhn, 1996; Zhang et al., 2008; Moore et al., 2010).

In this section, we define a *splicing module* as a set of exons that are regulated by the same splicing factors. The exons in a splicing module can belong to different genes, but they must exhibit correlated splicing patterns (in terms of being included or excluded in their respective transcripts) across different conditions, thus forming an exon co-splicing cluster. To study genome-wide splicing regulation, we identify co-splicing clusters that appear frequently across multiple conditions, and thus very likely represent splicing modules – the fundamental units of the splicing regulatory network.

3.3.1 Methods and Materials

We model each RNA-seq data set as an exon co-splicing network, where the nodes represent exons and the edges are weighted by the correlations between exon inclusion rate profiles. We identified 38 human RNA-seq data sets from the

NCBI Sequence Read Archive, each with at least six samples. All these data sets provide transcriptome profiling under multiple experimental conditions, such as diverse tissues or diseases. For each data set, we used Tophat (Trapnell et al., 2009) combined with Cufflinks (Trapnell et al., 2010) to estimate expressions for all transcripts with known UCSC transcript annotations (Fujita et al., 2011). We calculated the inclusion rate of each exon as the ratio between its expression (the sum of FPKM[1] over all transcripts that cover the exon) and the host gene's expression (the sum of FPKM over all transcripts of the gene). On the basis of these profiles, we constructed an exon co-splicing network from each RNA-seq data set using Pearson's correlation between the inclusion rate profiles of exon pairs. For details, refer to Dai et al. (2012).

We apply the tensor-based method presented in the previous section to the 38 co-splicing networks derived from human RNA-seq data sets. We then demonstrate that the co-splicing clusters identified by the method represent potential splicing modules, by validating against four biological knowledge databases.

3.3.2 Results

We adopt two empirical criteria for co-splicing clusters to be included in our results: "heaviness" ≥ 0.4 and cluster size ≥ 5 exons. The tensor-based method produces an atlas of 7,194/3,104/1,422/594 co-splicing clusters with recurrences $\geq 3/4/5/6$, respectively. To assess the biological significance of the clusters, we evaluate the extent to which these exon clusters represent splicing modules and transcriptional regulatory modules.

Splicing Regulatory Analysis

By construction, the exons in our identified co-splicing clusters have highly correlated inclusion rate profiles across different experimental conditions. Therefore, the clusters are likely to include exons that are co-regulated by the same splicing factors. It has been shown that splicing factors can affect alternative splicing by interacting with cis-regulatory elements in a position-dependent manner (Licatalosi et al., 2008). We collected experimental RNA target motifs (2220 RNA binding sites) for 62 splicing factors from the SpliceAid2 database (Piva et al., 2011). To identify possible splicing factors associated with a co-splicing cluster, for each exon in a given co-splicing cluster, we retrieved the internal exon region and its 50bp flanking intron region that are enriched in

---

[1] FPKM stands for "fragments per kilobase of exon per million fragments mapped," as defined in Trapnell et al. (2010).

the motifs of these 62 splicing factors by performing a BLAST search (E-score < 0.001). If the exons of a cluster are highly enriched in the targets of a splicing factor, we consider the cluster to be "splicing homogeneous." Although the collection of known splicing motifs is very limited, we still observed (based on hypergeometric test) that 4.9% of clusters with ≥ 5 exons and ≥ 6 recurrences are splicing homogenous at the p-value < 0.05 level. In contrast, only 1.6% of randomly generated patterns with the same size distribution are splicing homogenous. The enrichment fold ratio for co-splicing clusters identified using the tensor-based approach is therefore 3.0. The five most frequently enriched splicing factors are *hnRNP E2*, *9G8*, *hnRNP U*, *SRp75*, and *SRp30c*. Both *hnRNP E2* and *hnRNP U* belong to the heterogeneous nuclear ribonucleoprotein family, whose members generally suppress splicing through binding to an exonic splicing silencer (Matlin et al., 2005). Studies show that *hnRNP E2* can repress exon usage when present at high levels in vitro (Rothrock et al., 2005), and *hnRNP U* binds to pre-mRNA as well as nuclear mRNA and, so can play an important role in the processing and transport of mRNA (Spraggon et al., 2007). Factors *9G8*, *SRp75*, and *SRp30c* all belong to the *SR* family of splicing regulators. Factor *9G8* protein, excluding other *SR* factors, can rescue the splicing activity of a *9G8*-depleted nuclear extract, indicating that *9G8* plays a crucial role in splicing (Cavaloc et al., 1994). *SRp75* is present in messenger ribonucleoproteins in both cycling and differentiated cells, and it shuttles between nucleus and cytoplasm, implicating its widespread roles in splicing regulation (Änkö et al., 2010). *SRp30c* can function as a repressor of $3'$ splice site utilization, so *SRp30c-CE9* interactions may contribute to the control of *hnRNP A1* alternative splicing (Simard and Chabot, 2002).

Transcriptional and Epigenomic Analysis

To evaluate how co-splicing is affected by transcriptional regulation, we used the same ChIP-seq data and analysis procedure as described in Section 3.2.2. If the host genes of an exon cluster are highly enriched in the targets of any regulatory factor, we consider the cluster to be transcription homogenous. At the significance level p-value < 0.01, 74.9% of clusters with recurrences ≥ 3 are transcription homogenous, compared to only 21.2% of randomly generated clusters with the same sizes. As expected, the enrichment fold ratio increases with recurrence. This result suggests a strong association between transcription and splicing. The four most frequently enriched regulatory factors are *TAF8*, *GABP*, *FOS*, and *NFYB*. *TAF8* is a subunit of transcription initiation factor *TFIID*, which is required for accurate and regulated initiation by RNA polymerase II (Horikoshi et al., 1992). As an ETS transcription factor, *GABP* plays a key role in regulating genes that are intimately involved in cell cycle control, protein synthesis, and cellular metabolism (Rosmarin et al., 2004). *FOS* can dimerise with *c-Jun* to form *AP-1* transcription factor, which upregulates transcription of a wide range of genes involved in proliferation and differentiation to defense against invasion and cell damage (Kovács, 1998). *NFYB* is a subunit of an ubiquitous heteromeric transcription factor *NF-Y*, which regulates 30% of mammalian promoters (Fossati et al., 2011).

Co-Splicing Clusters Reveal Novel Functions That Are Not Identified by Coexpression Clusters

Studies have shown that genes that are co-regulated transcriptionally do not necessarily overlap with those that are co-spliced (Pan et al., 2004). Therefore, the co-splicing clusters can reveal functionally related genes that could not be discovered from transcription analysis. To identify novel functions associated with co-splicing but not coexpression, we complement the preceding analysis by constructing a gene coexpression network from each RNA-seq data set. The nodes of these networks represent genes, and the edges are weighted by Pearson's correlation between two gene expression profiles. We then apply our tensor-based pattern mining algorithm to identify frequent coexpression clusters in the 38 coexpression networks, using the same heaviness (≥ 0.4) and size (≥ 5) criteria as for the identifying co-splicing clusters. We found that 98.8% of co-splicing clusters with recurrence ≥ 3 have low expression correlations (the average correlations ≤ 0.2). Therefore, many of the functions associated with post-transcriptional regulation are enriched in co-splicing clusters but not in coexpression clusters. These functions include maintenance of protein location, regulation of protein catabolic process, cytoplasmic sequestering of protein, regulation of intracellular protein transport, regulation of ubiquitin-protein ligase activity, ribonucleoprotein complex assembly, RNA splicing, via transesterification reactions, and RNA export from nucleus.

Exons Can Dynamically Participate in Different Pathways upon Different Co-Splicing Mechanisms

Alternatively skipping or including a cassette exon can change the functions of a protein by deleting or inserting a protein domain. In other words, protein isoforms alternatively spliced from the same gene may participate in different pathways. In our results, we observed that 70.3% of exons are members of at least two clusters (recurrence ≥ 3) with different functions. For example, exon8 of the gene *Rela* appears in four co-splicing clusters with recurrences ≥ 3, which are enriched with the following distinct functions: "ER-associated protein catabolic process" (p-value $= 2.20\text{E-}5$), "response to extracellular stimulus" (p-value $= 3.80\text{E-}5$), "regulation of gene-specific transcription"

(p-value $= 8.89$E-5), and "positive regulation of intracellular protein kinase cascade" (p-value $= 2.49$E-5). *Rela* encodes the transcription factor *p65*, which is an important subunit of the *NF-κB* complex that affects several hundred genes by *NF-κB* signaling. Recent research has identified several alternative splice variants of *Rela*, e.g. *p65△*, *p65△2* and *p65△3*. In fact, *p65△* arises by the use of an alternative splice site located 30 nucleotides into exon8, and *p65△3* was identified as a splice variant lacking exon7 and exon8 (Leeman and Gilmore, 2008). These facts are consistent with our finding that exon8 is dynamically included in multiple co-splicing clusters.

3.4 Coupling Mechanisms Between Transcription and Splicing

Transcription and splicing regulate gene expression differently, yet in a coordinated manner (Fong and Zhou, 2001; Kornblihtt et al., 2004). Thus far, it is not clear to what extent, by what mechanisms, or for which cellular functions, transcription-splicing coupling takes place. To study such coupling mechanisms, we propose to identify coupled transcription-splicing modules in a set of paired gene coexpression and exon co-splicing networks, each pair being derived from an RNA data set. The concept of our approach is illustrated in Figure 3.3. A set of co-expressed genes (i.e., genes that are heavily interconnected in the coexpression networks) is likely to be co-regulated by the same transcription factor and thus may represent a *transcription module*. Similarly, a set of co-spliced exons (i.e., heavily interconnected in the exon co-splicing networks) is likely to be co-spliced by the same splicing factor and thus may represent a *splicing module*. Therefore, when we find a gene transcription module, whose exons (or a subset thereof) form a splicing module, it is likely that the transcription and splicing processes are coupled. We call such patterns coupled modules. In particular, if a coupled module recurrently appears among multiple paired gene and exon networks, then the enhanced signal-to-noise ratio increases the likelihood that this frequent coupled cluster (FCC) represents a coupled module than would those coupled clusters derived from a single data set.

3.4.1 Methods and Materials

We used the same $L = 38$ human RNA-seq data sets as described in Section 3.3. Processing these data sets yields a collection of 38 paired gene coexpression networks and exon co-splicing networks. These networks can be represented as two third-order tensors, $\mathcal{G} = (g_{ijk})_{N \times N \times L}$ and $\mathcal{E} = (e_{ijk})_{M \times M \times L}$. We also define a binary relation matrix between the N genes and M exons, $R = (r_{ij})_{N \times M}$. Each element g_{ijk} (e_{ijk}) in the tensor is the nonnegative weight of the edge between

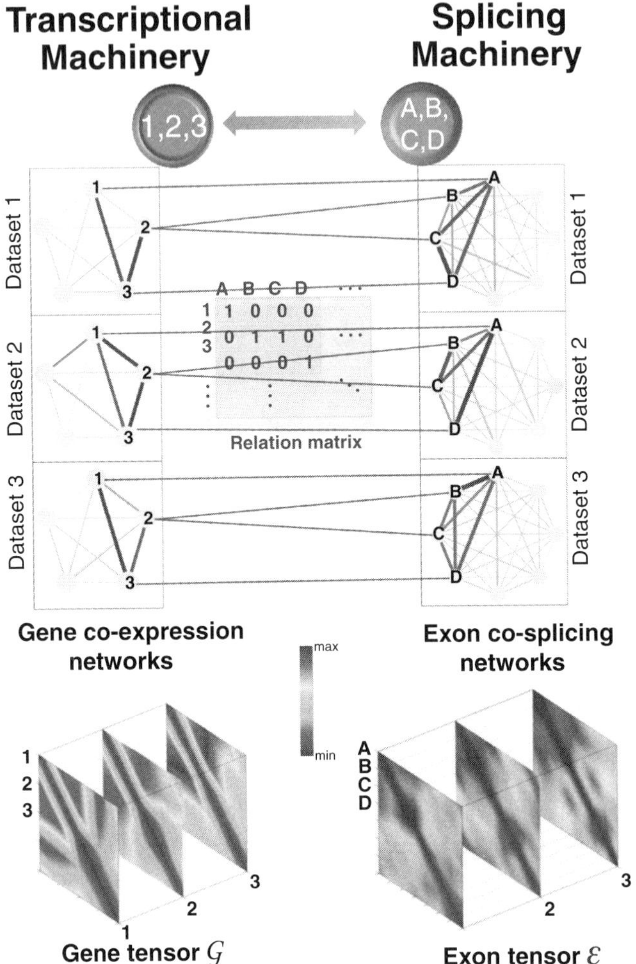

Figure 3.3 Illustration of a *frequent coupled cluster* (FCC). In a collection of three paired gene coexpression and exon co-splicing networks, a subset of genes {1,2,3} is heavily interconnected and their exons {A,B,C,D} are also heavily interconnected. The two subsets form an FCC: a coupled transcription-splicing module. The gene and exon clusters intuitively correspond to the heavy subtensors in $\mathcal{G}$ and $\mathcal{E}$.

genes (exons) i and j in the kth gene coexpression (exon co-splicing) network. As each exon belongs to only one gene but a gene may have at least one exon, the relation matrix R has two characteristics: (1) $r_{ij} = 1$ when gene i contains exon j, $r_{ij} = 0$ otherwise; and (2) each column vector has only one nonzero element. Therefore, R is very sparse, with exactly M nonzeros. On the basis of the preceding tensor data and vector variables, an FCC is defined as a set of genes G, a set of exons E, and a set of data sets D that satisfy the following two criteria: (1) heavy subgraph: the genes of G are heavily connected to each other in each data set of D, and the exons of E are heavily connected to each other in the same data sets and, (2) relation: each gene of G contains ≥ 1 exons of E, whereas each exon of E is contained by ≤ 1 genes of G.

Any FCC is described by three membership vectors: (1) the *gene membership vector* $\mathbf{x} = (x_1, \ldots, x_N)^T$, where $x_i = 1$ if gene i belongs to the gene set G of the cluster and $x_i = 0$ otherwise; (2) the *exon membership vector* $\mathbf{y} = (y_1, \ldots, y_M)^T$, where $y_j = 1$ if exon j belongs to the exon set E of the cluster and $y_j = 0$ otherwise; and (3) the *data set membership vector* $\mathbf{w} = (w_1, \ldots, w_L)^T$, where $w_k = 1$ if the cluster appears in the data set k of D (we call these the "active data sets" of the cluster) and $w_k = 0$ otherwise. Using these three membership vectors, the heavy subgraph criterion defined earlier can be formulated by maximizing the "heaviness" functions of the gene and exon subgraphs in the active data sets D: $H_G(\mathbf{x}, \mathbf{w}) = \sum_{i=1}^{N} \sum_{j=1}^{N} \sum_{k=1}^{L} g_{ijk} x_i x_j w_k$ and $H_{\mathcal{E}}(\mathbf{y}, \mathbf{w}) = \sum_{i=1}^{M} \sum_{j=1}^{M} \sum_{k=1}^{L} e_{ijk} x_i x_j w_k$. The relation criterion can be formulated by using the idea of the linear assignment problem formulation in operations science (Burkard et al., 2009). Let the relation variables $Z = (z_{ij})_{N \times M}$ indicate the matching between genes and exons, where $z_{ij} = 1$ if gene i contains exon j and both belong to the FCC and 0 otherwise. Then the relation criterion can be formulated by maximizing the objective function $O_R(Z) = \sum_{r_{ij} \neq 0} z_{ij} r_{ij} x_i y_j$ under the constraint $\sum_{j=1}^{M} z_{ij} \leq K_2$ for all $i = 1, \ldots, N$, where K_2 is the number of exons in the cluster). Therefore, maximizing $O_R(Z)$ becomes a process of finding a set of genes and exons with one-to-many relationships between each other, and also having large associated weights x_i and y_j, because the special characteristic of R already guarantees that each exon belongs to only ONE gene.

The three criteria are combined into one objective function: $O(\mathbf{x}, \mathbf{y}, \mathbf{w}, Z) = H_G(\mathbf{x}, \mathbf{w}) + \lambda H_{\mathcal{E}}(\mathbf{y}, \mathbf{w}) + \mu O_R(\mathbf{x}, \mathbf{y}, Z)$, where $\lambda, \mu > 0$ are constant weights of individual criterion. This function is defined on the discrete variables and thus the task of maximizing this function is an NP-hard problem and not solvable in reasonable time even for small data sets. We instead solve a continuous optimization problem with the same objective by relaxing the integer constraints to continuous constraints. The new problem is formally expressed as follows:

$$\max_{\mathbf{x}, \mathbf{y}, \mathbf{w}, Z \in \mathbb{R}^+} \quad O(\mathbf{x}, \mathbf{y}, \mathbf{w}, Z) = H_G(\mathbf{x}, \mathbf{w}) + \lambda H_{\mathcal{E}}(\mathbf{y}, \mathbf{w}) + \mu O_R(\mathbf{x}, \mathbf{y}, Z)$$

$$\text{subject to} \begin{cases} \text{Constraint I:} & \|\mathbf{x}\|_f = 1, & 1 \leq f < 2 \\ \text{Constraint II:} & \|\mathbf{y}\|_g = 1, & 1 \leq g < 2 \\ \text{Constraint III:} & \|\mathbf{w}\|_h = 1, & h \geq 2 \\ \text{Constraint IV:} & \sum_{j=1}^{M} z_{ij} \leq 1, & i = 1, 2, \ldots, N \end{cases}$$

$$(3.2)$$

Constraints I and II give "sparse" solutions where only a few genes (exons) are selected in the module. Constraint III gives a "smooth" solution, so that the discovered cluster occurs in as many data sets as possible. Equation (3.2) defines a tensor-based optimization problem of identifying FCCs. By solving Equation (3.2), users can easily identify the top-ranking data sets (after sorting elements of $\mathbf{w}$ in non-increasing order) and top-ranking genes/exons (after sorting elements $\mathbf{x}/\mathbf{y}$ in non-increasing order) contributing to the objective function. After rearranging the tensors in this manner, the optimum FCC occupies a corner of the 3D tensors $\mathcal{G}$ and $\mathcal{E}$. We then mask the edges of this cluster in the gene and exon networks with zeros and optimize Equation (3.2) again to find the next module. We developed an iterative algorithm to obtain the solution to Equation (3.2). For details, refer to Li et al. (2012).

We define an FCC extracted from the top-ranking genes, exons and data sets as a triplet (G', E', D') that satisfies three empirically determined thresholds: (1) each gene subgraph formed by G' and exon subgraph formed by E' in each active data set of D' has an average edge weight ("heaviness") $\geq$ a predefined threshold, (2) the fraction of host genes of E' that are included in G', $\text{coverage}_{\text{gene}}(G', E') = \frac{|G' \cap \text{hostgene}(E')|}{|G'|} \geq$ a predefined threshold, and (3) the fraction of exons of G' that are included in E' $\text{coverage}_{\text{exon}}(G', E') = \frac{|E' \cap \text{exon}(G')|}{|E'|} \geq$ a predefined threshold. The first criterion gives a certain degree of how heavy and frequent the cluster should be, and the second and third criteria measure the degree of "coupling" between the gene set and exon set. There may be more than one triplet that satisfies these criteria for a given optimization result, for example, a small cluster that appears in many data sets and a larger cluster that appears in just three data sets. A group of FCCs derived from the same membership vectors $\mathbf{x}$, $\mathbf{y}$, and $\mathbf{w}$ is called a *family of FCCs*. The code of this algorithm is available at `http://zhoulab.usc.edu/tensor/`.

3.4.2 Results

We identified 8,667 FCC families containing a total of 43,580 FCCs. Each FCC contains ≥ 5 member genes, ≥ 5 member exons, appears in ≥ 2 RNA-seq data sets, has a "heaviness" ≥ 0.4, $\text{coverage}_{\text{gene}} \geq 0.7$ and $\text{coverage}_{\text{exon}} \geq 0.7$. The average gene/exon size of these patterns is 12.61/12.64, and the average recurrence is 2.04. Because FCCs within the same family have extensive overlap, we treat a family of FCCs as a unit in the following analyses.

Transcription and Splicing Module Analysis

Because the genes of an FCC are strongly co-expressed in multiple data sets generated under different conditions, they are likely to represent a transcription module. To assess this possibility, we used the ChIP-seq data and our definition of a "transcriptionally homogenous module" from Section 3.2.2. Among all identified FCC families, 61.2% of them are transcriptionally homogenous with an enrichment q-value < 0.05, compared to 8.9% of the randomly generated patterns with the same size distribution. The five most frequently enriched regulators are *YY1*, *E2F4*, *c-Myc*, *MAX*, and *TAF*, all of which have been implicated in cancer pathogenesis or progression (Gordon et al., 2005; Suzuki et al., 2010).

Because the exons of an FCC have highly correlated inclusion rate profiles across different experimental conditions, they are likely to be co-regulated by the same splicing factors. We used the targets of splicing factors and our definition of a "splicing homogenous module" from Section 3.3.2. Although the collection of known splicing data is very limited, we still observed that 8.9% of the FCC families are splicing homogenous, compared to 5.1% of randomly generated patterns with the same size distribution. The five most frequently enriched splicing regulators are *PSF*, *hnRNP-D*, *hnRNP-C1*, *SLM-2*, and *HuB*.

Exploring the Mechanisms of Transcription-Splicing Coupling

Our catalogs of coupled transcription-splicing modules provide a unique opportunity to systematically assess the mechanisms of the coupled recruitment of transcription and splicing factors. In particular, we evaluate the hypothesis that functionally coupled recruitment is mediated by direct or indirect protein-protein interactions between the two kinds of factors. We examined the available protein-protein interactions (PPIs) between enriched transcription and splicing factors in coupled module families. We consider not only the direct PPIs between transcription and splicing factors, but also indirect interactions through a mediator protein (one-hop interactions), because experiments have shown that transcription and splicing factors can be recruited by a common third protein (Blencowe et al., 1999). To achieve broad coverage, we considered the 109 human transcription factors in the ENCODE (Thomas et al., 2007) and JASPAR (Sandelin et al., 2004) databases. In addition, with 10,278 DNA binding motifs downloaded from Tabach et al. (2007), we performed BLAST searches (E-score < 0.05) in 1000 bp regions of the transcription start sites of all genes, then used the hyper-geometric test (q-value < 0.05) to identify transcription factors for the genes of a given FCC. The putative splicing factors for each FCC were identified based on the description in Section 3.3.2. The total number of PPIs between transcription and splicing factors within the same FCC families (including direct and one-hop interactions) is 105, compared to only 14.8 in random families with the same gene and exon size distribution. The enhancement fold ratio is 7.1. This evidence supports the hypothesis that PPI-mediated association can be an important mechanism for transcription-splicing coupling.

3.5 Conclusion

Biological network data are rapidly accumulating for a wide range of organisms under various conditions. The integrative analysis of multiple biological networks is a powerful approach to unraveling the complex and dynamic mechanisms of gene regulation. In this review, we proposed to discover three types of patterns in multiple biological networks. These patterns are robust to experimental bias and noise, and their varied nature provides insight into the complexity of gene regulation. Modeling a collection of networks as a third-order tensor enables us to formulate the pattern discovery problem as an optimization task, and draws on a wealth of tensor-based numerical methods in the established fields. Although we used coexpression and co-splicing networks (derived from RNA-seq data) throughout this chapter, our tensor-based computational models are well suited to other types of biological networks for integrative analysis.

Acknowledgment

This work was supported by NIH grants NHLBI MAPGEN U01HL108634 and NIGMS R01GM105431 and by NSF grant 0747475.

References

Änkö, Minna-Liisa, Morales, Lucia, Henry, Ian, Beyer, Andreas, and Neugebauer, Karla M. 2010. Global analysis reveals SRp20- and SRp75-specific mRNPs in cycling and neural cells. *Nature Structural and Molecular Biology*, **17**(8), 962–970.

Blencowe, B.J., Bowman, J.A.L., McCracken, S., and Rosonina, E. 1999. SR-related proteins and the processing of messenger RNA precursors. *Biochemistry and Cell Biology*, **77**(4), 277–291.

Breitkreutz, B.J., Stark, C., Reguly, T., Boucher, L., Breitkreutz, A., Livstone, M., Oughtred, R., et al. 2008. The BioGRID Interaction Database: 2008 update. *Nucleic Acids Research*, **36**(Database issue), D637–D640.

Burkard, Rainer, Dell'Amico, Mauro, and Martello, Silvano. 2009. *Assignment Problems*. SIAM.

Cavaloc, Y., Popielarz, M., Fuchs, J. P., Gattoni, R., and Stévenin, J. 1994. Characterization and cloning of the human splicing factor 9G8: a novel 35 kDa factor of the serine/arginine protein family. *The EMBO Journal*, **13**(11), 2639–2649.

Dai, Chao, Li, Wenyuan, Liu, Juan, and Zhou, Xianghong J. 2012. Integrating many co-splicing networks to reconstruct splicing regulatory modules. *BMC Systems Biology*, **6**(Suppl 1), S17.

Fong, Y.W., and Zhou, Q. 2001. Stimulatory effect of splicing factors on transcriptional elongation. *Nature*, **414**(6866), 929–933.

Fossati, Andrea, Dolfini, Diletta, Donati, Giacomo, and Mantovani, Roberto. 2011. NF-Y recruits Ash2L to impart H3K4 trimethylation on CCAAT promoters. *PloS One*, **6**(3), e17220.

Fujita, Pauline A., Rhead, Brooke, Zweig, Ann S., Hinrichs, Angie S., Karolchik, Donna, Cline, Melissa S., Goldman, Mary, et al. 2011. The UCSC Genome Browser database: update 2011. *Nucleic Acids Research*, **39**(Database issue), D876–882.

Gordon, S., Akopyan, G., Garban, H., and Bonavida, B. 2005. Transcription factor YY1: structure, function, and therapeutic implications in cancer biology. *Oncogene*, **25**(8), 1125–1142.

Hedley, M.L., and Maniatis, T. 1991. Sex-specific splicing and polyadenylation of dsx pre-mRNA requires a sequence that binds specifically to tra-2 protein in vitro. *Cell*, **65**(4), 579–586.

Hentze, M.W., and Kuhn, L.C. 1996. Molecular control of vertebrate iron metabolism: mRNA-based regulatory circuits operated by iron, nitric oxide, and oxidative stress. *Proceedings of the National Academy of Sciences of the United States of America*, **93**(16), 8175–82.

Horikoshi, M., Bertuccioli, C., Takada, R., Wang, J., Yamamoto, T., and Roeder, R. G.. 1992. Transcription factor TFIID induces DNA bending upon binding to the TATA element. *Proceedings of the National Academy of Sciences of the United States of America*, **89**(3), 1060–1064.

Jernej Ule, A., Ule, J.S., Alan Williams, J.S.H., Melissa Cline, H.W., Tyson Clark, C.F., Matteo Ruggiu, B.R.Z., David Kane, J.N.W., and John Blume, R.B.D. 2005. Nova regulates brain-specific splicing to shape the synapse. *Nature Genetics*, **37**(8), 844–852.

Kornblihtt, A.R., de la Mata, M., Fededa, J.P., Munoz, M.J., and Nogues, G. 2004. Multiple links between transcription and splicing. *RNA*, **10**(10), 1489–1498.

Kovács, K. J. 1998. c-Fos as a transcription factor: a stressful (re)view from a functional map. *Neurochemistry International*, **33**(4), 287–297.

Leeman, Joshua R., and Gilmore, Thomas D. 2008. Alternative splicing in the NF-kappaB signaling pathway. *Gene*, **423**(2), 97–107.

Li, Wenyuan, Liu, Chun-Chi, Zhang, Tong, Li, Haifeng, Waterman, Michael S., and Zhou, Xianghong Jasmine. 2011. Integrative analysis of many weighted co-expression networks using tensor computation. *PLoS Computational Biology*, **7**(6), e1001106.

Li, Wenyuan, Dai, Chao, Liu, Chun-Chi, and Zhou, Xianghong Jasmine. 2012. Algorithm to identify frequent coupled modules from two-layered network series: application to study transcription and splicing coupling. *Journal of Computational Biology*, **19**(6), 710–730.

Licatalosi, Donny D., Mele, Aldo, Fak, John J., Ule, Jernej, Kayikci, Melis, Chi, Sung Wook, Clark, Tyson A., et al. 2008. HITS-CLIP yields genome-wide insights into brain alternative RNA processing. *Nature*, **456**(7221), 464–469.

Matlin, Arianne J., Clark, Francis, and Smith, Christopher W.J. 2005. Understanding alternative splicing: towards a cellular code. *Nature Reviews Molecular Cell Biology*, **6**(5), 386–398.

Moore, M.J., Wang, Q., Kennedy, C.J., and Silver, P.A. 2010. An alternative splicing network links cell-cycle control to apoptosis. *Cell*, **142**(4), 625–636.

Nagoshi, R.N., McKeown, M., Burtis, K.C., Belote, J.M., and Baker, B.S. 1988. The control of alternative splicing at genes regulating sexual differentiation in D. melanogaster. *Cell*, **53**(2), 229–236.

Newman, M.E.J. 2004. Analysis of weighted networks. *Physical Reviews E*, **70**(5), 056131.

Pan, Qun, Shai, Ofer, Misquitta, Christine, Zhang, Wen, Saltzman, Arneet L, Mohammad, Naveed, Babak, Tomas, et al. 2004. Revealing global regulatory features of mammalian alternative splicing using a quantitative microarray platform. *Molecular Cell*, **16**(6), 929–941.

Piva, Francesco, Giulietti, Matteo, Burini, Alessandra Ballone, and Principato, Giovanni. 2011. SpliceAid 2: A database of human splicing factors expression data and RNA target motifs. *Human Mutation*, Sept.

Rosmarin, Alan G., Resendes, Karen K., Yang, Zhongfa, McMillan, John N., and Fleming, Shawna L. 2004. GA-binding protein transcription factor: a review of GABP as an integrator of intracellular signaling and protein-protein interactions. *Blood Cells, Molecules and Diseases*, **32**(1), 143–154.

Rothrock, Caryn R., House, Amy E., and Lynch, Kristen W. 2005. HnRNP L represses exon splicing via a regulated exonic splicing silencer. *The EMBO Journal*, **24**(15), 2792–2802.

Sandelin, Albin, Alkema, Wynand, Engström, Pär, Wasserman, Wyeth W., and Lenhard, Boris. 2004. JASPAR: an open-access database for eukaryotic transcription factor binding profiles. *Nucleic Acids Research*, **32**(Database issue), D91–94.

Simard, Martin J., and Chabot, Benoit. 2002. SRp30c is a repressor of 3′ splice site utilization. *Molecular and Cellular Biology*, **22**(12), 4001–4010.

Spraggon, L., Dudnakova, T., Slight, J., Lustig-Yariv, O., Cotterell, J., Hastie, N., and Miles, C. 2007. hnRNP-U directly interacts with WT1 and modulates WT1 transcriptional activation. *Oncogene*, **26**(10), 1484–1491.

Suzuki, H., Igarashi, S., Nojima, M., Maruyama, R., Yamamoto, E., Kai, M., Akashi, H., et al. 2010. IGFBP7 is a p53-responsive gene specifically silenced in colorectal cancer with CpG island methylator phenotype. *Carcinogenesis*, **31**(3), 342.

Tabach, Yuval, Brosh, Ran, Buganim, Yossi, Reiner, Anat, Zuk, Or, Yitzhaky, Assif, Koudritsky, Mark, Rotter, Varda, and Domany, Eytan. 2007. Wide-scale analysis of human functional transcription factor binding reveals a strong bias towards the transcription start site. *PloS One*, **2**(8), e807.

Takahashi, Kyoko, Hayashi, Natsuko, Shimokawa, Toshibumi, Umehara, Nagayoshi, Kaminogawa, Shuichi, and Ra, Chisei. 2008. Cooperative regulation of Fc receptor γ-chain gene expression by multiple transcription factors, including Sp1, GABP, and Elf-1. *Journal of Biological Chemistry*, **283**(22), 15134–15141.

Thomas, Daryl J., Rosenbloom, Kate R., Clawson, Hiram, Hinrichs, Angie S., Trumbower, Heather, Raney, Brian J., Karolchik, Donna, et al. 2007. The ENCODE Project at UC Santa Cruz. *Nucleic Acids Research*, **35**(Database issue), D663–D667.

Trapnell, Cole, Pachter, Lior, and Salzberg, Steven L. 2009. TopHat: discovering splice junctions with RNA-Seq. *Bioinformatics*, **25**(9), 1105–1111.

Trapnell, Cole, Williams, Brian A., Pertea, Geo, Mortazavi, Ali, Kwan, Gordon, van Baren, Marijke J., Salzberg, Steven L., Wold, Barbara J., and Pachter, Lior. 2010. Transcript assembly and quantification by RNA-Seq reveals unannotated transcripts and isoform switching during cell differentiation. *Nature Biotechnology*, **28**(5), 511–515. PMID: 20436464.

Xu, M., Kao, M.C.J., Nunez-Iglesias, J., Nevins, J.R., West, M., and Zhou, X.J. 2008. An integrative approach to characterize disease-specific pathways and their coordination: a case study in cancer. *BMC Genomics*, **9**(Suppl. 1), S12.

Zhang, C., Zhang, Z., Castle, J., Sun, S., Johnson, J., Krainer, A.R., and Zhang, M.Q. 2008. Defining the regulatory network of the tissue-specific splicing factors Fox-1 and Fox-2. *Genes and Development*, **22**(18), 2550–2563.

Zhang, Tong. 2010. Analysis of multi-stage convex relaxation for sparse regularization. *Journal of Machine Learning Research*, **11**, 1081–1107.

Zhou, X.J., Kao, M.C.J., Huang, H., Wong, A., Nunez-Iglesias, J., Primig, M., Aparicio, O.M., Finch, C.E., Morgan, T.E., Wong, W.H., et al. 2005. Functional annotation and network reconstruction through cross-platform integration of microarray data. *Nature Biotechnology*, **23**, 238–243.

4

Network Integration of Genetically Regulated Gene Expression to Study Complex Diseases

ZHIDONG TU, BIN ZHANG, AND JUN ZHU

Abstract

Understanding the molecular mechanisms underlying the connections between genotype and phenotype in human is one of the most important biological questions. By integrating genomic and genetic information to constructing network models, we demonstrate that potential gene regulatory mechanisms and key genes can be identified. We review multiple network algorithms that have been applied to discovering the key pathways linking the genetic and phenotype. Using two application examples, we show that novel hypotheses can be formed and experimentally validated to help us to better understand human complex diseases.

4.1 Introduction

A major goal in current biomedical research is to understand the mechanisms of various diseases to allow the development of novel or improved treatments. For complex diseases (e.g., cancers, diabetes, and cardiovascular diseases), multiple genetic factors interact with environmental factors to determine the disease development and progression (Hunter, 2005; Schadt, 2009). With recent large efforts in genome-wide association studies (GWAS), thousands of genetic variants have been identified for the association with various disease phenotypes. However, for most of these variants, little is known of their mechanisms of causing disease. This poses a major challenge for complex disease research in the post-GWAS era. For variants in the protein coding region that lead to nonsynonymous changes, it is reasonable to suspect that disease phenotypes are the consequence of the aberrant changes from the corresponding proteins' function. However, more than 80% of these single nucleotide polymorphisms (SNPs) are located in noncoding regions, most of them are likely to function through gene expression regulation (Schaub et al., 2012). This is supported by the findings that GWAS SNPs are enriched within DNase I hypersensitive

(DHS) sites (Maurano et al., 2012) and more likely to be eQTL (expression quantitative trait locus) SNPs (Nicolae et al., 2010). With the advancement of high-throughput technologies to allow quantitatively measuring the whole transcriptome, studying the genetics of genome-wide gene expression regulation became possible and was first performed in yeast (Brem et al., 2002) and then quickly applied to other species like mouse and human (Cheung et al., 2003; Schadt et al., 2003). The study of the genetics of gene expression has recently reached a new milestone with the launch of the Genotype of Tissue Expression (GTEx) project, which plans to profile individuals' genotypes and transcriptomes from nearly 1000 donors in more than 30 tissues using genotype array and RNA sequencing (RNAseq) technology (GTEx Consortium 2013).

With these large efforts in identifying the genetic architecture of tissue-specific gene expression, hundreds of thousands eQTLs have been identified or will emerge soon. One key challenge is to identify the regulatory mechanisms of these eQTLs and use such information to aid human disease research.

In the case of *cis*-eQTLs, the genetic variants are in proximity to the associated genes; it is intuitive to combine additional regulatory motif information (such as transcription factor binding) to corroborate the regulatory function of such SNPs. With the large body of data generated from projects like ENCODE (Encode Project Consortium et al., 2012), integrated analyses have been done and demonstrated that lead SNPs based on GWAS are enriched for DNase peaks and ChIP-seq peak (Schaub et al., 2012). Additional discussion is provided in Chapter 1. Conversely, it remains a challenge to interpret the mechanisms of *trans*-regulated gene expression, in which case the genetic variants are either far away from the regulated genes (e.g., greater than 1Mb) or on a different chromosome. It is generally believed that such regulations have multiple steps and are often mediated by transcription factors (TFs). Similar to studying *cis*-regulation, additional orthogonal data can help to reveal the potential regulatory mechanisms of *trans*-eQTLs. One approach is to rely on the protein-protein interaction (PPI) network to model the relay of signals from the genetic variants to the target genes.

In the first half of this chapter, we review a few computational approaches that model the flow of genetic information to gene expression in biological networks. We then provide an example in which we consider the information flow and the epistatic interaction between multiple genetic variants. This case study is to demonstrate how meaningful disease genes can be prioritized through an integrated network analysis. In the second half of this chapter, we review methods studying the differential connectivity in the coexpression network and discuss an example based on such an approach to help to identify disease causal genes.

4.2 Modeling Genetic Information Flow in Network

To discover the potential mechanisms of genetically regulated gene expression variation, we developed a random walk approach relying on a gene network comprising protein-protein interactions and protein-DNA interactions (Tu et al., 2006). The random walk basically simulates signal transductions occurring in the real biological interaction network, where most interactions are happening through proteins bumping into each other in a fluid environment consisting of various biological molecules and cellular components. Specifically, the network of proteins and transcription factor–DNA interactions are modeled as a graph G. At the kernel of this approach, we try to find paths in the network that connect genes located within eQTLs to their target genes (genes that are transcriptionally mapped to an eQTL).

Given a target gene and its eQTL, we initiate "walks" in the network starting from TFs that bind to the promoter region of the target gene. Once a node is reached, the next node to visit is chosen stochastically by favoring genes whose expressions are highly correlated with target gene g_t (Figure 4.1). Some walks will eventually arrive at genes within eQTL, and these genes will be visited at different frequencies when walks are repeated multiple times. The algorithm can be formalized as follows.

For a target gene g_t, the set of transcription factors binding to its promoter region are denoted as $T_{g_t} = (t_1, \ldots, t_n)$, and the candidate causal genes in the eQTL regions are denoted as $C_{g_t} = (g_{c_1}, \ldots, g_{c_m})$. The gene network is represented as a graph G in which the protein-protein interactions are represented as undirected edges, whereas protein phosphorylation and TF-DNA bindings are represented as directed edges. For each $t_k \in T_{g_t}$, we start a stochastic search procedure, as shown in Figure 4.1.

We denote all the neighbors of a particular gene in the gene network as $Nei(\cdot)$, so that $b \in Nei(a) \Leftrightarrow e_{ba} \in G$, where e_{ba} represents a directed edge from b to a. Starting from t_k, we estimate for each $g_i \in Nei(t_k)$ the "likelihood" that g_i is causative for the expression variation of the target gene g_t. We estimate such causal effect by the absolute value of the Pearson correlation coefficient of g_i and g_t expression levels, denoted as $|\rho_{(g_i, g_t)}|$. Intuitively, a gene with strong expression correlation with the target gene is more likely to be involved in the same pathway. However, as not all genes on the pathway necessarily correlate with the target gene because of other posttranslational regulation mechanisms, we give noncorrelated genes a residual probability for being on

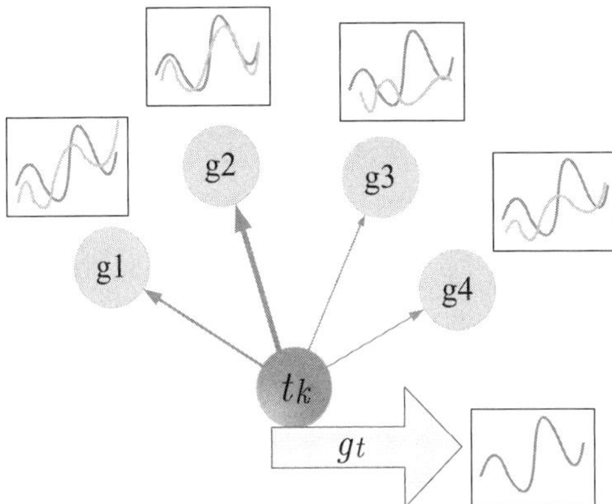

Figure 4.1 Initial step for "walking" in the network. TFs for the target gene are identified based on TF-DNA binding data and are taken as initiating points for the walk. The next gene to visit is selected stochastically from all neighbors based on expression correlation with the target gene. Genes with stronger correlation have a better chance to be selected as the next node, as indicated by the thickness of the arrows.

the regulatory pathway by defining the casual effect of g_i with respect to g_t as $\xi(g_i, g_t) = \max\{|\rho(g_i, g_t)|, \varepsilon\}$, where $0 < \varepsilon < 1$ is the default causal effect that a non-correlated gene could have on g_t.

We denote a path as $P(g_0, g_1, \ldots, g_z)$, where $g_0, g_1, \ldots, g_z$ are nodes in the graph and cycles are prohibited in the path, that is, $g_i \neq g_j$ for any g_i, g_j on the path. To ensure paths are noncyclic, a set $\mathbf{U}$ is introduced that contains only unvisited genes. We stochastically select $g_{i \in Nei(t_k)} \cap \mathbf{U}$ and transit from t_k to g_i. The transition probability is determined by Eq. (4.1).

Unvisited neighbor genes will be randomly drawn according to this transition probability. The chosen gene will be removed from $\mathbf{U}$ thereafter.

$$\Pr\{g_i | t_k, g_i \in Nei(t_k) \cap \mathbf{U}\} = \frac{\xi(g_i, g_t)}{\displaystyle\sum_{g_s \in Nei(t_k) \cap \mathbf{U}} \xi(g_s, g_t)} \qquad (4.1)$$

After we arrive at g_i, the same procedure is repeated. We select $g_i' \in Nei(g_i) \cap \mathbf{U}$ based on similar transition probability as described by Eq. (4.2).

$$\Pr\{g_i' | g_i, g_i' \in Nei(g_i) \cap \mathbf{U}\} = \frac{\xi(g_i', g_t)}{\displaystyle\sum_{g_s \in Nei(g_i) \cap \mathbf{U}} \xi(g_s, g_t)} \qquad (4.2)$$

It is worth noting that the algorithm always calculates the causal effect of a gene g_i with respect to g_t, which is different from many transcription regulatory network inference algorithms. In this procedure, the objective is not to identify the relationship between connected genes (i.e., g_i and g_i') but to find connected genes that are likely to be causative for the expression variation of the target gene g_t.

The proceeding procedure stops when it reaches any gene $g_i \in C_{g_t}$ or when it enters a dead end (i.e., $Nei(g_i) \cap \mathbf{U} = \emptyset$). We also set an upper bound for the total number of transitions allowed to ensure a stop. The upper bound is chosen to be high for many known pathways. Suppose we stop at $g_c \in C_{g_t}$ after one round of the procedure; the path can then be written as $P(t_k, \ldots, g_i, \ldots, g_c)$. The causal effect of g_c on g_t through $P(t_k, \ldots, g_i, \ldots, g_c)$ can be calculated by Eq. (4.3)

$$p(g_c, t_k, P(t_k, \ldots, g_c)) = \xi(t_k, g_t) \cdot \ldots \cdot \xi(g_c, g_t) \tag{4.3}$$

Eq. (4.3) measures the causal effect of g_c on g_t with respect to a specific potential pathway, and the general causal effect of g_c considering the whole gene network can be estimated by Eq. (4.4), where $P_{t_k}^{g_c}$ denotes all the paths starting from t_k and ending at g_c:

$$p(g_c, t_k) = \sum_{P_{t_k}^{g_c}} p(g_c, t_k, P(t_k, \ldots, g_i, \ldots, g_c)) \tag{4.4}$$

To calculate $p(g_c, t_k)$, each gene $g_i \in G$ is associated with a counter $V_{t_k}(g_i)$ to record the times it's been visited. We iterate the whole procedure N times, and N is set to be large enough so that Eq. (4.5) can be approximated, where $V_{t_k}(g_c)$ denotes the visit times for $g_c \in C_{g_t}$.

$$\lim_{N \to +\infty} V_{t_k}(g_c)/N = p(g_c, t_k) \tag{4.5}$$

If the target gene has more than one TF, each TF is assigned a weight based on its causal effect on the target gene and is linearly combined as shown by Eq. (4.6). The probability that g_c is the casual gene in the eQTL considering all the TFs for the target gene g_t is estimated by Eq. (4.7).

$$V_T(g_c) = \frac{\displaystyle\sum_{k=1}^{m} \xi(t_k, g_t) V_{t_k}(g_c)}{\displaystyle\sum_{k=1}^{m} \xi(t_k, g_t)} \tag{4.6}$$

$$\widehat{\Pr(g_c)} = \frac{V_T(g_c)}{\displaystyle\sum_{g_s \in C(g_t)} V_T(g_s)} = \frac{\displaystyle\sum_{k} p(g_c, t_k^c)}{\displaystyle\sum_{s: g_s \in C(g_t)} \sum_{k} p(g_s, t_k^s)} \tag{4.7}$$

Assuming there is only one causal gene in each eQTL, the gene with the largest posterior probability is reported as the causal gene based on Eq. (4.8).

$$g_c^* = \arg\max_{g_s \in C_{g_t}} \widehat{\Pr(g_s)} \tag{4.8}$$

To identify the underlying pathway, we start from g_c^* and trace backwards. We find from $Nei(g_c^*)$ the gene with the largest visit count and move to that gene (not stochastically). We repeat until we arrive at t_k. By this way, we find the most probable pathway which links g_c^* and t_k. The linear pathway generated by this approach is mainly for simplicity consideration. As indicated by Eq. (4.4), there could be multiple paths connecting g_c^* and t_k, and all of them contribute to the causal effect of g_c^* in a network manner.

4.3 Related Approaches

The random walk approach is further modeled as an electric circuit by Suthram et al. (2008), who showed that their algorithm, named eQED (eQTL electrical diagrams) can better capture the signal flow in the network; a possible reason is that the random walk approach counts the "dead end" paths, and such paths are not supposed to play a role in biological functions, and not counting them may actually improve performance. Kim et al. (2011) further developed the circuit flow algorithm and applied it to a glioblastoma multiforme data set. On the basis of that algorithm, they uncovered candidate causal genes and causal paths that are potentially responsible for the altered expression of disease genes in glioblastoma.

Similarly, Vanunu et al. (2010) proposed an approach called PRINCE to associate genes and protein complexes with disease via network propagation. The main idea is to have some prior information on gene-disease association, then let such information propagate in the protein-protein interaction network to achieve a new estimate on the gene-disease association in the global network. For a protein-protein interaction network $G = (V, E, w)$, where V is the node representing protein, E is the edge for protein-protein interactions and w is the weight of edges representing the reliability of such interactions. A prior knowledge $Y(v) \in [0, 1]$ represents the probability that protein v is associated with query disease. The algorithm is calculated based on a prioritization function $F : V \to \Re$ so that $F(v)$ is a statistics representing the strength of association between protein v and the query disease. F is defined by Eq. (4.9)

$$F(v) = \alpha \left[\sum_{u \in N(v)} F(u)w'(v, u) \right] + (1 - \alpha)Y(v) \tag{4.9}$$

so that $F(v)$ is a linear combination of prior information on v's disease association and the F scores of v's neighbors; $w'(v, u)$ is a normalized edge weight $w(v, u)$. Considering F and Y as vector of length $|V|$ and W' as a matrix of dimension $|V| \times |V|$, F can be computed using matrix algebra as in the following equation, where I is the identity matrix

$$F = (I - \alpha W')^{-1}(1 - \alpha)Y \qquad (4.10)$$

In practice, Vanunu et al. computed F using an iterative approach:

$$F^t = \alpha W' F^{t-1} + (1 - \alpha)Y \qquad (4.11)$$

where F^1 is set to be Y.

Poirel et al. (2013) developed an approach called LINKER, which combines teleporting random walks and k shortest path computations to identify pathways involved in a particular cellular process. Given a network $G = (V, E, w)$, which is a directed network with edge weight described by w, the goal is to identify a connected, low-weight subnetwork of G that connects the source set S to target set T (S and T are subsets of V). The LINKER solves this problem in two stages. In the first stage, a teleporting random walk known as PageRank is used to rank nodes in G with respect to a node in S. In the second stage, they used the node visitation probabilities to compute the k most probable paths from S to T.

Different from the PRINCE algorithm, where the signal has to rely on real interactions to propagate, the PageRank allows teleportation. For a node s_u in the source set S, with probability qs_x, it can teleport to any node $x \in V$, including itself. The starting probability of s_x is defined as $s_x = 1/|S|$ for $s_x \in S$ and $s_x = 0$ otherwise. In addition to teleporting, it can also "walk". It can move from its current node u to its direct neighbors $v \in N(u)$ with probability proportional to $(1 - q)w_{uv}$, where w_{uv} is normalized so that $\sum_{x \in N(u)} w_{ux} = 1$. Similar to PRINCE, PageRank also defines a transition matrix among nodes in G. The stationary visitation probability p_v for every node $v \in V$ can be computed by either solving the corresponding linear system or using an iterative method.

In the second stage, LINKER searches the interpretable paths that connect the source node to the target nodes. This is basically to find the paths from s_u to any node in T so the product of the visitation probability is maximized. By performing this two-stage calculation, LINKER is able to identify the "most likely" paths connecting the source nodes and target nodes.

Another approach reported to be useful is to identify the optimal subnetwork for a given set of seed genes by solving a Steiner tree or prize collecting Steiner tree (PCST) problem (Bienstock et al., 1993). To follow-up Ideker et al.'s (2002) very early work on identifying the dense submodules in the protein interaction network, Dittrich et al. (2008) revisited the problem by providing an optimal solution based on integer-linear programming and the PCST problem.

Given a connected undirected vertex and edge-weighted graph $G = (V, E, c, p)$ with vertex prize $p : V \rightarrow \Re^{\geq 0}$ and edge costs $c : E \rightarrow \Re^{\geq 0}$, the goal is to find a connected subgraph $T = (V_T, E_T)$ of G, $V_T \subseteq V$, $E_T \subseteq E$, such that the following term is maximized

$$\max \left[\sum_{v \in V_T} p(v) - \sum_{e \in E_T} c(e) \right] \tag{4.12}$$

The algorithm developed from Ljubie et al. Ljubic et al. (2006) is used to solve the the PCST problem. A number of preprocessing steps are first applied to simplify and transform the input network, and then a cut-based ILP model is solved in a branch-and-cut framework to produce an optimal solution.

Bailly-Bechet et al. (2011) worked on the same PCST problem, but their solution is based on statistical physical modeling and belief propagation. All these work and recent developments (Sadeghi and Fröhlich, 2013; Tuncbag et al., 2013) are providing new solutions for finding the signaling and regulatory subnetworks in the biological systems.

4.4 Application 1: Prioritize Diabetes Genes

Modeling the genetic information flow in biological networks can help us to better understand the disease mechanisms. One objective of such modeling is to identify the disease causal genes or genes on the information flow paths that could be suitable for drug development. In this part, we discuss in detail an application of such modeling to serve this goal.

We developed a network-based model to identify key genes that regulate plasma insulin levels in a B6 × BTBR obese F2 cross. The F2 inter-cross was generated between diabetes-resistant (B6) and diabetes-susceptible (BTBR) mouse strains, made genetically obese in response to the Lepob mutation Keller et al. 2008. The cross consisted of >500 mice, evenly split between males and females. A comprehensive set of ~5000 genotype markers was used to genotype each F2 mouse (~2000 informative SNPs were used for analysis), and the expression levels of ~40,000 transcripts (corresponding to 25,901 unique genes) were monitored in five tissues (adipose, liver, pancreatic islets, hypothalamus, and gastroc (gastrocnemius muscle)) that were harvested from each mouse at 10 weeks of age. In addition to gene expression, several key T2D-related traits were determined for each mouse.

At a genome-wide *p*-value less than 0.05, plasma insulin shows significant linkage to multiple loci in male mice, including chromosomes 2 at 59.5–82.5 cM, 6 at 0–33.66 cM, and 19 at 25–35.38 cM with LOD scores of $\sim$6.5, 4.4, and 5.4, respectively. Using linear regression modeling, the loci on chromosomes 2, 6, and 19 explain 10.6%, 6.0%, and 8.4% of the variation of plasma insulin, respectively. The top two loci at chromosomes 2 and 19 jointly explain 16.8% of the variance. The top two loci are significant at a genome-wide *p*-value less than 0.01, and are consistent with the results of another independent F2 cross from the same two strains (Stoehr et al., 2000).

To elucidate the gene-gene interactions underlying the heritability of plasma insulin, we examined gene expression profiles in several key tissues and extended the causality method to construct a protein interaction network for genes with expression quantitative trait loci, or eQTLs linked to insulin loci. To simplify the interaction models, we modeled the effects of two loci. Further analysis focused on interactions of genes with eQTLs linked to the top two plasma insulin loci, at chromosomes 2 and 19. We hypothesized that the joint regulation of plasma insulin at the two QTLs is mediated by gene-gene interactions whose expression variations are linked to the two loci. We further developed a novel ranking algorithm to infer candidate genes for regulation of plasma insulin.

Treating gene expression as a phenotypic trait, we computed eQTLs, for all genes expressed in pancreatic islets, white adipose tissue, liver, hypothalamus, and gastroc muscle of each male F2 mouse. We hypothesized that genes with eQTLs that colocalized with insulin QTLs are coregulated by common genetic factors (Schadt et al., 2005). We identified eQTLs within each tissue that had LOD profile peaks on chromosomes 2 and 19, the same genomic regions containing the peak insulin linkages. Among genes physically located within the insulin QTLs on chromosome 2 or 19, 89 genes have *cis*-eQTLs (gene expression QTLs are mapped to within 10 Mb of the genomic location of the genes) in islet, 66 in white adipose tissue, 52 in liver, 51 in hypothalamus, and 5 in gastroc. Clearly genes with *cis*-eQTLs may play a significant role in modulating insulin, and methods have been developed to identify the causal genes with *cis*-eQTLs for various phenotypic traits. However, each gene with a *cis*-eQTL can only explain the variance in the trait linked to its location. Here we considered a complementary strategy where we focused on genes with *trans*-eQTLs and interactions among them that integrate perturbations from multiple loci. As a greater number of genes showed *trans*-linkage, it is worth

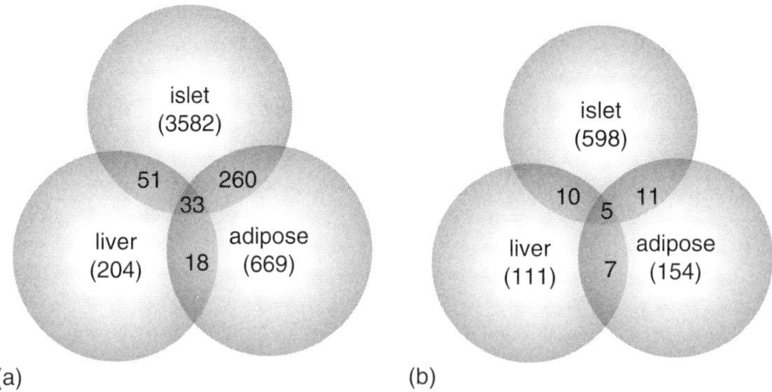

Figure 4.2 Number of gene transcripts in islet, liver, and white adipose tissue containing eQTLs that overlap with the insulin QTLs on (A) chromosome 2 and (B) chromosome 19. Numbers in parentheses show tissue-specific eQTLs.

studying the potential mechanisms by which these genes jointly mediate the phenotypic variation.

The expression traits that overlapped with the insulin QTLs were tissue specific, and are enriched in different GO biological pathways. The largest number of these traits was from pancreatic islets (Figure 4.2). In addition, islets contained the largest proportion of eQTLs that showed linkage to both loci on chromosomes 2 and 19, indicating that, similar to traditional complex traits (e.g., insulin), gene expression is also regulated by multiple genetic loci. Co-localization of gene eQTLs and plasma insulin QTLs does not imply that the gene is related to plasma insulin regulation. To filter out genes that, though linked to the same QTL region as insulin QTLs, are likely independent of plasma insulin regulation, we applied a genetic causality test developed by Schadt et al. (2005) to further narrow our list of candidate regulatory genes. Given the known feedback loop (islets–insulin levels–peripheral tissues, and with glucose levels), genes supported as either causal (QTL → gene → insulin) or reactive with respect to insulin levels (QTL → insulin → gene) were identified for consideration as insulin regulation genes.

A model considering the two loci on chromosomes 2 and 19 accounts for a greater part of the variation in plasma insulin than a single locus model. Several models of various degrees of complexity could explain the joint regulation of a common trait by multiple loci. As plotted in Figure 4.3A, the simplest case (M1) would be that the two loci directly regulate the same gene and that such a gene is responsible for modulating the trait. A slightly more complex case (M2) would be when each locus regulates a different gene, which could collaborate

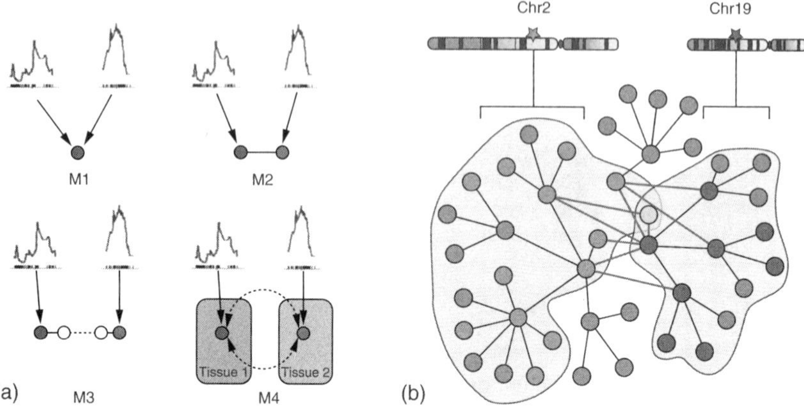

Figure 4.3 (A) Mechanistic models for joint regulation by two loci. In model 1, two loci directly regulate same gene. In model 2, each locus regulates one gene, and the two genes have physical interaction. Model 3 is similar to model 2, but there are multiple steps between the two genes. In model 4, each locus regulates one gene in a single tissue, and the cross-tissue interaction leads to a joint effect on phenotypic variation. (B) Interactive model for the additive effects between two genetic loci that regulate plasma insulin. Genetic variation at chromosome 2 changes expression for some nodes in the network (orange), while variation at chromosome 19 changes expression of other nodes (green), including genes regulated by chromosome 2 (yellow node). The blue nodes represent other nodes in the global protein-protein interaction network. Genes bound by gray curves are genes sharing the same eQTLs. Nodes involved in an interaction between the two sub-networks (shaded in light orange and green) are connected by bold red lines. We hypothesize that these nodes would be more influenced by genetic variation at both loci than a single locus. Genes involved in these cross-group interactions may be key regulators of plasma insulin.

directly through protein-protein physical interaction to influence the trait. In model M3, genes regulated by different loci interact indirectly and multiple steps exist before the perturbation signals merge on the common trait. In model M4, multiple tissues and their interactions are involved in regulating the trait. Here we developed an approach that seeks to combine the first two models to identify those components of the network underlying insulin regulation that are modulated by the chromosome 2 and 19 genetic loci and that may be physically interacting. As shown in Figure 4.3B, genetic variation at a single locus results in perturbations of biological functions that are reflected in the transcripts, or nodes, linked to that locus (orange or green nodes). Genetic variation at two loci could result in a larger functional influence on nodes showing linkage to both loci (yellow nodes) or nodes interacting across the two sub-networks (nodes connected by red edges).

Here we only consider the situation in which single genetic variations are synergistic, although antagonistic relationships may certainly occur. We hypothesize that nodes mediating the interaction between the two sub-networks are critical points for integrating the effects of multiple loci. To identify and rank these critical nodes, we developed an algorithm for assessing a gene's potential for being such an integrator in each of five tissues. We first collected a mouse protein-protein interactome by combining information from various databases, as previously described (Tu et al., 2009), where most interactions were experimentally derived and manually curated. We then extracted tissue-specific networks by mapping genes with eQTLs overlapping insulin QTLs on chromosome 2 or 19 onto this interactome and considering only interactions across the two eQTL gene groups. Islets contain the greatest number of genes involved in a cross-group interaction between the sub-networks showing linkage to chromosome 2 or 19.

Many genes are supported as being involved in cross-group interactions and could conceivably play a critical role in regulating plasma insulin levels. To assess their potential in mediating cross-group interactions and regulating clinical traits, we designed a novel ranking algorithm that integrates trait, protein-protein interaction, and gene expression (referred to as the TIE score) to identify those genes most likely to play a critical role in insulin regulation. Instead of focusing on the property of individual genes, the TIE score incorporates an interaction potential (IP) between a protein pair. Because PPI data are not assayed in a relevant physiological context, we leverage the expression data, which are assayed in a relevant context, to weigh whether a protein interaction pair is relevant to our context of interest (i.e., interactions between diabetes-relevant tissues in the cross population). Many post-transcriptional and post-translational modifications may impact protein-protein interactions and their resulting functional activities. However, these modifications cannot be inferred by gene expression profiling. For a pair of proteins known to interact, we make the simplifying assumption that their protein activities and binding affinity only rely on their expression levels. We further assume a strong IP if both genes are highly expressed in a mouse relative to the two genes in other F2 mice (because protein-protein association rates depend on protein concentrations in a simple diffusion model); if one or the other gene has relatively low expression, the IP will be weaker. We then calculate the correlation between the IP and trait, yielding a trait-IP correlation (TIPC). The TIPC represents the potential of an interaction (instead of individual genes) in regulating a clinical trait. Using TIPC as a weight for each interaction in the protein-protein interaction network, the TIE score is then computed for each node based on the small subnetwork formed by the node and its direct neighbors. A node receives a high TIE score if it has numerous interactions with large TIPC values. TIE scores are context dependent so that a given gene may have different TIE scores in different tissues, given that the levels of its own expression and that of its interaction partners' may be different between tissues. In contrast to other network-based approaches, the TIE score enables us to identify key nodes within a network that have multiple protein interactions with neighboring nodes, where such interactions are supported as exerting a strong influence on the particular clinical trait under investigation (in our case, plasma insulin levels).

For two interacting proteins i and j, we define IP for the protein pair in an individual mouse k as

$$IP_k(i, j) = \frac{e_i^k - e_i^{\min}}{e_i^{\max} - e_i^{\min}} \cdot \frac{e_j^k - e_j^{\min}}{e_j^{\max} - e_j^{\min}} \tag{4.13}$$

where e_i^k is gene i's expression in mouse k, so is e_j^k for gene j; $e_i^{\max}$ and $e_i^{\min}$ are the maximum and minimum tissue-specific expression observed across the entire F2 panel for gene i, respectively. The calculation assumes that the variation of protein abundance is approximated by its gene expression and that IP is proportional to the relative levels of the two proteins. Thus a reduction in gene expression would lead to reduced protein-protein interaction for a given pair, and vice versa. The predictive power of $IP(i, j)$ for a specific trait value (such as plasma insulin in our case) can be calculated as the TIPC. We define $TIPC((i, j), t) = |corr_{k \in K}(IP_k(i, j), t_k)|$, where t is the trait under consideration, $corr$ is the function for calculating the correlation coefficient, and K is the union of all the mice. We consider both network topology and TIPC scores to rank genes. Assume protein i interacts with a set of proteins J; the TIE score $S(i)$ is then computed as

$$S_t(i) = \frac{\sum_{j \in J} TIPC((i, j), t)}{|J|} \ln |J| \tag{4.14}$$

where $j \in J$ and $|J|$ is the number of proteins contained in J. A high TIE score indicates that a gene has large number of interactions, and for its direct neighbors, the average correlation between interaction potential and trait is also high.

For ranking purposes, we limit our calculation to genes with five or more interactions and set the TIE score to zero for the rest of the genes, given that genes with too few interactions may spuriously influence the TIE score and thus lead to unreliable results. We also permutated gene expression for the nodes in the network a thousand times and calculated the empirical distribution of TIE scores to assess the significance of derived TIE scores.

Table 4.1 *Top five genes in adipose and islet tissues ranked by TIE scores whose gene expression-insulin correlation > 0.2*

Rank	Gene symbol	No. interactions	TIE score	P-Value	Corr. ins
Adipose					
1	*Cdkn1a*	14	0.23	6.7×10^{-5}	0.29
2	*Aldoa*	18	0.052	0.48	0.46
3	*Src*	22	0.031	1	−0.22
4	*Tpi1*	16	0.027	1	0.51
5	*Pcx*	16	0.016	1	0.30
Islet					
1	*App*	102	0.78	≈ 0	−0.62
2	*Gria3*	15	0.52	≈ 0	−0.57
3	*Grb10*	7	0.52	≈ 0	−0.52
4	*Calca*	13	0.49	$<10^{-16}$	−0.39
5	*Ins1*	47	0.49	$<10^{-16}$	0.55

Pancreatic islets had the greatest number of genes with a significant nonzero TIE score. Adipose tissue had many fewer genes with a nonzero TIE score than islets, and the other three tissues had no genes with nonzero TIE scores, suggesting that protein-protein interactions in these tissues may not be the mechanism underlying the cross-locus regulation of plasma insulin involving chromosomes 2 or 19. The top five ranked genes for both adipose and islet tissues are given in Table 4.1. Our result suggests that pancreatic islets contribute the most to variation in plasma insulin in our F2 cross between B6 and BTBR mice. Circulating levels of plasma triglyceride, an indicator of insulin resistance, showed no significant genetic linkage, suggesting that factors controlling insulin resistance may be distinct from those controlling plasma insulin in our B6XBTBR-ob/ob F2 cross.

Compared to the top scores in adipose tissue, the top scores for genes in the islet network are much higher (Table 4.1), suggesting that a greater number of genes in islet make a larger contribution to insulin variation.

The top ranked gene in the islet network is *App* (Amyloid β Precursor Protein). Successive proteolytic processing of *App* by β- and γ-secretase enzymes generates the amyloid-β peptide, a primary component of amyloid plaques, which are thought to be central to the etiology of Alzheimer's disease (AD) (Palop and Mucke, 2010). Although this gene has been heavily studied by AD researchers, its relevance in type 2 diabetes is much less known. In addition to *App*, several other genes with high TIE scores could potentially be involved in regulating plasma insulin. Wang et al. (2007) showed that peripheral-tissue-specific knockout of *Grb10* results in enhanced insulin sensitivity in vivo that could be due to the loss of *Grb10*-mediated degradation of the insulin receptor. More recently, disrupting *Grb10* has been shown to increase pancreatic beta cell mass and reduce beta cell apoptosis in mice (Zhang et al., 2012). The enrichment of genes previously shown to participate in the regulation of circulating insulin in the top islet gene list, such as *Grb10* and insulin *Ins1*, supports our use of TIE in identifying novel regulators of insulin.

The expression of *App* in islets strongly negatively correlates with plasma insulin levels in the F2 cross (Pearson correlation coefficient $r = -0.68$, p-value $\ll 0.01$). Previous work has shown that compared to wild-type mice, whole-body *App* knockout mice (*App–/–*) have reduced plasma glucose and elevated insulin secretion in response to an intravenous glucose injection (Needham et al., 2008). Given that *App* is expressed in multiple tissues, including the brain, where it may regulate neurogenesis (Zhou et al., 2011), we sought to determine if the changes observed in *App–/–* mice reflect direct or indirect effects of *App* on islet function and/or health.

To assess whether the loss of *App* has a direct impact on islet function, we monitored insulin secretion ex vivo from pancreatic islets collected from either wild-type or *App–/–* mice. At 10 weeks of age, insulin secretion from wild-type versus *App-/-* mice was not different. However, at 17 weeks of age, insulin secretion was elevated approximately two to three fold from *App–/–* islets in response to glucose ($p < 0.05$), a depolarizing concentration of KCl ($p < 0.01$), or a membrane-permeant analogue of cAMP ($p < 0.01$). The amount of insulin per islet (insulin content) was not significantly different between wild-type and *App–/–* mice. Further, insulin secretion in response to basal glucose concentration (1.7 mM) at either 10 or 17 weeks of age was not significantly different between wild-type and *App–/–* mice. These results suggest that *App* directly functions as a negative regulator of insulin secretion in islets, and this only occurs in older mice.

4.5 Differential Connectivity in Coexpression Network

One important approach of studying human diseases is to compare the difference in the molecular data between healthy and disease states. Differential gene expression analysis is the most frequently used analysis in gene expression experiments, and numerous analytical methods have been developed (Tusher et al., 2001; Smyth, 2005). More recently, much attention has been drawn to examining the differential connectivity in the coexpression networks. For example, Van Nas et al. (2009) found gender-specific coexpression modules and Southworth et al. (2009) found that aging in mice was associated with a general decrease in coexpression. Although differential gene expression and differential coexpression analyses are related, they do not depend on each other. It has been reported that in some studies, very few differentially expressed genes are identifiable, whereas a large number of differentially coexpressed modules can be found (Zhang et al., 2013). In this part of the chapter, we review a few current methods for studying differential coexpression networks and discuss an application at the end.

Differential coexpression methods can be roughly divided into two categories: one category is based on predefined gene modules and is therefore called the "targeted" approach, whereas the second one is "untargeted" approach, which aims at grouping genes into modules based on their differential coexpression pattern (Tesson et al., 2010). Choi and Kendziorski (2009) focused on analyzing modules based on known Gene Ontology categories and therefore their approach belongs to the targeted approach category. Ihmels et al. (2005) developed a differential clustering algorithm (DCA) to systematically characterize both similarities and differences in the fine structure of coregulation patterns between fungal pathogen *Candida albicans* and model yeast *Saccharomyces cerevisiae*. The DCA is applied to a set of orthologous genes that are present in both organisms. In the first step, the pairwise correlations between these genes are measured in each organism separately, defining two pairwise correlation matrices (PCMs) of the same dimension (i.e., the number of orthologous genes). Next, the PCM of the primary ("reference") organism is clustered, assigning genes into subsets that are coexpressed in this organism but not necessarily in the second ("target") organism. Finally, the genes within each coexpressed subgroup are reordered by clustering according to the PCM of the target organism. This procedure is performed twice, reciprocally, such that each PCM is used once for primary and once for secondary clustering, yielding two distinct orderings of the genes. The results of the DCA are presented in terms of the rearranged PCMs. Because these matrices are symmetric and refer to the same set of orthologous genes, they can be combined into a single matrix without losing information. Specifically, we join the two PCMs into one composite matrix such that the lower-left triangle depicts the pairwise correlations in the reference organism, while the upper-right triangle depicts the correlations in the target organism.

Inspection of the rearranged composite PCM allows for an intuitive extraction of the differences and similarities in the coexpression pattern of the two organisms. An automatic scoring method is then applied to classify clusters into one of the four conservation categories: full, partial, split, or no conservation of coexpression.

Tesson et al. (2010) developed an unsupervised, therefore "untargeted" method called DiffCoEx to identify differentially coexpressed gene modules. The method was built on WGCNA (Zhang and Horvath, 2005), a widely applied coexpression network analysis package. It has five major steps and it starts with the calculation of the pairwise correlation $c_{ij}^{[k]} = \mathrm{cor}(i, j)$ in each condition k for all the gene pairs (i, j). In the second step, the matrix of adjacency difference D is computed, and each element in this matrix is defined by

$$d_{ij} = \left(\sqrt{\frac{1}{2} |\mathrm{sign}\left(c_{ij}^{[1]}\right) * \left(c_{ij}^{[1]}\right)^2 - \mathrm{sign}\left(c_{ij}^{[2]}\right) * \left(c_{ij}^{[2]}\right)^2 |} \right)^{\beta} \tag{4.15}$$

d_{ij} therefore represents the difference of correlation between gene pair (i, j) in condition 1 versus condition 2. The parameter β is a positive integer and is similar to the soft power in the WGCNA package to transform the original difference so that large difference is emphasized over smaller difference. In the next step, the topological overlap-based dissimilarity matrix T is calculated based on D:

$$T : t_{ij} = 1 - \frac{\sum\limits_{k \in N(i,j)} (d_{ik}d_{kj}) + d_{ij}}{\min\left(\sum\limits_{k \in N(i)} d_{ik}, \sum\limits_{k \in N(j)} d_{jk}\right) + 1 - d_{ij}} \tag{4.16}$$

The use of topological overlap is an advantage over simply considering the correlation between a single pair of genes as the correlation is "smoothed" out based on the neighboring genes around the gene pair that is under consideration. In the next two steps, DiffCoEx performs dynamicTreeCut to cluster modules based on the matrix T and assess the statistical significance of the differential coexpression findings. Amar et al. (2013) proposed a comparable method called "DICER" to solve a similar problem.

4.6 Application 2: LOAD Brain Study

In this part of the chapter, we discuss an example of applying the differential coexpression connectivity analysis to study AD. Zhang et al. (2013) processed 1647 autopsied tissues from dorsolateral prefrontal cortex (PFC), visual cortex (VC), and cerebellum (CB) from 549 brains of 376 late-onset Alzheimer's disease (LOAD) patients and 173 nondemented healthy controls. Each tissue sample was genotyped and profiled for the whole transcriptome. Multitissue coexpression networks consisting of the top one-third most variable gene-expression traits per brain region in individuals from all the three regions are constructed in both the LOAD patients and controls. To analytically compare

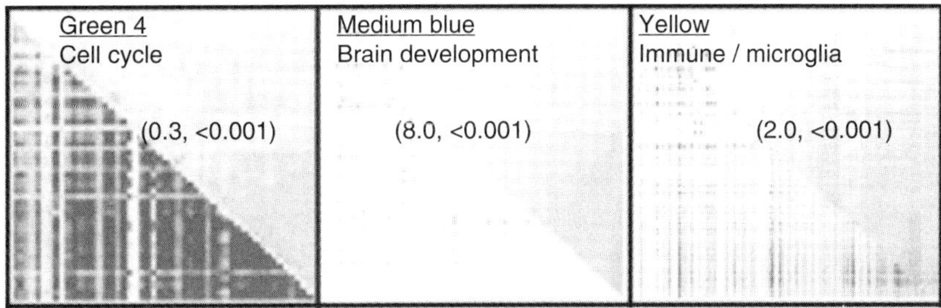

Figure 4.4 Example of gain of connectivity and loss of connectivities as calculated by the MDC for modules in LOAD (the upper right triangle of each module) versus that in the nondemented state (the lower left triangle of each module). Differential connectivity (MDC) and FDR estimates are specified in each panel in parenthese (MDC, FDR).

the coexpression structure difference between LOAD and nondemented individuals, modular differential connectivity (MDC) is calculated. To compare the MDC of a module Ω that contains N genes as defined by the common genes between two networks x and y, the ratio of the average connectivity among the N genes in the network x to that among the same gene set in network y is specified by

$$\text{MDC}: \delta_\Omega(x, y) = \frac{\sum\limits_{i=1}^{N-1} \sum\limits_{j=i+1}^{N} k_{ij}^x}{\sum\limits_{i=1}^{N-1} \sum\limits_{j=i+1}^{N} k_{ij}^y} \tag{4.17}$$

MDC > 1 indicates gain of connectivity (GOC) or enhanced coregulation between genes, whereas MDC < 1 indicates loss of connectivity (LOC) or reduced coregulation between genes. The statistical significance of the MDC metrics was computed through the false discovery rate (FDR) procedure. An example of loss connectivity in dementia patients is shown by the green 4 module in Figure 4.4, while two modules (medium blue and yellow modules) are illustrated as examples for gain of connectivity in the same figure. Coregulation structures like the glutathione transferase (GST) module showed extreme cases gain of connectivity in LOAD while structures like the nerve myelination module showed loss of connectivity. Thus, new modules are created in LOAD, whereas in other cases, a portion of the network is completely disrupted. The authors then constructed Bayesian networks for each coexpression module. Detailed information regarding Bayesian network construction has been previously reported (Zhu et al., 2004, 2012). A gene *TYROBP* was selected based on key driver analysis, MDC, and additional information as a top causal regulator for experimental validation with successful outcome.

4.7 Discussion

Gene expression measured at transcriptome scale provides an enormous amount of information and opens a door to the biological systems for us to examine its complexity at the molecular level. Such information when combined with genotype and phenotype data has proven to be very helpful for providing insights into the connections and regulations within the system. However, we need be aware of a few limitations.

For methods relying on the predefined networks, such as protein-protein interactions, caution needs be taken regarding the incompleteness of such information. In many cases, the accuracy of the annotated interactions needs to be improved. For methods that do not consider protein interactions but mainly use gene expression information, such as coexpression or the Bayesian network approach, this is not a concern. However, the findings based on gene expression analysis usually do not directly map to protein physical interactions, therefore it could be more difficult to experimentally verify such indirect interactions.

Another important aspect of biological systems is the dynamic property of the systems. All the data and associated modeling we discussed so far are capturing a single time point, a snap-shot of the system. Obviously, important information could be lost when we omit the dynamic properties of the system. Studying the dynamic properties is clearly more demanding because a greater number of samples needs to be collected and profiled at multiple time points, which can be practically challenging for human studies. Nevertheless, a great amount of research work has been done in this area, and the dynamic properties of complex systems will continue to attract more research attention (Zhu et al., 2010; Gitter et al., 2013; Wang et al., 2013).

With the fast development of new high-throughput technologies, we are facing a deluge of biological data. The scale and heterogeneity of the data (generally called "big data") pose a great challenge for developing novel analytic methods. Although the case studies we discussed tend to demonstrate that it is possible to computationally reveal the underlying biological mechanisms buried in the big data, there is clearly much more to conquer as we are still at this early stage of understanding complex human disease systems.

References

Amar D, Safer H, Shamir R. 2013. Dissection of regulatory networks that are altered in disease via differential co-expression. *PLoS computational biology* **9**(3): e1002955.

Bailly-Bechet M, Borgs C, Braunstein A, Chayes J, Dagkessamanskaia A, François J, Zecchina R. 2011. Finding undetected protein associations in cell signaling by belief propagation. *Proceedings of the National Academy of Sciences* **108**(2): 882–887.

Bienstock D, Goemans M, Simchi-Levi D, Williamson D. 1993. A note on the prize collecting traveling salesman problem. *Mathematical programming* **59**: 413–420.

Brem R, Yvert G, Clinton R, Kruglyak L. 2002. Genetic dissection of transcriptional regulation in budding yeast. *Science (New York, NY)* **296**(5568): 752–755.

Cheung V, Conlin L, Weber T, Arcaro M, Jen K-Y, Morley M, Spielman R. 2003. Natural variation in human gene expression assessed in lymphoblastoid cells. *Nature genetics* **33**(3): 422–425.

Choi Y, Kendziorski C. 2009. Statistical methods for gene set co-expression analysis. *Bioinformatics (Oxford, England)* **25**(21): 2780–2786.

Dittrich M, Klau G, Rosenwald A, Dandekar T, Müller T. 2008. Identifying functional modules in protein-protein interaction networks: an integrated exact approach. *Bioinformatics (Oxford, England)* **24**(13): i223–i231.

Encode Project Consortium, Bernstein B, Birney E, Dunham I, Green E, Gunter C, Snyder M. 2012. An integrated encyclopedia of DNA elements in the human genome. *Nature* **489**(7414): 57–74.

Gitter A, Carmi M, Barkai N, Bar-Joseph Z. 2013. Linking the signaling cascades and dynamic regulatory networks controlling stress responses. *Genome research* **23**(2): 365–376.

GTEx Consortium. 2013. The Genotype-Tissue Expression (GTEx) project. *Nature genetics* **45**(6): 580–585.

Hunter D. 2005. Gene-environment interactions in human diseases. *Nature Reviews Genetics* **6**(4): 287–298.

Ideker T, Ozier O, Schwikowski B, Siegel A. 2002. Discovering regulatory and signalling circuits in molecular interaction networks. *Bioinformatics* **18**(Suppl 1): S233–S240.

Ihmels J, Bergmann S, Berman J, Barkai N. 2005. Comparative gene expression analysis by differential clustering approach: application to the Candida albicans transcription program. *PLoS genetics* **1**(3): e39.

Keller M, Choi Y, Wang P, Davis D, Rabaglia M, Oler A, Stapleton D, Argmann C, Schueler K, Edwards S, et al. 2008. A gene expression network model of type 2 diabetes links cell cycle regulation in islets with diabetes susceptibility. *Genome research* **18**(5): 706–716.

Kim Y-A, Wuchty S, Przytycka T. 2011. Identifying causal genes and dysregulated pathways in complex diseases. *PLoS computational biology* **7**(3): e1001095.

Ljubic I, Weiskircher R, Pferschy U, Klau G, Mutzel P, Fischetti M. 2006. An algorithmic framework for the exact solution of the prize-collecting steiner tree problem. *Mathematical Programming* **105**: 427–449.

Maurano M, Humbert R, Rynes E, Thurman R, Haugen E, Wang H, Reynolds A, Sandstrom R, Qu H, Brody J, et al. 2012. Systematic localization of common disease-associated variation in regulatory DNA. *Science (New York, NY)* **337**(6099): 1190–1195.

Needham B, Wlodek M, Ciccotosto G, Fam B, Masters C, Proietto J, Andrikopoulos S, Cappai R. 2008. Identification of the Alzheimer's disease amyloid precursor protein (APP) and its homologue APLP2 as essential modulators of glucose and insulin homeostasis and growth. *Journal of pathology* **215**(2): 155–163.

Nicolae D, Gamazon E, Zhang W, Duan S, Dolan M, Cox N. 2010. Trait-associated SNPs are more likely to be eQTLs: annotation to enhance discovery from GWAS. *PLoS Genetics* **6**(4): e1000888.

Palop J, Mucke L. 2010. Amyloid-beta-induced neuronal dysfunction in Alzheimer's disease: from synapses toward neural networks. *Nature neuroscience* **13**(7): 812–818.

Poirel C, Rodrigues R, Chen K, Tyson J, Murali T. 2013. Top-down network analysis to drive bottom-up modeling of physiological processes. *Journal of computational biology* **20**(5): 409–418.

Sadeghi A, Fröhlich H. 2013. Steiner tree methods for optimal sub-network identification: an empirical study. *BMC bioinformatics* **14**(1): 144.

Schadt E, Monks S, Drake T, Lusis A, Che N, Colinayo V, Ruff T, Milligan S, Lamb J, Cavet G, et al. 2003. Genetics of gene expression surveyed in maize, mouse and man. *Nature* **422**(6929): 297–302.

Schadt E, Lamb J, Yang X, Zhu J, Edwards S, Guhathakurta D, Sieberts S, Monks S, Reitman M, Zhang C, et al. 2005. An integrative genomics approach to infer causal associations between gene expression and disease. *Nature genetics* **37**(7): 710–717.

Schadt E. 2009. Molecular networks as sensors and drivers of common human diseases. *Nature* **461**(7261): 218–223.

Schaub M, Boyle A, Kundaje A, Batzoglou S, Snyder M. 2012. Linking disease associations with regulatory information in the human genome. *Genome research* **22**(9): 1748–1759.

Smyth G. 2005. Limma: linear models for microarray data. In *Bioinformatics and Computational Biology Solutions Using R and Bioconductor*, pp. 397–420. Springer.

Southworth L, Owen A, Kim S. 2009. Aging mice show a decreasing correlation of gene expression within genetic modules. *PLoS genetics* **5**(12): e1000776.

Stoehr J, Nadler S, Schueler K, Rabaglia M, Yandell B, Metz S, Attie A. 2000. Genetic obesity unmasks nonlinear interactions between murine type 2 diabetes susceptibility loci. *Diabetes* **49**(11): 1946–1954.

Suthram S, Beyer A, Karp R, Eldar Y, Ideker T. 2008. eQED: an efficient method for interpreting eQTL associations using protein networks. *Molecular systems biology* **4**: 162.

Tesson B, Breitling R, Jansen R. 2010. DiffCoEx: a simple and sensitive method to find differentially coexpressed gene modules. *BMC bioinformatics* **11**: 497.

Tu Z, Wang L, Arbeitman M, Chen T, Sun F. 2006. An integrative approach for causal gene identification and gene regulatory pathway inference. *Bioinformatics* **22**(14): e489–e496.

Tu Z, Argmann C, Wong K, Mitnaul L, Edwards S, Sach I, Zhu J, Schadt E. 2009. Integrating siRNA and protein-protein interaction data to identify an expanded insulin signaling network. *Genome research* **19**(6): 1057–1067.

Tuncbag N, Braunstein A, Pagnani A, Huang S-SC, Chayes J, Borgs C, Zecchina R, Fraenkel E. 2013. Simultaneous reconstruction of multiple signaling pathways via the prize-collecting steiner forest problem. *Journal of computational biology* **20**(2): 124–136.

Tusher V, Tibshirani R, Chu G. 2001. Significance analysis of microarrays applied to the ionizing radiation response. *Proceedings of the National Academy of Sciences* **98**(9): 5116–5121.

Van A, GuhaThakurta D, Wang S, Yehya N, Horvath S, Zhang B, Ingram-Drake L, Chaudhuri G, Schadt E, Drake T, et al. 2009. Elucidating the role of gonadal hormones in sexually dimorphic gene coexpression networks. *Endocrinology* **150**(3): 1235–1249.

Vanunu O, Magger O, Ruppin E, Shlomi T, Sharan R. 2010. Associating genes and protein complexes with disease via network propagation. *PLoS computational biology* **6**(1): e1000641.

Wang J, Chen B, Wang Y, Wang N, Garbey M, Tran-Son-Tay R, Berceli S, Wu R. 2013. Reconstructing regulatory networks from the dynamic plasticity of gene expression by mutual information. *Nucleic acids research* **41**(8): e97.

Wang L, Balas B, Christ-Roberts CY, Kim RY, Ramos FJ, Kikani CK, Li C, Deng C, Reyna S, Musi N, et al. 2007. Peripheral disruption of the Grb10 gene enhances insulin signaling and sensitivity in vivo. *Molecular and cellular biology* **27**(18): 6497–6505.

Zhang B, Horvath S. 2005. A general framework for weighted gene co-expression network analysis. *Statistical applications in genetics and molecular biology* **4**: Article 17.

Zhang B, Gaiteri C, Bodea L-G, Wang Z, McElwee J, Podtelezhnikov A, Zhang C, Xie T, Tran L, Dobrin R, et al. 2013. Integrated systems approach identifies genetic nodes and networks in late-onset Alzheimer's disease. *Cell* **153**(3): 707–720.

Zhang J, Zhang N, Liu M, Li X, Zhou L, Huang W, Xu Z, Liu J, Musi N, DeFronzo R, et al. 2012. Disruption of growth factor receptor-binding protein 10 in the pancreas enhances β-cell proliferation and protects mice from streptozotocin-induced β-cell apoptosis. *Diabetes* **61**(12): 3189–3198.

Zhou Z, Chan C, Ma Q, Xu X, Xiao Z, Tan E. 2011. The roles of amyloid precursor protein (APP) in neurogenesis: implications to pathogenesis and therapy of Alzheimer disease. *Cell adhesion and migration* **5**(4): 280–292.

Zhu J, Lum P, Lamb J, GuhaThakurta D, Edwards S, Thieringer R, Berger J, Wu M, Thompson J, Sachs A, et al. 2004. An integrative genomics approach to the reconstruction of gene networks in segregating populations. *Cytogenetic and genome research* **105**(2–4): 363–374.

Zhu J, Chen Y, Leonardson A, Wang K, Lamb J, Emilsson V, Schadt E. 2010. Characterizing dynamic changes in the human blood transcriptional network. *PLoS computational biology* **6**(2): e1000671.

Zhu J, Sova P, Xu Q, Dombek K, Xu E, Vu H, Tu Z, Brem R, Bumgarner R, Schadt E. 2012. Stitching together multiple data dimensions reveals interacting metabolomic and transcriptomic networks that modulate cell regulation. *PLoS biology* **10**(4): e1001301.

5

Integrative Analysis of Multiple ChIP-X Data Sets Using Correlation Motifs

HONGKAI JI AND YINGYING WEI

Abstract

Genome-wide chromatin immunoprecipitation experiments including ChIP-seq and ChIP-chip, jointly referred to as ChIP-X, are high-throughput technologies to map protein-DNA interactions in the genome. When multiple related ChIP-X data sets are available, separately analyzing each data set is not optimal because it may lack power to detect consistent but relatively weak signals in multiple studies. Jointly analyzing all data sets may allow one to borrow information across studies to improve signal detection. However, a common problem in data integration is the difficulty in handling data set–specific signals that cannot be dealt with by simply assuming that the signal status for each genomic locus is the same across all studies. An integration model that naively enumerates all possible study specificity patterns, conversely, has exponential complexity because there are 2^D possible combinatorial signal presence and absence patterns for D studies. Correlation motifs provide a useful solution to this problem. By introducing a small number of latent probability vectors called correlation motifs, this approach can describe the major correlation structure among multiple data sets, which can then be used to guide information sharing across data sets. The correlation motif approach is capable of improving signal detection. At the same time, it does not have the problem of exponential model complexity and is flexible enough to handle all possible data set–specific signal configurations.

5.1 Introduction

Chromatin immunoprecipitation (ChIP) coupled with high-throughput sequencing (ChIP-seq) (Johnson et al., 2007; Barski et al., 2007) and ChIP coupled with genome tiling array hybridization (ChIP-chip) (Ren et al., 2000; Cawley et al., 2004) are two high-throughput technologies capable of mapping protein-DNA interactions (PDIs) genome-wide. These two technologies, collectively referred to as "ChIP-X," have been widely used to produce information on transcription factor binding sites (TFBSs) (Johnson et al., 2007;

Robertson et al., 2007), histone modifications (HMs) (Barski et al., 2007; Mikkelsen et al., 2007), nucleosome positioning (Schmid and Bucher, 2007), allele-specific PDIs (Mikkelsen et al., 2007; Rozowsky et al., 2011), variations of PDIs among the population (McDaniell et al., 2010; Kasowski et al., 2010), and the evolution of gene regulation (Schmidt et al., 2010). Today, ChIP-X has become an indispensable tool used by both individual investigators and large consortium projects such as the Encyclopedia of DNA Elements (ENCODE), modENCODE, and Roadmap Epigenomics consortia to annotate the human genome and epigenome (Consortium, 2012; Celniker et al., 2009; Bernstein et al., 2010).

ChIP-X data deposited in public databases grow rapidly. For instance, the recently developed database hmChIP (Chen et al., 2011a) has compiled 2000+ consistently processed ChIP-X samples representing 170+ different human and mouse transcription factors (TFs) and histone modifications. The Gene Expression Omnibus (GEO) (Barrett et al., 2009) contains 13,000+ data sets that involves ChIP-seq or ChIP-chip. The growing data opens the door to better utilizing the data. By examining large amounts of data, one can obtain insights on how to better separate signals from noises. These insights are not always available by looking at each individual data set. For instance, Judy and Ji (2009) have shown that by using a large number of ChIP-chip samples in GEO, one can estimate and better remove systematic probe effects from ChIP-chip data generated by Affymetrix tiling arrays. Choi et al. (2009), Wu and Ji (2010), Chen et al. (2011b), and Zeng et al. (2013) have shown that jointly analyzing two or more related ChIP-X data sets can improve the sensitivity and specificity of peak calling. Ji et al. (2013) have shown that for detecting differential PDIs, integrative analysis of multiple ChIP-seq data sets is more reliable than analyzing each data set separately. Wei et al. (2012) have demonstrated that jointly analyzing multiple ChIP-seq data sets also improves the analysis of allele-specific binding (ASB).

Existing ChIP-X data analysis tools are mostly designed for analyzing data collected in one experiment (Laajala et al., 2009; Wilbanks and Facciotti, 2010). New computational tools are required to deal with the emerging needs in data integration. However, data integration in the ChIP-X analyses is non-trivial. One has to address numerous challenges including, for instance, storing and accessing big data, building scalable statistical models, and developing computationally efficient algorithms. Furthermore, different applications and data types may require different data integration methods, making it difficult to find a common solution to all data integration problems. Despite this, some emerging ideas are broadly useful in many applications. This chapter discusses one such idea, the "correlation motif approach," which is applicable to a broad class of problems commonly seen in ChIP-X data analyses. The correlation motif approach considers the problem of detecting signals from multiple ChIP-X data sets. Here signal is a general term. Its meaning depends on the specific application in question. For example, if the purpose of the analysis is to detect TFBSs, then signals are the peaks in the ChIP-X data. For detecting allele-specific binding, signal refers to genomic sites where protein binding intensities on the paternal and maternal alleles are different.

5.2 General Problem Setting and Motivations

Consider D ChIP-X data sets from which one wants to detect signals. The data sets are biologically related, but different data sets may measure different signals. For instance, they can measure binding sites of different TFs, measure binding sites of the same TF in different cell types, or represent data produced by different labs to study the same TF in the same cell type. Each data set may contain multiple replicate samples. Depending on the experiment design, they may also contain control samples (e.g., input control).

The analysis goal is to identify signals in each and every data set. A simple solution to this problem is to analyze each data set separately. This approach, however, ignores the correlation among different data sets. As a result, it may not be the most efficient way to use the data. This can be demonstrated using the example in Figure 5.1. This example studies binding sites of transcription factors GLI1 and GLI3 in different biological contexts. Both GLI1 and GLI3 belong to the same TF family and recognize the same DNA motif. The figure shows the fold enrichments between the normalized ChIP and control probe intensities in four ChIP-chip experiments conducted in different cell types. Whereas there are clear data set–specific binding patterns, the binding sites in different data sets tend to occur at the same genomic loci. In light of this observation, when there is a strong peak in data set 1 and a relatively weak peak in data set 2, one can be more confident that the peak in data set 2 is a real binding event instead of random noise. In this example, the second data set (Gli1 limb) has a low signal-to-noise ratio overall, which makes it hard to discriminate peaks from noise. If one analyzes different data sets separately, one may not be able to confidently identify the weak GLI binding site in data set 2 around the Ptch1 promoter region, which indeed is a true GLI1 binding site (Vokes et al., 2008). In contrast, an integrative analysis would allow one to pool information from the other three data sets to help with finding real peaks in data set 2 with higher accuracy.

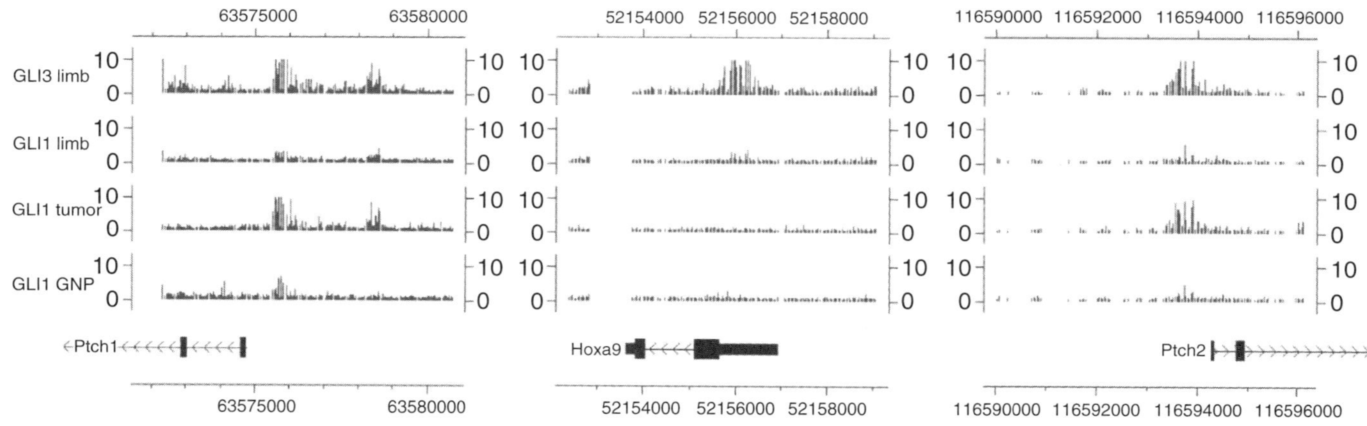

Figure 5.1 Related ChIP-X data sets share information. Peaks in four related ChIP-chip data sets for GLI have similar locations, although different loci may exhibit different data set–specific binding intensity patterns.

In the peak calling example, a naive approach to performing integrative analysis is to assume that each genomic locus is either bound in all studies or not bound in any study. This concordance model may allow one to identify loci with weak but consistent binding signals in all data sets. However, it ignores the reality that there are also many loci that are bound by GLI1 but not GLI3, or bound by GLI only in limb but not in other cell types, and so on. The concordance model cannot identify these data set–specific signals.

A more general approach is to consider all possible combinatorial signal occurrence patterns. In the example, each locus can either be bound or not bound in each data set. Given D data sets, there are 2^D possible signal patterns. One can describe the data using a mixture of 2^D models. Each model corresponds to a signal pattern. Once the relative frequency of each model is given, the correlation structure among the data sets is specified. Using this approach, one can both deal with data set specificity and use the correlation among data sets to develop a principled solution to sharing information across data sets. However, an obvious limitation of this approach is that it does not scale well with the increasing D. As the number of data sets increases, the model complexity will increase exponentially. This motivated development of the correlation motif approach aims at providing a flexible and scalable solution to this type of problem.

5.3 Method

5.3.1 The Correlation Motif Model

Suppose there are G genomic loci. Let $\mathbf{x}_{gd}$ be the observed data for locus g in data set d. Depending on the applications, the observed data could be probe intensities in ChIP-chip or read counts in ChIP-seq. Each $\mathbf{x}_{gd}$ can be a vector that contains data from all replicate ChIP samples and/or control samples in data set d; $\mathbf{x}_{gd}$s are assumed to be normalized and appropriately transformed. The collection of data for locus g is $\mathbf{x}_g = \{\mathbf{x}_{gd} : d = 1, \ldots, D\}$. The collection of all observed data is $\mathbf{X} = \{\mathbf{x}_g : g = 1, \ldots, G\}$.

For simplicity, assume that the signal status of a locus in a data set can be indicated by a binary variable a_{gd}. Let $a_{gd} = 1$ indicate that locus g carries a signal of interest in study d, and $a_{gd} = 0$ otherwise. This section introduces the correlation motif approach based on this two-signal-states assumption. The correlation motif approach, however, is general, and it is straightforward to extend the model to deal with multiple signal states (e.g., up-, down-, and no change in the differential binding analysis).

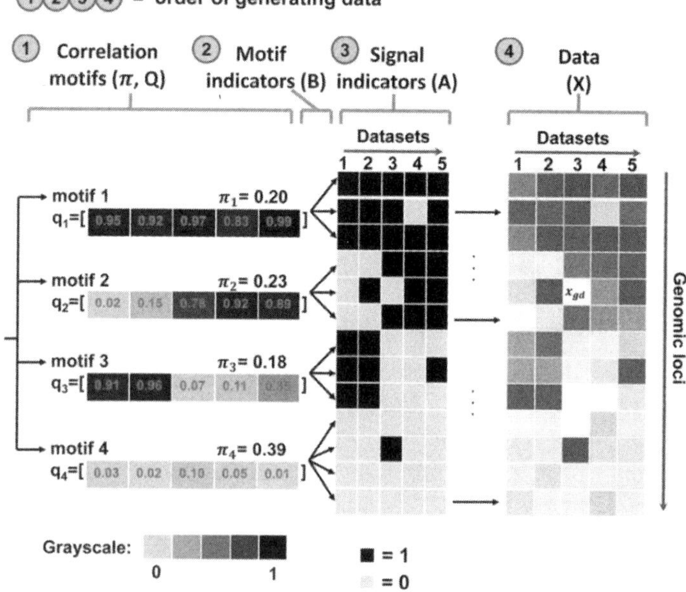

Figure 5.2 The data generative process assumed by the correlation motif model.

Let the vector $\mathbf{a}_g = [a_{g1}, a_{g2}, \ldots, a_{gD}]$ contain the signal states of locus g in all D data sets. Each locus can carry signal in none, some, or all of the D data sets. The collection of all signal indicators is $\mathbf{A} = \{\mathbf{a}_1, \ldots, \mathbf{a}_G\}$. With D studies, each $\mathbf{a}_g$ has 2^D possible configurations. The relative frequency of each configuration gives the joint distribution of $[a_{g1}, \ldots, a_{gD}]$ and thus the correlation structure among data sets. However, modeling correlation in this way requires $O(2^D)$ parameters and therefore is not scalable when the number of data sets D increases.

The correlation motif approach circumvents this limitation by introducing a hierarchical mixture model shown in Figure 5.2. Genomic loci are assumed to belong to K different classes or "motifs" ($K \ll 2^D$). The observed data $\mathbf{X}$ are assumed to be generated by the following generative probabilistic model.

- First, each locus g is randomly assigned a class label b_g. A priori, the probability that a locus belongs to class k is $\pi_k \equiv Pr(b_g = k)$. Here $\sum_{k=1}^{K} \pi_k = 1$.
- Second, given a locus's class label b_g, its signal states $\mathbf{a}_g$ are stochastically generated according to the following probabilities: $q_{kd} \equiv Pr(a_{gd} = 1 | b_g =$

k). Conditional on b_g, a_{gd}s in different data sets are assumed to be independent. Under these assumptions, the prior probability that a locus carries signal in data set d is $\sum_k \pi_k q_{kd}$. The joint prior probability that a locus carries signal in all data sets is $\sum_k \pi_k(\prod_d q_{kd})$. Clearly this probability is different from $\prod_d(\sum_k \pi_k q_{kd})$, the joint probability one would expect if signals were to occur independently in different data sets. This shows that, despite the conditional independence assumption, the a_{gd}s are not independent marginally after integrating out the latent class label b_g. This is why the hierarchical mixture model used here can model the correlation among data sets.

- Third, given a locus's signal states $\mathbf{a}_g$, its observed data are generated independently for each data set based on $f_{gd1}(\mathbf{x}_{gd}) \equiv f(\mathbf{x}_{gd}|a_{gd} = 1)$ and $f_{gd0}(\mathbf{x}_{gd}) \equiv f(\mathbf{x}_{gd}|a_{gd} = 0)$. Here f_{gd1} and f_{gd0} are probability distributions for signal and noise, respectively. They are usually constructed based on the data type in question. In what follows, f_{gd1} and f_{gd0} are assumed to be known.

Besides the preceding assumptions, the model also assumes that the class label b_g, signal states $\mathbf{a}_g$, and data $\mathbf{x}_g$ for different loci are generated independently.

Let $\boldsymbol{\pi} = [\pi_1, \ldots, \pi_K]$ and $\mathbf{q_k} = [q_{k1}, \ldots, q_{kD}]$. Introduce indicator function $\delta(\cdot)$: $\delta(\cdot) = 1$ if its argument is true, and $\delta(\cdot) = 0$ otherwise. Using these notations, the joint probability distribution of signal states $\mathbf{A}$, class labels $\mathbf{B}$, and observed data $\mathbf{X}$ conditional on model parameters $\boldsymbol{\pi}$ and $\mathbf{Q}$ can be written as

$$Pr\,(\mathbf{A}, \mathbf{B}, \mathbf{X}|\boldsymbol{\pi}, \mathbf{Q})$$

$$= \prod_{g=1}^{G}\prod_{k=1}^{K}\left\{\pi_k \prod_{d=1}^{D}[q_{kd}f_{gd1}(\mathbf{x}_{gd})]^{a_{gd}}[(1 - q_{kd})f_{gd0}(\mathbf{x}_{gd})]^{1-a_{gd}}\right\}^{\delta(b_g=k)} \tag{5.1}$$

This model has several important features. First, each loci class k is associated with a vector $\mathbf{q}_k$ that specifies the prior signal occurrence probability in each and every data set. If the elements q_{kd} in this vector are all nonzero, then the class can generate all 2^D configurations of $\mathbf{a}_g$ with nonzero probabilities. Thus, the model is capable of handling all possible data set–specific patterns. Second, even though each class could allow all 2^D configurations of $\mathbf{a}_g$, different configurations have different probabilities of occurring. Loci with similar configurations will tend to be grouped into the same class. Therefore the model essentially couples two tasks, signal detection and loci clustering, together. Clustering genomic loci into multiple classes provides the key to model the complex correlation structure among data sets. Each loci class represents a signal cooccurrence pattern across data sets. With multiple classes, the model is able to describe multiple different signal cooccurrence patterns. The correlation structure among data sets is encoded through this mixture of signal cooccurrence patterns. For this reason, each $\mathbf{q}_k$ is called a correlation motif. Subsequently, the correlation motifs will also provide the key to integrating information across data sets. Third, this model only requires $O(KD)$ rather than $O(2^D)$ parameters for modeling correlation through π and $\mathbf{Q}$. In real data analysis, K is typically much smaller than 2^D. Thus, the model allows one to avoid the exponential complexity problem. To summarize, the hierarchical mixture model can accommodate all possible signal configurations, provides the flexibility to model the complex correlation structure, and is also scalable to increasing data set number.

5.3.2 Model Fitting and Statistical Inference

In the preceding model, only $\mathbf{X}$ is observed. Both π and $\mathbf{Q}$ are unknown parameters. $\mathbf{A}$ and $\mathbf{B}$ can be viewed as unobserved missing data. Suppose the class number K is given. To infer the signal states $\mathbf{A}$, one can employ an empirical Bayes approach with the following steps.

Step 1

Introduce a Dirichlet prior $Dir(\alpha_1, \ldots, \alpha_K)$ for π and a Beta prior $B(\beta_1, \beta_2)$ for q_{kd}. The joint posterior distribution of unknown parameters and missing indicators can be shown to be

$$Pr(\pi, \mathbf{Q}, \mathbf{A}, \mathbf{B}|\mathbf{X}) \propto \prod_{k=1}^{K} \pi_k^{\alpha_k-1} \prod_{k=1}^{K}\prod_{d=1}^{D} q_{kd}^{\beta_1-1}(1-q_{kd})^{\beta_2-1}$$

$$* \prod_{g=1}^{G}\prod_{k=1}^{K} \left\{ \pi_k \prod_{d=1}^{D}[q_{kd}f_{gd1}(\mathbf{x}_{gd})]^{a_{gd}}[(1-q_{kd})f_{gd0}(\mathbf{x}_{gd})]^{1-a_{gd}} \right\}^{\delta(b_g=k)} \tag{5.2}$$

Step 2

Estimate π and $\mathbf{Q}$ using the posterior mode of $Pr(\pi, \mathbf{Q}|\mathbf{X})$ where $\mathbf{A}$ and $\mathbf{B}$ are integrated out. This can be achieved by running an expectation-maximization (EM) algorithm (Dempster et al., 1977). The EM algorithm is an iterative algorithm that iterates between two steps: an E-step and an M-step. Assume that $\pi^{(t)}$ and $\mathbf{Q}^{(t)}$ are the parameter estimates at iteration t. The E-step and M-step at the $(t+1)$th iteration are given in the following.

E-step: This step computes the following Q-function.

$$E_{\text{old}}[\log Pr(\boldsymbol{\pi}, \mathbf{Q}, \mathbf{A}, \mathbf{B}|\mathbf{X})] = \sum_{k=1}^{K} \left\{ \sum_{g=1}^{G} E_{\text{old}}[\delta(b_g = k)] + \alpha_k - 1 \right\} \log(\pi_k)$$

$$+ \sum_{k=1}^{K} \sum_{d=1}^{D} \left\{ \sum_{g=1}^{G} E_{\text{old}}[a_{gd}\delta(b_g = k)] + \beta_1 - 1 \right\} \log(q_{kd})$$

$$+ \sum_{k=1}^{K} \sum_{d=1}^{D} \left\{ \sum_{g=1}^{G} E_{\text{old}}[(1 - a_{gd})\delta(b_g = k)] + \beta_2 - 1 \right\} \log(1 - q_{kd})$$

$$+ C \tag{5.3}$$

Here the expectation E_{old} is computed with respect to $Pr(\mathbf{A}, \mathbf{B}|\boldsymbol{\pi}^{(t)}, \mathbf{Q}^{(t)}, \mathbf{X})$. C is a constant that does not involve the unknown parameters. Note that

$$E_{\text{old}}[\delta(b_g = k)] = Pr(b_g = k|\boldsymbol{\pi}^{(t)}, \mathbf{Q}^{(t)}, \mathbf{X})$$

$$= \frac{\pi_k^{(t)} \prod_{d=1}^{D}[q_{kd}^{(t)} f_{gd1}(\mathbf{x}_{gd}) + (1 - q_{kd}^{(t)}) f_{gd0}(\mathbf{x}_{gd})]}{\sum_l \left\{ \pi_l^{(t)} \prod_{d=1}^{D}[q_{ld}^{(t)} f_{gd1}(\mathbf{x}_{gd}) + (1 - q_{ld}^{(t)}) f_{gd0}(\mathbf{x}_{gd})] \right\}} \tag{5.4}$$

and

$$E_{\text{old}}[a_{gd}\delta(b_g = k)] = Pr(a_{gd} = 1, b_g = k|\boldsymbol{\pi}^{(t)}, \mathbf{Q}^{(t)}, \mathbf{X})$$

$$= Pr(b_g = k|\boldsymbol{\pi}^{(t)}, \mathbf{Q}^{(t)}, \mathbf{X}) Pr(a_{gd} = 1|b_g = k, \boldsymbol{\pi}^{(t)}, \mathbf{Q}^{(t)}, \mathbf{X}) \tag{5.5}$$

In the preceding formula, $Pr(b_g = k|\boldsymbol{\pi}^{(t)}, \mathbf{Q}^{(t)}, \mathbf{X})$ is computed using Eq. (5.4), and $Pr(a_{gd} = 1|b_g = k, \boldsymbol{\pi}^{(t)}, \mathbf{Q}^{(t)}, \mathbf{X})$ is computed using

$$Pr(a_{gd} = 1|b_g = k, \boldsymbol{\pi}^{(t)}, \mathbf{Q}^{(t)}, \mathbf{X}) = \frac{q_{kd}^{(t)} f_{gd1}(\mathbf{x}_{gd})}{q_{kd}^{(t)} f_{gd1}(\mathbf{x}_{gd}) + (1 - q_{kd}^{(t)}) f_{gd0}(\mathbf{x}_{gd})} \tag{5.6}$$

M-step: This step searches for $\boldsymbol{\pi}$ and $\mathbf{Q}$ that maximizes the Q-function. The results will serve as $\boldsymbol{\pi}^{(t+1)}$ and $\mathbf{Q}^{(t+1)}$ for the next iteration of the EM algorithm.

It can be shown that

$$\pi_k^{(t+1)} = \frac{\sum_{g=1}^{G} Pr(b_g = k | \boldsymbol{\pi}^{(t)}, \mathbf{Q}^{(t)}, \mathbf{X}) + \alpha_k - 1}{G + \sum_{k'=1}^{K}(\alpha_k' - 1)} \tag{5.7}$$

$$q_{kd}^{(t+1)} = \frac{\sum_{g=1}^{G} Pr(a_{gd} = 1, b_g = k | \boldsymbol{\pi}^{(t)}, \mathbf{Q}^{(t)}, \mathbf{X}) + \beta_1 - 1}{\sum_{g=1}^{G} Pr(b_g = k | \boldsymbol{\pi}^{(t)}, \mathbf{Q}^{(t)}, \mathbf{X}) + \beta_1 + \beta_2 - 2} \tag{5.8}$$

In Eqs. (5.7) and (5.8), $\alpha_k - 1$ and $\beta_l - 1$ can be viewed as prior pseudo counts. They are combined with the observed data to estimate the posterior of $\boldsymbol{\pi}$ and $\mathbf{Q}$; α_k and β_k are usually set to 2 so that the actual pseudo counts $\alpha_k - 1$ and $\beta_l - 1$ used in the posterior mode calculation are equal to 1.

Step 3

The EM stops when $\boldsymbol{\pi}^{(t+1)}$ and $\mathbf{Q}^{(t+1)}$ are sufficiently close to $\boldsymbol{\pi}^{(t)}$ and $\mathbf{Q}^{(t)}$. Using the estimated $\hat{\boldsymbol{\pi}}$ and $\hat{\mathbf{Q}}$, one can compute the posterior probability that locus g belongs to class k using Eq. (5.4) after replacing $\boldsymbol{\pi}^{(t)}$ and $\mathbf{Q}^{(t)}$ using the final estimates of $\boldsymbol{\pi}$ and $\mathbf{Q}$.

Step 4

Next, one can compute the posterior probability that locus g carries signal in data set d:

$$Pr(a_{gd} = 1 | \hat{\boldsymbol{\pi}}, \hat{\mathbf{Q}}, \mathbf{X}) = \sum_{k} Pr(a_{gd} = 1 | b_g = k, \hat{\boldsymbol{\pi}}, \hat{\mathbf{Q}}, \mathbf{X}) Pr(b_g = k | \hat{\boldsymbol{\pi}}, \hat{\mathbf{Q}}, \mathbf{X})$$

$$\tag{5.9}$$

Here $Pr(b_g = k | \hat{\boldsymbol{\pi}}, \hat{\mathbf{Q}}, \mathbf{X})$ and $Pr(a_{gd} = 1 | b_g = k, \hat{\boldsymbol{\pi}}, \hat{\mathbf{Q}}, \mathbf{X})$ can be computed using Eqs. (5.4) and (5.6), respectively, after replacing $\boldsymbol{\pi}^{(t)}$ and $\mathbf{Q}^{(t)}$ using the final estimates of $\boldsymbol{\pi}$ and $\mathbf{Q}$.

When the posterior probability passes a user-specified cutoff (e.g., 0.5), one can claim that locus g carries a signal of interest in data set d. $Pr(a_{gd} = 1 | \hat{\boldsymbol{\pi}}, \hat{\mathbf{Q}}, \mathbf{X})$ can also be used to prioritize genomic loci in each data set. The ranked genomic loci lists can be used to guide the choice of follow-up targets.

Choose K

If the motif number K is unknown, one has to determine its value. This is a model selection problem. One can use the Bayesian information criterion (BIC) to choose K. For each given K, the BIC is computed as follows:

$$BIC(K) = -2 * \log Pr(\mathbf{X}|\boldsymbol{\pi}, \mathbf{Q}) + (K - 1 + KD)\log G$$

$$= -2 * \sum_{g=1}^{G} \log \left[\sum_{k=1}^{K} \left\{ \pi_k \prod_{d=1}^{D} \left[q_{kd} f_{gd1}(\mathbf{x}_{gd}) + (1 - q_{kd}) f_{gd0}(\mathbf{x}_{gd}) \right] \right\} \right]$$

$$+ (K - 1 + KD) * \log G \tag{5.10}$$

Here $K - 1$ is the number of parameters for $\boldsymbol{\pi}$, and KD is the number of parameters for $\mathbf{Q}$. BIC can be computed for different K, and the K with the smallest BIC can be chosen to perform the final analysis.

5.3.3 Why the Method Works

The correlation motif approach improves the signal detection through two-way information pooling.

First, according to Eq. (5.9), $Pr(a_{gd} = 1|\hat{\boldsymbol{\pi}}, \hat{\mathbf{Q}}, \mathbf{X})$ is computed by combining $Pr(a_{gd} = 1|b_g = k, \hat{\boldsymbol{\pi}}, \hat{\mathbf{Q}}, \mathbf{X})$ and $Pr(b_g = k|\hat{\boldsymbol{\pi}}, \hat{\mathbf{Q}}, \mathbf{X})$. Here $Pr(b_g = k|\hat{\boldsymbol{\pi}}, \hat{\mathbf{Q}}, \mathbf{X})$ is determined using locus g's information from all data sets (see Eq. (5.4)), and $Pr(a_{gd} = 1|b_g = k, \hat{\boldsymbol{\pi}}, \hat{\mathbf{Q}}, \mathbf{X})$ contains information specific to data set d (see formula 5.6). Thus, the model pools information across all data sets to strengthen the information in each specific data set.

Second, calculation of both $Pr(a_{gd} = 1|b_g = k, \hat{\boldsymbol{\pi}}, \hat{\mathbf{Q}}, \mathbf{X})$ and $Pr(b_g = k|\hat{\boldsymbol{\pi}}, \hat{\mathbf{Q}}, \mathbf{X})$ involves prior probabilities $\hat{\boldsymbol{\pi}}$ and $\hat{\mathbf{Q}}$, which are estimated by examining data from all loci. Thus, the model also borrows information across loci to help with inference.

Because information is borrowed across both data sets and loci, one can use the data more efficiently to improve the separation between signal and noise. This is especially useful in data sets with relatively low signal-to-noise ratios.

5.4 Application 1: ChIP-Chip Peak Calling

JAMIE (Wu and Ji, 2010) is a recently developed method for jointly detecting protein-DNA binding sites in multiple ChIP-chip data sets. This method is a special case of the correlation motif approach.

In the ChIP-chip analysis, the observed data $\mathbf{X}$ are probe intensities. The probes are approximately evenly distributed along the genome. Protein-DNA binding sites can be found by identifying "peaks" or "bumps." Each peak is a consecutive run of probes with high probe intensities.

Suppose there are D ChIP-chip data sets. To identify peaks in each data set, JAMIE assumes that an L base pair (bp) long window arbitrarily retrieved from the genome can fall into two categories: background or potential binding regions (PBR). Let $b_g = 1$ denote that window g is a background window. Let $b_g = 2$ denote that window g is a PBR. If a window is background, it is inactive (i.e., contains no peaks) in all data sets. In this case, $a_{gd} = 0$ for all d. If a window is a PBR, it can become active (i.e., contain a peak) in each data set with nonzero probability. In this case, each a_{gd} can either be 1 or 0. Consider window g. If it is active in data set d, then its data $\mathbf{x}_{gd}$ are assumed to follow distribution $f_{gd1}(.)$. If it is inactive in data set d, then $\mathbf{x}_{gd}$ follow $f_{gd0}(.)$. This is the data generative model used by JAMIE. The model is essentially a correlation motif model with $K = 2$. Here windows are viewed as genomic loci. Background and PBR are two loci classes. In the background class ($k = 1$), all q_{kd}s are forced to be zero. In the PBR class ($k = 2$), q_{kd}s can be nonzero. These q_{kd}s are unknown and need to be estimated.

Specifying $f_{gd1}(.)$ and $f_{gd0}(.)$ is the key to tailoring the general correlation motif framework to specific applications. In ChIP-chip peak calling, each genomic window can contain multiple probes. To construct $f_{gd1}(.)$ and $f_{gd0}(.)$, one can first build models for each probe. For probe i, let $p_{id1} \equiv p_{id1}(\mathbf{x}_{id})$ be the probability of generating the observed probe intensities when the probe is covered by a peak, and let $p_{id0} \equiv p_{id0}(\mathbf{x}_{id})$ be the probability of generating the probe intensities when the probe is not within a peak. JAMIE uses two normal distributions with shifted means to build p_{id1} and p_{id0}. The parameters of these two distributions can be empirically estimated from the data and are assumed to be fixed and known in JAMIE. Readers are referred to Wu and Ji (2010) for details.

Once the probe-level models p_{id1} and p_{id0} are specified, they can be used to construct the window-level model f_{gd1} and f_{gd0}. Let T_g denote the set of probes covered by window g. If window g is inactive in data set d (i.e., $a_{gd} = 0$), then

$$f_{gd0}(\mathbf{x}_{gd}) = \prod_{l \in T_g} p_{ld0} \tag{5.11}$$

In other words, all probes in window g and data set d are assumed to generate their intensities independently according to the background model p_{id0}.

If window g is active in data set d (i.e., $a_{gd} = 1$), then it should contain a peak in that data set. Instead of assuming that the peak expands across the whole window, JAMIE assumes that the peak begins at a randomly selected probe in the window and has length W_{gd} ($\leq L$). A priori, the peak length W_{gd} is randomly chosen from a set of allowable lengths $\mathbf{W}$ (e.g., $\{500, 600, \ldots, 1000\}$

bps). The peak start position and peak length are chosen subject to the constraint that the peak should be fully contained within the window. All possible peak configurations that satisfy this constraint are assumed to have equal prior probability to occur. In JAMIE, both $\mathbf{W}$ and the window size L are specified by users. L should be chosen so that a PBR is able to cover typical ChIP-chip peaks in all D data sets. For instance, one can first run an initial peak calling on each data set separately using a standard peak calling tool. L can then be set to be $1\sim2$ times the third quartile of the ChIP-chip peak lengths obtained from this initial analysis. On the basis of our experience, within this range, the algorithm is not too sensitive to the choice of L.

On the basis of these assumptions, a PBR can be active in multiple data sets. However, its peaks in different data sets are not required to align. They may have different start positions and lengths. This is illustrated in Figure 5.3A. These assumptions make the model flexible so that experiment-specific data characteristics (e.g., different peak lengths in different data sets) can be taken care of. The model also allows one to jointly analyze binding sites of different TFs that bind to the same cis-regulatory module but not to the same DNA motif site. To compare with PBR ($b_g = 2$), Figure 5.3B illustrates a typical background window ($b_g = 1$).

Let u_{gd} denote the probe that starts the binding peak within an active window, $T_g^{W_{gd}, u_{gd}}$ be probe indices covered by a W_{gd} bp peak starting at u_{gd}, and $T_g \backslash T_g^{W_{gd}, u_{gd}}$ be the remaining probe indices in window g. Let $T_g(W_{gd})$ be the set of probe indices that can be used to start a peak of length W_{gd} that is fully covered by the window. Denote the number of elements in a set Z as $|Z|$. By integrating out all possible peak start u_{gd} and peak length W_{gd}, one can derive

$$f_{gd1}(\mathbf{x}_{gd}) = \frac{1}{\sum_{W_{gd} \in \mathbf{W}} |T_g(W_{gd})|}$$

$$\sum_{W_{gd} \in \mathbf{W}} \sum_{u_{gd} \in T_g(W_{gd})} \left\{ \prod_{l \in T_g^{W_{gd}, u_{gd}}} p_{ld1} \prod_{l \in T_g \backslash T_g^{W_{gd}, u_{gd}}} p_{ld0} \right\} \tag{5.12}$$

With $f_{gd0}(.)$ and $f_{gd1}(.)$ specified, the JAMIE algorithm can be derived using the general correlation motif framework. One minor difference is that in the JAMIE model, the q_{kd}s for the background class are fixed to zero and do not need to be estimated. This can be accommodated by slightly modifying the EM algorithm introduced in the previous section. These modifications are straightforward and are skipped here in the interest of space.

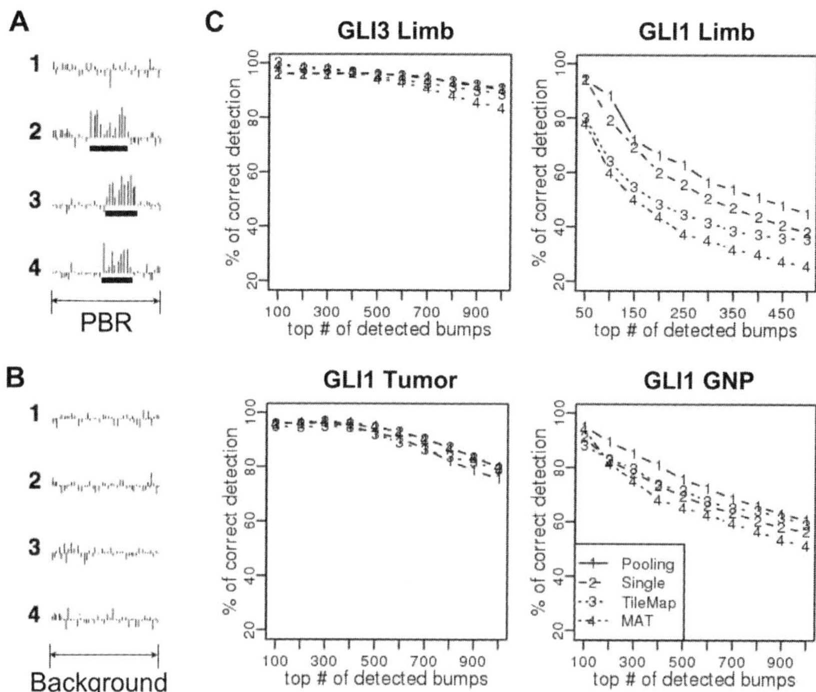

Figure 5.3 Joint peak calling by JAMIE. (A) Illustration of the model for PBR. The four tracks correspond to four ChIP-chip data sets. Each vertical bar corresponds to a probe. The bar height represents the log2 fold enrichment between ChIP and control samples. A PBR can be either active or inactive in each data set. If it is active in a data set, it will contain a peak. The peak region is underlined using a thick horizontal line. Peaks in different data sets are not required to align. (B) Illustration of the model for background window. A background window does not contain any peaks. (C) Performance of JAMIE in the four GLI ChIP-chip data sets. In each data set, JAMIE (pooling) and several single-data set-based algorithms including JAMIE single (single), MAT, and TileMap, are used for peak calling. For each algorithm, the percentages of the top reported peaks that are gold standard GLI binding sites are shown. The figure is reproduced from Wu and Ji (2010) by permission of Oxford University Press.

JAMIE first divides the genome into non-overlapping L bp windows. It then applies the EM algorithm to these windows to estimate the model parameters π and $\mathbf{Q}$. Given the estimated parameters, JAMIE will use an L bp sliding window to scan the genome again to call PBRs and peaks. Details of the JAMIE algorithm can be found in Wu and Ji (2010).

In Wu and Ji (2010), JAMIE has been applied to analyze the four GLI ChIP-chip data sets in Figure 5.1. These data were collected by two different labs using Affymetrix Mouse Promoter 1.0R arrays. The ChIP-chip experiments were designed to map GLI binding sites in different cell types, including GLI3 binding in developing limbs of mouse embryos ("GLI3 Limb"), GLI1 binding in mouse developing limb ("GLI1 Limb"), medulloblastoma ("GLI1 Tumor"), and granule neuron precursor cells ("GLI1 GNP"). GLI1 and GLI3 belong to the same TF family and recognize the same DNA motif. A subset of their binding targets in different cell types is shared. Conversely, each TF and cell type is expected to have TF-specific and cell-type-specific targets. Thus, these four data sets are expected to have distinct but correlated binding profiles. Each data set contains three replicate ChIP samples and three replicate control samples.

To demonstrate how data integration improves peak calling, Wu and Ji (2010) compared JAMIE with two state-of-the-art peak calling methods, MAT (Johnson et al., 2006) and TileMap (Ji and Wong, 2005), designed for single data set analysis. Both MAT and TileMap were run on each data set to produce a ranked list of peaks. In addition, the JAMIE model can also be modified to analyze each data set separately. This is done by first assuming that windows in each data set are either active or inactive and then using an EM algorithm to fit this two-component mixture to each data set. JAMIE applied to each data set separately is called "JAMIE single." JAMIE single uses the same probe intensity model $f_{gd0}(.)$ and $f_{gd1}(.)$ as in JAMIE. It only differs from JAMIE in that JAMIE single does not analyze multiple data sets jointly. Therefore, JAMIE single was also applied to analyze each GLI data set to provide a well-controlled comparison for evaluating the net gain of data integration.

In this real data example, people do not know all the true binding sites. For evaluation purpose, a gold standard peak list was constructed for each individual data set using the common peaks reported by all three single data set algorithms: MAT, TileMap, and JAMIE single. Next, one replicate sample was dropped from each data set to create a reduced data set. All four algorithms, MAT, TileMap, JAMIE single, and JAMIE, were then applied to the reduced data. Figure 5.3C compares the peak calling accuracies of different algorithms in the reduced data. For each data set and each algorithm, the figure shows the percentage of top reported peaks that overlap with the gold standard peaks. On the basis of the results, JAMIE performed either better than or comparable to the competing algorithms. Among the four data sets, the second ("GLI1 Limb") had low signal-to-noise ratio. In that data set, substantial improvement was observed.

The GLI analysis is only one illustrative example that demonstrates JAMIE. In Wu and Ji (2010), JAMIE was also compared with other algorithms using a number of other real data sets as well as simulated data. The evaluation based on other criteria was also conducted. Extensive analysis in that study shows that the joint peak calling by JAMIE robustly outperforms the single-data set analysis methods.

JAMIE has been implemented as an R package. The package is freely available at http://www.biostat.jhsph.edu/~hji/jamie. The core functions are implemented in C to make computation efficient. In a test consisting of four data sets, each with three ChIP and three control samples and 3.8 million probes, it took about 15 minutes to run on a PC with 2.2 GHz CPU and 4 GB RAM.

5.5 Application 2: Analysis of Allele-Specific Binding in ChIP-Seq

In ChIP-seq, protein-bound DNAs are directly read out by massively parallel sequencing. For this reason, the technology not only allows one to map genome-wide protein-DNA binding sites but also makes it possible to study allele-specific protein-DNA binding (ASB). Suppose ChIP-seq data are collected for a genotyped individual. After mapping sequence reads to the genome, the number of reads mapped to each allele can be counted at each heterozygote SNP in this individual. The allelic read counts provide the basis for evaluating whether the two alleles are bound at the same level.

Detecting ASB in a single ChIP-seq data set often has very low statistical power. This is because only sequence reads mapped to heterozygote SNPs are informative for discriminating two alleles. Whereas each ChIP-seq sample may contain tens of millions of reads, only a small fraction are aligned to heterozygote SNPs. The small allelic read counts at each SNP lead to low power.

With increasing use of ChIP-seq and the rapid growth of ChIP-seq data in public databases, it is now not unusual to have multiple ChIP-seq data sets collected for the same individual and in the same cell type. These data sets may measure binding of different TFs and/or different HMs, but they are all collected for the same individual for which genotype does not change. When faced with such a scenario, integrative analysis of all data sets may improve the ASB detection. This is because ASB of different proteins may be correlated. For instance, proteins often form complexes. If two proteins A and B have to form a protein complex in order to bind to DNA, then the allelic imbalances of A and B are expected to cooccur. As a result, weak evidence for allelic imbalances for both proteins A and B may be combined to provide a much stronger evidence for ASB.

The iASeq method (Wei et al., 2012) is a new method for jointly analyzing multiple ChIP-seq data sets to detect ASB. The iASeq model is another example of the correlation motif approach. In this model, heterozygote SNPs are assumed to belong to K different classes. Different classes have different ASB cooccurence patterns across data sets. Because ASB can occur in two directions depending on which allele is active, each SNP g can have three possible states in each data set d: skewed to the reference allele (SR), skewed to the nonreference allele (SN), or not allele specific (NS). Here reference allele is defined as the allele consistent with the reference genome. To accommodate these three states, iASeq uses two binary indicators a_{1gd} and a_{2gd} to indicate SR and SN, respectively. This is slightly different from the two-state correlation motif model introduced earlier, which only requires one indictor a_{gd}. Similarly, instead of using one q_{kd} to denote the prior probability of carrying signal, iASeq uses two probabilities q_{1kd} and q_{2kd} to denote the prior probability of being SR and SN, respectively, in data set d for SNPs in class k. Similar to JAMIE, one class (here class K) is assumed to be a class with pure background. For this class, q_{1Kd}s and q_{2Kd}s are all forced to be zero. Once the data models $f_{gd1}(\mathbf{x}_{gd})$, $f_{gd2}(\mathbf{x}_{gd})$, and $f_{gd0}(\mathbf{x}_{gd})$ for the three states SR, SN, and NS are specified, the iASeq model will be fully specified. Although the model is slightly different from the two-state model presented in Section 5.3, all the model fitting and inference techniques remain almost the same and only require minor modifications.

To specify the data model $f_{gdl}(\mathbf{x}_{gd})$, recall that each data set may contain multiple replicate ChIP samples. For SNP g, data set d and replicate sample j, let x_{gdj} and y_{gdj} be the read count for the reference allele and nonreference allele, respectively. Let $n_{gdj} = x_{gdj} + y_{gdj}$ be the total read count. For SNP g and data set d, if the SNP is SR (i.e., $a_{1gd} = 1$), then x_{gdj} in each replicate sample (d, j) is assumed to be generated according to a probability distribution $f_{gdj1}(x_{gdj}) \equiv Pr(x_{gdj}|n_{gdj}, a_{1gd} = 1, a_{2gd} = 0)$. Data in different replicate samples are assumed to be generated independently. Similarly, if the SNP is SN (i.e., $a_{2gd} = 1$), then x_{gdj}s are generated according to $f_{gdj2}(x_{gdj}) \equiv Pr(x_{gdj}|n_{gdj}, a_{1gd} = 0, a_{2gd} = 1)$. If the SNP is NS (i.e., $a_{1gd} = 0$ and $a_{2gd} = 0$), then x_{idj}s are generated according to $f_{gdj0}(x_{gdj}) \equiv Pr(x_{gdj}|n_{gdj}, a_{1gd} = 0, a_{2gd} = 0)$.

To specify f_{gdjk}, x_{gdj} is assumed to follow a binomial distribution $x_{gdj}|n_{gdj}, p_{gdj} \sim \text{Binom}(n_{gdj}, p_{gdj})$, where p_{gdj} is the probability that a read generated at SNP g in sample (d, j) represents the reference allele. Next, p_{gdj} is modeled depending on the values of a_{1gd} and a_{2gd}.

If SNP g is NS in data set d (i.e., $a_{1gd} = 0$ and $a_{2gd} = 0$), p_{gdj} is assumed to follow a beta distribution $\text{Beta}(\alpha_{dj}, \beta_{dj})$ with mean $p_{dj0} = \alpha_{dj}/(\alpha_{dj} + \beta_{dj})$. Here one uses a beta distribution for the background p_{gdj} instead of setting it to a constant because it was observed that a constant p_{dj0} is insufficient for capturing the background variation (Skelly et al., 2011). When specifying the beta distribution, if there is no read mapping biases, p_{dj0} should be equal to 0.5.

However, owing to various mapping biases (e.g., it can be easier to map reads to the reference allele than to the non-reference allele), p_{dj0} may be different from 0.5 in real data. Therefore, in iASeq, α_{dj} and β_{dj} are estimated empirically from the data based on allelic read counts from all heterozygote SNPs in a data set. Once estimated, they are treated as fixed and known (see Wei et al., 2012, for details). After integrating out all possible values of p_{gdj}, the distribution of x_{gdj} conditional on $a_{1gd} = 0$ and $a_{2gd} = 0$ can be derived as

$$
\begin{aligned}
f_{gdj0}(x_{gdj}) &= \int_0^1 Pr(x_{gdj}|n_{gdj}, p_{gdj}, a_{1gd} = 0, a_{2gd} = 0) \\
&\quad * f(p_{gdj}|a_{1gd} = 0, a_{2gd} = 0)dp_{gdj} \\
&= \frac{C_{n_{gdj}}^{x_{gdj}} B(x_{gdj} + \alpha_{dj}, n_{gdj} - x_{gdj} + \beta_{dj})}{B(\alpha_{dj}, \beta_{dj})}
\end{aligned}
\tag{5.13}
$$

Here C_n^k is the binomial coefficients "n choose k", and $B(.,.)$ is the beta function.

If SNP g is SR in data set d (i.e., $a_{1gd} = 1$ and $a_{2gd} = 0$), p_{gdj} is assumed to follow a uniform distribution $U[p_{dj0}, 1]$. Here $p_{dj0} = \alpha_{dj}/(\alpha_{dj} + \beta_{dj})$ is defined before. After integrating out p_{gdj}, one has

$$
\begin{aligned}
f_{gdj1}(x_{gdj}) &= \int_0^1 Pr(x_{gdj}|n_{gdj}, p_{gdj}, a_{1gd} = 1, a_{2gd} = 0) \\
&\quad * f(p_{gdj}|a_{1gd} = 1, a_{2gd} = 0)dp_{gdj} \\
&= \frac{C_{n_{gdj}}^{x_{gdj}}}{1 - p_{dj0}} \int_{p_{dj0}}^1 p^{x_{gdj}}(1 - p)^{n_{gdj} - x_{gdj}} dp
\end{aligned}
\tag{5.14}
$$

Similarly, if SNP g is SN in data set d (i.e., $a_{1gd} = 0$ and $a_{2gd} = 1$), p_{gdj} is assumed to follow a uniform distribution $U[0, p_{dj0}]$. One has

$$
\begin{aligned}
f_{gdj2}(x_{gdj}) &= \int_0^1 f(x_{gdj}|n_{gdj}, p_{gdj}, a_{1gd} = 0, a_{2gd} = 1) \\
&\quad * f(p_{gdj}|a_{1gd} = 0, a_{2gd} = 1)dp_{gdj} \\
&= \frac{C_{n_{gdj}}^{x_{gdj}}}{p_{dj0}} \int_0^{p_{dj0}} p^{x_{gdj}}(1 - p)^{n_{gdj} - x_{gdj}} dp
\end{aligned}
\tag{5.15}
$$

Once the data models for each sample are given, the models for each data set can be obtained by combining replicates in the same data set. Under the assumption that replicate samples are collected independently, one has $f_{gdl}(\mathbf{x}_{gd}) = \prod_j f_{gdjl}(x_{gdj})$. Here $l = 0$, 1 or 2. The iASeq model is now fully specified. Upon minor modifications, one can use the EM algorithm introduced before to estimate model parameters and perform statistical inference. Details of the iASeq algorithm can be found in Wei et al. (2012).

In Wei et al. (2012), iASeq was used to analyze 41 ChIP-seq data sets collected in the ENCODE project. These data consist of a total of 78 samples and 94,519 heterozygote SNPs from the GM12878 cell line. Based on BIC, the optimal K was 3 (Figure 5.4A). Besides the background class, the algorithm discovered two other SNP classes representing two ASB occurrence patterns (Figures 5.4B and 5.4C). Figures 5.4B and 5.4C show heatmaps of q_{1kd}s and q_{2kd}s, respectively. It can be seen that the two nonbackground motif patterns are in opposite directions. Within each motif, most of the analyzed TFs and HMs are highly correlated. They tend to be skewed to the same direction. A few data sets have less dramatic allelic imbalance in these two motifs and do not exhibit strong correlation with others. A detailed examination reveals that these data sets correspond to several H3K27me3 experiments conducted by different labs. This indicates that the allele specificity of the repressive histone modification H3K27me3 may behave differently from the allele specificity of most other TFs and HMs examined here.

Using the posterior probability of ASB (i.e., $Pr(a_{1gd} = 1|\hat{\pi}, \hat{\mathbf{Q}}, \mathbf{X}) + Pr(a_{2gd} = 1|\hat{\pi}, \hat{\mathbf{Q}}, \mathbf{X}))$, SNPs were ranked in each data set. When the iASeq ranking was compared with the rankings provided by different methods that analyze each data set separately, iASeq was found to significantly improve the ASB detection accuracy. For instance, Figures 5.4D and 5.4E show the ranking results for two representative data sets. In each data set and for each ASB detection method, the figure shows the number of chromosome X (chrX) SNPs identified among the top ranked SNPs. Because GM12878 is a female lymphoplast line, most non-pseudoautosomal-region chrX SNPs are expected to be allele specific due to X-inactivation. For this reason, the number of chrX SNPs among the detected SNPs can serve as a benchmark to evaluate an algorithm's performance. Figures 5.4D and 5.4E show that iASeq clearly has the highest power for detecting ASB compared to the five single-data-set-based methods. These single-data-set-based methods were described in detail in Wei et al. (2012). For each data set and algorithm, the area under the performance curve (AUC) can be computed. The relative improvement of iASeq over the best competing algorithm is defined as $\mathrm{dAUC} = \frac{\mathrm{AUC}_{\mathrm{iASeq}} - \mathrm{AUC}_{\mathrm{best}}}{\mathrm{AUC}_{\mathrm{best}}}$, where $\mathrm{AUC}_{\mathrm{best}}$ is the AUC of the best algorithm other than iASeq; $\mathrm{dAUC} > 0$ indicates that iASeq outperforms all other algorithms. Figure 5.4F shows the histogram of dAUC in all 41 data sets. Clearly iASeq improves the ASB analysis in all data sets.

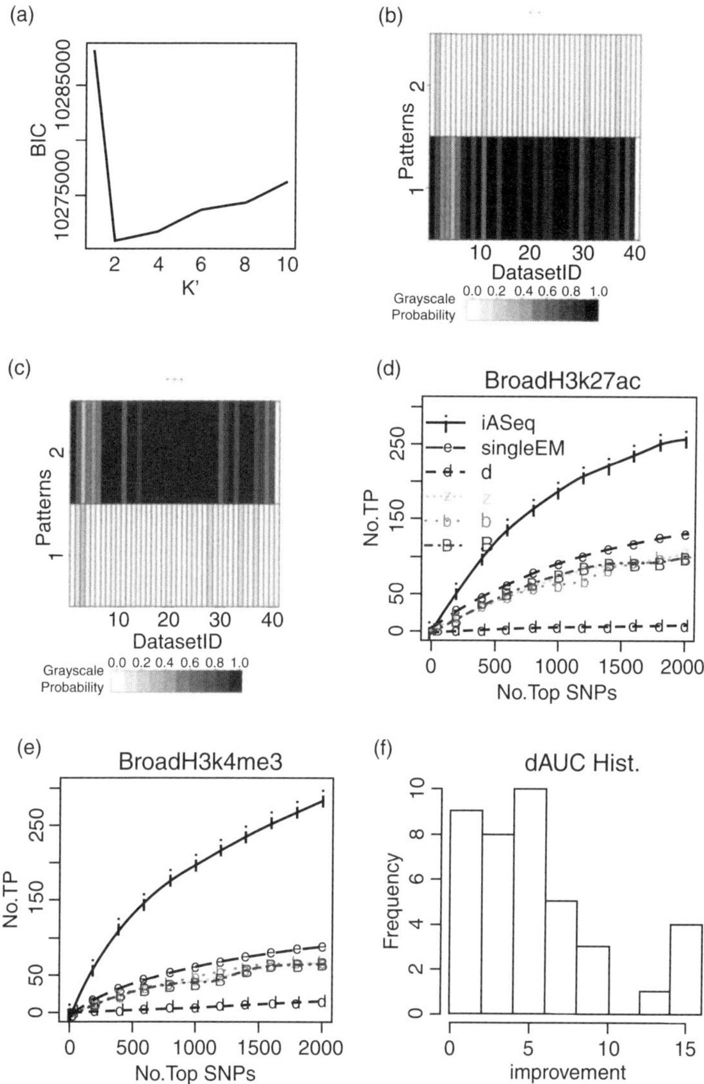

Figure 5.4 ASB analysis by iASeq. (A) BIC versus $K' = K - 1$. K' is the number of non-background SNP classes. $K' = 2$ (or $K = 3$) is associated with the smallest BIC. (B) Heatmap showing the estimated q_{1kd}s. Each row is a motif, and each column is a data set. The background motif is not shown because all its q_{1kd}s are equal to 0. (C) Heatmap showing the estimated q_{2kd}s. (D–E) Performance of iASeq in two representative data sets. The number of chromosome X SNPs among the top reported SNPs are shown for iASeq and several other single data set based algorithms. iASeq is the most accurate algorithm. (F) Performance of iASeq in all 41 GM12878 data sets. For each data set, dAUC is computed. dAUC > 0 indicates that iASeq performs better than all other algorithms. The histogram shows the distribution of dAUCs in all 41 data sets. The figure is reproduced from Wei et al. (2012) with permission.

In Wei et al. (2012), iASeq was also evaluated using other criteria as well as simulation data. These analyses collectively show that iASeq robustly outperforms the single-data-set analysis methods (see Wei et al., 2012, for details).

The iASeq software is freely available as an R package in bioconductor at http://www.bioconductor.org/packages/release/bioc/html/iASeq.html. For analyzing the GM12878 data with 78 samples and 94,519 SNPs, iASeq took 16 hours to run the EM algorithm to fit a single model with $K = 10$ on a machine with 2.7 GHz CPU and 4GB RAM.

5.6 Discussion

To summarize, the correlation motif approach is a flexible and scalable statistical framework for integrative analyses of ChIP-X data. This framework can be applied to a variety of applications. In principle, it can also be applied to data types other than ChIP-X, such as gene expression data and DNA methylation data. To apply this framework to a new application, one only needs to tailor $f_{gdl}(.)$, the probability distributions of the observed data in the signal and noise states, to fit the data type in question.

Inferring the signal state a_{gd} can be viewed as a hypothesis testing problem. Identifying signals in a single high-throughput genomic data set is often framed as a multiple testing problem. From this angle, the correlation motif approach can be viewed as an empirical Bayes solution to conducting multiple testing in multiple data sets. This approach couples the multiple testing with the loci clustering. Clustering allows one to identify the correlation structure among multiple data sets, which in turn can help with improving the accuracy of the hypothesis testing.

The correlation motif approach models the correlation of the discrete hidden signal states. The current model does not describe the correlation at the raw data level. Unlike the discrete signal states, the raw data can be continuous. Thus, modeling the correlation at the raw data level represents an alternative solution to data integration. The recently developed differential principal component analysis (dPCA) (Ji et al., 2013) represents an initial effort toward that direction. Despite the initial efforts exemplified by dPCA, whether modeling the raw data correlation provides additional gain over the correlation motif model is still under investigation. If the answer is yes, a natural future research direction would be how to extend the correlation motif approach to further incorporate the raw data correlation that cannot be explained by the hidden signal states.

Acknowledgements

This work was supported by NIH grants R01HG006282 and R01HG006841.

References

Barrett, T., Troup, D. B., Wilhite, S. E., Ledoux, P., Rudnev, D., Evangelista, C., Kim, I. F., A., Soboleva, Tomashevsky, M., Marshall, K. A., Phillippy, K. H., Sherman, P. M., Muertter, R. N., and Edgar, R. 2009. NCBI GEO: archive for high-throughput functional genomic data. *Nucleic Acids Res.*, **37**, D885–890.

Barski, A., Cuddapah, S., Cui, K., Roh, T. Y., Schones, D. E., Wang, Z., Wei, G., Chepelev, I., and Zhao, K. 2007. High-resolution profiling of histone methylations in the human genome. *Cell*, **129**, 823–837.

Bernstein, B. E., Stamatoyannopoulos, J. A., Costello, J. F., Ren, B., Milosavljevic, A., Meissner, A., Kellis, M., Marra, M. A., Beaudet, A. L., Ecker, J. R., Farnham, P. J., Hirst, M., Lander, E. S., Mikkelsen, T. S., and Thomson, J. A. 2010. The NIH Roadmap Epigenomics Mapping Consortium. *Nat. Biotechnol.*, **28**, 1045–1048.

Cawley, S., Bekiranov, S., Ng, H. H., Kapranov, P., Sekinger, E. A., Kampa, D., Piccolboni, A., Sementchenko, V., Cheng, J., Williams, A. J., Wheeler, R., Wong, B., Drenkow, J., Yamanaka, M., Patel, S., Brubaker, S., Tammana, H., Helt, G., Struhl, K., and Gingeras, T. R. 2004. Unbiased mapping of transcription factor binding sites along human chromosomes 21 and 22 points to widespread regulation of noncoding RNAs. *Cell*, **116**, 499–509.

Celniker, S. E., Dillon, L. A., Gerstein, M. B., Gunsalus, K. C., Henikoff, S., Karpen, G. H., Kellis, M., Lai, E. C., Lieb, J. D., MacAlpine, D. M., Micklem, G., Piano, F., Snyder, M., Stein, L., White, K. P., Waterston, R. H., and modENCODE Consortium. 2009. Unlocking the secrets of the genome. *Nature*, **459**, 927–930.

Chen, L., Wu, G., and Ji, H. 2011a. hmChIP: a database and web server for exploring publicly available human and mouse ChIP-seq and ChIP-chip data. *Bioinformatics*, **27**, 1447–1448.

Chen, Y., Meyer, C. A., Liu, T., Li, W., Liu, J. S., and Liu, X. S. 2011b. MM-ChIP enables integrative analysis of cross-platform and between-laboratory ChIP-chip or ChIP-seq data. *Genome Biol.*, **12**, R11.

Choi, H., Nesvizhskii, A. I., Ghosh, D., and Qin, Z. S. 2009. Hierarchical hidden Markov model with application to joint analysis of ChIP-chip and ChIP-seq data. *Bioinformatics*, **25**, 1715–1721.

Consortium, ENCODE Project. 2012. An integrated encyclopedia of DNA elements in the human genome. *Nature*, **489**, 57–74.

Dempster, A. P., Laird, N. M., and Rubin, D. B. 1977. Maximum likelihood from incomplete data via the EM algorithm. *J. R. Stat. Soc. B*, **39**, 1–38.

Ji, H., and Wong, W. H. 2005. TileMap: create chromosomal map of tiling array hybridizations. *Bioinformatics*, **21**, 3629–3636.

Ji, H., Li, X., Wang, Q. F., and Ning, Y. 2013. Differential principal component analysis of ChIP-seq. *Proc. Natl. Acad. Sci. USA*, **110**, 6789–6794.

Johnson, D. S., Mortazavi, A., Myers, R. M., and Wold, B. 2007. Genome-wide mapping of in vivo protein-DNA interactions. *Science*, **316**, 1497–1502.

Johnson, W. E., Li, W., Meyer, C. A., Gottardo, R., Carroll, J. S., Brown, M., and Liu, X. S. 2006. Model-based analysis of tiling-arrays for ChIP-chip. *Proc. Natl. Acad. Sci. USA*, **103**, 12457–12462.

Judy, J. T., and Ji, H. 2009. TileProbe: modeling tiling array probe effects using publicly available data. *Bioinformatics*, **25**, 2369–2375.

Kasowski, M., Grubert, F., Heffelfinger, C., Hariharan, M., Asabere, A., Waszak, S. M., Habegger, L., Rozowsky, J., Shi, M., Urban, A. E., Hong, M. Y., Karczewski, K. J., Huber, W., Weissman, S. M., Gerstein, M. B., Korbel, J. O., and Snyder, M. 2010. Variation in transcription factor binding among humans. *Science*, **328**, 232–235.

Laajala, T. D., Raghav, S., Tuomela, S., Lahesmaa, R., Aittokallio, T., and Elo, L. L. 2009. A practical comparison of methods for detecting transcription factor binding sites in ChIP-seq experiments. *BMC Genomics*, **10**, 618.

McDaniell, R., Lee, B. K., Song, L., Liu, Z., Boyle, A. P., Erdos, M. R., Scott, L. J., Morken, M. A., Kucera, K. S., Battenhouse, A., Keefe, D., Collins, F. S., Willard, H. F., Lieb, J. D., Furey, T. S., Crawford, G. E., Iyer, V. R., and Birney, E. 2010. Heritable individual-specific and allele-specific chromatin signatures in humans. *Science*, **328**, 235–239.

Mikkelsen, T. S., Ku, M., Jaffe, D. B., Issac, B., Lieberman, E., Giannoukos, G., Alvarez, P., Brockman, W., Kim, T. K., Koche, R. P., Lee, W., Mendenhall, E., O'Donovan, A., Presser, A., Russ, C., Xie, X., Meissner, A., Wernig, M., Jaenisch, R., Nusbaum, C., Lander, E. S., and Bernstein, B. E. 2007. Genome-wide maps of chromatin state in pluripotent and lineage-committed cells. *Nature*, **448**, 553–560.

Ren, B., Robert, F., Wyrick, J. J., Aparicio, O., Jennings, E. G., Simon, I., Zeitlinger, J., Schreiber, J., Hannett, N., Kanin, E., Volkert, T. L., Wilson, C. J., Bell, S. P., and Young, R. A. 2000. Genome-wide location and function of DNA binding proteins. *Science*, **290**, 2306–2309.

Robertson, G., Hirst, M., Bainbridge, M., Bilenky, M., Zhao, Y., Zeng, T., Euskirchen, G., Bernier, B., Varhol, R., Delaney, A., Thiessen, N., Griffith, O. L., He, A., Marra, M., Snyder, M., and Jones, S. 2007. Genome-wide profiles of STAT1 DNA association using chromatin immunoprecipitation and massively parallel sequencing. *Nat. Methods*, **4**, 651–657.

Rozowsky, J., Abyzov, A., Wang, J., Alves, P., Raha, D., Harmanci, A., Leng, J., Bjornson, R., Kong, Y., Kitabayashi, N., Bhardwaj, N., Rubin, M., Snyder, M., and Gerstein, M. 2011. AlleleSeq: analysis of allele-specific expression and binding in a network framework. *Mol. Syst. Biol.*, **7**, 522.

Schmid, C. D., and Bucher, P. 2007. ChIP-seq data reveal nucleosome architecture of human promoters. *Cell*, **131**, 831–832.

Schmidt, D., Wilson, M. D., Ballester, B., Schwalie, P. C., Brown, G. D., Marshall, A., Kutter, C., Watt, S., Martinez-Jimenez, C. P., Mackay, S., Talianidis, I., Flicek, P., and Odom, D. T. 2010. Five-vertebrate ChIP-seq reveals the evolutionary dynamics of transcription factor binding. *Science*, **328**, 1036–1040.

Skelly, D. A., Johansson, M., Madeoy, J., Wakefield, J., and Akey, J. M. 2011. A powerful and flexible statistical framework for testing hypotheses of allele-specific gene expression from RNA-seq data. *Genome Res.*, **21**, 1728–1737.

Vokes, S. A., Ji, H., Wong, W. H., and McMahon, A. P. 2008. A genome-scale analysis of the cis-regulatory circuitry underlying sonic hedgehog mediated patterning of the mammalian limb. *Genes Dev.*, **22**, 2651–2663.

Wei, Y., Li, X., Wang, Q. F., and Ji, H. 2012. iASeq: integrative analysis of allele-specificity of protein-DNA interactions in multiple ChIP-seq datasets. *BMC Genomics*, **13**, 681.

Wilbanks, E. G., and Facciotti, M. T. 2010. Evaluation of algorithm performance in ChIP-seq peak detection. *PLoS ONE*, **5**, e11471.

Wu, H., and Ji, H. 2010. JAMIE: joint analysis of multiple ChIP-chip experiments. *Bioinformatics*, **26**, 1864–1870.

Zeng, X., Sanalkumar, R., Bresnick, E. H., Li, H., Chang, Q., and Keles, S. 2013. jMOSAiCS: joint analysis of multiple ChIP-seq datasets. *Genome Biol.*, **14**, R38.

VERTICAL INTEGRATIVE ANALYSIS (GENERAL METHODS)

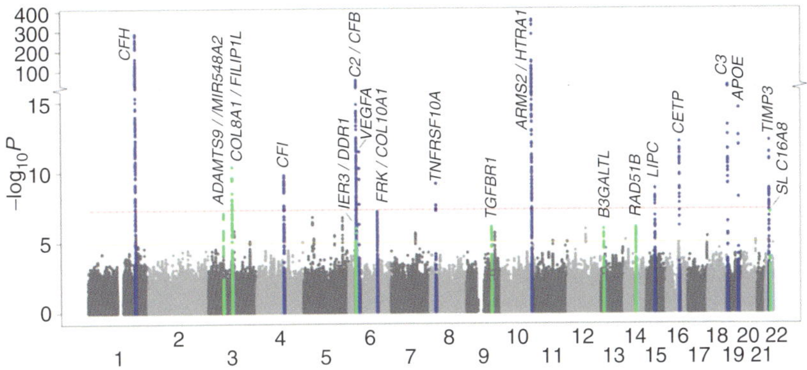

Plate 1.3 Meta-analysis of age-related macular degeneration.

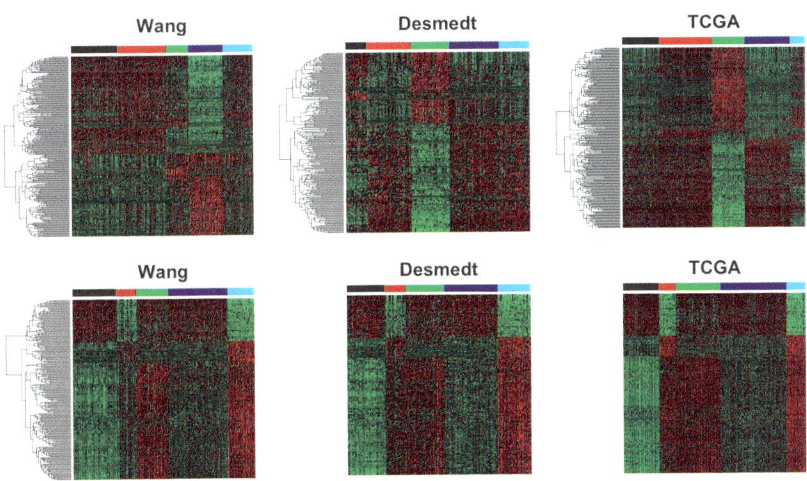

Plate 2.5 Meta Sparse K-means result for breast cancer data of three studies from different platforms. Each row represents a gene that is concordant across studies. In each study, patients are totally divided into five clusters, represented by five unique colors in the color bar at the top of the data matrix of each study. The top three figures are individual clustering results. The bottom three figures are meta-clustering results.

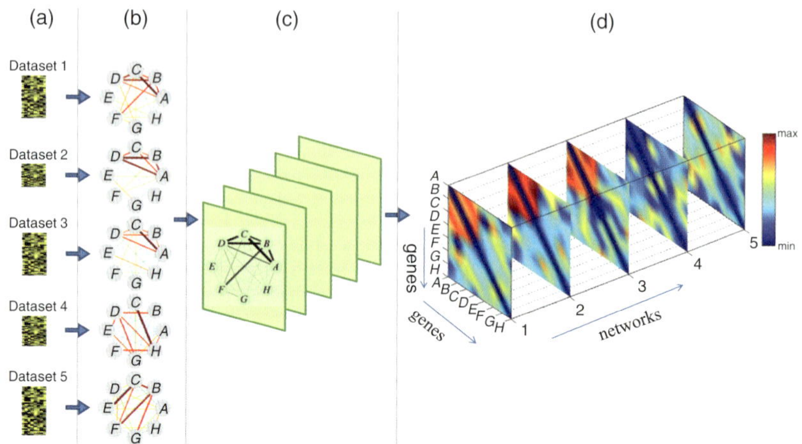

Plate 3.1 Illustration of the tensor representation for multiple networks and a recurrent heavy subgraph. (A) Microarray data sets are modeled as (B) a collection of coexpression networks. (C) These coexpression networks can be "stacked" together into (D) a third-order tensor such that each slice represents the adjacency matrix of one network. The weights of edges in the coexpression networks and their corresponding tensor elements are indicated by the color scale to the right of the figure. In (D), after reordering the tensor using the gene and network membership vectors, it becomes clear that the subtensor in the top-left corner of the tensor (formed by genes A, B, C, D in networks $1, 2, 3$) corresponds to a recurrent heavy subgraph.

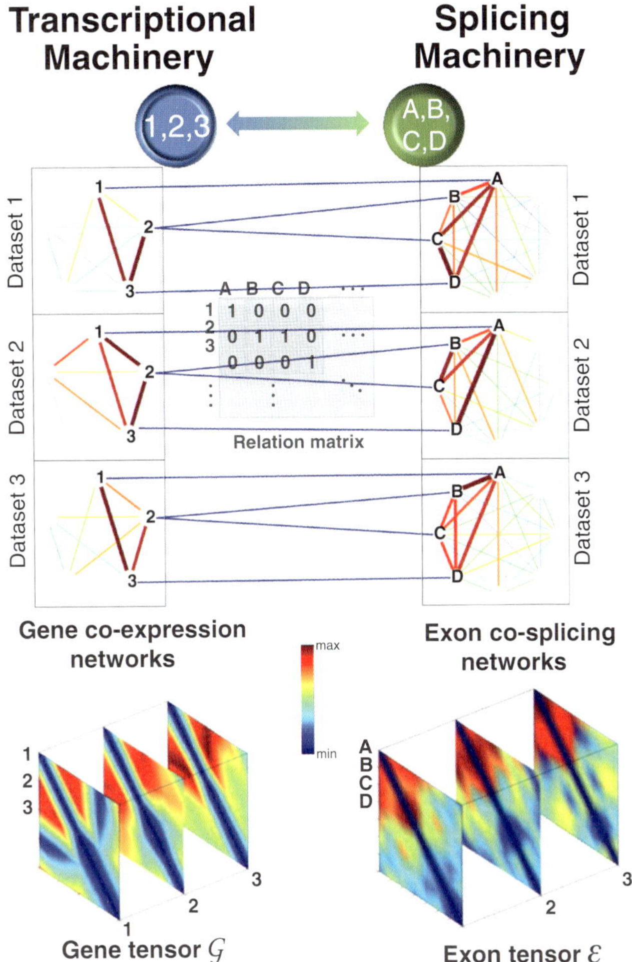

Plate 3.3 Illustration of a *frequent coupled cluster* (FCC). In a collection of three paired gene coexpression and exon co-splicing networks, a subset of genes {1,2,3} is heavily interconnected and their exons {A,B,C,D} are also heavily interconnected. The two subsets form an FCC: a coupled transcription-splicing module. The gene and exon clusters intuitively correspond to the heavy subtensors in $\mathcal{G}$ and $\mathcal{E}$.

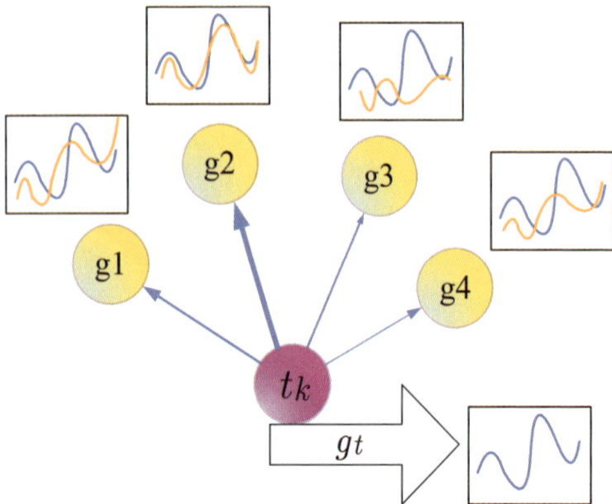

Plate 4.1 Initial step for "walking" in the network. TFs for the target gene are identified based on TF-DNA binding data and are taken as initiating points for the walk. The next gene to visit is selected stochastically from all neighbors based on expression correlation with the target gene. Genes with stronger correlation have a better chance to be selected as the next node, as indicated by the thickness of the arrows.

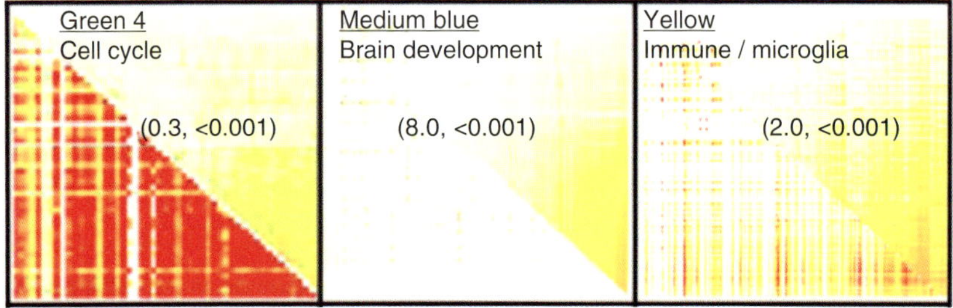

Plate 4.4 Example of gain of connectivity and loss of connectivities as calculated by the MDC for modules in LOAD (the upper right triangle of each module) versus that in the nondemented state (the lower left triangle of each module). Differential connectivity (MDC) and FDR estimates are specified in each panel in parenthese (MDC, FDR).

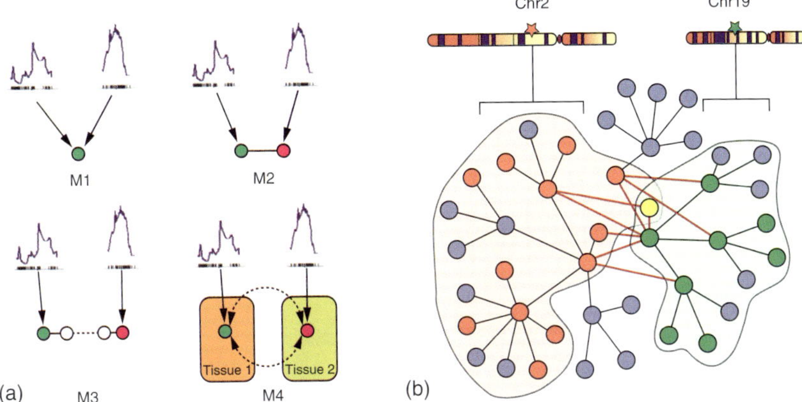

Plate 4.3 (A) Mechanistic models for joint regulation by two loci. In model 1, two loci directly regulate same gene. In model 2, each locus regulates one gene, and the two genes have physical interaction. Model 3 is similar to model 2, but there are multiple steps between the two genes. In model 4, each locus regulates one gene in a single tissue, and the cross-tissue interaction leads to a joint effect on phenotypic variation. (B) Interactive model for the additive effects between two genetic loci that regulate plasma insulin. Genetic variation at chromosome 2 changes expression for some nodes in the network (orange), while variation at chromosome 19 changes expression of other nodes (green), including genes regulated by chromosome 2 (yellow node). The blue nodes represent other nodes in the global protein-protein interaction network. Genes bound by gray curves are genes sharing the same eQTLs. Nodes involved in an interaction between the two sub-networks (shaded in light orange and green) are connected by bold red lines. We hypothesize that these nodes would be more influenced by genetic variation at both loci than a single locus. Genes involved in these cross-group interactions may be key regulators of plasma insulin.

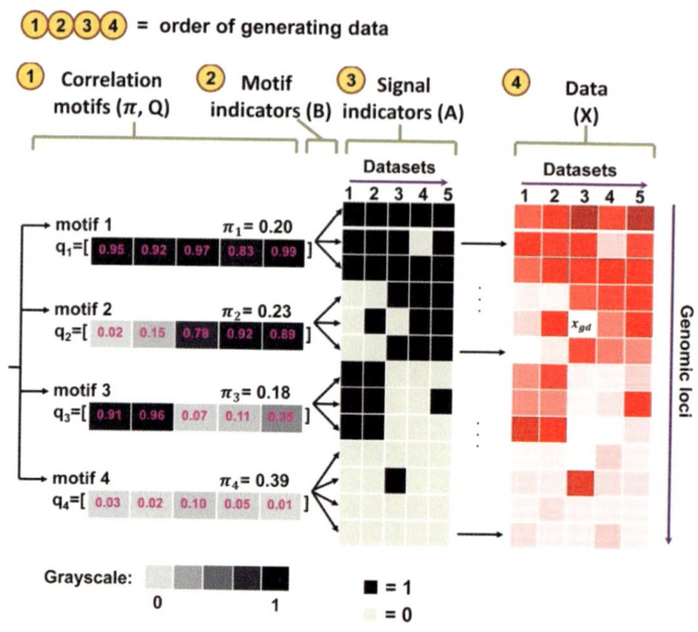

Plate 5.2 The data generative process assumed by the correlation motif model.

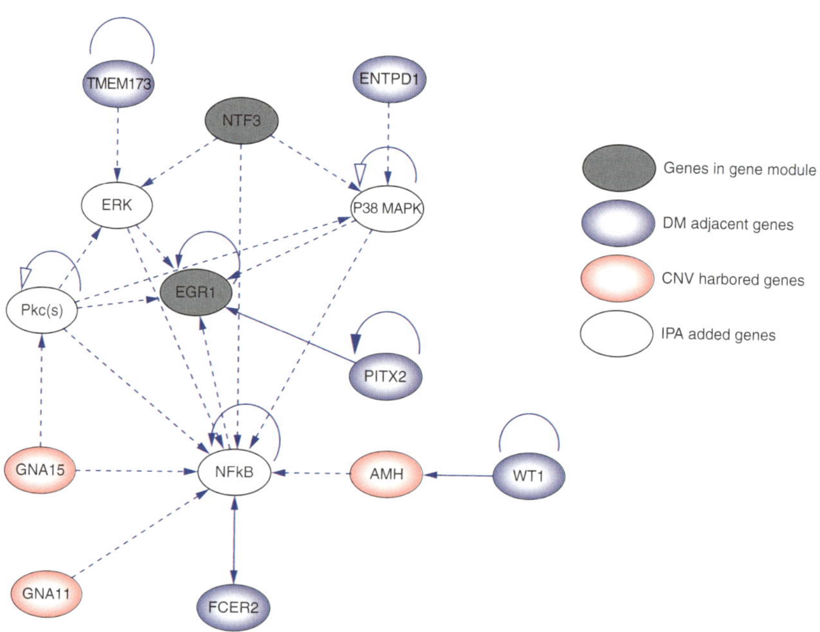

Plate 6.3 The molecular interaction network (constructed by IPA) centered on the gene *EGR1*. The solid lines represent direct interactions, and the dashed lines represent indirect interactions.

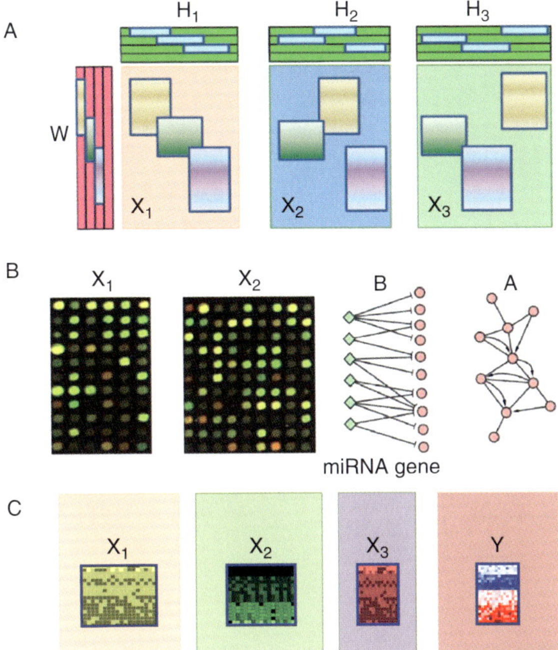

Plate 6.1 Illustration of the data resources and major features of the three methods: (A) joint nonnegative matrix factorization (NMF), (B) the network-regularized joint NMF, and (C) the sparse multiple-block PLS.

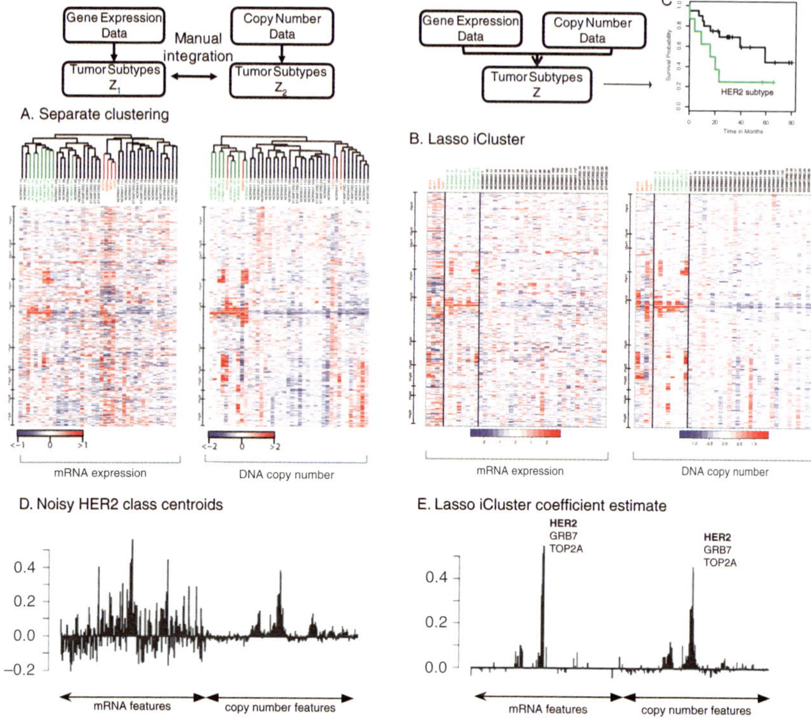

Plate 7.2 A motivating example using the Pollack data set to demonstrate that a joint analysis using the lasso iCluster method outperforms the separate clustering approach in subtype analysis given DNA copy number and mRNA expression data. (A) Heatmap with samples ordered by separate hierarchical clustering. Rows are genes and samples are columns. Samples labeled in red are breast cancer cell line samples. Samples labeled in green are HER2 breast tumors. (B) Heatmap with samples ordered by integrative clustering using the lasso iCluster method. (C) Kaplan-Meier plot indicates the HER2 subtype has poor survival outcome. (D) Standard cluster centroid estimates. (E) Sparse coefficient estimates under the lasso iCluster model.

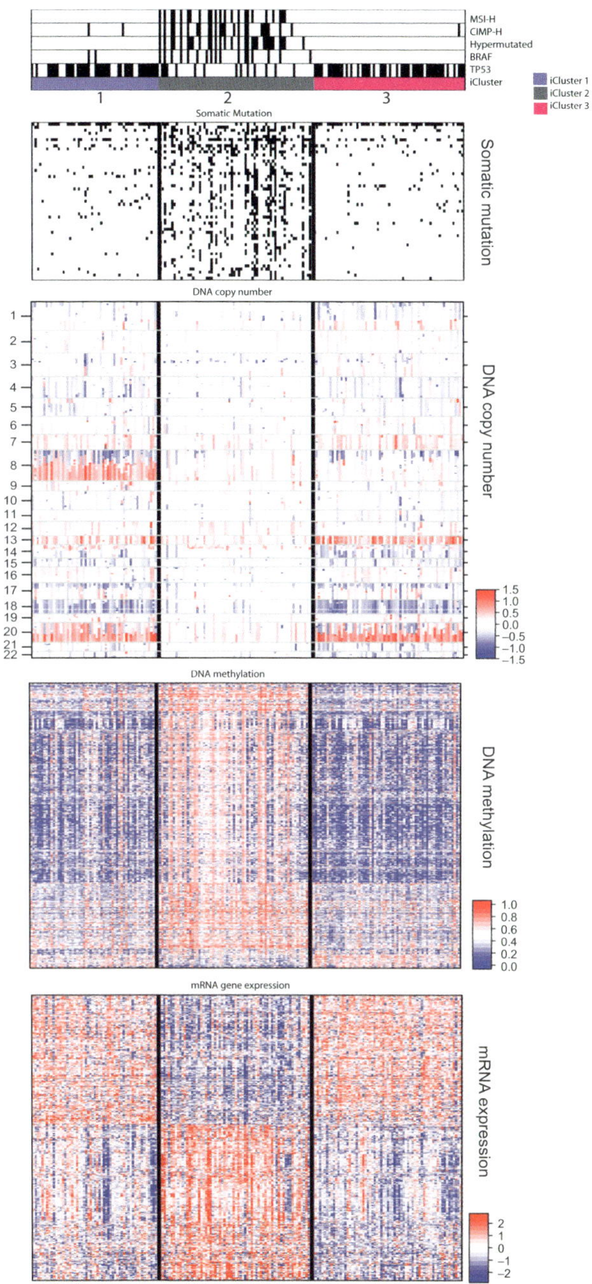

Plate 7.4 Three molecular clusters identified in 189 TCGA colorectal cancer samples. Heat map display of lasso-selected cluster-discriminant features. Rows are features and columns are tumor samples. The first panel shows genes that are mutated (black) or not mutated (white) in each cluster; the second panel shows genomic regions amplified (red) or deleted (blue); the third panel shows genes hypermethylated (red) or hypomethylated (blue); and the fourth panel shows genes overexpressed (red) or underexpressed (blue) in each cluster.

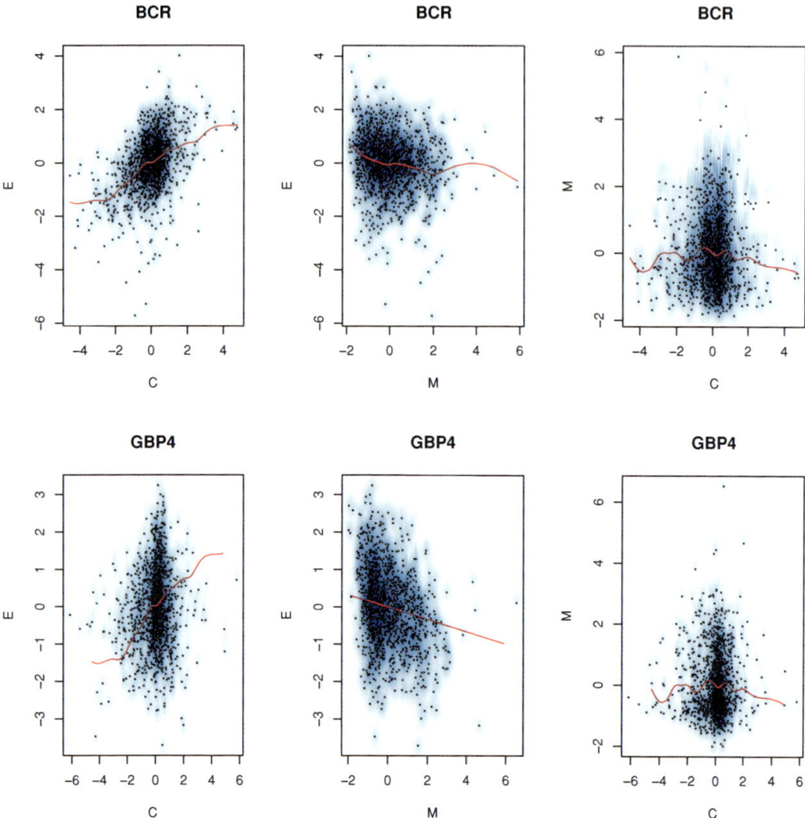

Plate 9.3 Smooth scatter plots of pairwise relationship among features C, M and E for genes *BCR* and *GBP4*. The red line in each smooth scatter plot is the lowess smoother. Dots correspond to the raw expression measurements from the TCGA data with 1,448 samples in Table 9.2.

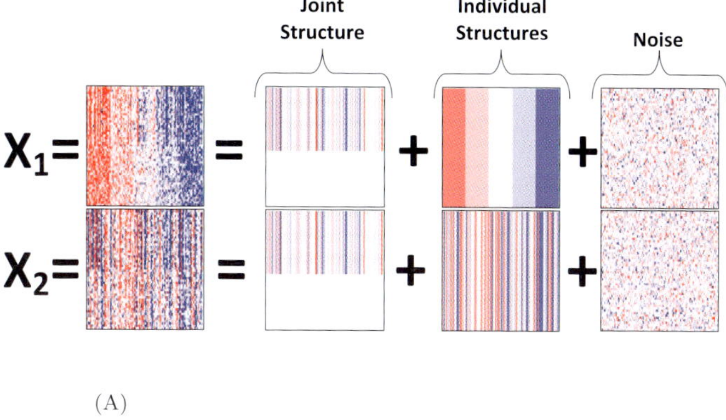

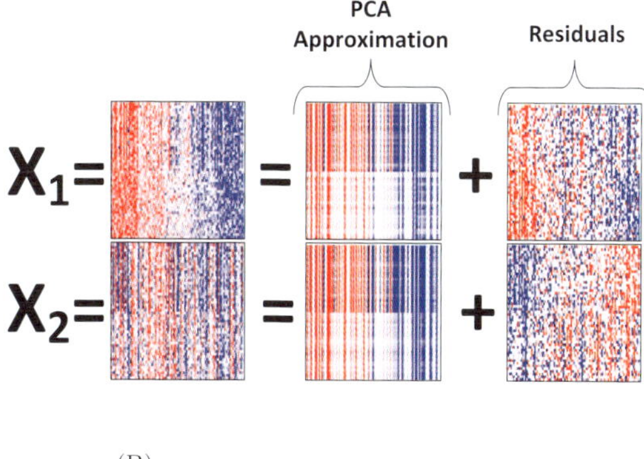

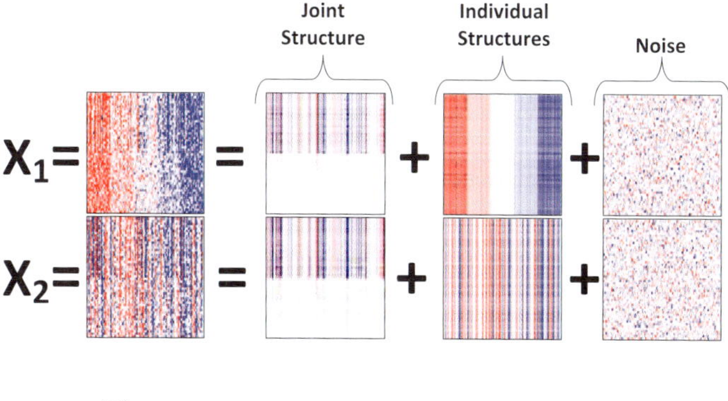

Plate 11.1 (A) Simulated data. $\mathbb{X}_1$ and $\mathbb{X}_2$ as the sum of simulated joint structure, individual structure, and noise. (B) PCA estimate. Approximation resulting from the first consensus principal component. (C) JIVE estimates. Blue corresponds to negative values, red to positive values.

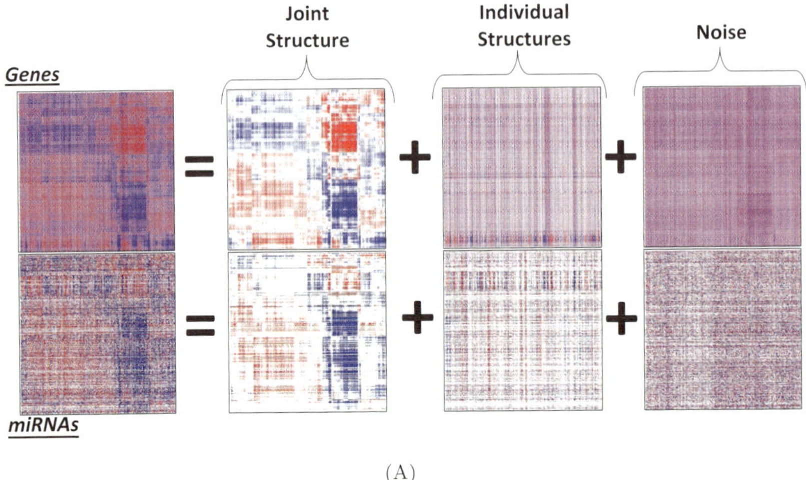

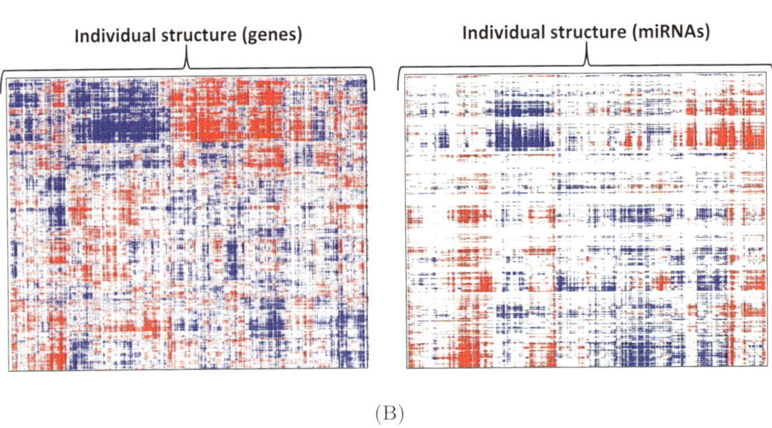

Plate 11.3 (A) Heatmaps of low-rank estimates for joint structure, individual structure, and residual noise in the gene expression and miRNA data; columns have the same order in all heatmaps. (B) Another view of the low-rank estimates for individual structure, in which columns are ordered separately for gene expression and miRNA data. Blue corresponds to negative values, red to positive values.

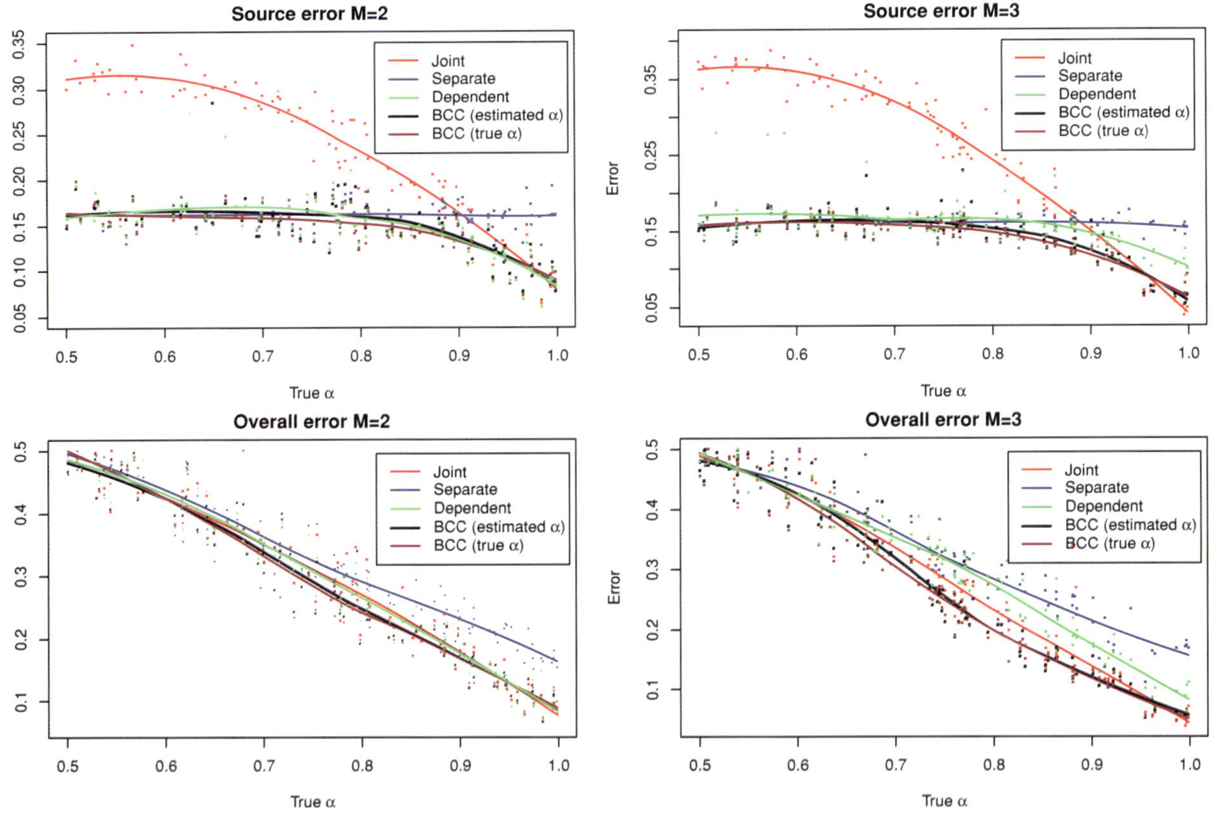

Plate 11.5 Source-specific and overall clustering error for 100 simulations with $M = 2$ and $M = 3$ data sources, shown for joint clustering, separate clustering, dependent clustering, BCC, and BCC using the true α. A LOESS curve displays clustering error as a function of α for each method.

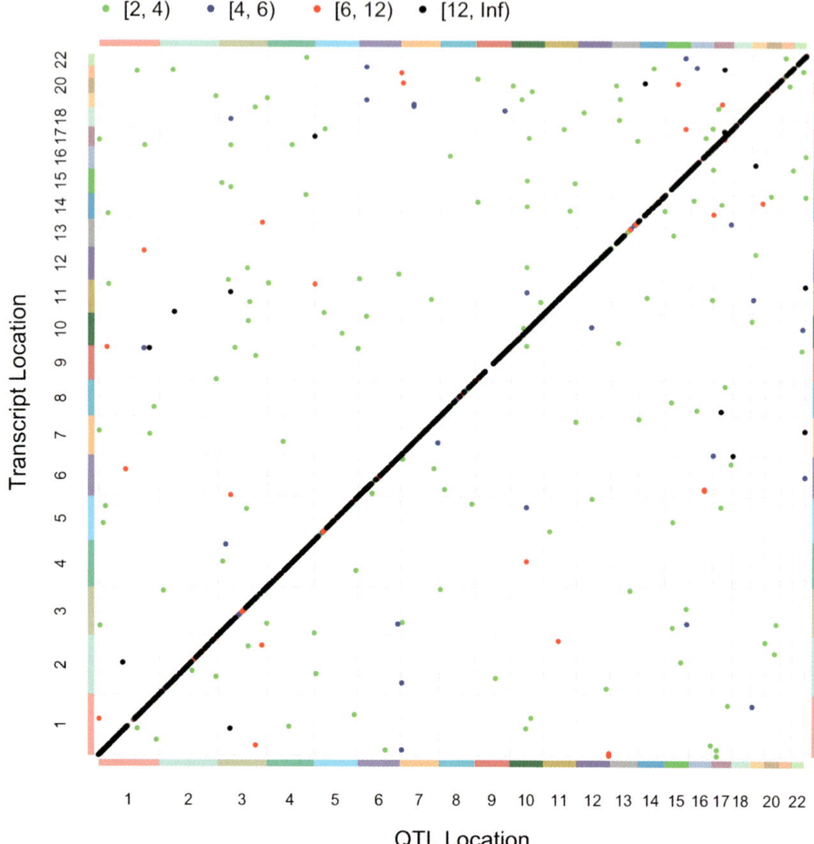

Plate 12.1 eQTL results of gene expression and SNP genotypes measured from whole blood of 1263 unrelated individuals. Each point indicates a significant association between one gene and one SNP. The color of a point reflects the range of $-\log_{10}(q\text{-value})$, which is labeled at the top of the figure.

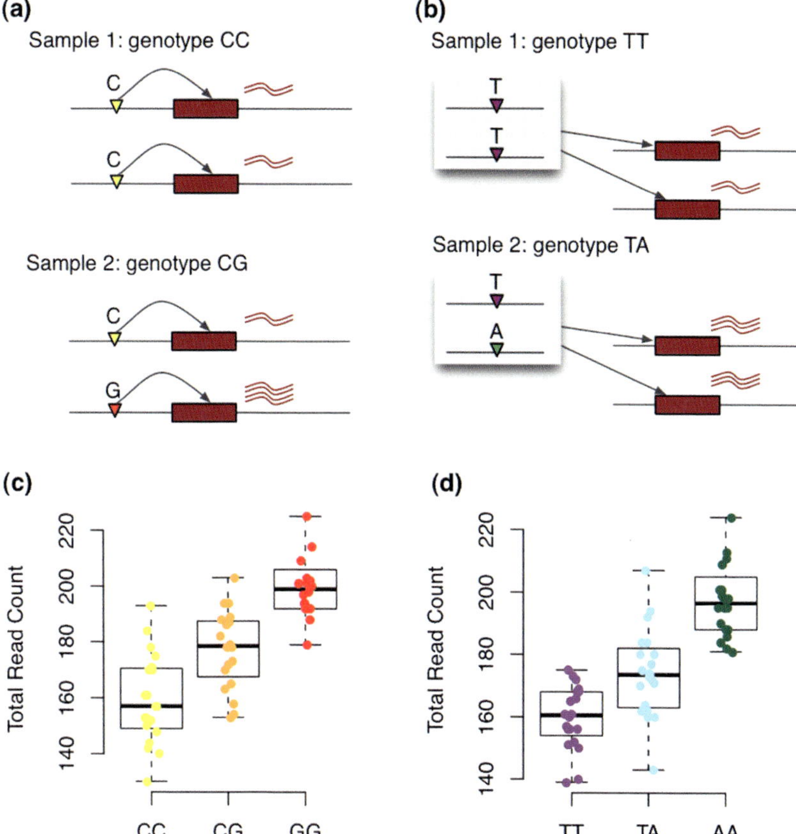

Plate 12.2 (A) An example of a *cis*-eQTL in two samples. In sample 2, where the candidate eQTL (the SNP for which we test association) has a heterozygous genotype CG, the expressions of the two alleles are different. (B) An example of a *trans*-eQTL in two samples. In sample 2, where the candidate eQTL has a heterozygous genotype TA, the expressions of the two alleles are the same. (C) Simulated data for a *cis*-eQTL across 60 samples with 20 samples within each genotype class. (D) Simulated data for a *trans*-eQTL across 60 samples with 20 samples within each genotype class. This figure is adapted from Sun and Hu (2013, Figure 12.1).

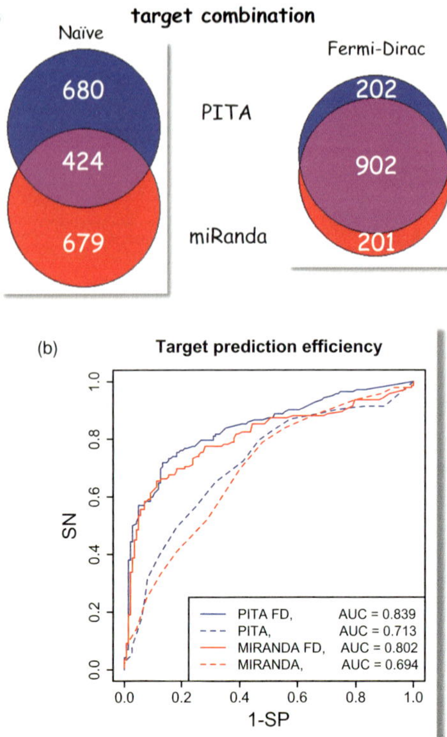

Plate 13.1 Using Fermi-Dirac to improve multiple target prediction. (A) Target overlap between miRanda and PITA using (left) the naive target combination method and (right) Fermi-Dirac. (B) ROC curves of miRanda and PITA on predicting mRNA targeted by multiple miRNAs. Dotted lines are naive combination of targets; solid lines are Fermi-Dirac combination of targets.

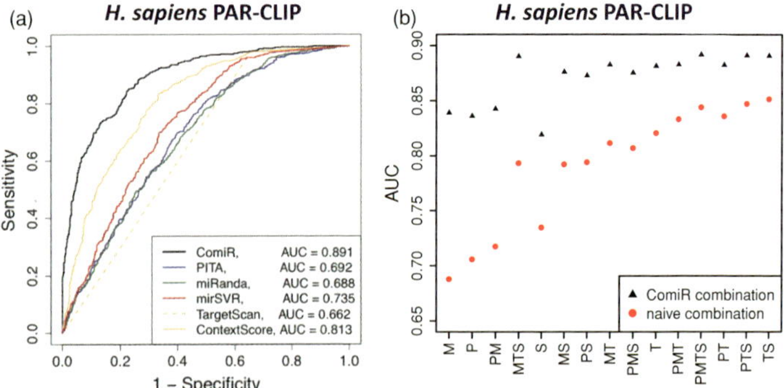

Plate 13.2 ComiR shows significantly improved performance in predicting targets of multiple miRNAs. (A) ROC curves of ComiR (black line) compared to TargetScan (dotted orange line, without conservation score; solid orange line, context score), miRanda (green line), PITA (blue line), and mirSVR (red line). All differences in area under the curve (AUC) are statistically significant. (B) Comparison of SVMs integrating different tools. Regardless of the tools used, the ComiR combination of scores (black triangles) outperforms the naive score combination (red circles). The largest improvement of ComiR scoring is for miRanda and PITA. *M*, miRanda; *P*, PITA; *T*, TargetScan; *S*, mirSVR.

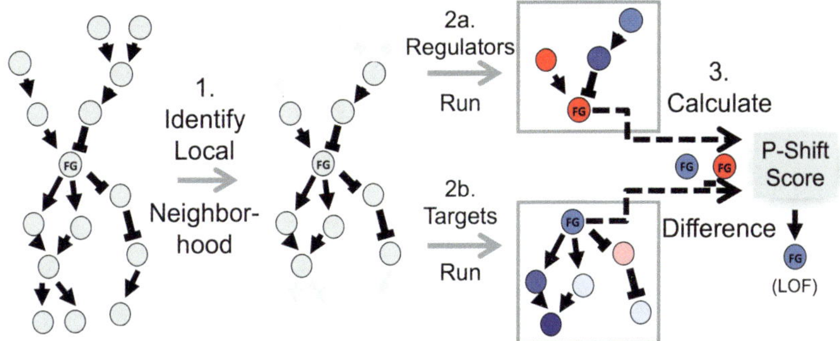

Plate 14.3 To calculate shift scores, PARADIGM-SHIFT goes through the following steps: it (1) undergoes feature selection to determine which features to include in the upstream (regulators) and downstream (targets) models, (2) calculates inferred activities for upstream and downstream networks using PARADIGM, and (3) computes the shift score as the difference between the downstream and upstream PARADIGM runs.

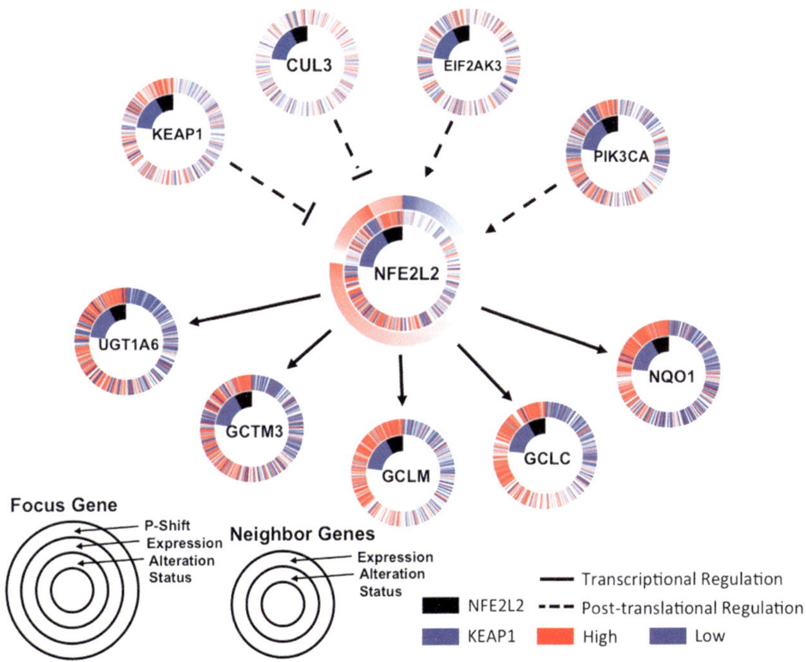

Plate 14.4 PARADIGM-SHIFT analysis of *NFE2L2* and *KEAP1* mutation on the Nrf2 signaling pathway. Circlemap display of mutation neighborhood selected around *NFE2L2*. Solid lines indicate transcriptional regulation and dashed lines indicate protein regulation. Samples were sorted first by the *NFE2L2* and *KEAP1* mutation status, then by shift score.

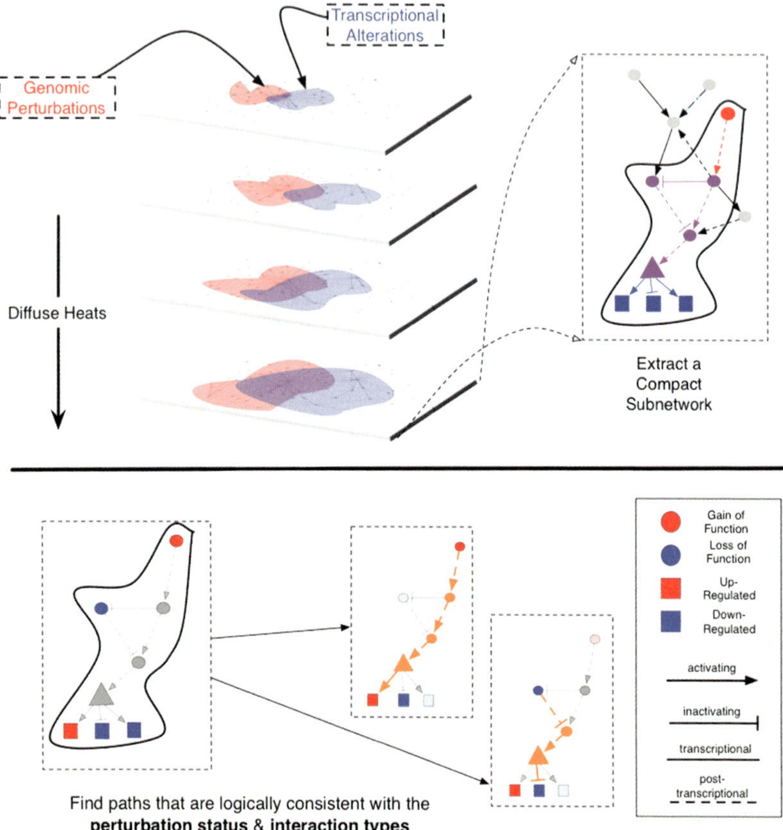

Plate 14.5 (top) Relevant genes from two distinct sets are shown by dyes diffusing on a pathway from a source set (e.g., genomically altered genes; red nodes) and a target set (e.g., transcription factors; blue nodes). "Linke" genes are shown as purple nodes, placed between the source and target sets; multiple time slices of the diffusion process are shown as stacked layers of the same network. (bottom) Subnetworks are extracted following the diffusion process. The algorithm finds all paths that connect genomic alterations to transcriptional alterations where the edge-interation logic is consistent with the sign of the source and target nodes (i.e., gain or loss of function; up- or down-regulation). After this filtering step, the union of all edges in the validated paths defines the resulting subnetwork.

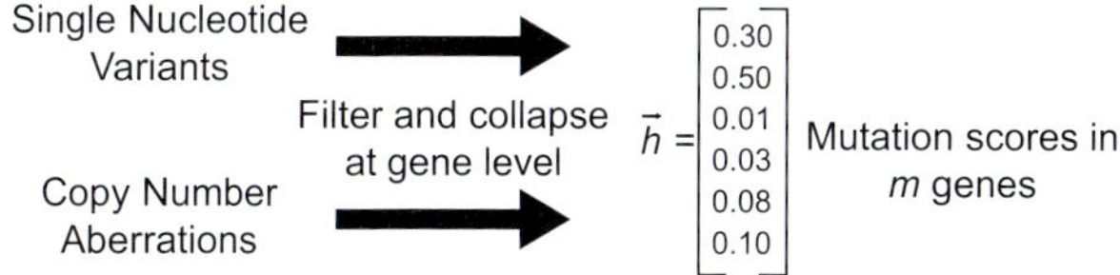

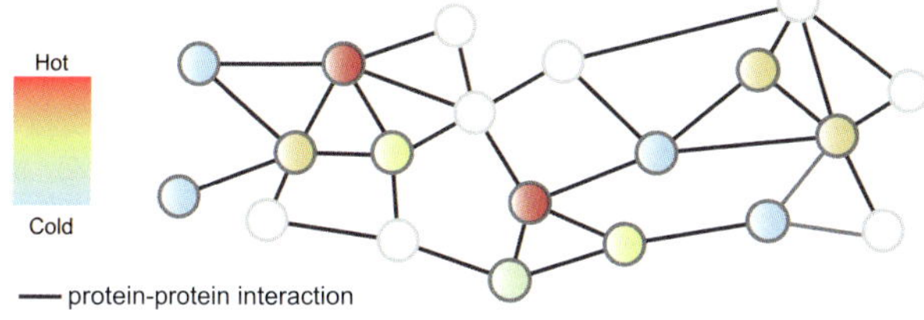

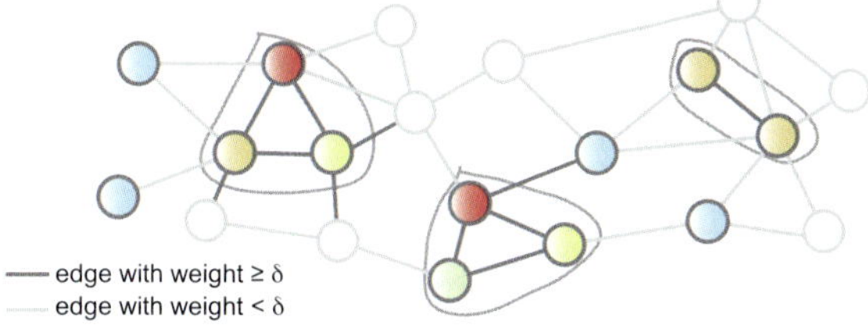

Plate 15.2 Overview of the Hotnet algorithm. (A) HotNet generates a heat score for each protein in a PPI network using single nucleotide variants and copy number aberrations from a cohort of tumors. (B) HotNet applies the heat to each protein in the network and allows it to diffuse for time t to create an edge-weighted graph. (C) HotNet partitions the graph by removing edges with weight less than a parameter δ, selected as described in the text. HotNet outputs the connected components of the partitioned graphs as significantly mutated subnetworks and evaluates their significance using a permutation test.

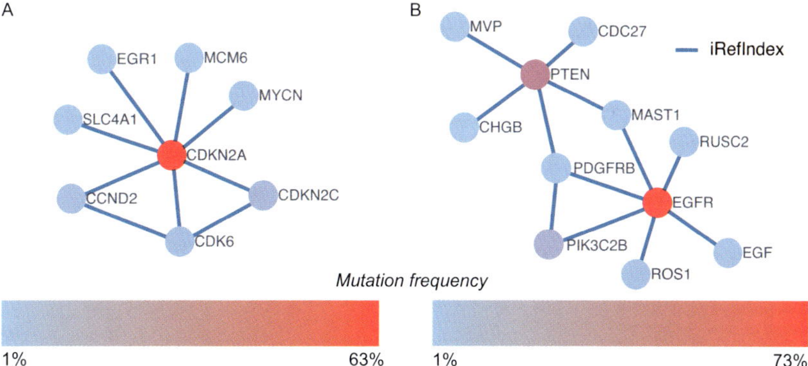

Plate 15.3 Subnetworks identified by HotNet on the TCGA GBM data set. (A) A subnetwork that includes multiple members of the Rb signaling pathway. (B) A subnetwork that includes members of the RTK and PI(3)K signaling pathways.

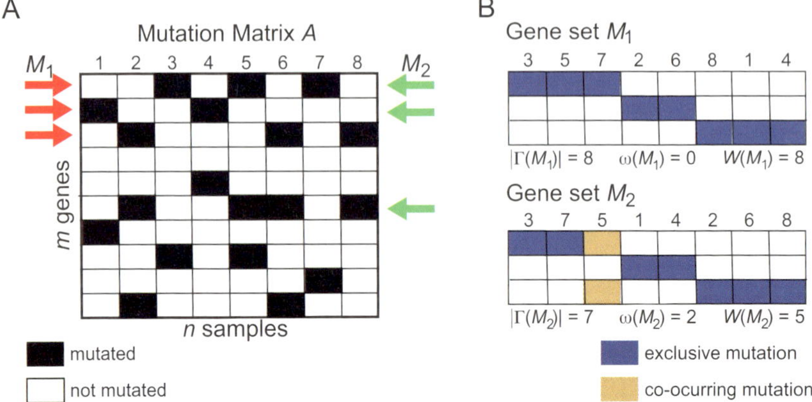

Plate 15.4 Overview of the Dendrix algorithm. (A) Dendrix takes as input a mutation matrix A (here shown at the level of individual genes) and finds gene sets with high weight W (i.e., approximately exclusive mutations and high coverage). The red and the green arrows point to the two highest-weight gene sets in A of size $k = 3$, respectively. Note that the gene sets overlap by two genes. (B) Mutation matrices for gene sets M_1 and M_2, with samples sorted independently in each set to illustrate exclusivity. Below each set are the coverage $|\Gamma|$, coverage overlap ω, and weight W.

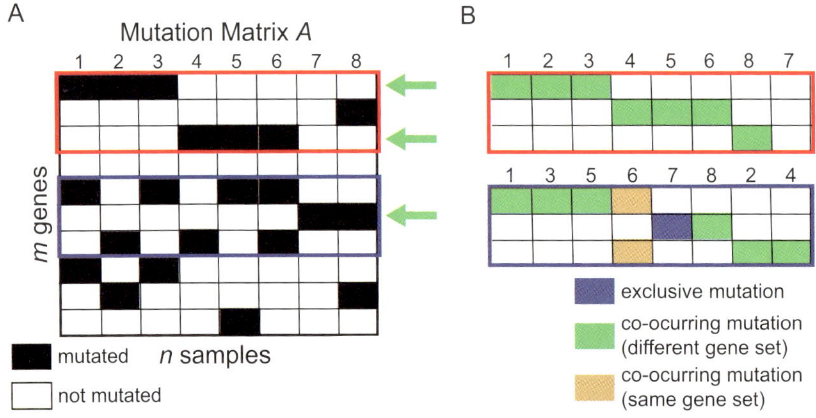

Plate 15.5 Overview of the Multi-Dendrix algorithm. (A) Multi-Dendrix identifies multiple gene sets simultaneously from a mutation matrix A. Shown in blue and red are the two gene sets of size $k = 3$ with maximum total weight. Methods that identify multiple gene sets *iteratively* would report the gene set with maximum weight indicated by the green arrows and thus would not find the optimal collection of gene sets. (B) The collection of exclusive gene sets identified by Multi-Dendrix from mutation matrix A. The samples (columns) in panel B are ordered to show the exclusivity of the mutations in the both gene sets. The mutations in each gene set largely exclusive (blue and green), with many co-occurring mutations between gene sets (green) and only one co-occurring mutation within the same gene set (orange).

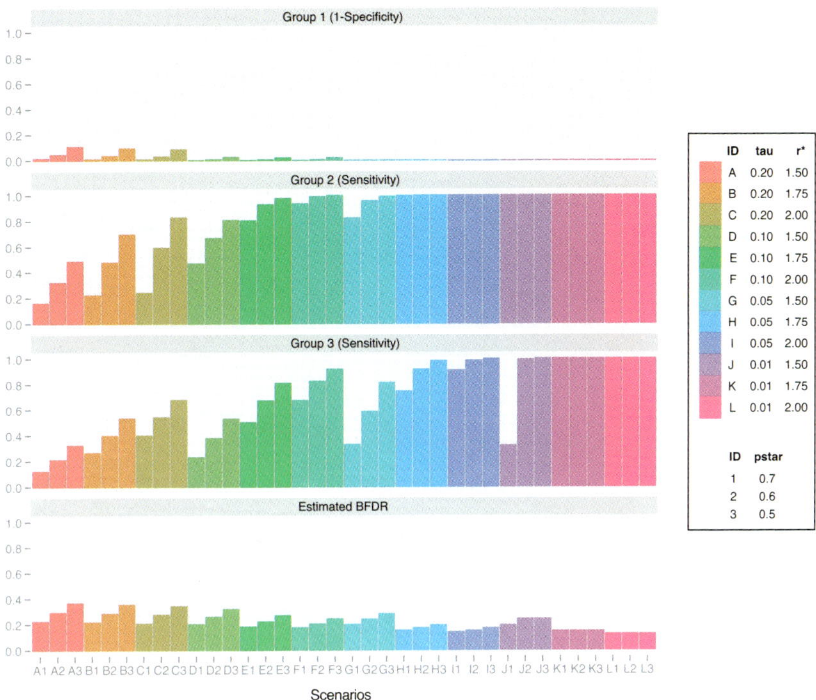

Plate 16.1 Simulation results. Proteins selected from group 1 across all time points are false positives. Proteins selected from group 2 at h_1 and group 3 at (h_2, h_3) are true positives.

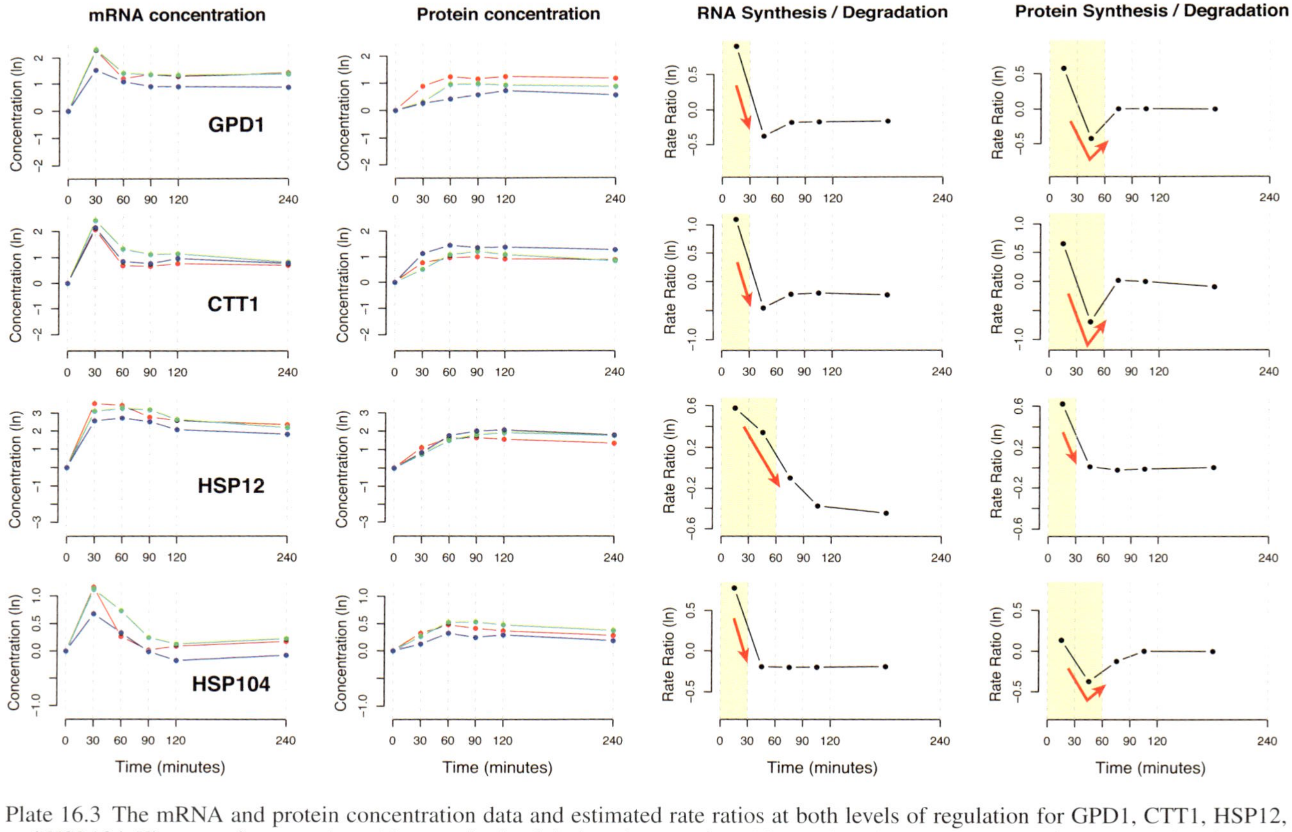

Plate 16.3 The mRNA and protein concentration data and estimated rate ratios at both levels of regulation for GPD1, CTT1, HSP12, and HSP104. These are four proteins with osmotic shock-induced expression. Blue, red, and green curves are time course data for each biological replicate. Yellow background indicates the time intervals during which the rate ratios deviated from the average range across the time course. Red arrows indicate significant regulation change at the RNA and protein level in each gene.

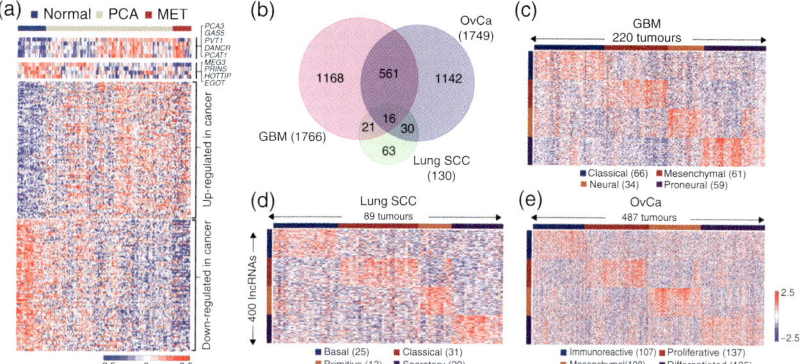

Plate 18.3 (A) The expression level of lncRNA that showed significantly differential expression between cancer and normal prostate tissues shown in heatmap across 29 normal prostate samples and 131 primary and 19 metastatic prostate tumor samples. Several known cancer-related lncRNA or lncRNA with established function in a noncancer context were highlighted. (B) Venn diagram representing the number of subtype-specific lncRNA in three cancers. The expression profile of the top 100 lncRNA that exhibited significantly higher expression in one subtype than the others for (C) GBM, (D) OvCa, and (E) Lung SCC shown in heatmap. (Note: the rank was based on the ascending order of the p-value.) Tumor samples were hierarchically clustered within each subtype.

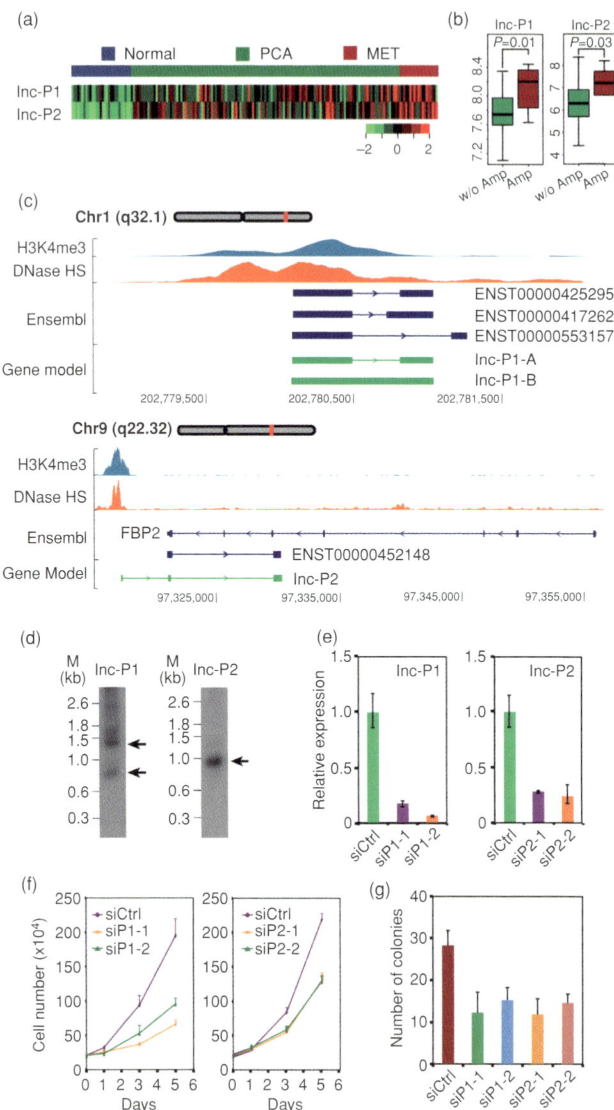

Plate 18.5 Experimental validation of lnc-P1 and lnc-P2 function. (A) Heatmap showing the expression of lnc-P1 and lnc-P2 in normal prostate tissue, primary and metastatic prostate cancer. (B) Box plot of lnc-P1 and lnc-P2 expression in tumors with genomic amplification and in the tumors without genomic amplification. (C) Transcript structure of lnc-P1 and lnc-P2 from Ensembl annotation and determined by 5′ and 3′ RACE experiments in LNCaP cell. In addition, the H3K4me3 and DNase I hypersensitive region profiles in the same cell line are shown. (D) The Northern blot of lnc-P1 and lnc-P2 transcripts. (E) Relative expression level of lnc-P1 and lnc-P2 upon knockdown by two different siRNA (purple and orange) and upon control siRNA treatment (green). (F) Growth curves of LNCaP cell with or without targeted siRNA-mediated knockdown of lnc-P1 or lnc-P2. The growth curves of control siRNA-treated cells and the growth curves of two targeted siRNA-treated cells plotted in purple, orange, and green, respectively. (G) Number of soft-agar colony formation of LNCaP cell with or without targeted siRNA-mediated knockdown of lnc-P1 or lnc-P2.

6

Identify Multi-Dimensional Modules from Diverse Cancer Genomics Data

SHIHUA ZHANG, WENYUAN LI, AND
XIANGHONG JASMINE ZHOU

Abstract

The rapid development of high-throughput biotechnology has made it possible to perform high-resolution genome profiling on several platforms simultaneously, resulting in an abundance of multidimensional genomic data. Such data provide unique and unprecedented opportunities to explore the coordination and cooperation between regulatory mechanisms on multiple levels. In the field of computational cancer genomics, integrating multiple types of genomic data for the discovery of combinatorial patterns is becoming a valuable and challenging issue. This chapter reviews recent progress in this direction, focusing on three methods developed by the authors. Specifically, we introduce a joint matrix factorization method, a network-regularized joint matrix factorization method, and a partial least squares regression method. The methods address the problem of integrating multiple data sets in the unsupervised, semi-supervised, or supervised manners. We also describe their applications in specific biological contexts. The methods described herein reveal biologically relevant patterns that would have been overlooked with only a single type of data, and uncover new associations between the different layers of cellular activities.

6.1 Introduction

Cellular systems are characterized by multiple levels of organization and complicated interactions between levels. The different levels (e.g., epigenetic status, transcriptions, and translations) must coordinate precisely to maintain the function and robustness of the cell. Gene expression, a crucial part of the cellular system, is a very complex process influenced by epigenetic, transcriptional, and posttranscriptional regulation, among other factors. In healthy cells, the dynamic interplay between these regulatory levels acts to maximize the efficiency and specificity of gene expression. Abundant studies support this view that gene regulation is governed by multiple, complex, and extensively coupled networks. However, studies of the coordination between cellular activities on

135

different levels have been hindered by a lack of appropriate data. Therefore, most genomic research focuses on global profiling of only one level.

The rapid development of high-throughput genomics technologies in the past decade, especially sequencing technologies, has significantly facilitated the characterization of cellular systems at multiple levels simultaneously. Such data have enabled researchers to obtain a global view of the principles underlying gene regulation. Microarray and sequencing technologies can measure genome-wide gene expressions, DNA variations, epigenetic regulations, and posttranscriptional regulations. For example, the Cancer Genome Atlas (TCGA) project is generating multi-dimensional maps of key genomic changes (single nucleotide polymorphism (SNP), copy number variation (CNV), DNA methylation (DM), gene expression (GE), and microRNA expression (ME)) measured on the same set of tumor samples for more than 20 cancer types (Cancer Genome Atlas Research Network, 2008). The NCI60 project has profiled 60 human cancer cell lines in terms of drug responses, gene expression, protein expression, microRNA expression, and other variables (Weinstein et al., 1997). As sequencing costs continue to decline, the multidimensional characterization of samples will soon become standard practice. Multidimensional genomics data sets provide unprecedented opportunities to discover the connections between different layers of gene regulation and different cellular systems.

However, most genome-wide studies still restrict their analysis to only one aspect of regulation, such as gene expression profiles (e.g. Alter et al., 2000; Omberg et al., 2007; Tamayo et al., 2007). This is because multidimensional genomic data pose new challenges for analysis and call for novel computational methods. Furthermore, the different types of genomics data have different scales and units, so it is not possible to aggregate the direct measurements of different data sets. There are ways around this problem: for example, various eQTL approaches have been developed to identify regulatory SNPs in a specific type of data set consisting of SNP and gene expressions (Zhang et al., 2010). However, eQTL cannot be applied to data sets with more than two dimensions. Multivariate regression is another method applicable to two-dimensional genomics data sets, to infer correlative relationships. Recently, Kutalik et al. (2008) proposed a powerful approach called the Ping-Pong algorithm to uncover *comodules* in gene expression and drug response data. All the preceding studies have identified important relationships between pairwise genomics variables.

Several methods have also been developed to analyze genomic data sets with more than two dimensions. For example, the multivariate model developed

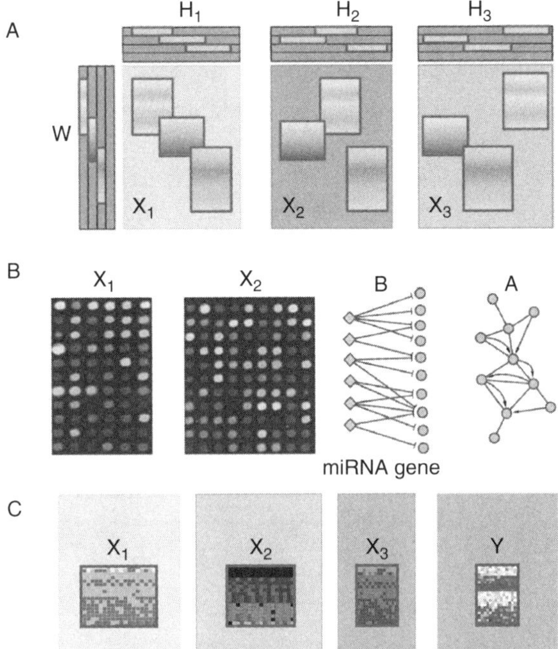

Figure 6.1 Illustration of the data resources and major features of the three methods: (A) joint nonnegative matrix factorization (NMF), (B) the network-regularized joint NMF, and (C) the sparse multiple-block partial least squares.

by Mankoo et al. (2011) and the sparse regression method proposed by Witten and Tibshirani (2009) can both learn from multidimensional genomic data in a supervised manner. Another relevant method, cMonkey, is a biclustering method that can analyze GE matrices from different species (Waltman et al., 2010). In addition, several papers have designed multiple kernel learning methods for integrating heterogeneous genomic data (Alpaydin, 2011; Hamid et al., 2012; Yu et al., 2010). In these methods, each data type is transformed into a kernel, and then these are "fused" into a single kernel to be used for prediction, regression, or feature selection.

In this chapter, we introduce three methods that we recently developed to discover combinatorial patterns by integrating multiple types of high-throughput genomics data (Figure 6.1). The first is a powerful joint matrix factorization framework to identify correlative modules in multi-dimensional genomics data (Zhang et al., 2012). We test its performance using data from the TCGA project, specifically the DNA methylation, miRNA, and gene expression profiles of 385 ovarian cancer samples. These three types of genomics variables are known to be highly interdependent. The method identifies subsets of mRNAs, miRNAs, and methylation markers for which all or a subset of the samples exhibit correlated profiles across different types of measurements (Figure 6.1A). These subsets are termed multi-dimensional modules (*mdmodules*). Second, we developed a network-regularized joint matrix factorization framework for reconstructing miRNA regulatory modules based on multiple genomic data sources (Zhang et al., 2011) (Figure 6.1B). We applied the method to three types of data: predicted miRNA-gene interactions, the expression profiles of miRNAs and genes, and the gene-gene interaction network. We found biologically appropriate solutions of the membership functions by applying sparsity penalties to enhance the signal-noise separation. Third, we designed a partial least squares regression method for supervised module discovery in multidimensional genomic data (Li et al., 2012) (Figure 6.1C). In the application, a regulatory module contains sets of regulatory factors from different layers that are likely to contribute jointly to a local gene expression factory. This approach facilitates reconstruction of the regulatory network across different layers.

6.2 The Joint NMF Method

Nonnegative matrix factorization (NMF) is increasingly being used to analyze high-dimensional genomics data. NMF factors a matrix $X_{M \times N}$ into two nonnegative matrices $X = WH$, where W is an M by K matrix containing the basis vectors, and H is a K by N matrix containing the coefficient vectors. Each element in W and H must be ≥ 0. The dimension K is much smaller than either M or N and chosen by the researcher before factorization. In this way, the matrices W and H represent the most important features of X in fewer dimensions. Following this idea, we proposed the joint NMF framework for integrative genomics analysis (Zhang et al., 2012). Let X_1, X_2, X_3 be $M \times N_1$, $M \times N_2$, and $M \times N_3$ matrices representing three types of genomic profiling of the same samples, for example, the methylation profiles of N_1 DNA markers and the expressions of N_2 genes and N_3 miRNAs of M samples. To extract mdmodules across the three data matrices, we decompose the three data matrices into a common basis matrix W and three coefficient matrices H_I ($I = 1, 2, 3$) such that each source matrix $X_I \approx WH_I$ and all the factors are nonnegative ($W \geq 0$, $H_I \geq 0$). W is an $M \times K$ matrix, and each column of W represents a basis vector of the reduced system. H_I is a matrix of size $K \times N_I$, and each row of H_I represents a coefficient vector. The most widely used error function in NMF is the squared Euclidean error function. We also adopt this function to define the objective function of the joint factorization: $\sum_{I=1}^{3} \| X_I - W H_I \|_F^2$.

We use "multiplicative update" equations to minimize the error function. Specifically, given a desired rank K, the algorithm starts by randomly initializing the matrices W and H_1, H_2, and H_3, then iteratively updates the four matrices as follows:

$$W_{ia} = W_{ia} \frac{(X_1 H_1^T + X_2 H_2^T + X_3 H_3^T)_{ia}}{(W(H_1 H_1^T + H_2 H_2^T + H_3 H_3^T))_{ia}},$$

$$(H_I)_{a\mu} = (H_I)_{a\mu} \frac{(W^T X_I)_{a\mu}}{(W^T W H_I)_{a\mu}}, \quad I = 1, 2, 3.$$

We repeat this procedure 50 times with different initial matrices. The factorization that minimizes the objective function value is used as the final solution for further analysis. The time complexity of our joint NMF decomposition is $O(t K (M + N_1 + N_2 + N_3)^2)$, similar to that of the original NMF model, where t is the number of iterations.

The result of NMF is that the three data matrices are projected into a common coordinate system, so that we can explore correlative relationships among the three types of variables. The coefficient matrices H_1, H_2, and H_3 can be used to identify membership vectors consisting of DM markers, miRNAs, and genes in multi-dimensional modules. We calculated z-scores for each element of the matrices H, relative to the row k in which the element resides: $z_{ij} = \frac{x_{ij} - \mu_i}{\sigma_i}$, where μ_i is the average value for feature j (DM markers/miRNA/gene) in H_I ($I = 1, 2, 3$), and σ_i is the standard deviation. We then assign feature j to the module of row k if z_{ij} is greater than a given threshold T. In this way, we identify a functional module for each row k of the factorized matrix. Each DM marker/miRNA/gene may be assigned to multiple mdmodules, which allows the identification of multiple functional activities of DM markers/miRNAs/genes.

6.2.1 Results

We tested the joint NMF algorithm using the TCGA ovarian cancer data set, which consists of gene expression, DNA methylation, and miRNA expression profiles across 385 samples (patients). After parameter optimization, we chose to factor the three large matrices into $K = 200$ building blocks, from which 200 mdmodules were derived. We know that dimension reduction preserves most of the information embedded in the original data, because the average sample wise correlations between the reconstructed data (based on W and H_I)

and the original data were 0.90, 0.92, and 0.91 in the methylation, miRNA, and gene expression dimensions, respectively. Each mdmodule comprises a set of genes, methylation markers, and miRNAs. In total, the 200 mdmodules cover 2985 genes, 2008 DNA methylation markers, and 270 miRNAs. The average module sizes in the gene, methylation marker, and miRNA dimensions are 239.6, 162.3, and 13.8, respectively.

Multidimensional Modules Reveal Vertical Associations and Cooperative Functional Effects

To assess the biological relevance of the multidimensional modules, we tested the functional homogeneity of members within each dimension. A set of genes is defined to be functionally homogenous if it is enriched in at least one gene ontology (GO) biological process category, with a q-value less than 0.05 (the q-value is the p-value after a false discovery rate multiple testing correction). Among the 200 mdmodules, 80% were functionally homogenous in the gene expression (GE) dimension with respect to member genes; 62.7% were functionally homogenous in the DNA methylation (DM) dimension, with respect to those genes directly adjacent to the member DNA methylation markers; finally, 12.5% were functionally homogenous in the miRNA expression (ME) dimension with respect to member miRNAs. The functions of miRNAs were predicted based on the functions of their target genes. All three values are significantly higher than the proportions observed in modules obtained after randomization (5%, 13.1%, and 3.9% for GE, DM, and ME, respectively).

Although all three dimensions showed significant enrichment in developmental processes that are known to be tightly associated with cancer pathogenesis, this preference is most obvious in the DM dimension, with additional strong participation in embryonic development. This result is consistent with a previous report that polycomb complex targets in the embryonic stem cell are predisposed to cancer-specific hypermethylation (Widschwendter et al., 2007). The most frequently activated biological processes in the GE dimension are responses to external stimuli (e.g., chemotaxis, locomotor behavior, and inflammatory responses). This observation points out that gene expression programs are flexible enough to adapt to external perturbations. The ME dimension shows a distinct preference for participation in transcriptional regulation (as expected) and cell differentiation.

Although the individual dimensions of these modules exhibit a significant level of functional homogeneity, combining them reveals an even stronger functional synergy. Ninety-three percent of the mdmodules were functionally homogenous when all three variables were taken into account, compared to only 7.9% after randomization. This result demonstrates the power of multidimensional data to identify different genomic variables that are involved in the same functional pathways.

The ability of the modules to capture multilevel synchronicity was also observed relative to perturbed KEGG pathways. For example, simply by combining multiple dimensions, we observed that nine modules showed significant perturbations in at least one KEGG pathway ($p < 0.05$) that were not shown otherwise. These pathways include TGF-β signaling, hedgehog signaling, bladder cancer, and cytokine-cytokine receptor interaction pathways, all of which have been confirmed to be closely associated with ovarian cancer. For 11.5% of the mdmodules, the pathway enrichment for combined members from all three dimensions is more significant than that for any individual dimension.

According to the model principle, a mdmodule should capture vertical associations between variables of different dimensions (e.g., GE and DM). Indeed, compared to randomly permuted modules, the Pearson's correlation coefficients between variables of any two dimensions are significantly higher ($p < 0.05/200$) in 65.5% of the modules. This result indicates that the probability of identifying these modules by chance is close to zero. The strong statistical correlations across different dimensions imply the coordinated activities of genes, methylations, and miRNAs.

To explore further the biological implications of these vertical correlations, we tested whether the genes in an mdmodule were likely to be located close to the methylation markers in the same module or/and targeted by miRNAs in the same module. At a significance level of 0.1, we found that 75 of the 200 mdmodules showed significant overlap between expressed genes and genes adjacent to methylation markers. This result confirms the strong influence of DNA methylation on the expression of adjacent genes. Likewise, 146 modules with $p < 0.1$ show significant overlap between genes targeted by miRNAs and expressed genes within the same mdmodule. Because the targeting relationship between miRNAs and genes is far from complete, our assessment of the overlap can only be an underestimate. These data show that mdmodule can elucidate the vertical association mechanisms between different layers of gene regulation.

In 44 modules, the genes from the GE and DM dimensions are enriched in protein-protein interactions ($p < 0.05$). In 18 of these 44 modules, one protein belongs to the GE dimension and another belongs to the DM dimension ($p < 0.05$ with right-tailed Fisher's exact test). This finding highlights the different regulatory effects on closely adjacent molecules of the same pathway.

Through two examples, we have also shown that the multidimensional modules capture the associations among epigenetic regulation, gene expression, and posttranscriptional regulation on various parts of the pathway. Such synchronized effects would be very difficult to identify by any means other than

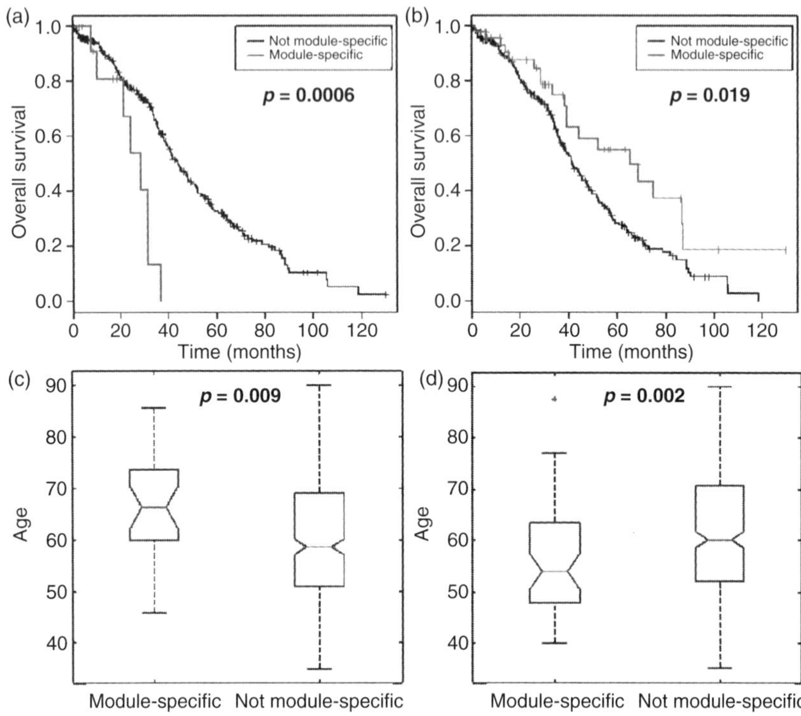

Figure 6.2 Kaplan-Meier survival analysis for patients associated with (A) module 166 or (B) module 3 compared to other patients. The *p*-values of the log-rank test were $p = 0.0006$ and $p = 0.019$, respectively. Median survivals for patients in module 166 or module 3, compared to other patients, were 26.4 vs. 36.1 years and 38.2 vs. 33.8 years, respectively. Boxplots for the ages of patients associated with (C) module 28 or (D) module 78 compared to other patients. The *p*-values of the rank-sum test were $p = 0.009$ and $p = 0.002$, respectively. The median ages for patients in module 28 or module 78 compared to other patients were 66.3 vs. 58.7 years and 54.1 vs. 60.2 years, respectively.

the integrative analysis of multidimensional data. Our mdmodules facilitate the discovery of abnormal functions at multiple regulatory levels. Thus, this method can aid the development of a holistic approach to drug intervention, one that simultaneously corrects the effects of various types of dysfunctions.

Clinical Associations of the Multi-Dimensional Modules

In the NMF framework, the decomposed component vector (i.e., a column of the W matrix) can provide information on the associations between a given sample/patient and the individual modules. This information, combined with the available clinical characterizations of each patient, can aid in the discovery of phenotype-specific mdmodule. An mdmodule that stratifies patients into clinically distinct groups can shed light on the molecular mechanisms of the respective clinical phenotypes.

On the basis of the information from the W matrix, we compared the survival times of ovarian cancer patients who are strongly associated with a specific mdmodule and those who are not. We found several mdmodules whose patients showed significantly shorter or longer median survival times (log-rank test $p < 0.05$). For example, 13 patients strongly associated with mdmodule 166 show significantly worse outcomes, with a median survival of 26.4 months, compared to 34.1 months for other patients ($p = 0.0006$, log-rank test) (Figure 6.2A). In fact, all three dimensions of this mdmodule have distinct characteristics not apparent in the rest of the patients. The module contains numerous cell cycle checkpoint genes (e.g., *BUB1B, CENPF, MAD2L1, CCNB1, BUB1, CCNA2, CHEK1*, and *TTK*) and is significantly enriched in genes from the "nuclear division" functional category ($p < 10^{-8}$). In another case, the patients in mdmodule three have an improved median survival time of 38.2 months versus 33.8 months in the remaining patients ($p < 0.02$, log-rank test). This module has a significant perturbation of the endometrial cancer pathway, with several key genes related to tumorigenesis, for example, *EGFR, CTNNA2*, and *ARAF*.

We identified 20 mdmodules that contain patients with significantly different age characteristics. For example, patients in module 28 had an older median age compared to other patients (66.3 years vs. 58.7 years; $p = 0.009$, rank-sum test) (Figure 6.2C), and mdmodule 78 was associated with significantly younger patients (median age of 54.1 years vs. 60.2 years; $p = 0.002$, rank-sum test) (Figure 6.2D).

6.3 The Network-Regularized Joint NMF Method

In this section, we describe our NMF framework for identifying miRNA-gene comodules (Figure 6.2B). We designed an objective function with three components. The first is based on the nonnegative miRNA and gene expression matrices X_1 and X_2, using the joint NMF framework discussed in the last section. The second component considers the effects of gene-gene interactions. The last component considers the effects of predicted miRNA-gene interactions.

We assume that there is a common basis matrix W for the miRNA and gene expression matrices X_1 and X_2, which have dimensions $s \times m$ and $s \times n$, respectively. We adopt the joint NMF framework to factor X_1 and X_2 as described in the previous section. This is done by optimizing the following objective function: $\sum_{I=1,2} \|X_I - WH_I\|_F^2$, where H_1 and H_2 have dimensions $k \times m$ and $k \times n$, respectively. The parameter k is chosen prior to optimization.

The solution to the joint NMF problem is often not unique and may be sensitive to noise in the expression data. Both of these limitations may confound the module discovery process. For these reasons, we guide the optimization process toward reasonable biological solutions by incorporating prior knowledge into the objective function. The essence of our semisupervised learning method is to define constraints that favor the placement of biologically linked variables in the same comodule. In addition to improving the biological relevance of the results, such constraints can greatly facilitate the discovery of comodules by narrowing down the large search space.

Let A (with dimensions $n \times n$) denote the adjacency matrix of a gene interaction network and B (with dimensions $m \times n$) denote the adjacency matrix of a bipartite miRNA-gene network. We enforce "must-link" constraints by including the following term in the objective function: $\mathcal{O}_1 = \sum_{ij} a_{ij}(h_i^2)^T h_j^2 = Tr(H_2 A H_2^T)$. This ensures that genes with known interactions have similar coefficient profiles. Similarly, known interactions between genes and miRNAs can be encoded by including the term $\mathcal{O}_2 = \sum_{ij} b_{ij}(h_i^1)^T h_j^2 = Tr(H_1 B H_2^T)$.

To discover miRNA-gene regulatory comodules, we combine the three objectives into a single optimization problem: $\sum_{I=1,2} \|X_I - W H_I\|_F^2 - \lambda_1 Tr(H_2 A H_2^T) - \lambda_2 Tr(H_1 B H_2^T)$. The parameters λ_1 and λ_2 are weights for the constraints defined in A and B. The first term favors modules with miRNA and gene expression profiles that are correlated in the common basis matrix W. The second term, $Tr(H_2 A H_2^T)$, summarizes all the must-link constraints in the gene-gene network. The third term, $Tr(H_1 B_{12} H_2^T)$, summarizes all the must-link constraints in the miRNA-gene network. We call this approach the network-regularized multiple NMF (NRNMF) framework.

To ensure that the discovered comodules have a clear biological interpretation, we want the coefficient matrices H_1 and H_2 to be relatively sparse. To this end, we extend the objective function as follows: $\sum_{I=1,2} \|X_I - W H_I\|_F^2 - \lambda_1 Tr(H_2 A H_2^T) - \lambda_2 Tr(H_1 B H_2^T) + \gamma_1 \|W\|_F^2 + \gamma_2(\sum_j \|h_j\|_1^2 + \sum_{j'} \|h_{j'}\|_1^2)$. The variables h_j and $h_{j'}$ denote the jth and j'th columns of H_1 and H_2 respectively. The term $\gamma_1 \|W\|_F^2$ limits the growth of W, whereas $\gamma_2(\sum_j \|h_j\|_1^2 + \sum_{j'} \|h_{j'}\|_1^2)$ encourages sparseness of the column vectors. This variant of the algorithm is called sparse network-regularized NMF (SNMRMF).

The NRNMF algorithm efficiently converges to a local minimum by iteratively updating the matrix decomposition. The objective function is guaranteed not to increase when the decomposition is updated. Furthermore, the objective function remains invariant if and only if W, H_1, and H_2 are at a stationary point. This behavior can be proved in the same way as for the classical NMF algorithm. Note that H_1 and H_2 are updated simultaneously, based on their current values at each iteration. The time complexity of the proposed algorithm is $O(tk(s + m + n)^2)$, where t is the number of iterations. The coefficient matrices H_1 and H_2 produced by the algorithm can be used to identify comodules.

6.3.1 Results

We have applied the SNMNMF method to identify miRNA-gene comodules by integrating multiple independent data sources (Zhang et al., 2011). The 49 miRNA-gene comodules identified in this study have an average of 3.8 miRNAs and 78 genes per module. Based on a distribution of correlations derived from randomized miRNA-gene comodules, the (anti-)correlations between miRNA and gene expressions are statistically significant in 69.4% of the identified modules (permutation test with p-value less than 0.05/50), indicating that the probability of finding similarly (anti-)correlated comodules by chance is close to zero.

The miRNA-gene comodules discovered by our method may shed light on the combinatorial regulation of miRNAs. Eleven of the identified modules are significantly enriched in at least one miRNA cluster, defined as a set of miRNAs that are located within 50 kb of each other in the genome ($q < 0.05$ after multiple testing correction). For example, comodule 48 contains nine miRNAs (mir-506, mir-507, mir-508-3p, mir-509-3p, mir-509-3-5p, mir-509-5p, mir-513b, mir-513c, mir-514), all of which belong to a miRNA cluster on chromosome Xq27.3.

On a basis of a literature survey, we found that spatially clustered miRNAs often have similar functions or play complementary roles. An abundant literature supports the biological significance of the comodules identified in this study. For example, in comodule 10, two of the four member miRNAs (mir-449a and 449b) belong to a miRNA cluster on chromosome 5q11.2, while the other two (miR-34b* and 34c-5p) belong to a cluster on chromosome 11q23.11. A recent study reported miR-449a and 449b to have a tumor-suppressing function by regulating *Rb/E2F1* activity(Yang et al., 2009). In addition, miR-34b* and 34c-5p are both targeted by *p53* and cooperatively control cell proliferation in ovarian cancer (Corney et al., 2007).

To take another example, three of the seven miRNAs in module 16 (miR-96, miR-182*, miR-183) are clustered on chromosome 7q32.2 and are dysregulated in various cancers. These miRNAs (along with others) cooperatively repress *FOXO1*, affecting cell cycle controls and apoptotic responses in endometrial cancer. The differential expressions of these miRNAs appear to depend on the mismatch repair status, a behavior characteristic of undifferentiated proliferative states in colon cancer. In addition, these miRNAs have been identified as important biomarkers in the detection and prognosis of prostate cancer. All this evidence shows that our comodules can indeed group miRNAs with cooperative roles and provide insights into their functional mechanisms.

We have calculated the enrichments of the comodules in GO biological process terms and KEGG pathways, using the hypergeometric test to assess their biological relevance. Twenty-six (53.1%) of the gene modules have at least one overrepresented GO biological process term with an FDR-corrected q-value less than 0.05. Taken together, the modules are enriched in 367 different GO biological processes and 57 KEGG pathways. The most frequently enriched biological processes are nuclear division, immune system process, microtubule-based processes, inflammatory response, response to external stimulus, cell cycle, and cell adhesion. When we performed the same test on a set of random modules, only 3.0% (2.4%) were enriched in any GO biological process. These observations demonstrate the power of our method to group genes that participate in the same processes or pathways.

To verify that the comodules are related to cancer, we also used a cancer benchmark data set consisting of 147 miRNAs from a review article (Koturbash et al., 2011). Each of these miRNAs was reported in the literature to be dysregulated in one or more cancers. Among them, 41 are relevant to ovarian cancer. This data set does not include any information from the TCGA ovarian cancer data that we used to identify comodules. Our comodules involve 117 different miRNAs, 52 of which belong to the benchmark set. This ratio is highly significant ($p = 1.1 \times 10^{-6}$). Even more important, 21 of the 52 shared miRNAs are related to ovarian cancer, with an enrichment significance of $p = 7.2 \times 10^{-6}$.

Furthermore, 69.4% of the modules contain at least two miRNAs that are known to be cancer related. For example, module 42 has seven miRNAs, five of which belong to the benchmark. Four of them (mir-199a-5p, mir-199b-3p, mir-127-3p, mir-214) are also reported to play roles in ovarian cancer. Further supporting this interpretation, the genes of this comodule are enriched in numerous cancer-related pathways, such as hedgehog signaling pathway, cell differentiation, TGFβ signaling pathway, and Wnt signaling pathway.

We explored cancer gene enrichment in the comodules using the large-scale, human-curated database of the Ingenuity Pathway Analysis (IPA) system. Most of the modules (63.3%) are highly enriched in cancer genes (multiple test corrected $p < 0.05$, as reported by the IPA system). Moreover, 10 modules are significantly enriched in ovarian cancer genes. For example, the 129 genes in module 23 include 64 cancer genes and 13 ovarian cancer genes. This module is overrepresented in several cancer-related pathways, including cell communication, TGFβ signaling pathway, and PPAR signaling pathway. These observations confirm that the miRNA-gene comodules discovered in this study play important roles in various cancers, especially ovarian cancers.

Based on the principles of our method, the genes in a comodule are likely to function together as a network, and the miRNAs in a comodule are likely to cooperatively target groups of networked genes. Using the IPA system and its database of molecular interactions, we found that the genes of a comodule are often organized into highly connected networks. Specifically, based on the IPA system, we found that 67.4% of the comodules are significantly connected, meaning they form at least one highly significant scored network with a cut-off larger than 30. These results, including the high levels of enrichment in known cancer miRNAs, cancer genes, and cancer-related processes and interactions, shed light on a miRNA-gene regulatory circuit that plays an important functional role in ovarian cancer and possibly other cancers.

6.4 The Sparse Multiple Block Partial Least Squares Method

In this section, we develop a method to identify subsets of three types of variables (CNV, DM, and ME) that jointly explain the expression of a subset of genes, in all or a subset of the samples. The union of the four subsets is termed a multidimensional regulatory module (MDRM) (Figure 6.1C). This approach captures the association between different types of variables (CNV-DM-ME) in terms of their joint impact on GE and facilitates the reconstruction of the regulatory network across different layers.

Let three input blocks or matrices $X_i \in \mathbb{R}^{K \times N_i}$ $(i = 1, 2, 3)$ represent three types of genomic profiling performed on the same K samples; that is, their column dimensions are $N_1/N_2/N_3$ for the CNV/DM/ME features. Let a response block/matrix $Y \in \mathbb{R}^{K \times M}$ represent the M gene expressions across the K samples. A multi-dimensional module is defined by satisfying the following criterion: "*the profiles extracted from n_i columns across k rows of X_i ($i = 1, 2, 3$) have strong associations with those from m columns across the same k rows of Y.*" Two columns have a "strong association" if they have similar and coherent patterns across their elements. To identify multi-dimensional modules, we developed a new method called sparse multi-block partial least squares (sMBPLS) regression.

Each input matrix X can be summarized by a single vector $\mathbf{t} = X\mathbf{w}$, which is a linear combination of all columns of X. The vector $\mathbf{w}$ contains the weights of the input columns. Similarly, $\mathbf{u} = Y\mathbf{q}$ is a "summary" vector of the columns of Y. Thus, the larger the covariance between the two "summary" vectors $\mathbf{t}$ and $\mathbf{u}$, the more similar the two matrices and the stronger their association. This measure of association can be extended to multiple blocks of input variables.

We use another weighted sum, $\mathbf{t} = \sum_{i=1}^{3} b_i \mathbf{t}_i$, to combine the three "summary vectors" $\mathbf{t}_i = X_i \mathbf{w}_i$ ($i = 1, 2, 3$) corresponding to the three sets of input variables. The weights $b_1, b_2, b_3 > 0$ indicate the contribution of each data block to the covariance structure of the input data. Therefore, the covariance between $\mathbf{t}$ and $\mathbf{u} = Y\mathbf{q}$ measures the strength of the association between all three input data blocks and the response data block. Maximizing the covariance between $\mathbf{t}$ and $\mathbf{u}$ can reveal biological associations between X_1, X_2, X_3 and Y, in turn leading to the discovery of a multi-dimensional module.

For this formulation to discover multidimensional modules containing a reasonable subset of the data columns, we need to add sparsity penalties to the objective function. We adopt the widely used lasso penalization (Tibshirani, 1996), which has been successfully applied in many fields. The lasso regularization of any vector $\mathbf{x}$, denoted P_λ, is defined as $P_\lambda(\mathbf{x}) = \sum_i p_\lambda(x_i) = \sum_i 2\lambda |x_i|$. The problem is formally expressed as follows:

$$\max_{\mathbf{w}_i, \mathbf{q}, \mathbf{t}_i, \mathbf{u}} \Omega(\mathbf{t}, \mathbf{u}, \mathbf{w}_i, \mathbf{q}, \mathbf{b}) = \mathrm{cov}(\mathbf{t}, \mathbf{u}) - \sum_{i=1}^{3} P_{\lambda_i}(\mathbf{w}_i) - P_{\lambda_4}(\mathbf{q})$$

$$\text{with } \mathbf{t}_i = X_i \mathbf{w}_i, \ \mathbf{u} = Y\mathbf{q}, \text{ and } \mathbf{t} = \sum_{i=1}^{3} b_i \mathbf{t}_i$$

$$\text{subject to } \|\mathbf{w}_i\|^2 = 1, \ \|\mathbf{q}\|^2 = 1, \ \|\mathbf{b}\|^2 = 1 \tag{6.1}$$

where the objective function $\Omega(\cdot)$ contains sparsity penalties for the weight vectors $\mathbf{w}_i$ ($i = 1, 2, 3$) and $\mathbf{q}$. To solve this problem, we propose the sparse multi-block PLS (sMBPLS) regression algorithm (Li et al., 2012). After we identify a module, we deflate the matrix by subtracting the signal from the current set of vectors that optimizes the objective function. The next module can be obtained by maximizing the objective function on the deflated matrix, and so on.

6.4.1 Results

We applied sMBPLS to the multidimensional TCGA ovarian cancer data and discovered 100 modules. On average, each module contains 30 samples, 45 CNV loci, 42 methylation marks, 5 microRNAs, and 44 genes. To evaluate the biological relevance of the multidimensional modules, we first test the functional homogeneity of each dimension individually. A set of genes is defined as functionally homogenous if it is enriched in at least one GO category with a q-value less than 0.05 (the q-value is the p-value after a false discovery rate multiple testing correction). The GE dimension is functionally homogenous in 36% of the modules, the CNV dimension is functionally homogenous with respect to CNV-harbored genes in 24% of the modules, and the DM dimension is functionally homogenous with respect to genes adjacent to methylation marks in 13% of the modules. The ME dimension is functionally homogenous with respect to microRNA in 9% of the modules, where microRNA function is predicted based on the functions of their target genes. These values are all significantly higher than the rates found in randomized modules, which are 1.24%, 1.48%, 2.32%, and 0.35%, respectively.

MDRMs have Synergistic Functions Across Multiple Dimensions

Although the individual dimensions of the discovered modules already exhibit significant levels of functional homogeneity, combining all four dimensions reveals even stronger synergy. When we analyzed all genes involved with the four variables, 48 of the 100 modules were found to be functionally homogenous, compared to 3.8% of randomized modules. This result highlights the power of multidimensional modules to group functionally relevant factors from different regulatory layers. Many of the identified modules are enriched in biological processes such as cell cycle, cell activation, and immune system, all of which indicate a possible involvement in cancer. Also, they contain many important genes (or microRNAs) known to be related to ovarian cancer. For example, module 37 includes four *HOX* family genes (*HOXB2*, *HOXB4*, *HOXB6*, and *HOXB7*) that have been extensively reported to be related to ovarian cancer (Widschwendter et al., 2009). The same module also contains a DNA methylation mark adjacent to *HOXA9*, which was reported to be significantly hypermethylated in ovarian cancer patients (Widschwendter et al., 2009). Finally, its member genes *FGF19*, *GAS2*, *BMP7*, *TNFSF11*, and *FGFR3* are all known to play important roles in tumor genesis and progression.

Next, we tested for transcriptional homogeneity in each dimension of the multidimensional modules. We used the 191 ChIP-seq profiles generated by the Encyclopedia of DNA Elements (ENCODE) Consortium. These data provide a set of potential targets for regulatory factors. A set of genes is defined as "transcriptional homogenous" if it is enriched in the targets for any regulatory factor with a q-value less than 0.05. We achieved similar results as those of functional homogeneity analysis. These modules are enriched of the transcription factors such as *SRF*, *STAT1*, and *H3K27me3*. *SRF* regulates the activity of many immediate-early genes and thereby participates in cell cycle regulation, apoptosis, cell growth, and cell differentiation. *STAT1* enhances inflammation and innate and adaptive immunity, triggering in most instances antiproliferative and pro-apoptotic responses in tumor cells. Particularly, *STAT1* negatively regulates the cell cycle by inducing *p21 WAF1/CIP1* in ovarian

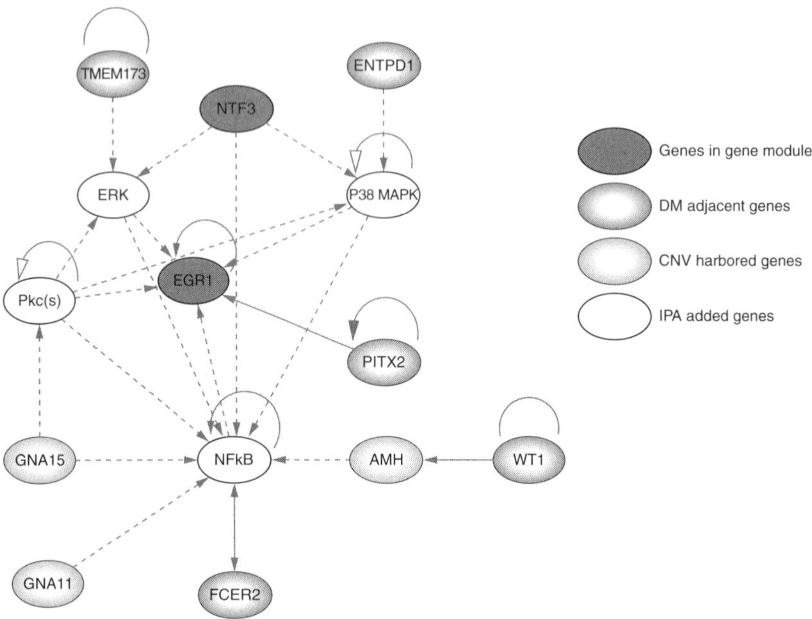

Figure 6.3 The molecular interaction network (constructed by IPA) centered on the gene *EGR1*. The solid lines represent direct interactions, and the dashed lines represent indirect interactions.

cancer (Burke et al., 1999). *H3K27me3* has been evaluated as a prognostic indicator for clinical outcome in patients with breast and ovarian cancers (Wei et al., 2008).

MDRMs Facilitate Regulatory Analysis

Our method has discovered sets of genomic features from different regulatory layers that are likely to have synergistic impacts on gene expression. To further elaborate on the relationships between the different features, we built molecular interaction networks using the IPA system. From each multi-dimensional module, we formed a set of genes involved in the CNV, DM, ME, and GE dimensions. Given this gene set, IPA constructs possible molecular networks based on literature-derived relationships between genes (or microRNAs) and computes a ranking score $-log(p)$ for each network. The p-value indicates the likelihood that the input genes would be found together due to random chance. All of the multidimensional modules produce highly statistically significant interaction networks ($p < 1.0\text{e-}20$), which again indicates the strength of their associations.

Now we provide in-depth descriptions of the heterogeneous regulatory networks that affect a key tumor suppressor gene (*EGR1*) in ovarian cancer. *EGR1* is a cancer-suppressing gene known to be down-regulated in ovarian cancer. The network derived from module 4 reveals a set of complex connections between heterogeneous factors that control the expression of *EGR1* (Figure 6.3). A direct regulation on *ERG1* comes from *PITX2*, a gene adjacent to a DNA methylation mark in our module and known to be essential for the expression of *EGR1* in rats (Suh et al., 2002). There are also multiple indirect influences on *EGR1* transmitted by the transcription factor *NFkB*, by *GNA15* and *GNA11* (genes hosted by CNVs in the module), and by *FCER2* (genes adjacent to the methylation marks in the module). In particular, *WT1* (regulated by DNA methylation) is known to positively regulate *AMH* (Nachtigal et al., 1998) (which is also regulated by a CNV), which in turn positively regulates *EGR1* via *NFkB*. In addition, *EGR1* is regulated by both *TMEM173* and *ENTD1* (adjacent to methylations) via *ERK* and *P38 MARK*, respectively. The disruption of multiple neighbor nodes to *EGR1* by different regulatory mechanisms highlights the complex nature of the controls on this key suppressor gene for ovarian cancer.

6.5 Discussion

Recent technology has enabled the simultaneous, multiplatform genomic profiling of biological samples, resulting in abundant multidimensional genomic data. Such data, coupled with other biological knowledge, are a rich resource in the exploration of cellular mechanisms. However, tools for the systematic analysis of such data are currently lacking and challenging to develop. A large number of tools have been designed to integrate just one or two types of data, and many of these have been applied to genomic data analysis with good results. In this chapter, we summarized recent progress on tools capable of integrating more than two data types, by describing three methods recently developed by the authors.

The first data analysis technique, joint NMF, discovers sophisticated modular structures (mdmodules) embedded in multi-dimensional genomics data. It breaks down massive data sets represented as matrices into small building blocks that exhibit similar patterns across certain rows and columns. This procedure provides two major advantages. First, by representing the raw data in terms of a smaller set of coherent features that hold across multiple data sets, we greatly reduce the complexity of the data and facilitate a global overview of its inherent structure. More importantly, this approach captures associations among different types of variables (mRNA, miRNA, and methylation). The mdmodules can identify associations between multiple regulatory levels, and reveal significantly disrupted pathways that would be ignored if only a single dimension of data were used. In addition, the mdmodules can stratify patients (samples) into clinically distinct groups. This analysis facilitates identification of the complex molecular mechanisms that underlie different clinical phenotypes.

Second, we developed a flexible and effective framework that integrates miRNA and gene expression profiles from the same patients, miRNA-gene networks, and gene interaction networks to identify miRNA-gene regulatory comodules. The comodules are helpful for exploring the combinatorial regulations and cooperative mechanisms that occur between miRNAs and genes. Testing the method on human ovarian cancer samples from the TCGA database and gene networks demonstrates that the comodules represent real cooperation between miRNAs and genes in several functions and phenotypes of cellular systems. The method can therefore provide new insights into the transcription and posttranscription regulatory organization of ovarian cancer. As genomic data sources increase in volume and diversity, our framework could provide new avenues for the systematic interpretation of combinatorial regulatory mechanisms. Moreover, the method is equally useful for many biological problems requiring the integration of several types of inputs. In particular, it is suitable for problems involving multidimensional genomic data (profiling multiple variables on the same set of samples) and independent priors identifying known relationships between the variables (e.g., miRNA-gene and gene-gene relationships).

Third, we proposed the sMBPLS regression method to identify regulatory modules in diverse types of genomic data measured on the same set of samples. Classical eQTL analysis can only be applied to relate one type of genomic marker (e.g., SNP) to GE. In contrast, sMBPLS can identify combinations of multiple types of genomic markers that jointly impact the expression of a set of genes. We have applied the sMBPLS method to a suite of genomic profiles from 230 ovarian cancer samples, including CNV, DM, microRNA, and GE data. The algorithm identified 100 modules, many of which display a high degree of functional homogeneity in at least one genomic dimension. If all dimensions of data are considered together, the modules exhibit an even greater degree of functional synergy. A detailed network view of individual modules reveals many genomic features that appear to be isolated if we only consider one type of data. By combining diverse types of data, sMBPLS discovers more coherent and connected regulatory networks. Furthermore, our method derives weights for the CNV, methylation, and microRNA dimensions in each module, indicating their relative contributions to the expressions of the genes.

We have demonstrated that multiple heterogeneous factors in a module can have combinatorial effects on GE. We should note that these modules (1) do not necessarily reflect direct causal mechanisms for GE but can be a good starting point for further study the underlying mechanisms and that (2) sMBPLS outperforms most existing algorithms in analyzing more than two data blocks.

In summary, we expect that there will soon be a rapid increase of multidimensional data and that developing methods for such data coupled with accumulated biological knowledge will become an active research area. Identifying coordinated patterns across multiple regulatory layers is a vital step towards revealing the real organizational principles of complex gene regulatory systems. In this chapter, we have reviewed three methods designed to reveal the coordinated patterns involving the epigenetic, transcription, and posttranscription levels, yet there are many other levels of regulatory controls that we have not yet attempted to include. Interpreting such complex patterns is still a major challenge, given our limited knowledge of multilayer coordination in biological systems. However, the rapid accumulation of multidimensional data and the discovery of increasingly complex networks will definitely drive a positive cycle of knowledge discovery. The methods described in this chapter can serve as powerful tools for the simultaneous integration of diverse data sources to discover multidimensional regulatory patterns. In future studies, it will be worthwhile to apply the proposed methods to more data sources simultaneously, uncovering more sophisticated "factories" that comprise many layers of regulatory factors.

Acknowledgment

This project was supported by NIH grants NHLBI MAPGEN U01HL108634, NIGMS R01GM105431 and NSF grant 0747475 to X.J.Z. and by National Natural Science Foundation of China grants 11001256 and 61379092, 61422309 to S.Z.

References

Burke, F., Smith, P.D., Crompton, M.R., Upton, C., and Balkwill, F.R. 1999. Cytotoxic response of ovarian cancer cell lines to IFN-gamma is associated with sustained induction of IRF-1 and p21 mRNA. *British Journal of Cancer*, **80**(8), 1236–1244.

Cancer Genome Atlas Research Network. 2008. Comprehensive genomic characterization defines human glioblastoma genes and core pathways. *Nature*, **455**, 1061–1068.

Corney, D.C., Flesken-Nikitin, A., Godwin, A.K., Wang, W., and Nikitin, A.Y. 2007. MicroRNA-34b and MicroRNA-34c are targets of p53 and cooperate in control of cell proliferation and adhesion-independent growth. *Cancer Research*, **67**(18), 8433–8438.

Koturbash, I., Zemp, F.J., Pogribny, I., and Kovalchuk, O. 2011. Small molecules with big effects: the role of the microRNAome in cancer and carcinogenesis. *Mutation Research*, **722**(2), 94–105.

Kutalik, Z., Beckmann, J.S., and Bergmann, S. 2008. A modular approach for integrative analysis of large-scale gene-expression and drug-response data. *Nat Biotechnol.*, **26**, 531–539.

Li, W., Zhang, S., Liu, C.C., and Zhou, X.J. 2012. Identifying multi-layer gene regulatory modules from multi-dimensional genomic data. *Bioinformatics*, **28**, 2458–2466.

Nachtigal, M.W., Hirokawa, Y., Enyeart-VanHouten, D.L., Flanagan, J.N., Hammer, G.D., and Ingraham, H.A. 1998. Wilms' tumor 1 and Dax-1 modulate the orphan nuclear receptor SF-1 in sex-specific gene expression. *Cell*, **93**(3), 445–454.

Suh, H., Gage, P.J, Drouin, J., and Camper, S.A. 2002. Pitx2 is required at multiple stages of pituitary organogenesis: pituitary primordium formation and cell specification. *Development (Cambridge, England)*, **129**(2), 329–337.

Tibshirani, R. 1996. Regression shrinkage and selection via the Lasso. *Journal of the Royal Statistical Society, Series B*, **58**(1), 267–288.

Wei, Y., Xia, W., Zhang, Z., Liu, J., Wang, H., Adsay, N.V., Albarracin, C., et al. 2008. Loss of trimethylation at lysine 27 of histone H3 is a predictor of poor outcome in breast, ovarian, and pancreatic cancers. *Molecular Carcinogenesis*, **47**(9), 701–706.

Weinstein, J.N., Myers, T.G., O'Connor, P.M., Friend, S.H., Fornace, A.J. Jr., Kohn, K.W., Fojo, T., et al. 1997. An information-intensive approach to the molecular pharmacology of cancer. *Science*, **275**, 343–349.

Widschwendter, M., Fiegl, H., Egle, D., Mueller-Holzner, E., Spizzo, G., Marth, C., Weisenberger, D.J., et al. 2007. Epigenetic stem cell signature in cancer. *Nature genetics*, **39**(2), 157–158.

Widschwendter, M., Apostolidou, S., Jones, A.A., Fourkala, E.O., Arora, R., Pearce, C.L., Frasco, M.A., et al. 2009. HOXA methylation in normal endometrium from premenopausal women is associated with the presence of ovarian cancer: a proof of principle study. *International Journal of Cancer*, **125**(9), 2214–2218.

Yang, X., Feng, M., Jiang, X., Wu, Z., Li, Z., Aau, M., and Yu, Q. 2009. miR-449a and miR-449b are direct transcriptional targets of E2F1 and negatively regulate pRb-E2F1 activity through a feedback loop by targeting CDK6 and CDC25A. *Genes & development*, **23**(20), 2388–2393.

Zhang, S., Li, Q., Liu, J., and Zhou, X.J. 2011. A novel computational framework for simultaneous integration of multiple types of genomic data to identify microRNA-gene regulatory modules. *Bioinformatics*, **27**, i401–i409.

Zhang, S., Liu, C.C., Li, W., Shen, H., Laird, P.W., and Zhou, X.J. 2012. Discovery of multi-dimensional modules by integrative analysis of cancer genomic data. *Nucleic Acids Res.*, **40**, 9379–9391.

Zhang, W., Zhu, J., Schadt, E.E., and Liu, J.S. 2010. A Bayesian partition method for detecting pleiotropic and epistatic eQTL modules. *PLoS Comput Biol*, **6**, e1000642.

7

A Latent Variable Approach for Integrative Clustering of Multiple Genomic Data Types

RONGLAI SHEN

Abstract

Clustering analysis is an unsupervised learning method that aims to group data into subsets based on the similarity among the data points. In gene expression microarray studies, clustering analysis has been used to identify biologically meaningful disease subtypes (samples in the same subtype share similar gene expression profiles), or to discover gene expression modules co-regulated through a similar mechanism. Recent technology advances have facilitated integrated genomic profiling across multiple platforms simultaneously including next-generation sequencing and high throughput array platforms. With the rapid accumulation of multidimensional datasets, there is an increasing need for robust and scalable statistical and computational methods for the analysis of such datasets. This book covers a wide range of topics on information integration of omics datasets. In this Chapter, we briefly review the recent advances in integrative clustering methods with a focus on introducing a latent variable approach developed by the authors and its extensions to perform variable selection, and to account for both discrete and continuous data types in the joint model. We also discuss several important questions in clustering analysis including how to determine the number of clusters and assess cluster stability. Finally, we demonstrate the application of the method to the TCGA colorectal cancer (CRC) dataset which includes whole-exome DNA-sequencing, Affymetrix SNP6.0 array, and RNA-sequencing in 276 CRC samples.

7.1 Introduction

Cancer is a heterogeneous disease. Identifying clinically relevant tumor subtypes that correlate with patient outcome (e.g., treatment response, survival) is an important yet difficult task. Over the past years, molecular classification based on microarray gene expression data has led to important discoveries of novel cancer subtypes (Perou et al., 1999; Alizadeh et al., 2000; Sorlie et al., 2001; Lapointe et al., 2003; Hoshida et al., 2003). However, the biological and therapeutic implications of most cancer expression subtypes remain largely

155

unknown due to the lack of understanding of the underlying disease mechanisms. In addition, expression changes may be related to cellular activities independent of tumorigenesis, and therefore leading to subtypes that may not be directly relevant for diagnostic and prognostic purposes.

In recent years, large scale comprehensive caner genome studies including the NCI-NHGRI Cancer Genome Atlas (TCGA) project is generating a large amount of data of multiple "omic" dimensions, where in addition to gene expression, genome-wide assessment of somatic mutations (changes in the DNA sequence), DNA copy number aberrations, DNA methylation are simultaneously obtained in the same biological samples. In isolation, none of the individual data type alone can completely capture the complexity of the cancer genome. Collectively, however, true oncogenic mechanisms may emerge as a result of joint analysis of multiple genomic data sources. As a result, integrative data analysis of multiple types of genomic alterations carry more power to characterize, classify, and predict outcomes than the conventional study involving a single genomic data type.

In this Chapter, we focus on class discovery problem given multiple omic data sets. Lock and Dunson (2013) introduced a Bayesian consensus clustering that permits a separate clustering of each data source that adhere loosely to an overall clustering. Yuan et al. (2011) applied a nonparametric Bayesian model for subtype discovery by integrating gene expression and copy number variation data. Among other applications, Alter and Golub (2004) introduced a joint matrix decomposition method to discover novel correlation between DNA replication initiation and RNA transcription during the yeast cell cycle by the integration of yeast genome-scale protein DNA-binding data with cell cycle mRNA expression.

In our previous work, we introduce a latent variable approach called iCluster for integrative clustering (Shen et al., 2009). Two extensions were introduced in our later publications. In Shen et al. (2012a), we introduced penalized latent variable methods for variable selection in the integrative clustering framework. In Mo et al. (2013), we introduced a hybrid Bayesian Expectation-Maximization approach for joint modeling both discrete (e.g., somatic mutation status) and continuous (e.g., copy number log-ratio, gene expression) data types, which broadened the scope of integration, and greatly increased our ability to harness the full potential of large-scale multidimensional cancer genomic datasets. The method has several successful applications to identify novel integrated subtypes of breast, colorectal, lung, endometrial, stomach cancers, and glioblastoma (Curtis et al., 2012; Mo et al., 2013; Cancer Genome Atlas Network, 2013a, 2012, 2013b, 2014; Shen et al., 2012b), through a comprehensive integration of somatic mutation, DNA copy number, DNA methylation, mRNA expression, and microRNA expression data. The R software package iCluster-Plus is available in Bioconductor (http://www.bioconductor.org/).

7.2 Methods

We first describe briefly some classic multivariate methods including K-means, Gaussian mixture model, principal component analysis (PCA), latent variable analysis, and the connections between them. This will provide a clear perspective for our proposal of using a latent variable approach for integrative clustering.

7.2.1 *K-means, Gaussian Mixture model, and Principal Component Analysis*

In standard K-means, given an initial set of K cluster assignments and the corresponding cluster centers, the procedure iteratively moves the centers to minimize the total within-cluster variance. For purposes of exposition, we assume the data are gene expression, although they could be any type of continuous genomic measurements. Let $\mathbf{X}$ denote the mean-centered expression data of dimension $p \times n$ with rows being genes and columns being samples. Given a partition C of the column space of $\mathbf{X}$ and the corresponding cluster mean vectors $\{\mathbf{m}_1, \ldots, \mathbf{m}_K\}$, the sample vectors $\mathbf{X} = \{\mathbf{x}_1, \ldots, \mathbf{x}_n\}$ are assigned cluster membership such that the sum of within-cluster squared distances is minimized:

$$\min \sum_{k=1}^{K} \sum_{C(i)=k} \|\mathbf{x}_i - \mathbf{m}_k\|^2 . \tag{7.1}$$

The cluster centers are subsequently recalculated successively based on the current partition. The algorithm iterates until the assignments do not change. It is well-known that Gaussian mixture model (GMM) is a "soft" version of K-means by probabilistic assignments of data points to cluster centers at each iteration of the Expectation-Maximization (EM) algorithm in the GMM estimation procedure (Hastie et al., 2009). Overviews of applying Gaussian mixture models to clustering analysis can be found in McLachlan and Peel (2002) and Fraley and Raftery (2002).

One of the main criticisms of K-means clustering is that the algorithm is sensitive to the choice of starting points; it can iterate to local minima rather than the global maximum. However, it has been recently shown that an approximate solution to the K-means loss function exists through eigenvalue decomposition (Zha et al., 2001). To see this, let $\mathbf{D} = (\mathbf{d}_1, \ldots, \mathbf{d}_K)'$ denote the cluster indicator matrix with the kth row being the indicator vector of cluster k normalized to have unit length:

$$\mathbf{d}'_k = (0, \ldots, 0, \underbrace{\frac{1}{\sqrt{n_k}}, \ldots \frac{1}{\sqrt{n_k}}}_{n_k}, 0, \ldots, 0), \tag{7.2}$$

where n_k is the number of samples in cluster k and $\sum_{k=1}^{K} n_k = n$. The objective is to obtain an optimal solution of the cluster assignment matrix $\mathbf{D}$ such that the within-cluster variance is minimized. Let $\mathbf{S} = \mathbf{X}'\mathbf{X}$ be the Gram matrix of the samples. The K-means loss function in (7.1) can be expressed as

$$trace(\mathbf{X}'\mathbf{X}) - trace(\mathbf{DX}'\mathbf{XD}'),$$

which is the total variance minus the between-cluster variance. Since the total variance is a constant given the data, it follows that minimizing (7.1) is equivalent to maximizing the between-cluster variance

$$trace(\mathbf{DX}'\mathbf{XD}'). \tag{7.3}$$

Now if we introduce $\mathbf{Z} = (\mathbf{z}_1, \ldots, \mathbf{z}_K)'$ as a continuous-valued version of $\mathbf{D}$ with the same orthonormality assumption, and rewrite the above maximization problem as

$$\max_{\mathbf{ZZ}'=I_K} trace(\mathbf{ZX}'\mathbf{XZ}'), \tag{7.4}$$

then this is equivalent to the eigenvalue decomposition of $\mathbf{S}$. Therefore the principal component analysis (PCA) provides a continuous analog to the solution of the cluster membership indicators in K-means clustering. The intuition behind this connection between PCA and K-means is that the PCs lie in a low-dimensional latent space where the original data are projected onto each of the first K principal directions such that the total variance is maximized. As such, any distinct subgroup structures will be automatically embedded in this set of orthogonal directional vectors. A later publication by Ding and He (2004) pointed out the redundancy in $\mathbf{Z}$ such that the K-means solution can be defined by the first $K - 1$ principal components (PCs).

7.2.2 PCA and Latent Variable Model

Principal component analysis can be considered a special case of factor analysis (Jolliffe, 2002). The basic idea is that a vector of observed random variables $\mathbf{x}$ of length p (in our case, gene expression) can be expressed as a linear function of a vector of hypothetical random variables $\mathbf{z}$ of length $K(\ll p)$. Specifically,

$$\mathbf{x} = \boldsymbol{\mu} + \mathbf{W}\mathbf{z} + \mathbf{e},$$

in which $\mathbf{W}$ is a factor loading matrix of dimension $p \times K$, and $\mathbf{e}$ is an independent error term with $E[\mathbf{e}] = \mathbf{0}$.

In the case of multivariate Gaussian case, PCA is equivalent to the maximum likelihood estimation (Tipping and Bishop, 1999; Lynn and McCulloch, 2000). Specifically, assume the prior distribution for $\mathbf{z}$ is $\mathbf{z} \sim N(\mathbf{0}, \mathbf{I})$, and the conditional distribution of the genomic data given the latent variables is

$$\mathbf{x}|\mathbf{z} \sim N(\boldsymbol{\mu} + \mathbf{W}\mathbf{z}, \sigma^2\mathbf{I}),$$

in which σ^2 is the variance component of the independent error term $\mathbf{e}$. Then the marginal distribution is

$$\mathbf{x} \sim N(\boldsymbol{\mu}, \boldsymbol{\Sigma}),$$

where $\boldsymbol{\Sigma} = \mathbf{W}\mathbf{W}^T + \sigma^2\mathbf{I}$. The posterior distribution $\mathbf{z}|\mathbf{x}$ is then

$$\mathbf{z}|\mathbf{x} \sim N(\boldsymbol{\Sigma}^{-1}\mathbf{W}^T(\mathbf{x} - \boldsymbol{\mu}), \sigma^2\boldsymbol{\Sigma}^{-1}).$$

The authors showed that the likelihood function is maximized when

$$\hat{\mathbf{W}} = \mathbf{U}(\boldsymbol{\Lambda} - \sigma^2\mathbf{I})^{1/2}\mathbf{R},$$

$$\hat{\sigma}^2 = \frac{1}{n - K} \sum_{l=K+1}^{n} \lambda_l,$$

where columns of $\mathbf{U}$ dare the first K eigenvectors of $\mathbf{S}$, $\boldsymbol{\Lambda}$ is a diagonal matrix with the diagonal being the corresponding eigenvalues $\lambda_1, \ldots, \lambda_K$, and $\mathbf{R}$ is an arbitrary $K \times K$ orthogonal rotation matrix. For simplicity, the authors choose $\mathbf{R} = \mathbf{I}$. Plugging in these maximum likelihood estimates, the posterior mean of $\mathbf{z}$ is

$$E[\mathbf{z}|\mathbf{x}] = (\boldsymbol{\Lambda} - \sigma^2\mathbf{I})^{1/2}\boldsymbol{\Lambda}^{-1/2}\mathbf{U}.$$

As $\sigma^2 \to 0$, the posterior mean reduces to $\mathbf{U}$, the first K eigenvectors of $\mathbf{S}$. The connections between K-means, PCA, and latent variable model motivate a subtype analysis using latent variables and has an immediate generalization to integrate multiple data types. We will illustrate this in the next two sections.

7.2.3 Subtype Analysis as a Latent Variable Analysis

The general assumption of latent variable models is that the observable phenomena are influenced by some underlying and unobserved causes. The use of latent variable models has been extensively used in economics, sociology and psychology to understand the underlying mechanisms that may have produced the observed data. For example, Dawkins (1989) analyzed the data on completion times of track races in different countries as a linear function of some latent athletic ability variable. Conway et al. (2002) used latent variables to study the relationship between working memory capacity, short term memory capacity, processing speed, and intelligence.

In cancer subtype analysis, we propose that latent variables represent the latent oncogenic processes underlying cancer genomic phenotypes (Shen et al., 2009). More specifically, it is assumed the diverse molecular phenotypes as measured by multiple genomic platforms (mutation, DNA copy number, gene expression) in a tumor are due to a set of common latent variables representing the underlying oncogenic mechanisms. For example, in colon cancer, genetic instability exists at two distinct levels: microsatellite instability (MIN) and chromosomal instability (CIN). In MIN tumors, the instability is observed at the nucleotide level and results in high rates of mutations in the DNA sequence. In most other tumors, the instability is observed at the chromosome level, resulting in losses and gains of whole chromosomes or large portions. A review of genetic instabilities in colon cancer can be found in Lengauer et al. (1998). The underlying cause of MIN lies in defective DNA mismatch repair (MMR) due to mutations in MMR genes. The molecular basis of CIN is less well understood. CIN is often associated with inactivating mutation of the tumor suppressor genes *TP53*, but not believed to be the primary cause. In Section 1.3, we will illustrate the use of a latent variable model for studying these distinct genetic pathways in colon cancer.

The connections between K-means, PCA and latent variable model discussed in Section 1.2.2 motivates the use of latent variables for subtype analysis, and provide an immediate generalization for integrative clustering of multiple omics data types. In the next section, we will show how this can be achieved.

7.2.4 Integrative Clustering Analysis Using Latent Variable Models

Let x_{ijt} denote the genomic variable associated with the jth ($j \in \{1, \ldots, p_t\}$) genomic feature in the ith ($i \in \{1, \ldots, n\}$) sample of the tth ($t \in \{1, \ldots, m\}$) data type. A genomic feature can be either a protein-coding gene (e.g., *TP53*) or a non-gene-centric element of interest (e.g., copy number region, CpG site, microRNA, etc.) depending on the data type. Let $\mathbf{z}_i = (z_{i1}, \ldots, z_{ik})'$ be a column-vector consists of k unobserved latent variables. We consider a class of generalized linear models (GLMs)

$$g(\mu_{ijt}) = \alpha_{jt} + \boldsymbol{\beta}_{jt}\mathbf{z}_i, \tag{7.5}$$

where $\mu_{ijt} = E(x_{ijt})$ and $g(\cdot)$ is the link function in GLM, α_{jt} is an intercept term, and $\boldsymbol{\beta}_{jt}$ is a length-k row vector of coefficients that determine the weights genomic variable j contributes to the latent variables. For a single data type $m = 1$, the Gaussian case of (7.5) corresponds to the Gaussian latent variable model discussed in Section 1.2.2. Shen et al. (2009) proposed a Gaussian case of (1.5) called iCluster for the integration of continuous data types. Specifically, the independent error term is assumed to follow a Gaussian distribution $\mathbf{e}_t \sim N(\mathbf{0}, \boldsymbol{\Psi}_t)$ for $t = 1, \ldots m$, where $\boldsymbol{\Psi} = \mathrm{diag}(\sigma_{1t}^2, \ldots, \sigma_{p_tt}^2)$. The Gaussian case is appropriate for integrating continuous data types (e.g., microarray gene expression, log of total copy number (logR) from SNP arrays).

In a follow-up paper, Mo et al. (2013) accounted for both continuous and discrete data types in the same model. The extended method called iCluster+ broadened the scope of integrative analysis for the inclusion of somatic mutation data by massively parallel sequencing. Suppose the *tth* data type contains mutation information. More specifically, $x_{ijt} = 1$ if gene j is mutated in tumor i and $x_{ijt} = 0$ if gene j is wild type in tumor i. we consider the following logistic regression

$$\mathrm{logit}\,P(x_{ijt} = 1|\mathbf{z}_i) = \alpha_{jt} + \boldsymbol{\beta}_{jt}z_i,$$

where $P(x_{ijt} = 1|\mathbf{z}_i)$ is the probability of gene j mutated in patient i given the value of the latent variable vector $\mathbf{z}_i$.

The joint log-likelihood of (x_{ijt}, z_i) can be written as

$$\ell(x_{ijt}, z_i; \alpha_{jt}, \boldsymbol{\beta}_{jt}) = \sum_{i=1}^{n}\sum_{t=1}^{m}\sum_{j=1}^{p_t} \left\{ \log f(x_{ijt}|z_i; \alpha_{jt}, \boldsymbol{\beta}_{jt}) + \log f(z_i) \right\},$$

where the summation is due to the conditional independence assumption of x_{ijt} given z_i. Here $f(z_i)$ is the density function of the standard multivariate normal distribution $N(\mathbf{0}, \boldsymbol{I}_k)$, and the conditional density $f(x_{ijt}|z_i; \alpha_{jt}, \boldsymbol{\beta}_{jt})$ has the form of normal, Bernoulli, multinomial, or Poisson density function depending on the type of genomic variable.

However, z_i is not observed in our model. A Markov chain simulation is particularly suitable for the latent variable formulation (Tanner and Wong, 2010). The basic idea is to replace the expression in the parameter updates shown above by its expectation with respect to z_i given x_{ijt} by repeatedly sampling (typically we use 1,000 draws) the latent variable z_i from its joint posterior distribution

$$\mathbf{z}_i |\cdot \propto f(z_i) \prod_{t=1}^{m} \prod_{j=1}^{p_t} f\left(x_{ijt} \mid \alpha_{jt}, \boldsymbol{\beta}_{jt}, z_i\right), \quad i = 1, \ldots, n,$$

using a random walk Metropolis-Hasting algorithm (Robert and Casella, 2004; Liu, 2001). Then we calculate parameters updates by their sample averages over the repeated draws. Sample clusters are assigned by the values of the latent variables. We use K-means clustering to divide the n samples into g clusters using k latent variables where $g = k + 1$. In the null model case where $k = 0$ (intercept only), this implies that all samples belong to one cluster.

Variable Selection in Integrative Clustering

In high-dimensional data, sparsity in the coefficient vector $\boldsymbol{\beta}_{jt} = (\beta_{jt1}, \ldots, \beta_{jtk})$ greatly impact the interpretability of the latent variables and our ability to identify the important genomic features that have significant contributions to the latent variables. By sparsity, we mean some of these coefficients will be estimated to be exactly zero. For example, a zero value in the rth element of $\boldsymbol{\beta}_{jt}$ means the jth genomic feature belonging to data type t has no weight on the rth latent variable. If the entire vector is estimated to be zero, then the corresponding genomic feature has no contribution to the latent variables and is considered non-informative. In order to identify genomic variables that make important contribution to the latent variables, we apply the L_1-norm penalty (Tibshirani, 1996) and consider the following penalized likelihood estimation:

$$\max_{\alpha_{jt}, \boldsymbol{\beta}_{jt}} \ell(x_{ijt}, z_i; \alpha_{jt}, \boldsymbol{\beta}_{jt}) - \sum_{t=1}^{m} \sum_{j=1}^{p_t} \lambda_t \|\boldsymbol{\beta}_{jt}\|_1$$

where $\|\boldsymbol{\beta}_{jt}\|_1 = |\beta_{jt1}| + \cdots + |\beta_{jtk}|$ is the L_1-norm penalty and λ_ts are non-negative tuning parameters. Due to the singularity of the L_1-norm penalty at $\beta_{jtr} = 0$, some estimated β_{jtr} will be exactly zero. If the entire vector $\boldsymbol{\beta}_{jt}$ is zero, then the corresponding genomic variable is effectively removed from the model. Allowing the tuning parameter (λ_t) to vary by t is motivated by the fact that different data types have different degrees of sparsity (number of informative features). In addition to variable selection, the Lasso-type procedures have also been shown to have good prediction ability in both finite samples and asymptotic situations (Bickel et al., 2009; Greenshtein and Ritov, 2004).

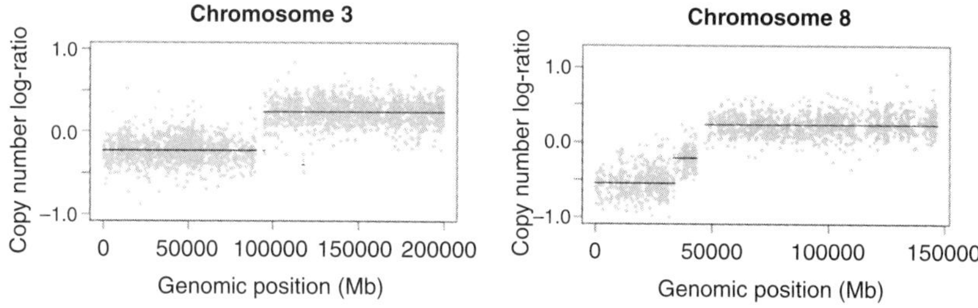

Figure 7.1 Illustration of copy number probe-level data from a lung tumor sample (Chitale et al., 2009). Log-ratios of copy number (tumor versus normal) on chromosome 3 and 8 are displayed. Log-ratio great than zero indicates copy number gain and log-ratio below zero indicates loss. Black line indicates the segmented value using the circular binary segmentation method (Olshen et al., 2004; Venkatraman and Olshen, 2007).

A Monte-Carlo Newton-Raphson algorithm can be used for model estimation (see Mo et al. (2013)).

In Shen et al. (2012a), we considered additional sparsity constraint including elastic net (Zou and Hastie, 2005) and fused lasso (Tibshirani et al., 2005). The elastic net penalty takes the following form

$$\sum_{t=1}^{m}\sum_{j=1}^{p_t}\lambda_{1t}\|\boldsymbol{\beta}_{jt}\|_1 + \sum_{t=1}^{m}\sum_{j=1}^{p_t}\lambda_{2t}\|\boldsymbol{\beta}_{jt}\|_2^2 \tag{7.6}$$

where $\|\boldsymbol{\beta}_{jt}\|_2^2 = \beta_{jt1}^2 + \cdots + \beta_{jtk}^2$ is the L_2 norm. The combination of L_1 (lasso) and L_2 (ridge) has an additional advantage by shrinking coefficients of correlated features toward each other, and thus encourages a grouping effect toward selecting highly correlated features together. The elastic net penalty also tends to be more numerically stable than lasso in practice. The use of fused lasso is motived by the fact that copy number aberrations tend to occur in contiguous regions along chromosomal positions (Figure 7.1). The fussed lasso takes the following form

$$\sum_{t=1}^{m}\sum_{j=1}^{p_t}\lambda_{1t}\|\boldsymbol{\beta}_{jt}\|_1 + \sum_{t=1}^{m}\sum_{j=2}^{p_t}\lambda_{2t}\|\boldsymbol{\beta}_{jt} - \boldsymbol{\beta}_{(j-1)t}\|_1, \tag{7.7}$$

in which first term encourages sparseness while the second term encourages smoothness along index j (genome position). The fused Lasso penalty is particularly suitable for DNA copy number data where contiguous regions of a chromosome tend to be altered in the same fashion (Tibshirani and Wang, 2008).

7.2.5 Model Selection

For model selection, we use a two-step approach. In the first step, given k (number of latent variables), we estimate the lasso penalty parameters $\{\lambda_t\}_{t=1}^{m}$ that minimize a modified Bayesian information criterion (BIC)

$$BIC = -2\ell + d \cdot log(n)$$

where d is the total number of nonzero parameter estimates (Wang and Zhu, 2008; Zou et al., 2007). It can be calculated directly after fitting the model with little extra computational effort. However, to determine the optimal combination of the penalty parameter values, a very large search space needs to be covered. We used an efficient sampling method based on a uniform design (UD) (Fang and Wang, 1994). A theoretical advantage of the uniform design over an exhaustive grid search is the uniform space-filling property that avoids wasteful computation at nearby sampling points. In the next step, we choose the best k based on a deviance ratio

$$\text{dev.ratio} = \frac{\ell_{k,\hat{\lambda}} - \ell_0}{\ell_{sat} - \ell_0},$$

where $\hat{\lambda} = \{\hat{\lambda}_t\}_{t=1}^{m}$ are the combination of penalty parameter values that give the minimal BIC under k, ℓ_0 is the log-likelihood under the null (intercept) model, and ℓ_{sat} denotes the log-likelihood under the saturated model (a model with a latent variable per sample). The deviance ratio metric can be interpreted as the percentage of variation explained by the current model, and k is thus chosen to achieve an optimal value of the deviance ratio.

In the model selection process, an exhaustive grid search for the optimal combination of the penalty parameters is inefficient and computationally prohibitive. We propose to use a uniform design the uniform design (UD) to generate good lattice points from the search domain, a similar strategy adopted by Wang et al. (2008). The uniform design seeks experimental points that scattered uniformly across the search domain for efficient sampling of tuning parameter combinations. A key theoretical advantage of UD over the traditional grid search is the uniform space filling property that avoids wasteful computation at close-by points. Let D be the search region. Using the concept of discrepancy that measures uniformity on $D \subset R^d$ with arbitrary dimension d, which is basically the Kolmogorov statistic for a uniform distribution on D, Fang and Wang (1994) point out that the discrepancy of the good lattice point set from a uniform design converges to zero with a rate of $O(n^{-1}(\log n)^d)$, where n (a prime number) denotes the number of generated points on D. They also point out that the sequence of equi-lattice points on D has a rate of $O(n^{-1/d})$ and the sequence of uniformly distributed random numbers on D has a rate of $O(n^{-1/2}(\log\log n)^{1/2})$. Thus the uniform design has an optimal rate for $d \geq 2$.

7.3 Applications

7.3.1 A Motivating Example

We will first show a motivating example where an integrated analysis of copy number and gene expression is far more insightful than separate analysis of each data type. Pollack et al. (2002) used customized microarrays to generate measurements of DNA copy number and mRNA expression in parallel for 37 primary breast cancer and 4 breast cancer cell line samples. Here x_{ij1} refers to the gene expression variable and x_{ij2} refers to the DNA copy number (log-ratio of tumor versus normal) corresponding to gene j and sample i. In this example, both data types have gene-centric measurement by design.

Here we present a simple example that involves the well-known HER2 breast cancer subtype characterized by a narrow (focal) DNA amplification event on chromosome 17 harboring the oncogene *HER2*. A heatmap of the genomic features on chromosome 17 is plotted in Figure 7.2. In the heatmap, rows are genes ordered by their genomic position and columns are samples ordered by separate hierarchical clustering (panels A) or by integrative clustering (panels B). There are two main subclasses in the 41 samples: the cell line subclass (samples labeled in red) and the HER2 tumor subclass (samples labeled in green). It is clear in Figure 7.2A that these subclasses cannot be distinguished well from separate hierarchical clustering analyses.

Separate clustering followed by manual integration as depicted in Figure 7.2A remains the most frequently applied approach to analyze multiple omics data sets in the current literature due to its simplicity and the lack of a truly integrative approach. However, Figure 7.2A clearly shows its lack of a unified system for cluster assignment and poor correlation of the outcome with biological and clinical annotation. In Shen et al. (2012a), we use simulation study to show that separate clustering can fail drastically in estimating the true number of clusters, classifying samples to the correct clusters, and selecting cluster-associated features.

The heatmap in Figure 7.2B demonstrates the superiority of integrative clustering in correctly identifying the subgroups (vertically divided by solid black lines). From left to right, cluster 1 (samples labeled in red) corresponds to the breast cancer cell line subgroup, distinguishing cell line samples from tumor samples. Cluster 2 corresponds to the *HER2* tumor subtype (samples labeled in green), showing concordant amplification in the DNA and overexpression in mRNA at the *HER2* locus (chr 17q12). This subtype is associated with poor

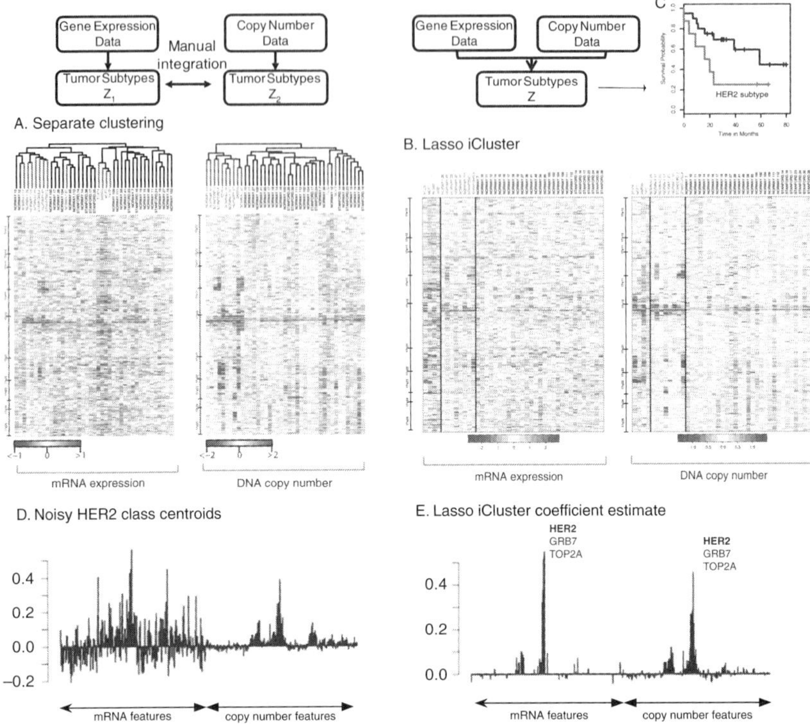

Figure 7.2 A motivating example using the Pollack data set to demonstrate that a joint analysis using the lasso iCluster method outperforms the separate clustering approach in subtype analysis given DNA copy number and mRNA expression data. (A) Heatmap with samples ordered by separate hierarchical clustering. Rows are genes and samples are columns. Samples labeled in red are breast cancer cell line samples. Samples labeled in green are HER2 breast tumors. (B) Heatmap with samples ordered by integrative clustering using the lasso iCluster method. (C) Kaplan-Meier plot indicates the HER2 subtype has poor survival outcome. (D) Standard cluster centroid estimates. (E) Sparse coefficient estimates under the lasso iCluster model.

survival as shown in Figure 7.2C. Cluster 3 (samples labeled in black) did not show any distinct patterns, though a pattern may have emerged if there were additional data types such as DNA methylation.

The motivation for variable selection by sparse constraints on the coefficient estimates is illustrated by Figure 7.2E. It clearly reveals the *HER2*-subtype genes (including *HER2, GRB7, TOP2A*). By contrast, the standard cluster centroid estimation is flooded with noise (Figure 7.2D), revealing an inherent problem with clustering methods without regularization.

7.3.2 Integrated Subtype Analysis of the TCGA Colorectal Cancer Dataset

Next we illustrate a whole-genome multiple-platform clustering using the TCGA colorectal cancer data set.

Data Pre-Processing and Filtering

In the TCGA colorectal carcinoma (CRC) study (TCGA Network, 2012), genome-scale analyses of 276 CRC samples were conducted and these included analysis of whole-exome sequencing, DNA copy number, promoter DNA methylation, and mRNA expression. Data processing methods are described in TCGA Network (2012); Mo et al. (2013) and are briefly described here. For whole-exome DNA sequencing, somatic mutations were called by the MuTect algorithm (Cibulskis et al., 2013) using paired tumor and matched normal sample pairs. The set of 461 most significantly mutated genes as identified by the MutSig algorithm was included for clustering. A gene by sample matrix of binary values (1-mutated, 0-WT) was generated for clustering. Copy number segmented data (germline CNV removed) based on Affymetrix SNP 6.0 array was used. Dimension reduction was performed for to obtain non-redundant copy number regions as described in Mo et al. (2013). Median absolute deviation was used to select the top 2,000 most variable CpG sites in methylation beta value for clustering. For RNA-sequencing, RSEM was used to quantify the transcript abundance of a gene. The top 2,000 most variable genes in mRNA expression were used for the analysis.

7.3.3 Model Selection

An integrative clustering analysis using iCluster+ was applied to a subset of 189 tumors that had all four data types available. A three-cluster solution was chosen (Figure 7.3A) by selection criteria described in Section 1.2.5.

Subsampling methods can be used to assess clustering stability. Here we generated 100 random subsamples (without replacement) of the original data with a sampling ratio of 0.8 to preserve the general structure (Ben-Hur et al., 2002). We ran iCluster+ for $k = 1 - 5$ (number of latent variables) on each of the subsampled datasets using the lasso parameter tuning procedure as described earlier. For a randomly chosen pair of subsampled datasets, a measure of agreement between the clustering is computed using the adjusted Rand index (Rand, 1971). When two partitions agree perfectly, the adjusted Rand index is 1. When the agreement is random (the degree of agreement equals the expected value under a model of randomness from the generalized hypergeometric distribution),

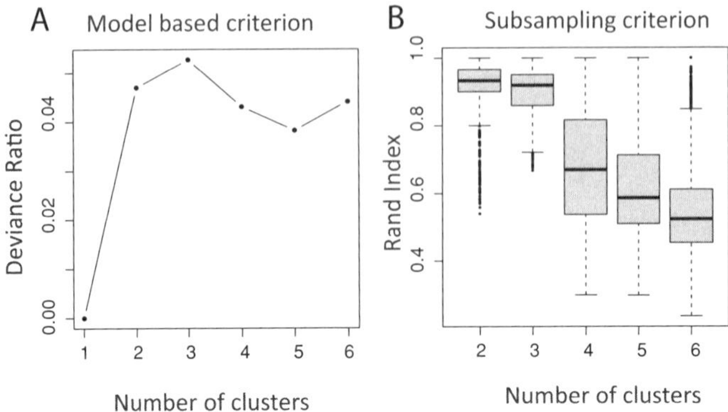

Figure 7.3 Determine the number of clusters in the colorectal cancer data analysis based on A. deviance ratio and B. subsampling.

the adjusted Rand index is 0. We exhausted all possible pairs and the box-plots of adjusted Rand Indices computed for the pairs for $g = 2 - 6$ (number of clusters) were shown in Figure 7.3B. The mean adjusted Rand index averaged over all possible pairs for $g = 2 - 6$ is 0.92 0.90, 0.68, 0.62, 0.54. For 3-cluster solutions, cluster assignments of any random subsamples of the colorectal cancer dataset are on average 90% concordant, suggesting stability of the results. Based on the stability plot, 3-cluster is the demarcating point which is consistent with our proposed criteria based on percent explained variation.

7.3.4 Integrated Subtypes of Colorectal Cancer

Figure 7.4 reveals that the integrated clusters are associated with the distinct types of genetic instability. The majority of subclass 2 tumors correspond to the hyper-mutated class previously described in TCGA Network (2012) which shows a substantially higher mutational burden than the rest of the tumors. These tumors are associated with microsatellite instability (MIN), CpG island methylator phenotype (CIMP) as defined in TCGA Network (2012), and *BRAF* mutation. Furthermore, tumors in this class have distinctly few *TP53* mutations ($p = 10^{-8}$) and are chromosomally stable. By contrast, clusters 1 and 3 display the classic chromosome instability (CIN) phenotype, with an average of over 26% of the genome altered in copy number. The differences is highly significant ($p < 10^{-15}$) and no significant differences in tumor purity as measured by the ABSOLUTE algorithm (Carter, 2012) was observed ($p = 0.13$). Among the TP53 mutated/CIN high group, patient samples were divided by chr8q amplification status (clusters 1 is clearly distinguished by chr8q amplification).

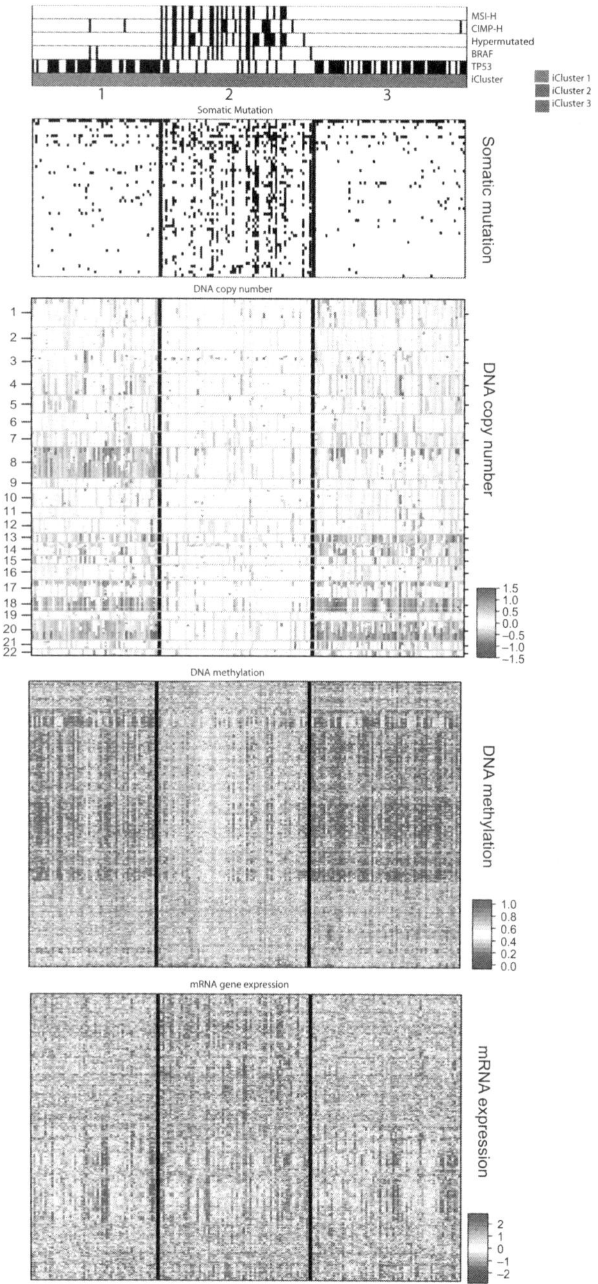

Figure 7.4 Three molecular clusters identified in 189 TCGA colorectal cancer samples. Heat map display of lasso-selected cluster-discriminant features. Rows are features and columns are tumor samples. The first panel shows genes that are mutated (black) or not mutated (white) in each cluster; the second panel shows genomic regions amplified (red) or deleted (blue); the third panel shows genes hypermethylated (red) or hypomethylated (blue); and the fourth panel shows genes overexpressed (red) or underexpressed (blue) in each cluster.

7.4 Discussion

The rapidly increasing size of integrated genomic datasets creates an unprecedented opportunity to study cancer biology and discover biomarkers and therapeutic targets in a novel way. In this Chapter, we presented a model-based approach for integrative clustering. A key aspect to the approach is the introduction of latent variables to represent the spectrum of the underlying disease driving factors, and that these latent variables facilitate the discovery of biological properties that lead to the phenotypic diversity observed with different cancer genomes.

A limitation to the current method lies in that statistical inference (significance test and confidence intervals for the "final" model selected) is not straightforward due to the computational complexity and the use of penalized regression. Statistical inference after model selection (post-model selection inference) is important yet very challenging even for simple linear regression. For a recent discussion on the topic, see Leeb and Potscher (2006); Berk et al. (2012). In Berk et al. (Berk et al., 2012), an intuitive and practical strategy has been proposed, but it fails when the dimension of the variables is high. To our knowledge, there is no clear solution in statistics for post-model selection inference problem for high-dimensional data. It is an interesting problem that may warrant future investigation.

References

Alizadeh, Ash A., Eisen, Michael B., Davis, Eric E., et al. 2000. Distinct types of diffuse large B-cell lymphoma identified by gene expression profiling. *Nature*, **403**(February), 503–511.

Alter, Orly, and Golub, Gene H. 2004. Integrative analysis of genome-scale data by using pseudoinverse projection predicts novel correlation between DNA replication and RNA transcription. *Proceedings of the National Academy of Sciences of the United States of America*, **101**(47), 16577–16582.

Ben-Hur, A., Elisseeff, A., and Guyon, I. 2002. A stability based method for discovering structure in clustered data. *Pacific Symposium on Biocomputing*.

Berk, R., Brown, L., Buja, A., Zhang, K., and Zhao, L. 2012. Valid post-selection inference. *Annals of statistics*, **41**, 802–837.

Bickel, P.J., Ritov, Y., and Tsybakov, A.B. 2009. Simultaneous analysis of lasso and dantzig selector. *Annals of statistics*, **37**, 1705–32.

Cancer Genome Atlas Network. 2012. Comprehensive genomic characterization of squamous cell lung cancers. *Nature*, **489**, 519–525.

Cancer Genome Atlas Network. 2013a. Comprehensive Molecular Profiling of Lung Adenocarcinoma. *Nature*, **In Press**.

Cancer Genome Atlas Network. 2013b. Integrated genomic characterization of endometrial carcinoma. *Nature*, **497**, 67–73.

Cancer Genome Atlas Network. 2014. Comprehensive Molecular Characterization of Gastric Adenocarcinoma. *Nature*, **In Press**.

Carter, S.L. et al. 2012. Absolute quantification of somatic DNA alterations in human cancer. *Nature*, **30**, 413–421.

Chitale, D., Gong, Y., Taylor, B.S., Broderick, S., Brennan, C., Somwar, R., Golas, B., Wang, L., Motoi, N., Szoke, J., Reinersman, J.M., Major, J., Sander, C., Seshan, V.E., Zakowski, M.F., Rusch, V., Pao, W., Gerald, W., and Ladanyi, M. 2009. An integrated genomic analysis of lung cancer reveals loss of DUSP4 in EGFR-mutant tumors. *Nature*, **28**(31), 2773–83.

Cibulskis, Kristian, Lawrence, Michael S., Carter, Scott L., Sivachenko, Andrey, Jaffe, David, Sougnez, Carrie, Gabriel, Stacey, Meyerson, Matthew, Lander, Eric S., and Getz, Gad. 2013. Sensitive detection of somatic point mutations in impure and heterogeneous cancer samples. *Nature Biotechnology*, **31**, 213–219.

Conway, A. R. A., Cowan, N., Bunting, M. F., Therriault, D. J., and Minkoff, S. R. B. 2002. A latent variable analysis of working memory capacity, short term memory capacity, processing speed, and general fluid intelligence. *Intelligence*, **30**, 163–183.

Curtis, Christina, Shah, Sohrab P., Chin, Suet-Feung, Turashvili, Gulisa, Rueda, Oscar M., Dunning, Mark J., Speed, Doug, Lynch, Andy G., Samarajiwa, Shamith, Yuan, Yinyin, Graf, Stefan, Ha, Gavin, Haffari, Gholamreza, Bashashati, Ali, Russell, Roslin, McKinney, Steven, Langerod, Anita, Green, Andrew, Provenzano, Elena, Wishart, Gordon, Pinder, Sarah, Watson, Peter, Markowetz, Florian, Murphy, Leigh, Ellis, Ian, Purushotham, Arnie, Borresen-Dale, Anne-Lise, Brenton, James D., Tavare, Simon, Caldas, Carlos, and Aparicio, Samuel. 2012. The genomic and transcriptomic architecture of 2,000 breast tumours reveals novel subgroups. *Nature*, **advance online publication**.

Dawkins, B. 1989. Multivariate Analysis of National Track Records. *The American Statistician*, **43**, 110–15.

Ding, Chris H. Q., and He, Xiaofeng. 2004. *K*-means clustering via principal component analysis. In: *ICML*. ACM International Conference Proceeding Series, vol. 69. ACM.

Fang, K.T., and Wang, Y. 1994. *Number theoretic methods in statistics*. London, UK: Chapman abd Hall.

Fraley, C., and Raftery, A. E. 2002. Model-based clustering, discriminant analysis, and density estimation. *Journal of American Statistical Association*, **97**, 611–631.

Greenshtein, E., and Ritov, Y. 2004. linear predictor-selection and the virtue of over-parametrization. *Bernoulli*, **10**, 971–988.

Hastie, T., Tibshirani, R., and Friedman, J. 2009. *elements of statistical learning 2nd edition*. Springer.

Hoshida, Y., Nijman, S.M., Kobayashi, M., Chan, J.A., Brunet, J.P., Chiang, D.Y., Villanueva, A., Newell, P., Ikeda, K., Hashimoto, M., Watanabe, G., Gabriel, S., Friedman, S.L., Kumada, H., Llovet, J.M., and Golub, T.R. 2003. Integrative transcriptome analysis reveals common molecular subclasses of human hepatocellular carcinoma. *Cancer Research*, **69**, 7385–92.

Jolliffe, I. T. 2002. *Principal Component Analysis*. New York, NY: Springer.

Lapointe, J., Li, C., Higgins, J.P., van de Rijn, M., Bair, E., Montgomery, K., Ferrari, M., Egevad, L., Rayford, W., Bergerheim, U., Ekman, P., DeMarzo, A.M., Tibshirani, R., Botstein, D., Brown, P.O., Brooks, J.D., and Pollack, J.R. 2003. Gene expression profiling identifies clinically relevant subtypes of prostate cancer. *Proceedings of the National Academy of Sciences*, **101**, 811–6.

Leeb, H., and Potscher, B.M. 2006. Can one estimate the conditional distribution of post-model-selection estimators? *Annals of statistics*, **34**, 2554–2591.

Lengauer, Christoph, Kinzler, Kenneth W., and Vogelstein, Bert. 1998. Genetic instabilities in human cancers. *Nature*, **396**(6712), 643–649.

Liu, J. 2001. *Monte Carlo Strategies in Scientific Computing*. New York, NY: Springer.

Lock, E. F., and Dunson, D. B. 2013. Bayesian consensus clustering. *Bioinformatics*, **29**(20), 2610–2616.

Lynn, H.S., and McCulloch, C. E. 2000. Using principal component analysis and correspondence analysis for estimation in latent variable models. *Journal of American Statistical Association*, **95**, 561–72.

McLachlan, G. J., and Peel, D. 2002. *Finite mixture models*. New York, NY: John Wiley & Sons.

Mo, Qianxing, Wang, Sijian, Seshan, Venkatraman E., Olshen, Adam B., Schultz, Nikolaus, Sander, Chris, Powers, R. Scott, Ladanyi, Marc, and Shen, Ronglai. 2013. Pattern discovery and cancer gene identification in integrated cancer genomic data. *Proceedings of the National Academy of Sciences*, **110**(11), 4245–4250.

Olshen, A.B., Venkatraman, E.S., Lucito, R., and Wigler, M. 2004. Circular binary segmentation for the analysis of array-based DNA copy number data. *Biostatistics*, **5**(4), 557–572.

Perou, Charles M., Jeffrey, Stefanie S., van de Rijn, Matt, et al. 1999. Distinctive gene expression patterns in human mammary epithelial cells and breast cancers. *Proceedings of the National Academy of Sciences*, **96**, 9212–9217.

Pollack, J. R., Sørlie, T., Perou, C. M., et al. 2002. Microarray analysis reveals a major direct role of DNA copy number alteration in the transcriptional program of human breast tumors. *Proceedings of the National Academy of Sciences*, **99**(October), 12963–12968.

Rand, W. M. 1971. Objective criteria for the evaluation of clustering method. *Journal of the American Statistical Association*, **66**, 846–885.

Robert, C. P., and Casella, G. 2004. *Monte Carlo Statistical Methods*. New York, NY: Springer.

Shen, R., Olshen, A.B., and Ladanyi, M. 2009. Integrative clustering of multiple genomic data types using a joint latent variable model with application to breast and lung cancer subtype analysis. *Bioinformatics*, **25**(22), 2906–2912.

Shen, R., Wang, S., and Mo, Q. 2012a. Sparse Integrative Clustering of Multiple Omics Data Sets. *Annals of Applied Statistics*, **7**, 269–294.

Shen, Ronglai, Mo, Qianxing, Schultz, Nikolaus, Seshan, Venkatraman E., Olshen, Adam B., Huse, Jason, Ladanyi, Marc, and Sander, Chris. 2012b. Integrative Subtype Discovery in Glioblastoma Using iCluster. *PLoS ONE*, **7**, e35236.

Sorlie, Therese, Perou, Charles M., Tibshirani, Robert, et al. 2001. Gene expression patterns of breast carcinomas distinguish tumor subclasses with clinical implications. *Proceedings of the National Academy of Sciences*, **98**(September), 10869–10874.

Tanner, M. A., and Wong, W. H. 2010. From EM to data augmentation: the emergence of MCMC Bayesian computation in the 1980s. *Statistical science*, **25**(4), 506–516.

TCGA Network. 2012. Comprehensive Molecular Characterization of Human Colon and Rectal Cancer. *Nature*, **In Press**.

Tibshirani, R. 1996. Regression shrinkage and selection via the lasso. *Journal of the Royal Statistical Society: Series B (Statistical Methodology)*, **58**, 267–288.

Tibshirani, R., and Wang, P. 2008. Spatial smoothing and hot spot detection for CGH data using the fused lasso. *Biostatistics*, **1**(9), 18–29.

Tibshirani, R., Saunders, M., Rosset, S., Zhu, J., and Knight, K. 2005. Sparsity and smoothness via the fused lasso. *Journal of the Royal Statistical Society Series B.*, **67**, 91–108.

Tipping, Michael E., and Bishop, Christopher M. 1999. Probabilistic Principal Component Analysis. *Journal of the Royal Statistical Society: Series B (Statistical Methodology)*, **61**(3), 611–622.

Venkatraman, E. S., and Olshen, A. B. 2007. A faster circular binary segmentation algorithm for the analysis of array CGH data. *Bioinformatics*, **23**(6), 657–663.

Wang, S., Nan, B., Zhu, J., and Beer, D. 2008. Doubly Penalized Buckley-James Method for Survival Data with High-Dimensional Covariates. *Biometrics*, **64**, 132–140.

Wang, Sijian, and Zhu, Ji. 2008. Variable selection for model-based high-dimensional clustering and its application to microarray data. *Biometrics*, **64**, 440–448.

Yuan, Yinyin, Savage, Richard S., and Markowetz, Florian. 2011. Patient-Specific Data Fusion Defines Prognostic Cancer Subtypes. *PLoS Comput Biol*, **7**(10), e1002227+.

Zha, Hongyuan, He, Xiaofeng, Ding, Chris H. Q., Gu, Ming, and Simon, Horst D. 2001. Spectral Relaxation for K-means Clustering. Pages 1057–1064 of: *NIPS*. MIT Press.

Zou, H., and Hastie, T. 2005. Regularization and variable selection via the elastic net. *Journal of the Royal Statistical Society: Series B (Statistical Methodology)*, **67**, 301–320.

Zou, H., Hastie, T., and Tibshirani, R. 2007. On the degrees of freedom of the lasso. *Annals of statistics*, **35**, 2173–2192.

8

Penalized Integrative Analysis of High-Dimensional Omics Data

JIN LIU, XINGJIE SHI, JIAN HUANG, AND SHUANGGE MA

Abstract

With omics data, results generated from single-dataset analysis are often unsatisfactory. Integrative analysis methods conduct the joint analysis of data from multiple independent studies or on multiple correlated responses, can effectively increase power, and outperform single-dataset analysis and meta-analysis. In this chapter, we review the penalized integrative analysis methods under both the homogeneity and heterogeneity models. Computation using the coordinate descent approach is described. We also discuss several important extensions. The analysis of a genome-wide association study demonstrates the applicability of reviewed methods.

8.1 Introduction

In the study of complex diseases such as cancer, cardiovascular diseases, and autoimmune diseases, profiling studies are now routinely conducted, generating *"large d, small n"* data, where the number of omics features profiled (genes, SNPs, methylation loci, etc.) d is much larger than the sample size n. Many different types of analyses can be conducted. For example, Chapters 3 and 4 were focused on identifying meaningful networks. In this chapter, our analysis goal is to identify a small subset of omics measurements that are associated with disease outcomes or phenotypes. Such measurements are also referred to as "markers" in the literature and in this chapter. Statistically, this is a variable selection problem. The development of integrative analysis methods has been partly motivated by the following examples.

8.1.1 Example 1

Consider the analysis of data generated in multiple independent studies with comparable designs. For example, in Ma et al. (2011), four pancreatic cancer data sets are collected and analyzed. The four data sets were generated in four independent studies, all having a case-control design, collecting mRNA gene expression measurements and searching for genes associated with the risk of pancreatic cancer. In high-dimensional omics studies, it has been recognized that the results generated in single-data-set analysis often have unsatisfactory properties such as low reproducibility. Among many possible contributing factors, the most important one is perhaps the small n. Multi-data-set analysis can effectively increase sample size and outperform single-data-set analysis (Guerra and Goldstein, 2009). This perspective has been explained in multiple chapters of this book. When the designs of multiple studies are "close enough", it can be reasonable to expect that they identify the same set of markers.

8.1.2 Example 2

Here the settings are similar to those of example 1. The difference is that the designs of multiple studies can be significantly different. As an example, Liu et al. (2014a) collected four independent studies on kidney, liver, prostate, and stomach cancers, respectively. All studies have a case-control design and mRNA gene expression measurements. The analysis goal is to identify markers associated with multiple types of cancers while allowing for cancer type-specific markers. Multi-cancer markers, such as those related to apoptosis, DNA repair, and cell cycle, represent the more essential features of cancer, whereas type-specific markers determine the unique characteristics of each cancer type. A significant difference between Example 1 and 2 is that here it is no longer sensible to expect the same set of markers across multiple studies.

8.1.3 Example 3

Consider the analysis of a single data set with multiple correlated outcome variables. An example is the genetic association study of complex traits in heterogeneous stock mice from the Wellcome Trust Case Control Consortium (WTCCC) (Valdar et al., 2006a,b). This data resource, which also includes pedigree information, was based on an advanced intercross mating among eight inbred strains over 50 generations of random mating, since the use of pseudo-random breeding for over 50 generations should result in an average distance between recombinants of <2 cM. The average linkage disequilibrium (LD), as measured by R^2 between adjacent markers, is 0.62 (Lorenz et al., 2011). Valdar et al. (2006a) conducted a genomewide association study to search for genetic markers associated with the phenotypes of immunological disorders. We are particularly interested in two outcome variables: CD4/CD8 ratio and

CD4:CD3. The CD4/CD8 ratio, which is also known as the T-Lymphocyte Helper/Suppressor Profile, is a basic laboratory test in which the percentage of CD3-positive lymphocytes in the blood positive for CD4 (T helper cells) and CD8 (a class of regulatory T cells) is calculated and compared. CD4:CD3 is another clinical index for immunological diseases. Both indices are related to the diagnosis of immunological diseases. The two outcome variables are statistically correlated, and the biological mechanisms behind them are expected to be related. More details on this data set and analysis results are provided in Section 8.4. The sets of markers associated with the two outcomes are expected to be similar, though not necessarily identical.

In the preceding examples, multi-data-set analyses need to be conducted – in example 3, each outcome variable combined with the omics measurements can be viewed as a separate "data set." In omics studies in general, the value of pooling information and conducting multi-data-set analysis has been well noted in the literature (Guerra and Goldstein, 2009) and in the previous chapters and is not reiterated here. In multi-data-set analysis, "classic" meta-analysis methods – as partly described in Chapters 1 and 2 – initially analyze each data set separately and then pool summary statistics (p-values, odds ratios, hazard ratios, etc.) across data sets. As an alternive, integrative analysis methods pool and analyze raw data from multiple studies. A few methods (e.g., intensity methods with gene expression data) search for transformations that make multiple data sets comparable and hence can be directly combined. However, such transformations do not necessarily exist, and even if they exist, they are computationally expensive and need to be conducted on a case-by-case basis. Thus we have been focusing on integrative analysis methods that do not require the direct comparability of different data sets. In a series of recent studies (Liu et al., 2014a; Ma et al., 2009, 2011), integrative analysis methods have been developed and shown to outperform single-data-set and meta-analysis methods with more accurate marker identification and better prediction. The intuition is that in meta-analysis, marker selection is conducted with each data set separately, and there is no "borrowing information" across data sets in this step. In contrast, integrative analysis pools raw data and takes multiple data sets simultaneously into consideration in marker selection and hence can be more effective.

In the rest of this chapter, we describe multiple integrative analysis methods that can be used to analyze the data set described in example 3, which has two correlated continuous outcome variables and high-dimensional SNP measurements. With very minor modifications, these methods can be used to conduct multi-data-set analysis as under examples 1 and 2. In a sense, example 3 is more complicated than examples 1 and 2 because of the correlations among multiple outcomes. With limited space, we refer the readers to published studies on examples 1 and 2 and omit them in this chapter. The proposed methods are directly applicable to others types of omics measurements, such as mRNA gene expression, methylation, microRNA, and others. Applications to other types of outcome variables (categorical, censored survival, etc.) are discussed in Section 8.5.1. We use the penalization technique for marker selection, which has been extensively adopted for omics data. In the literature, alternative techniques, such as regularized thresholding (Ma et al., 2009) and sparse boosting (Huang et al., 2012), have also been adopted.

8.2 Data and Model Settings

Let $M(> 1)$ be the number of outcome variables, n be the number of subjects, and d be the number of SNPs (also referred to as covariates). Denote $y^1, \ldots, y^M$ as the M outcome variables and x as the length-d vector of covariates. For $m = 1, \ldots, M$, assume that y^m is associated with x via the linear regression model $y^m = x'\beta^m + \epsilon^m$, where $\beta^m = (\beta_1^m, \ldots, \beta_d^m)'$ is the vector of regression coefficients and ϵ^m is the random error. An intercept term is needed if the data are not normalized. Here a linear model is assumed. Accommodating other models and other types of outcome variables is discussed in Section 8.5.1. For the heterogeneous stock mice data, the same set of covariates is shared by the two outcome variables. With minor revisions, the methods described here can also accommodate partially matched sets of covariates for different outcomes. For subject $i(= 1, \ldots, n)$, let $y_i = (y_i^1, \ldots, y_i^M)'$ be the length-M vector of response variables. Denote x_{ij} as the value of covariate j. Then the $M \times Md$ covariate matrix has the form $X_i = (U_{i1}, \ldots, U_{id})$, where $U_{ij} = x_{ij} I_M$ and I_M is the $M \times M$ identity matrix. The overall regression coefficient vector is then $\beta = (\beta_1', \ldots, \beta_d')'$, where $\beta_j = (\beta_j^1, \ldots, \beta_j^M)'$.

Consider the least squares-type loss function

$$R(\beta, \Omega) = \frac{1}{2n} \sum_{i=1}^{n} (y_i - X_i \beta)' \Omega (y_i - X_i \beta).$$

Ω is a $M \times M$ symmetric matrix that does not depend on β. The most essential requirement for Ω is that under the "classic" asymptotics with $n \gg d$, minimizing $R(\beta, \Omega)$ with respect of β leads to a consistent estimate. Consider the following specific examples. (1) $\Omega = \Omega_m$, which has its (m, m)th element equal to 1 and all others 0. Then R is simply the least squares loss function for the mth outcome. (2) $\Omega = \Omega_I$, which is the $M \times M$ identity matrix. Here the correlations among outcomes are ignored. This loss function is the most sensible under examples 1 and 2, which have independent data. It can still lead to consistent estimates under example 3, although there may be a loss of efficiency. (3) $\Omega = \Omega_\Sigma = \Sigma^{-1}$, where Σ is the covariance matrix for ϵ^ms. With normally distributed errors, this leads to the negative log-likelihood function.

8.2.1 Homogeneity and Heterogeneity Models

As evidenced in several chapters, integrative analysis has multiple aspects. In this chapter, we focus on marker selection. With multiple outcomes (or data sets), there are two possible scenarios (Liu et al., 2014a). The first is the *homogeneity model*, under which all outcomes share the same set of markers. The second is the *heterogeneity model*, under which different outcomes are allowed to have different sets of markers. For a SNP, under the homogeneity model, we only need to determine whether it is associated with outcomes at all. That is, only *one-dimensional* selection is needed. In contrast, under the heterogeneity model, we need to first determine whether a SNP is associated with any outcome at all. In addition, we also need to determine which outcome(s) it is associated with. That is, *two-dimensional* selection is needed. Under examples 1 and 3, both models can be sensible, depending on how "similar" the outcomes (data sets) are. Under example 2, only the heterogeneity model is sensible.

The heterogeneity model includes the homogeneity model as a special case and can be more flexible. At first sight it may seem that the homogeneity model is not needed. However, under sensible settings, the homogeneity model demands one fewer dimension of selection, which can lead to reduced computational complexity. In addition, reinforcing the same set of markers may lead to improved marker selection accuracy.

8.3 Penalized Marker Selection

For marker selection, we adopt penalization, which has been extensively used in the analysis of high-dimensional omics data. Regularized thresholding and sparse boosting have also been used to conduct similar analyses. We conjecture that other variable selection techniques, such as stochastic variable search selection, can also be adapted. As a specific example, we use the minimax concave penalty (MCP, Zhang (2010)), which is defined as $\rho(t; \lambda, \gamma) = \lambda \int_0^{|t|} \left(1 - \frac{x}{\lambda\gamma}\right)_+ dx$, wheres $\lambda > 0$ is the tuning parameter, as with many other penalties, and $\gamma > 0$ is the regularization parameter. Under simple data and model settings, MCP has been shown to outperform some penalties, such as lasso, while having comparable performance as, for example, SCAD.

8.3.1 Marker Selection under the Homogeneity Model

Consider the penalized estimate $\hat{\beta} = argmin_\beta\{R(\beta, \Omega) + P_{gMCP}(\beta)\}$. We adopt the estimation-based marker selection strategy, where a nonzero component of $\hat{\beta}$ indicates an association between the corresponding SNP and outcome. Stability- and inference-based identification have also been adopted in the literature but are not discussed here. P_{gMCP} is the *group MCP* (gMCP) penalty defined as

$$P_{gMCP}(\beta) = \sum_{j=1}^{d} \rho(||\beta_j||_2; \lambda, \gamma) \tag{8.1}$$

where $|| \cdot ||_2$ is the ℓ_2-norm. Denote $\Sigma^{-1/2}$ as the $M \times M$ symmetric matrix that satisfies $(\Sigma^{-1/2}\Sigma^{-1/2})\Sigma = I_M$. Denote $\tilde{U}_{ij} = \Sigma^{-\frac{1}{2}}U_{ij}$, $\tilde{U}_j = (\tilde{U}'_{1j}, \ldots, \tilde{U}'_{nj})'$, and $\Sigma_j = n^{-1}\tilde{U}'_j\tilde{U}_j$. Liu et al. (2013b) proposes replacing $||\beta_j||_2$ with $||\beta_j||_{\Sigma_j} = ||\beta'_j\Sigma_j\beta_j||_2$. This SNP-wise normalization can reduce computational complexity. It can also be achieved by orthogonalizing measurements corresponding to β_j first and then applying the penalty in (8.1). Under examples 1 and 2 with multiple data sets, it is possible that some omics covariates are not measured in all data sets. To accommodate such a scenario, λ in (8.1) needs to be revised as $\sqrt{d_j}\lambda$, where d_j is the number of data sets that measure the jth covariate.

The gMCP penalty in (8.1) is a sum of d individual penalties, with one for each SNP. For a specific SNP, the penalty is a *composite* of the outer MCP and inner ℓ_2-norm, which can be viewed as (the square root of) a ridge penalty. The outer penalty determines whether a SNP is associated with outcomes at all. As only one dimension of selection is needed, the inner penalty is ridge, which has the shrinkage but not selection property. The computational aspects of gMCP, such as computational algorithm and tuning parameter selection, are similar to but simpler than those described in Section 8.3.2 and are omitted here.

8.3.2 Marker Selection under the Heterogeneity Model

For the heterogeneity model, Liu et al. (2013b) describes a two-step method that first applies gMCP and then MCP to the markers selected by gMCP only. The methodological and computational aspects of the two-step method are relatively simple. In the following, we describe two alternative methods. The loss function is still $R(\beta, \Omega)$.

Composite Penalization

Consider the *1-norm gMCP*, where the penalty is defined as

$$P_{1-norm\ gMCP}(\boldsymbol{\beta}) = \sum_{j=1}^{d} \rho(\|\beta_j\|_1; \lambda, \gamma) \tag{8.2}$$

$\|\beta_j\|_1 = \sum_{m=1}^{M} |\beta_j^m|$ is the ℓ_1-norm of β_j. This is also a composite penalty. The outer MCP determines whether a SNP is associated with any outcome at all. For a selected SNP, the inner lasso penalty (ℓ_1-norm) determines which outcome(s) it is associated with. As with (8.1), under examples 1 and 2 with multiple data sets and missing measurements, λ needs to be revised.

In (8.2), lasso is adopted as the inner penalty. Unless under very strong conditions, lasso does not have the selection consistency property (Zhang and Huang, 2008). Motivated by such a consideration, we adopt the *composite MCP (cMCP)* method, where the penalty is

$$P_{cMCP}(\boldsymbol{\beta}) = \sum_{j=1}^{d} \rho \left(\sum_{m=1}^{M} \rho(\beta_j^m; \lambda_2, b); \lambda_1, a \right) \tag{8.3}$$

The inner and outer MCP penalties may have different tuning parameters and regularization parameters. The rationale is similar to that for 1-norm gMCP.

Computational algorithm for cMCP

Denote Y as the length-Mn vector composed of y_is and X as the $Mn \times Md$ design matrix composed of X_is. Denote $\Omega^{1/2}$ as the $M \times M$ symmetric matrix that satisfies $\Omega^{1/2}\Omega^{1/2} = \Omega$. With $\tilde{Y} = \text{diag}(\Omega^{1/2}, \ldots, \Omega^{1/2})Y$ and $\tilde{X} = \text{diag}(\Omega^{1/2}, \ldots, \Omega^{1/2})X$, the loss function can be rewritten as $R(\boldsymbol{\beta}, \Omega) = \frac{1}{2n}\|\tilde{Y} - \tilde{X}\boldsymbol{\beta}\|_2^2$.

Coordinate descent has been adopted in a large number of recent penalization studies and is used here. We expect that alternative computational algorithms, for example, gradient descent and parallel coordinate descent, are also applicable in penalized integrative analysis. For the penalty in (8.3), consider the group coordinate descent algorithm. This algorithm is iterative and optimizes over the regression coefficients of one SNP at a time. It cycles through all SNPs, and the overall iteration is repeated multiple times until convergence. The key to this algorithm is the update for a single group of coefficients that corresponds to a SNP. Unfortunately, the cMCP does not have a simple form for updating individual groups. To tackle this problem, we adopt an approximation approach. Consider update with the jth group. By taking the first-order Taylor series approximation about β_j and evaluating at the current estimate $\tilde{\beta}_j$, the penalty as a function of β_j^k is approximately proportional to $\tilde{\lambda}_{jk}|\beta_j^k|$, where

$$\tilde{\lambda}_{jk} = \rho'\left(\sum_{m=1}^{M}\rho(|\tilde{\beta}_j^m|;\lambda_2, b);\lambda_1, a\right)\rho'(|\tilde{\beta}_j^k|;\lambda_2, b) \tag{8.4}$$

For update with each β_j^k, we have an explicit solution:

$$\hat{\beta}_j^k = f_{cMCP}(z;\lambda) = S_1(z, \lambda) \tag{8.5}$$

with $S_1(z, \lambda) = sgn(z)(|z| - \lambda)_+$, and z and λ defined later in this chapter.

Consider the following algorithm. With fixed tuning and regularization parameters,

1. Initialize $s = 0$, the estimate $\boldsymbol{\beta}^{(0)} = (\beta_1^{(0)'}, \ldots, \beta_d^{(0)'})' = (0, \ldots, 0)'$, and the vector of residuals $r = \tilde{Y} - \tilde{X}\boldsymbol{\beta}^{(0)}$;
2. For $j = 1, \ldots, d$,
   1. calculate $\tilde{\lambda}_{jk}$ according to expression (8.4)
   2. calculate $z_j = n^{-1}\tilde{X}[j]'r + \beta_j^{(s)}$. $\tilde{X}[j]$ is the submatrix of $\tilde{X}$ that corresponds to β_j
   3. for $k = 1, \ldots, M$, update $\beta_j^{k(s+1)} \leftarrow f_{cMCP}(z_j^k; \tilde{\lambda}_{jk})$, where z_j^k is the kth element of z_j
   4. update $r \leftarrow r - \tilde{X}[j](\beta_j^{(s+1)} - \beta_j^{(s)})$
   5. update $s \leftarrow s + 1$
3. Repeat step 2 until convergence.

For this algorithm as well as those for the other estimates defined in this chapter, we use the ℓ_2-norm of the difference between two consecutive estimates smaller than 0.001 as the convergence criterion. In the data analysis in Section 8.4, convergence is achieved for all methods within 50 iterations. The loss function $R(\boldsymbol{\beta}, \Omega)$ has a simple least squares form. It is regular in the sense of Tseng (2001). The penalties (8.2) and (8.3) are separable. Thus the coordinate descent estimate converges to a coordinate-wise minimum of the loss function, which is also a stationary point. Stronger conditions are needed to achieve the global minimum.

Tuning and regularization parameters

Tuning and regularization parameters play an important role in determining properties of the estimates. As an example, consider penalty (8.3), which contains tuning parameters λ_1 and λ_2 and regularization parameters a and b. As under simpler settings, λ_1 and λ_2 control the sparsity, with larger values leading to fewer identified markers. Smaller values of a and b are better at retaining the unbiasedness of the MCP penalty for large coefficients, but they also have the risk of creating objective functions with a nonconvexity problem that are difficult to optimize and yield solutions that are discontinuous with respect to λ_1 and λ_2. It is therefore advisable to choose values of a and b that are big enough to avoid this problem, but not too big. As suggested in Breheny and Huang (2011) and Zhang (2010), we have experimented with a few values for a and b, particularly including 1.8, 3, 6, and 10.

In numerical study, we select tunings using V-fold cross-validation. As only simple calculations are involved in the coordinate descent algorithm and not many overall iterations are needed, cross-validation is in fact affordable. In addition, we only need to search over a small number of values for a and b, which significantly reduces computational cost. With (λ_1, λ_2), one may expect that their values cannot become very small because there are regions that are not locally convex (Breheny and Huang, 2009, 2011). The criteria over nonlocally convex regions may go up and down. To avoid the unexpectedness of such regions, we select (λ_1, λ_2) where the criterion first goes up. We refer to Breheny and Huang (2011) for related discussions.

Sparse Group Penalization

Consider the *sparse group MCP*, where the penalty is defined as

$$P_{sparse\ gMCP}(\boldsymbol{\beta}) = \sum_{j=1}^{d} \left\{ \rho(||\beta_j||_2; \lambda_1, a) + \sum_{k=1}^{M} \rho(|\beta_j^k|; \lambda_2, b) \right\}. \quad (8.6)$$

Notation has similar definitions as in the earlier sections. As described under the homogeneity model, $||\beta_j||_2$ can be replaced by $||\beta_j||_{\Sigma_j}$ to simplify calculation, and λ_1 may need to be revised as $\sqrt{d_j}\lambda_1$ under examples 1 and 2 to accommodate partially matched omics measurement sets. This method also uses two penalties to achieve two-dimensional selection. Different from (8.3), it uses the *sum* as opposed to composition. The first penalty determines whether a SNP is associated with any outcome, and the second penalty determines which outcome(s) it is associated with. This method has been motivated by the recent studies on sparse group lasso (Simon et al., 2013), which is a sum of group lasso and lasso penalties. We adopt MCP because of the concern over the variable selection inconsistency of lasso. In addition, the current multi-outcome (data set) settings are more complicated than those in Simon et al. (2013).

Computational algorithm

For computation, we again resort to coordinate descent. Using the same notations as for the cMCP, the overall objective function is

$$
\frac{1}{2n}||\tilde{Y} - \sum_{j=1}^{d} \tilde{X}[j]\beta_j||_2^2 + \sum_{j=1}^{d} \left\{ \rho(||\beta_j||_2; \sqrt{d_j}\lambda_1, a) + \sum_{k=1}^{M} \rho(|\beta_j^k|; \lambda_2, b) \right\}
\tag{8.7}
$$

Here we keep the $\sqrt{d_j}$ term to be thorough. Given the group parameter vectors β_l ($l \neq j$) fixed at their current estimates $\tilde{\beta}_l$, we seek to minimize the objective function with respect to β_j. Consider only the terms that involve β_j:

$$
\frac{1}{2n}||r_{-j} - \tilde{X}[j]\beta_j||_2^2 + \rho(||\beta_j||_2; \sqrt{d_j}\lambda_1, a) + \sum_{k=1}^{M} \rho(|\beta_j^k|; \lambda_2, b)
\tag{8.8}
$$

where $r_{-j} = \tilde{Y} - \sum_{l \neq j} \tilde{X}[l]\tilde{\beta}_l$. The first-order derivative of expression (8.8) is

$$
-\frac{1}{n}\tilde{X}[j]'r_{-j} + \frac{1}{n}\tilde{X}[j]'\tilde{X}[j]\beta_j
$$

$$
+ \frac{\beta_j}{||\beta_j||_2} \begin{cases} \sqrt{d_j}\lambda_1 - \frac{||\beta_j||_2}{a}, & \text{if } ||\beta_j||_2 \leq a\sqrt{d_j}\lambda_1 \\ 0, & \text{if } ||\beta_j||_2 > a\sqrt{d_j}\lambda_1 \end{cases} + t
\tag{8.9}
$$

where

$$
t = \left(sgn(\beta_j^1) \begin{cases} \lambda_2 - \frac{|\beta_j^1|}{b}, & \text{if } |\beta_j^1| \leq b\lambda_2 \\ 0, & \text{if } |\beta_j^1| > b\lambda_2 \end{cases}, \ldots, sgn(\beta_j^M) \right.
$$

$$
\left. \times \begin{cases} \lambda_2 - \frac{|\beta_j^M|}{b}, & \text{if } |\beta_j^M| \leq b\lambda_2 \\ 0, & \text{if } |\beta_j^M| > b\lambda_2 \end{cases} \right)'
$$

By setting expression (8.9) as zero, we have:

$$
-z_j + g(\beta_j)\beta_j + t = 0,
\tag{8.10}
$$

where $z_j = n^{-1}\tilde{X}[j]'r_{-j} = (z_j^1, \ldots, z_j^M)'$,

$$
g(\beta_j) = \left(1 + \frac{1}{||\beta_j||_2} \right) \begin{cases} \sqrt{d_j}\lambda_1 - \frac{||\beta_j||_2}{a}, & \text{if } ||\beta_j||_2 \leq a\sqrt{d_j}\lambda_1 \\ 0, & \text{if } ||\beta_j||_2 > a\sqrt{d_j}\lambda_1 \end{cases}.
$$

Denote z_j^k as the kth element of z_j. First, fix $g(\beta_j)$ at the current estimate $\tilde{\beta}_j$. We use g short for $g(\tilde{\beta}_j)$. The kth element in Eq. (8.10) can be rewritten as:

$$-\frac{z_j^k}{g} + \beta_j^k + sgn(\beta_j^k) \begin{cases} \frac{\lambda_2}{g} - \frac{|\beta_j^k|}{bg}, & \text{if } |\beta_j^k| \leq b\lambda_2 \\ 0, & \text{if } |\beta_j^k| > b\lambda_2 \end{cases} = 0 \qquad (8.11)$$

The solution to Eq. (8.11) is

$$\widehat{g\beta_j^k} = \begin{cases} \frac{S_1(z_j^k, \lambda_2)}{1 - \frac{1}{bg}}, & \text{if } |z_j^k| \leq b\lambda_2 g \\ z_j^k, & \text{if } |z_j^k| > b\lambda_2 g. \end{cases}$$

Here $S_1(z, \lambda) = sgn(z)(|z| - \lambda)_+$. For $k = 1, \ldots, M$, set $u_k = \widehat{g\beta_j^k}$ and $u = (u_1, \ldots, u_M)'$. Taking u back into its definition,

$$\beta_j + \frac{\beta_j}{\|\beta_j\|_2} \begin{cases} \sqrt{d_j}\lambda_1 - \frac{\|\beta_j\|_2}{a}, & \text{if } \|\beta_j\|_2 \leq a\sqrt{d_j}\lambda_1 \\ 0, & \text{if } \|\beta_j\|_2 > a\sqrt{d_j}\lambda_1 \end{cases} = u \qquad (8.12)$$

Solving expression (8.12) leads to

$$\hat{\beta}_j = \begin{cases} \frac{a}{a-1} S_2(u, \sqrt{d_j}\lambda_1), & \text{if } \|u\|_2 \leq a\sqrt{d_j}\lambda_1 \\ u, & \text{if } \|u\|_2 > a\sqrt{d_j}\lambda_1 \end{cases}, \qquad (8.13)$$

where $S_2(u, \lambda) = (1 - \lambda/\|u\|_2)_+ u$. Thus solving Eq. (8.10) amounts to iteratively calculating Eqs. (8.11) to (8.13) until convergence. The overall coordinate descent iteration and discussions on the tuning and regularization parameters are similar to those with cMCP and are not reiterated here.

Remarks

Composite penalization and sparse group penalization are two families of methods that can conduct two-dimensional selection. Although they have different forms, the intuitions behind them are similar. Theoretically and numerically, there is very limited direct comparison of their performance. In our data analysis in Section 8.4, they lead to different marker sets and estimates for the heterogeneous stock mice data. Both involve two penalties. For cMCP and sparse gMCP, both penalties are MCP. Conceptually, the MCP can be replaced by other penalties, such as members of the lasso family, SCAD, and others. It is also possible to "mismatch" penalties, for example, a composite or sum of MCP and bridge. However, we conjecture that the mismatched penalties can be computationally difficult. Computer programs written in R are available at http://works.bepress.com/shuangge/ for some of the proposed penalization methods. More portable software is still under development.

8.3.3 Accommodating the Interplay Among SNPs

When constructing the regression model, we assume the simple additive SNP effects. The penalties described previously are the sums of individual penalties, with one for each SNP. Overall, the SNP effects are interchangeable. Some SNPs (especially those "nearby") can have coordinated biological functions, whereas others have different functions. Statistically, some SNP measurements have high correlations. In single data set/outcome analysis, multiple statistical methods have been developed to accommodate the interplay among omics measurements. See, for example, the network-based analysis in Horvath (2011). In the following, we consider pathway and network analyses, which have been popular in omics data analysis. In the other contexts of integrating omics data, pathway and network analyses are also discussed in Chapters 3–5, 13, and 19 of this book.

Pathway analysis

Pathways are composed of genes with coordinated functions. With SNP data, we consider the "SNPs within pathways" hierarchical structure but note that a more comprehensive structure should be the two-level "SNPs within genes" and "genes within pathways" structure. In addition, we take a simplified perspective and view pathways as static clusters of SNPs.

Denote c^* as the number of pathways, and $c(j)$ as the pathway membership of SNP j. SNPs not annotated can be put in a "super pathway," excluded from analysis, or clustered using statistical approaches. For pathway $c = 1, \ldots, c^*$, denote $\beta_c = \{\beta'_j : c(j) = c\}$, that is, the regression coefficient vector representing all SNPs in this pathway. Consider (8.1) under the homogeneity model. It can be revised as $P_{gMCP}(\boldsymbol{\beta}) = \sum_{c=1}^{c^*} \rho(||\beta_c||_2; \lambda, \gamma)$. Under this penalty, all SNPs within the same pathway are selected as a whole, and the homogeneity model condition is reinforced. Now consider the heterogeneity model and cMCP (8.3). For pathway $c = 1, \ldots, c^*$, denote $\beta_c^m = \{\beta_j^m : c(j) = c\}$, which is the vector of regression coefficients representing all SNPs within pathway c for outcome m. Then penalty (8.3) can be revised as

$$P_{cMCP}(\boldsymbol{\beta}) = \sum_{c=1}^{c^*} \rho \left(\sum_{m=1}^{M} \rho(||\beta_m^c||_2; \lambda_2, b); \lambda_1, a \right) \qquad (8.14)$$

Under a scenario similar to example 2, we have recently studied this method (Liu et al., 2014b). Following a similar strategy, the sparse gMCP can be revised to conduct pathway analysis.

Under the penalties described earlier, SNPs within the same pathway are "all in" or "all out." This can be sensible for small pathways composed of very "similar" SNPs. However, for a large pathway, it is possible that some SNPs are associated with an outcome but others are not. The preceding penalties can be revised to accommodate such a scenario and generate sparser models. For example, penalty (8.14) can be revised as $P_{cMCP}(\beta) = \sum_{c=1}^{c^*} \rho\left(\sum_{m=1}^{M} \rho(||\beta_m^c||_1; \lambda_2, b); \lambda_1, a\right)$. That is, for β_m^c, a lasso penalty is imposed to conduct within-pathway selection. It is noted that partly because of computational complexity, this penalty has not been studied yet. Other penalties can be revised following a similar strategy.

Network analysis

In network analysis, a node represents a SNP. A network contains very rich information (see, e.g., Chapters 4 and 5). Here we focus on the adjacency measure, which quantifies how closely any two nodes are connected and is perhaps the most important characteristic of a network. It is often defined based on the notion of similarity between nodes, and the similarity can be defined biologically or statistically. In the literature, many approaches for constructing the adjacency measure have been developed (Horvath, 2011). We have described a few statistical approaches and their applications in integrative analysis in a recent study (Liu et al., 2013a).

For SNP j and k $(= 1, \ldots, d)$, denote a_{jk} as their adjacency measure. Denote r_{jk} as the absolute value of a correlation coefficient (Pearson, rank, etc.) or a linkage disequilibrium measure. Many commonly used adjacency measures take the following forms: (1) $a_{jk} = I(r_{jk} > r)$, where r is a data-dependent cutoff; (2) $a_{jk} = r_{jk}^{\alpha}$ with $\alpha > 0$ and can be determined using, for example, the scale-free topology criterion (Zhang and Horvath, 2005); and (3) $a_{jk} = r_{jk}^{\alpha} I(r_{jk} > r)$. For more detailed discussions, we refer to Horvath (2011), Liu et al. (2013a), and references therein. In our numerical study, we take adjacency measure (3), with r_{jk} being the canonical correlation and r calculated from the permutation that corresponds to the null that all SNPs are not associated with outcomes.

To accommodate the network structure, we propose imposing the following *Laplacian penalty* in addition to those described in Sections 8.3.1 and 8.3.2:

$$\frac{1}{2}\tau \times M \times \sum_{1 \leq j < k \leq d} a_{jk} \left(||\beta_j||_2 - ||\beta_k||_2\right)^2 \tag{8.15}$$

where $\tau \geq 0$ is a data-dependent tuning parameter. Denote $\theta = (\theta_1, \ldots, \theta_d)' = (||\beta_1||_2, \ldots, ||\beta_d||_2)'$. We can express the preceding penalty using a positive semi-definite matrix L, which satisfies $\theta'L\theta = \sum_{1 \leq j < k \leq d} a_{jk}(\theta_j - \theta_k)^2$. Let $A = (a_{jk}, 1 \leq j, k \leq d)$ and $G = \text{diag}(g_1, \ldots, g_d)$, where $g_j = \sum_{k=1}^{d} a_{jk}$. In a network where a_{jk} is the weight of edge (j, k), g_j is the degree of vertex j. The penalty in (8.15) can then be written in a "network-friendly" manner as $\frac{1}{2}\tau M\theta' L\theta$, where $L = G - A$. Here the Laplacian matrix is not normalized, meaning that the weight g_j is not standardized to 1. For discussions on normalization, we refer to Huang et al. (2011).

In integrative analysis, penalty (8.15) is first studied in Liu et al. (2013a), inspired by the single-data-set study in Huang et al. (2011). In omics data analysis (Horvath, 2011), it has been observed that tightly connected SNPs (with large a_{jk}s) tend to have similar biological functions and similar regression coefficients. The Laplacian penalty in (8.15) shrinks the difference between β_j and β_k and encourages them to be similar. The degree of shrinkage is adjusted by a_{jk}. In this spirit, it may seem more natural to revise (8.15) as $\frac{1}{2}\tau \times M \times \sum_{1 \le j < k \le d} a_{jk}||\beta_j - \beta_k||_2^2$. Our preliminary investigation (unpublished) suggests that formulation (8.15) is computationally more feasible.

8.4 Analysis of the Heterogeneous Stock Mice Data

This study has been briefly described in Section 8.1. We refer to (Valdar et al., 2006a,b) for more detailed information. The data set we analyze includes the fully phenotypic records on 2202 mice, and each was genotyped for 13,459 SNPs. We use fastPHASE to impute the missing SNP measurements. After deleting observations with missing phenotypes and alleles with minor allele frequency less than 0.05, there are 1514 mice and 9991 SNPs in 19 autosomes. Benchmark analyses, particularly including the "classic" one-SNP-at-a-time analysis and lasso penalized analysis on each outcome separately, are provided in Liu et al. (2012) and are not reiterated here. Multiple methods are described in Section 8.3. Here we present the results of five different methods: (M_1) apply MCP penalization to each outcome separately; this set of analyses serves as benchmark; (M_2) assume the homogeneity model, and apply the gMCP (8.1) with $\Omega = \Omega_I$; (M_3) assume the homogeneity model, and apply the gMCP (8.1) with $\Omega = \Omega_\Sigma$; (M_4) assume the heterogeneity model, and apply the sparse gMCP (8.6) with $\Omega = \Omega_I$; (M_5) account for the network structure, and apply gMCP (8.1) with $\Omega = \Omega_I$ + Laplacian penalty (8.15). For all methods, tuning parameters are selected using fivefold cross-validation. The five analysis methods cover both the single-outcome analysis and multioutcome integrative analysis, both the homogeneity model and heterogeneity model, with and without accounting for the correlation among outcomes, and individual-SNP and network-based analyses. M_2 and M_3 differ in that M_2 only accounts for the "connectedness" between two outcomes in marker selection, whereas M_3 also accounts for that in the loss function. A few other methods are also mentioned

Table 8.1 *Overlaps of different analysis methods*

	M_1	M_2	M_3	M_4	M_5	M_1	M_2	M_3	M_4	M_5
	CD4/CD8					CD4:CD3				
M_1	65	35	4	40	43	63	35	9	41	35
M_2		83	10	66	57		83	10	67	57
M_3			43	9	11			43	8	11
M_4				85	61				82	57
M_5					216					216

in Section 8.3. With the limited scope of this chapter, they will not be applied here.

In Table 8.1, we summarize the number of SNPs identified by each method and overlap. Under the heterogeneity model, the markers identified for the two outcomes are different, and so summaries are provided separately. More analysis results are presented in Tables 8.2 (M_3) and 8.3 (M_4) and in the appendix of Liu et al. (2013c) (Table 1.4 for M_1, 1.5 for M_2, and 1.6 for M_5).

Table 8.1 shows that different methods identify significantly different SNP sets. For example, for CD4/CD8, the five methods identify 65, 83, 43, 85, and 216 SNPs, respectively. There are only small to moderate overlaps. For CD4/CD8, M_2 and M_4 have 66 overlapping SNPs, whereas M_3 and M_4 have only 9 overlapping SNPs. When analyzing the two outcomes separately, although there are overlapping SNPs, the two identified SNP sets are largely different. In integrative analysis under the homogeneity model, it is reinforced that the same set of SNPs is identified. Under the heterogeneity model (Table 8.3), the degree of overlap is much higher than under M_1. The integrative analysis methods lead to more coherent sets of markers, which may simplify interpretations and downstream functional studies. M_5 identifies significantly more SNPs. This observation is reasonable as M_5 incorporates an ℓ_2 penalty, which in general leads to denser models. For SNPs that are correlated with those identified under M_2, method M_5 encourages them to have nonzero coefficients and to be identified.

With our limited knowledge on the susceptibility SNPs for immunology, we are not able to objectively evaluate the biological implications of identified SNPs. To provide more insights into different methods, we first evaluate prediction performance as follows: (1) randomly split data into a training set and a testing set with size 4:1; (2) analyze the training set; (3) use the model obtained in step 2 to make predictions for subjects in the testing set; (4) repeat step 1–3 100 times, and compute the average prediction performance measure. Here two measures are adopted. The first is the familiar mean prediction squared errors. In addition, considering that penalization leads to estimates

Table 8.2 *Analysis results of gMCP ($\Omega = \Omega_{\Sigma}$): identified SNPs, estimates, and OOI*

	CD4/CD8		CD4:CD3			CD4/CD8		CD4:CD3	
SNP	Est.	OOI	Est.	OOI	SNP	Est.	OOI	Est.	OOI
CEL-12_17860196	0.0104	0.53	0.0123	0.53	rs3677347	−0.0037	0.16	0.0025	0.16
mCV22813496	−0.2198	1.00	0.2971	1.00	rs3687193	−0.0027	0.18	−0.0023	0.18
mCV22965443	0.2075	1.00	−0.3229	1.00	rs3705078	0.0088	0.85	0.0264	0.85
UT_8_130.396331	0.0431	0.59	0.0315	0.59	rs3708635	0.1670	0.37	−0.0583	0.37
rs13475794	−0.0077	0.98	−0.0524	0.98	rs3709725	−0.0008	0.30	0.0003	0.30
rs13475847	0.1185	0.73	−0.0378	0.73	rs3711751	0.0041	0.13	−0.0045	0.13
rs13476239	−0.1340	0.48	0.1244	0.48	rs3714738	−0.0177	0.36	0.0079	0.36
rs13476913	−0.0212	0.74	−0.0177	0.74	rs3718812	0.0744	0.95	0.0034	0.95
rs13478276	−0.0426	0.86	0.0166	0.86	rs3722157	0.0022	0.14	0.0021	0.14
rs13478656	0.0288	0.97	0.0405	0.97	rs4135717	−0.0013	0.29	−0.0015	0.29
rs13479673	0.0024	0.25	0.0129	0.25	rs4139535	−0.0240	0.81	0.0369	0.81
rs13480135	−0.0117	0.59	−0.0177	0.59	rs4161151	−0.0121	0.22	−0.0130	0.22
rs13480367	−0.0096	0.39	−0.0174	0.39	rs4219905	−0.0005	0.49	−0.0147	0.49
rs13481412	0.0204	0.90	−0.0206	0.90	rs4224947	−0.0006	0.19	−0.0002	0.19
rs13483449	−0.0001	0.22	0.0048	0.22	rs4226699	−0.0022	0.42	−0.0050	0.42
rs3657482	0.1603	0.41	−0.0154	0.41	rs6171719	0.0355	0.75	0.0325	0.75
rs3662130	−0.0002	0.34	−0.0006	0.34	rs6328018	0.0092	0.53	0.0167	0.53
rs3665567	−0.0905	0.54	0.1629	0.54	rs6377831	−0.0156	0.50	0.0509	0.50
rs3666537	−0.0093	0.55	−0.0317	0.55	rs6393715	−0.0011	0.60	−0.0065	0.60
rs3670360	0.0484	0.92	−0.0219	0.92	rs6409750	−0.0626	0.89	0.1631	0.89
rs3671932	0.0137	0.68	−0.0089	0.68	rs8245237	−0.0071	0.10	0.0060	0.10
rs3675028	−0.0068	0.39	−0.0070	0.39					

Table 8.3 *Analysis results of sparse gMCP ($\Omega = \Omega_1$): identified SNPs, estimates, and OOI*

SNP	CD4/CD8		CD4:CD3		SNP	CD4/CD8		CD4:CD3	
	Est.	OOI	Est.	OOI		Est.	OOI	Est.	OOI
CEL-13_25470193	−0.0108	0.80	0.0312	0.84	rs3659070	−0.0111	0.24	0.0095	0.24
CEL-17_31069801	0.2476	1.00	−0.2418	1.00	rs3662979	−0.0277	0.61		
CEL-17_62934746	−0.0367	0.97	0.0075	0.95	rs3667348	0.0787	0.05	−0.0489	0.05
CEL-5_56325748	0.0530	0.50	−0.0111	0.43	rs3670360	0.0017	0.30		
CEL-9_17007263	−0.0032	0.54	0.0022	0.53	rs3674314	0.0080	0.57	−0.0042	0.54
gnf01.102.630	0.0283	0.92	−0.0259	0.92	rs3681853	−0.0097	0.53	0.0055	0.53
gnf06.096.938	0.0087	0.19	−0.0008	0.12	rs3701438	0.0435	0.91	−0.0403	0.91
gnf08.130.033	0.0095	0.20			rs3702472	−0.0173	0.72	0.0278	0.71
gnf11.058.315	−0.0227	0.42	0.0153	0.42	rs3705078	−0.0196	0.73	0.0799	0.86
mCV22965443	0.2064	1.00	−0.2531	1.00	rs3710419	−0.0763	0.79	0.0210	0.73
mCV25073238	−0.0096	0.54	0.0157	0.54	rs3711751	0.0193	0.34	−0.0134	0.34
UT_14_59.839857	−0.0401	0.64	0.0428	0.64	rs3715491	0.0001	0.10		
UT_15_103.451774	−0.0070	0.45	0.0066	0.45	rs3718829	−0.0043	0.36	0.0010	0.27
UT_3_90.000791	−0.0133	0.62	0.0386	0.84	rs4135996	0.0056	0.30		
rs13472132	0.0519	0.82	−0.0241	0.82	rs4174183	−0.0109	0.26	0.0152	0.26
rs13475847	0.0244	0.49			rs4184231	−0.0098	0.40		
rs13476251	−0.0871	0.80	0.0770	0.80	rs4219613	−0.0080	0.39	0.0594	0.50
rs13476267	0.0017	0.46	−0.0084	0.51	rs4222922	−0.0037	0.39	0.0214	0.65
rs13476350	−0.0096	0.18			rs4223701	−0.0007	0.39	0.0007	0.39
rs13476690	0.0053	0.38	−0.0015	0.36	rs4232449	−0.0444	0.71	0.0209	0.70
rs13476846	0.0231	0.63	−0.0014	0.28	rs6180306	−0.0275	0.70	0.0285	0.70

Marker					Marker				
rs13476913	−0.0005	0.32			rs6183014	0.0025	0.21	−0.0074	0.23
rs13477354	−0.0016	0.47	0.0023	0.48	rs6196764	0.0607	0.92	−0.0407	0.92
rs13477409	−0.0057	0.25			rs6217029	−0.0097	0.36		
rs13477511	−0.0025	0.14	0.0006	0.10	rs6245977	−0.0350	0.92	0.0265	0.91
rs13477584	−0.0189	0.49	0.0085	0.46	rs6280411	−0.0340	0.49	0.0389	0.48
rs13478446	−0.0016	0.40			rs6292954	−0.0029	0.24	0.0117	0.34
rs13478730	0.0190	0.27	−0.0172	0.27	rs6315002	−0.0094	0.32	0.0172	0.32
rs13479465	−0.0403	0.78	0.0084	0.63	rs6334723	−0.0679	0.99	0.0324	0.99
rs13479621	−0.0160	0.64	0.0096	0.63	rs6342158	−0.0251	0.46	0.0253	0.46
rs13479930	0.0343	0.60	−0.0310	0.60	rs6360080	0.0147	0.40	−0.0093	0.38
rs13480400	0.0113	0.57	−0.0106	0.57	rs6377183	−0.0088	0.39	0.0009	0.28
rs13480567	0.0051	0.32	−0.0103	0.32	rs6377831	−0.0955	1.00	0.1035	1.00
rs13480946	0.0091	0.40	−0.0003	0.13	rs6378343	−0.0812	0.56	0.0921	0.56
rs13481063	0.0095	0.13	−0.0080	0.13	rs6385968	0.0019	0.39	−0.0012	0.34
rs13481363	−0.0503	0.88	0.0591	0.88	rs6393715	0.0082	0.62	−0.0278	0.80
rs13481411	−0.0291	0.83	0.0121	0.80	rs6401555	−0.0254	0.38	0.0048	0.30
rs13481499	0.0066	0.25	−0.0020	0.24	rs6409750	−0.0860	0.99	0.1209	0.99
rs13481571	−0.0156	0.26	0.0081	0.25	rs13475794			−0.0034	0.26
rs13481618	0.0294	0.79	−0.0090	0.65	rs13476820			−0.0015	0.44
rs13481961	−0.0112	0.43	0.0092	0.43	rs13478656			0.0143	0.74
rs13482007	0.0058	0.40	−0.0107	0.41	rs13480138			0.0481	0.30
rs13482388	0.0212	0.16	−0.0013	0.08	rs13482668			0.0238	0.46
rs13482738	0.0009	0.34	−0.0164	0.68	rs3682465			0.0141	0.33
rs13483373	−0.0174	0.50	0.0084	0.50	rs3685111			−0.0013	0.14
rs13483459	−0.0061	0.22	0.0135	0.23	rs4220927			−0.0008	0.22
rs3657320	0.0795	0.89	−0.0476	0.89	rs6157163			−0.0005	0.41

shrinking to zero, we also compute the mean correlation coefficients between the observed and predicted outcomes. The mean prediction squared errors are 0.223, 0.214, 0.307, 0.233, and 0.208 for M_1 to M_5, respectively. The mean prediction correlation coefficients are 0.681, 0.682, 0.510, 0.670, and 0.697 for M_1 to M_5, respectively. It may seem counterintuitive that M_3, which accounts for the correlation between outcomes in the loss function, has the worst prediction performance. As discussed in Liu et al. (2012), Σ is the correlation matrix among residuals and needs to be estimated. It is different from the correlation matrix between the observed outcomes, which can be directly calculated. In our numerical study, we follow the strategy in Liu et al. (2012) and use a simple estimate, which may lead to the poor prediction performance. The other four methods have similar prediction performance, with M_5 having the best prediction. Prediction correlation coefficients close to 0.7 suggest satisfactory performance.

In addition, we also evaluate the stability of each method. Consider the procedure described earlier and the 100 sets of estimates from the 100 random splits. For a SNP, we compute the probability of it being identified across splits. This probability is referred to as the *observed occurrence index (OOI)* in Huang and Ma (2010). This evaluation is a by-product of the prediction evaluation and does not incur additional computational cost. For SNPs identified using the whole data set, we present the OOIs in the tables. The median OOIs (for SNPs identified using all observations) are 0.42 (M_1), 0.42 (M_2), 0.53 (M_3), 0.48 (M_4) and 0.54 (M_5), respectively. Compared to the benchmark M_1, stability can be improved with integrative analysis methods M_3 to M_5. It may seem that the integrative analysis OOIs are not "high enough". However it should be noted that for the SNPs not identified using the whole data set, the OOIs are much lower (e.g., for such SNPs under M_5, the median OOI is 0.03). Thus the proposed methods can lead to a good separation of important and unimportant SNPs.

This data set has also been analyzed in the literature. However, the published studies have been focused on marginal analysis, which analyzes one SNP/gene at a time. In addition, marker identification has been based on significance level, which differs significantly from that based on penalized estimation. The marker identification results generated using the proposed methods are significantly different from those published, partly because of the difference in analysis strategy and partly because of the new analysis technique.

8.5 Discussion

8.5.1 Conclusion

Profiling studies generate data with high-dimensional omics measurements and low sample sizes. Even with the cost of profiling falling drastically, as it is often difficult to collect samples, particularly human samples, the limitation in sample size may continue to exist in the foreseeable future. Pooling information across data sets can effectively increase sample size. In addition, for some studies, such as the identification of multi-cancer markers, multi-data-set analysis is inevitable. Compared to meta-analysis, integrative analysis provides a different strategy and has been less investigated. In this chapter, we have focused on identifying markers that are associated with outcome variables. Multiple penalization methods have been described, under different model assumptions and conducting different types of analyses. To avoid confusion in notation, we have used a data set with multiple correlated continuous outcomes as an example. Methods described in this chapter and related discussions are applicable to multi-data-set analysis with minor modifications. In our previous studies, categorical and censored survival data have also been analyzed. Examining the penalization methods suggests that the definition of penalties is relatively "independent" of the loss function. Thus, as long as loss functions can be properly constructed, for example, from likelihood functions or estimating equations, the proposed penalties are applicable. The computational algorithms and convergence results can be more challenging. When the loss functions are continuously differentiable, we can resort to the Taylor expansion or majorize-minimization algorithm, develop an iterative algorithm that has a least squares loss function in each iteration, and then adopt the existing computational algorithms. Some of the convergence results may still be derived using Tseng (2001). We limit this chapter to MCP-based penalties. In our previous publications, we have also studied penalties in the lasso family. We conjecture that other penalties, in particular SCAD, are also applicable, even though we have not pursued these penalties. We analyze the heterogeneous stock mice data to demonstrate the applicability of integrative analysis methods and show that they may lead to significantly different analysis results. The integrative analysis methods can lead to more coherent sets of identified markers with improved stability. Method M_5, which accommodates the network structure, has the best prediction performance and stability at the cost of more identified markers. In the literature, very limited research directly compares different integrative analysis methods. Our data analysis may provide some insights. A more comprehensive comparison demands systematic theoretical studies and large-scale simulations and is postponed until future study.

8.5.2 *Related and Possible Future Developments*

Several very important practical aspects have not been discussed. In the analysis of multiple data sets (outcomes), each data set needs to undergo rigorous data quality control and processing, as in single-data-set analysis. In addition, selecting proper data sets to pool is nontrivial. For example, the homogeneity model demands a high degree of "similarity" across data sets. It is not always more efficient to have more data sets if they are different. Meta-data need to be carefully evaluated to select similar data sets. Another practical problem comes from the incomparability of different platforms used in different studies. We acknowledge the importance of these practical problems but refer to other studies, such as Guerra and Goldstein (2009) and Chapter 2 of this book, for more discussion.

Penalized integrative analysis is still evolving, and more efficient methods still need to be developed. Consider, for example, example 1 where multiple data sets have been generated under comparable designs. Numerical studies in Liu et al. (2013b) and others suggest that sometimes with significant differences in experimental settings, the homogeneity model may be too restricted, and the heterogeneity model is needed. Conversely, it is expected that the marker sets identified in different data sets will be somewhat "similar." However, the methods described in Section 8.3.2 do not have a mechanism to *encourage* such similarity. As another example of possible methodological developments, consider the network-based analysis described in Section 8.3.3. The adjacency measures we consider and the Laplacian method can only accommodate *undirected* networks, which are computationally simple; they cannot accurately describe the regulations among SNPs/genes. It is still not clear to us how to accommodate directed networks in penalized integrative analysis.

Another limitation of the existing integrative analyses is that the theoretical aspect has not been thoroughly developed. For a few specific models and penalization methods (Liu et al., 2014b; Ma et al., 2012), we have shown that they enjoy the selection and estimation consistency properties. This result especially holds if $\log(d)/N \to 0$ as $N, d \to \infty$, where N is the combined sample size across data sets. However, we or others have not been able to show theoretically that integrative analysis methods can more accurately identify markers than meta-analysis methods. In the existing studies, the superiority of integrative analysis methods has been established using extensive simulation studies. More theoretical investigations are needed.

Acknowledgements

We thank the book editors for the invitation to contribute this chapter. This study has been partly supported by NIH awards CA142774, CA182984, and CA165923, National Social Science Foundation of China (13CTJ001), and by the National Bureau of Statistics Funds of China (2012LD001).

Appendix

Table 8.4 *Analysis results by applying MCP to each outcome separately: identified SNPs, estimates, and OOI*

SNP	CD4/CD8		CD4:CD3		SNP	CD4/CD8		CD4:CD3	
	Est.	OOI	Est.	OOI		Est.	OOI	Est.	OOI
CEL-13_25470193	−0.0057	0.16	0.0634	0.16	rs4184231	−0.0243	0.49		
CEL-17_31069801	0.3489	1.00	−0.3397	1.00	rs4223448	−0.1244	0.50		
CEL-17_62934746	−0.0572	0.83			rs4232449	−0.0594	0.40		
CEL-5_56325748	0.0735	0.42			rs6180306	−0.0120	0.16	0.0324	0.16
CEL-9_17007263	−0.0306	0.42			rs6196764	0.0761	0.71	−0.0541	0.71
gnf01.102.630	0.0267	0.58	−0.0387	0.58	rs6217029	−0.0220	0.36		
gnf11.058.315	−0.0296	0.58			rs6245977	−0.0602	0.89	0.0002	0.89
mCV22965443	0.3103	1.00	−0.3638	1.00	rs6280411	−0.0743	0.52		
mCV25073238			0.0209	0.18	rs6334723	−0.0789	0.98	0.0180	0.98
UT_14_59.839857			0.0469	0.22	rs6373215	−0.0195	0.43		
UT_15_103.451774			0.0219	0.12	rs6377831	−0.1309	1.00	0.1363	1.00
UT_3_90.000791			0.0500	0.00	rs6378343	−0.1079	0.34	0.1367	0.34
rs13472132	0.0583	0.84			rs6385968	0.0448	0.41		
rs13475867	−0.0405	0.60			rs6409750	−0.1263	0.88	0.1792	0.88
rs13476248	0.0874	0.67			rs13475794			−0.0033	0.00
rs13476846	0.0724	0.67			rs13476251			0.1278	0.33
rs13476892	−0.0125	0.34			rs13476267			−0.0265	0.13
rs13477584	−0.0274	0.47			rs13477354			0.0363	0.05
rs13477889	0.0073	0.30			rs13478546			−0.0002	0.02
rs13478446	−0.0064	0.33			rs13478656			0.0207	0.00
rs13478734	−0.0174	0.09	0.0132	0.09	rs13478947			0.0082	0.00
rs13478822	−0.0072	0.00			rs13479673			0.0218	0.00
rs13479376	0.0143	0.07			rs13480141			0.0956	0.00
rs13479465	−0.0823	0.73			rs13480400			−0.0265	0.14
rs13479621	−0.0088	0.34			rs13480567			−0.0194	0.05
rs13479930	0.0535	0.48	−0.0325	0.48	rs13481042			−0.0213	0.05

(continued)

Table 8.4 *(continued)*

| SNP | CD4/CD8 | | CD4:CD3 | | SNP | CD4/CD8 | | CD4:CD3 | |
---	Est.	OOI	Est.	OOI	---	Est.	OOI	Est.	OOI
rs13480045	−0.0212	0.13			rs13481817			−0.0476	0.05
rs13480277	0.0097	0.35			rs13482007			−0.0210	0.13
rs13481242	−0.0095	0.22			rs13482668			0.0492	0.00
rs13481353	0.0411	0.59			rs13482738			−0.0218	0.01
rs13481363	−0.0106	0.50	0.0965	0.50	rs13483456			0.0261	0.00
rs13481411	−0.0822	0.73			rs3655057			−0.0110	0.00
rs13481568	0.0111	0.44			rs3656890			−0.0210	0.00
rs13481618	0.0287	0.60			rs3659070			0.0124	0.20
rs13481961	−0.0084	0.34			rs3682465			0.0319	0.00
rs13482418	−0.0035	0.28			rs3692702			−0.0019	0.00
rs13483373	−0.0021	0.43			rs3705078			0.0956	0.00
rs3656583	0.0108	0.43			rs3709102			−0.0175	0.00
rs3657320	0.1566	0.89	−0.0354	0.89	rs3716179			−0.0039	0.00
rs3662979	−0.0330	0.40			rs3718405			0.0203	0.00
rs3672425	0.0703	0.47			rs3722316			−0.0107	0.00
rs3674314	0.0199	0.45			rs3723781			0.0124	0.00
rs3680871	0.0231	0.30			rs4174183			0.0180	0.07
rs3681655	0.0256	0.36			rs4219905			−0.1058	0.00
rs3681853	−0.0152	0.35			rs4220927			−0.0101	0.00
rs3697744	0.0007	0.27			rs4222922			0.0402	0.01
rs3701438	0.0423	0.83	−0.0432	0.83	rs4223428			−0.0516	0.10
rs3702472	−0.0070	0.49	0.0450	0.49	rs4223701			0.0071	0.08
rs3708393	−0.0044	0.18	0.0024	0.18	rs6157163			−0.0120	0.02
rs3709888	−0.0042	0.06			rs6183014			−0.0232	0.11
rs3710419	−0.1391	0.77			rs6239288			−0.0117	0.02
rs3711751	0.0325	0.17			rs6260804			0.0087	0.00
rs3722416	0.0250	0.40			rs6315002			0.0281	0.01
rs3724110	−0.0720	0.49			rs6315152			0.0403	0.17
rs3725706	−0.0035	0.20			rs6393715			−W0.0420	0.01

Table 8.5 *Analysis results of gMCP ($\Omega = \Omega_1$): identified SNPs, estimates, and OOI*

SNP	CD4/CD8		CD4:CD3		SNP	CD4/CD8		CD4:CD3	
	Est.	OOI	Est.	OOI		Est.	OOI	Est.	OOI
CEL-13_25470193	−0.0137	0.59	0.0259	0.59	rs13481961	−0.0281	0.45	0.0257	0.45
CEL-17_31069801	0.2480	1.00	−0.2435	1.00	rs13482007	0.0090	0.28	−0.0122	0.28
CEL-17_62934746	−0.0413	0.98	0.0196	0.98	rs13482388	0.0136	0.10	−0.0041	0.10
CEL-5_56325748	0.0628	0.43	−0.0286	0.43	rs13482668	−0.0024	0.10	0.0103	0.10
gnf01.102.630	0.0333	0.81	−0.0320	0.81	rs13482738	0.0006	0.51	−0.0017	0.51
gnf06.096.938	0.0059	0.14	−0.0019	0.14	rs13483373	−0.0212	0.38	0.0155	0.38
gnf11.058.315	−0.0353	0.58	0.0279	0.58	rs13483459	−0.0102	0.12	0.0151	0.12
gnf15.090.425	−0.0025	0.42	0.0051	0.42	rs3657320	0.0876	0.89	−0.0576	0.89
mCV22735181	−0.0216	0.13	0.0372	0.13	rs3659070	−0.0282	0.18	0.0248	0.18
mCV22965443	0.2025	1.00	−0.2520	1.00	rs3662979	−0.0186	0.45	0.0044	0.45
mCV25073238	−0.0289	0.50	0.0385	0.50	rs3670360	0.0023	0.15	−0.0005	0.15
UT_14_59.839857	−0.0440	0.57	0.0457	0.57	rs3674314	0.0074	0.36	−0.0047	0.36
UT_15_103.451774	−0.0112	0.51	0.0102	0.51	rs3681853	−0.0232	0.51	0.0183	0.51
UT_3_90.000791	−0.0127	0.65	0.0237	0.65	rs3701438	0.0467	0.82	−0.0460	0.82
UT_8_130.396331	0.0113	0.25	0.0082	0.25	rs3702472	−0.0319	0.69	0.0367	0.69
rs13469581	−0.0015	0.22	0.0021	0.22	rs3705078	−0.0393	0.68	0.0921	0.68
rs13472132	0.0488	0.71	−0.0282	0.71	rs3708411	−0.0100	0.19	0.0085	0.19
rs13475794	−0.0009	0.43	−0.0023	0.43	rs3710419	−0.0752	0.63	0.0338	0.63
rs13475847	0.0086	0.30	0.0003	0.30	rs3718829	−0.0087	0.31	0.0040	0.31
rs13475867	−0.0042	0.45	−0.0013	0.45	rs4174183	−0.0234	0.27	0.0233	0.27
rs13476251	−0.0864	0.49	0.0738	0.49	rs4219613	−0.0027	0.31	0.0084	0.31

(continued)

Table 8.5 *(continued)*

SNP	CD4/CD8		CD4:CD3		SNP	CD4/CD8		CD4:CD3	
	Est.	OOI	Est.	OOI		Est.	OOI	Est.	OOI
rs13476690	0.0192	0.41	−0.0116	0.41	rs4222922	−0.0094	0.47	0.0214	0.47
rs13476846	0.0161	0.34	−0.0068	0.34	rs4223428	0.0717	0.53	−0.0539	0.53
rs13476891	0.0006	0.14	−0.0005	0.14	rs4232449	−0.0429	0.68	0.0242	0.68
rs13477354	−0.0127	0.50	0.0168	0.50	rs6180306	−0.0273	0.54	0.0268	0.54
rs13477584	−0.0082	0.39	0.0049	0.39	rs6196764	0.0671	0.83	−0.0432	0.83
rs13478656	−0.0011	0.65	0.0217	0.65	rs6245977	−0.0378	0.81	0.0296	0.81
rs13478730	0.0223	0.19	−0.0205	0.19	rs6285803	−0.0325	0.10	0.0269	0.10
rs13479465	−0.0349	0.72	0.0115	0.72	rs6292954	−0.0151	0.15	0.0319	0.15
rs13479621	−0.0239	0.44	0.0182	0.44	rs6315002	−0.0148	0.28	0.0206	0.28
rs13480138	−0.0018	0.24	0.0182	0.24	rs6334723	−0.0763	0.96	0.0456	0.96
rs13480148	−0.0170	0.45	0.0269	0.45	rs6342158	−0.0251	0.26	0.0269	0.26
rs13480400	0.0155	0.41	−0.0144	0.41	rs6360080	0.0237	0.27	−0.0156	0.27
rs13480567	0.0109	0.20	−0.0153	0.20	rs6377183	−0.0019	0.24	0.0007	0.24
rs13480946	0.0216	0.37	−0.0088	0.37	rs6377831	−0.0975	1.00	0.1119	1.00
rs13481063	0.0189	0.08	−0.0153	0.08	rs6378040	−0.0065	0.01	0.0017	0.01
rs13481363	−0.0590	0.76	0.0603	0.76	rs6378343	−0.0763	0.37	0.0850	0.37
rs13481411	−0.0055	0.67	0.0024	0.67	rs6385968	0.0068	0.44	−0.0064	0.44
rs13481445	0.0126	0.13	−0.0122	0.13	rs6393715	0.0102	0.64	−0.0256	0.64
rs13481499	0.0060	0.20	−0.0040	0.20	rs6401555	−0.0293	0.29	0.0128	0.29
rs13481568	0.0132	0.46	−0.0103	0.46	rs6409750	−0.0846	1.00	0.1181	1.00
rs13481618	0.0277	0.40	−0.0165	0.40					

Table 8.6 *Analysis results of gMCP with the Laplacian penalty ($\Omega = \Omega_1$): identified SNPs, estimates, and OOI*

	CD4/CD8		CD4:CD3			CD4/CD8		CD4:CD3	
SNP	Est.	OOI	Est.	OOI	SNP	Est.	OOI	Est.	OOI
CEL-13_25470193	−0.0472	0.88	0.0703	0.88	rs3089070	0.0009	0.24	−0.0006	0.24
CEL-17_62934746	−0.0385	0.99	0.0171	0.99	rs3089636	0.0003	0.36	−0.0003	0.36
CEL-5_58483009	−0.0034	0.84	0.0014	0.84	rs3156741	−0.0007	0.47	0.0008	0.47
CEL-7_28971404	−0.0079	0.83	0.0043	0.83	rs3160730	0.0009	0.64	−0.0011	0.64
CEL-9_17007263	−0.0278	0.62	0.0219	0.62	rs3656403	0.0051	0.31	−0.0045	0.31
gnf01.102.630	0.0045	0.92	−0.0045	0.92	rs3656875	0.0009	0.71	−0.0010	0.71
gnf01.103.560	0.0045	0.92	−0.0045	0.92	rs3662979	−0.0001	0.31	0.0000	0.31
gnf01.103.869	0.0045	0.92	−0.0045	0.92	rs3666143	−0.0043	0.95	0.0017	0.95
gnf01.104.109	0.0045	0.92	−0.0045	0.92	rs3667348	0.0002	0.23	−0.0002	0.23
gnf04.098.802	−0.0023	0.46	0.0031	0.46	rs3670168	0.0071	0.76	−0.0100	0.76
gnf06.102.000	−0.0009	0.32	0.0011	0.32	rs3670360	0.0123	0.65	−0.0017	0.65
gnf08.108.032	−0.0025	0.68	0.0021	0.68	rs3672425	0.0239	0.75	−0.0233	0.75
gnf08.130.033	0.0121	0.18	0.0024	0.18	rs3673305	0.0101	0.91	−0.0118	0.91
gnf15.090.425	−0.0080	0.57	0.0147	0.57	rs3675335	0.0021	0.17	−0.0030	0.17
mCV22576656	0.0041	0.48	−0.0037	0.48	rs3676491	0.0001	0.41	−0.0002	0.41
mCV22602426	−0.0023	0.46	0.0031	0.46	rs3677807	−0.0025	0.68	0.0021	0.68
mCV22813496	−0.1108	0.46	0.1176	0.46	rs3678620	−0.0027	0.88	0.0026	0.88
mCV22907805	−0.0012	0.91	0.0014	0.91	rs3681853	−0.0062	0.48	0.0056	0.48
mCV22965443	0.1211	1.00	−0.0542	1.00	rs3682321	−0.0055	0.40	0.0044	0.40
mCV23375358	−0.0031	0.81	0.0013	0.81	rs3682465	−0.0097	0.33	0.0204	0.33
mCV23534950	−0.0003	0.08	0.0002	0.08	rs3682923	0.0038	0.41	−0.0864	0.41
mCV24217147	0.0118	0.77	−0.0151	0.77	rs3685305	0.0013	0.90	−0.0016	0.90
mCV24657328	−0.0011	0.26	0.0009	0.26	rs3686617	−0.0038	0.92	0.0037	0.92
mCV27580773	0.0006	0.56	−0.0007	0.56	rs3688504	0.0129	0.76	−0.0157	0.76
UT_11_95.85971	−0.0022	0.52	0.0018	0.52	rs3692826	−0.0043	0.95	0.0017	0.95
UT_11_95.889102	−0.0022	0.52	0.0018	0.52	rs3695790	−0.0038	0.92	0.0037	0.92

(continued)

Table 8.6 (*continued*)

SNP	CD4/CD8		CD4:CD3		SNP	CD4/CD8		CD4:CD3	
	Est.	OOI	Est.	OOI		Est.	OOI	Est.	OOI
UT_14_59.839873	0.0006	0.28	−0.0008	0.28	rs3699406	−0.0003	0.60	0.0004	0.60
UT_17_33.238924	−0.1108	0.46	0.1176	0.46	rs3699421	−0.0012	0.52	0.0011	0.52
UT_3_131.430324	0.0129	0.76	−0.0157	0.76	rs3699784	0.0008	0.70	−0.0003	0.70
UT_3_90.000791	−0.0418	0.95	0.0731	0.95	rs3701438	0.0254	0.66	−0.0252	0.66
UT_6_72.775257	0.0101	0.91	−0.0118	0.91	rs3702472	−0.0006	0.70	0.0007	0.70
UT_6_73.601283	0.0132	0.93	−0.0150	0.93	rs3703501	0.0006	0.56	−0.0007	0.56
UT_8_70.331085	0.0006	0.56	−0.0007	0.56	rs3704471	0.0002	0.23	−0.0002	0.23
UT_8_71.231527	0.0013	0.90	−0.0016	0.90	rs3704618	−0.0089	0.19	0.0111	0.19
UT_8_71.961718	0.0013	0.90	−0.0016	0.90	rs3704920	−0.0031	0.81	0.0013	0.81
UT_8_72.075442	0.0013	0.90	−0.0016	0.90	rs3705078	−0.0089	0.90	0.0299	0.90
rs13459103	−0.0012	0.91	0.0014	0.91	rs3708144	0.0062	0.26	−0.0047	0.26
rs13459163	−0.0729	0.89	0.0506	0.89	rs3708393	−0.0094	0.78	0.0093	0.78
rs13469581	−0.0067	0.38	0.0084	0.38	rs3708411	−0.0202	0.38	0.0139	0.38
rs13472132	0.0455	0.87	−0.0255	0.87	rs3710419	−0.0289	0.54	0.0090	0.54
rs13475794	−0.0023	0.71	−0.0158	0.71	rs3711099	0.0006	0.56	−0.0007	0.56
rs13475847	0.0079	0.61	0.0009	0.61	rs3712907	0.0018	0.63	−0.0004	0.63
rs13475867	−0.0065	0.59	−0.0012	0.59	rs3712953	0.0015	0.69	−0.0021	0.69
rs13475960	0.0007	0.44	−0.0001	0.44	rs3715491	0.0022	0.07	−0.0004	0.07
rs13475972	0.0007	0.44	−0.0001	0.44	rs3716288	0.0031	0.17	−0.0063	0.17
rs13475973	0.0007	0.44	−0.0001	0.44	rs3717803	0.0009	0.24	−0.0006	0.24
rs13476251	−0.0814	0.85	0.0669	0.85	rs3718829	−0.0091	0.45	0.0042	0.45
rs13476267	0.0022	0.63	−0.0042	0.63	rs3722665	−0.0007	0.47	0.0008	0.47
rs13476690	0.0174	0.70	−0.0114	0.70	rs3724110	−0.0253	0.78	0.0127	0.78
rs13476740	−0.0027	0.88	0.0026	0.88	rs4135996	0.0005	0.33	−0.0001	0.33
rs13476763	−0.0158	0.90	0.0083	0.90	rs4174183	−0.0095	0.45	0.0093	0.45

rs13476764	0.0060	0.99	−0.0028	0.99	rs4184231	−0.0001	0.37	0.0000	0.37
rs13476820	0.0001	0.41	−0.0002	0.41	rs4184702	−0.0032	0.57	0.0007	0.57
rs13476846	0.0196	0.55	−0.0081	0.55	rs4189683	−0.0053	0.57	0.0042	0.57
rs13476892	−0.0147	0.40	0.0115	0.40	rs4190470	−0.0039	0.43	0.0032	0.43
rs13476913	−0.0109	0.71	−0.0021	0.71	rs4198737	0.0046	0.44	−0.0018	0.44
rs13477269	−0.0034	0.37	0.0016	0.37	rs4199044	0.0069	0.63	−0.0025	0.63
rs13477354	−0.0138	0.42	0.0171	0.42	rs4219613	−0.0463	0.58	0.0772	0.58
rs13477396	0.0071	0.76	−0.0100	0.76	rs4220927	0.0003	0.28	−0.0025	0.28
rs13477397	0.0071	0.76	−0.0100	0.76	rs4222922	−0.0059	0.51	0.0138	0.51
rs13477584	−0.0117	0.53	0.0060	0.53	rs4223701	−0.0079	0.32	0.0070	0.32
rs13477889	0.0029	0.41	−0.0023	0.41	rs4227221	0.0006	0.56	−0.0007	0.56
rs13478276	−0.0034	0.84	0.0014	0.84	rs4227700	0.0030	0.18	−0.0067	0.18
rs13478285	−0.0043	0.95	0.0017	0.95	rs4232449	−0.0079	0.83	0.0043	0.83
rs13478286	0.0018	1.00	−0.0007	1.00	rs6154545	0.0008	0.70	−0.0003	0.70
rs13478446	−0.0064	0.44	0.0009	0.44	rs6164049	0.0004	0.24	−0.0004	0.24
rs13478656	0.0004	0.78	0.0228	0.78	rs6180306	−0.0025	0.68	0.0021	0.68
rs13478732	−0.0011	0.26	0.0009	0.26	rs6184947	−0.0033	0.53	0.0026	0.53
rs13478734	−0.0032	0.51	0.0026	0.51	rs6187409	0.0018	1.00	−0.0007	1.00
rs13478735	−0.0033	0.53	0.0026	0.53	rs6194426	0.0110	0.62	−0.0057	0.62
rs13478736	−0.0016	0.40	0.0013	0.40	rs6196764	0.0668	0.98	−0.0496	0.98
rs13478738	0.0002	0.45	−0.0002	0.45	rs6203570	0.0005	0.33	−0.0001	0.33
rs13478939	−0.0009	0.32	0.0011	0.32	rs6217029	−0.0129	0.45	0.0058	0.45
rs13478941	0.0004	0.52	−0.0004	0.52	rs6217662	0.0132	0.93	−0.0150	0.93
rs13479341	0.0002	0.14	−0.0020	0.14	rs6238771	−0.0011	0.26	0.0009	0.26
rs13479376	0.0091	0.16	−0.0053	0.16	rs6239339	0.0009	0.52	−0.0003	0.52
rs13479402	−0.0041	0.29	0.0034	0.29	rs6243819	0.0009	0.51	−0.0002	0.51
rs13479465	−0.0424	0.64	0.0163	0.64	rs6245801	−0.0015	0.33	0.0019	0.33
rs13479621	−0.0090	0.75	0.0055	0.75	rs6245977	−0.0277	0.92	0.0213	0.92
rs13479823	−0.0012	0.91	0.0014	0.91	rs6250696	0.0007	0.44	−0.0001	0.44
rs13479828	−0.0007	0.47	0.0008	0.47	rs6280404	−0.0029	0.69	0.0000	0.69
rs13479930	0.0437	0.29	−0.0399	0.29	rs6280411	−0.0149	0.33	0.0162	0.33

(continued)

Table 8.6 *(continued)*

SNP	CD4/CD8		CD4:CD3		SNP	CD4/CD8		CD4:CD3	
	Est.	OOI	Est.	OOI		Est.	OOI	Est.	OOI
rs13480141	−0.0007	0.39	0.0057	0.39	rs6293022	−0.0007	0.47	0.0008	0.47
rs13480148	−0.0035	0.50	0.0066	0.50	rs6296189	0.0009	0.64	−0.0011	0.64
rs13480277	0.0038	0.37	−0.0023	0.37	rs6301008	0.0009	0.52	−0.0003	0.52
rs13480400	0.0057	0.33	−0.0043	0.33	rs6315002	−0.0030	0.57	0.0042	0.57
rs13481042	0.0384	0.49	−0.0330	0.49	rs6315152	−0.0247	0.51	0.0306	0.51
rs13481242	−0.0208	0.31	0.0184	0.31	rs6317255	−0.0027	0.88	0.0026	0.88
rs13481276	−0.0022	0.59	0.0021	0.59	rs6317361	−0.0007	0.36	0.0005	0.36
rs13481353	0.0024	0.31	0.0000	0.31	rs6334723	−0.0506	0.97	0.0285	0.97
rs13481363	−0.0559	0.51	0.0629	0.51	rs6342158	−0.0255	0.38	0.0269	0.38
rs13481411	−0.0304	0.84	0.0170	0.84	rs6342799	−0.0002	0.40	0.0001	0.40
rs13481445	0.0022	0.59	−0.0021	0.59	rs6343634	0.0009	0.52	−0.0003	0.52
rs13481446	0.0022	0.59	−0.0021	0.59	rs6355384	−0.0079	0.83	0.0043	0.83
rs13481499	0.0020	0.56	−0.0018	0.56	rs6360080	0.0201	0.50	−0.0105	0.50
rs13481571	−0.0458	0.32	0.0344	0.32	rs6361420	0.0003	0.36	−0.0003	0.36
rs13481618	0.0134	0.40	−0.0039	0.40	rs6373215	−0.0005	0.61	0.0001	0.61
rs13481788	0.0003	0.52	−0.0001	0.52	rs6377183	−0.0002	0.30	0.0001	0.30
rs13481961	−0.0019	0.40	0.0014	0.40	rs6378040	−0.0004	0.44	0.0001	0.44
rs13482007	0.0118	0.87	−0.0148	0.87	rs6393715	0.0049	0.73	−0.0132	0.73
rs13482225	−0.0020	0.39	0.0010	0.39	rs6395241	−0.0013	0.45	0.0010	0.45
rs13482388	0.0147	0.53	−0.0032	0.53	rs6401555	−0.0118	0.71	0.0041	0.71
rs13482668	−0.0001	0.20	0.0007	0.20	rs6406122	0.0008	0.70	−0.0003	0.70
rs13482738	0.0059	0.68	−0.0138	0.68	rs6409750	−0.0846	0.98	0.1197	0.98
rs13482968	0.0687	1.00	−0.1134	1.00	rs6411422	0.0124	0.47	−0.0045	0.47
rs13483373	−0.0194	0.61	0.0132	0.61	rs8250941	0.0005	0.24	−0.0004	0.24
rs13483448	0.0019	0.35	−0.0040	0.35	rs8261820	0.0001	0.30	−0.0001	0.30
rs3023435	−0.0053	0.57	0.0042	0.57	rs8270116	0.0053	0.52	−0.0047	0.52

References

Breheny, P., and Huang, J. 2009. Penalized methods for bi-level variable selection. *Statistics and Its Interface*, **2**, 369–380.

Breheny, P., and Huang, J. 2011. Coordinate descent algorithms for nonconvex penalized regression, with applications to biological feature selection. *Annals of Applied Statistics*, **5**, 232–253.

Guerra, R., and Goldstein, D. R. 2009. *Meta-analysis and Combining Information in Genetics and Genomics*. Chapman and Hall/CRC.

Horvath, S. 2011. *Weighted Network Analysis: Applications in Genomics and Systems Biology*. Springer.

Huang, J., and Ma, S. 2010. Variable selection in the accelerated failure time model via the bridge method. *Lifetime Data Analysis*, **16**, 176–195.

Huang, J., Ma, S., Li, H., and Zhang, C. H. 2011. The sparse Laplacian shrinkage estimator for high dimensional regression. *Annals of Statistics*, **39**, 2021–2046.

Huang, Y., Huang, J., Shia, B. C., and Ma, S. 2012. Identification of cancer genomic markers via integrative sparse boosting. *Biostatistics*, **13**, 509–522.

Liu, J., Huang, J., and Ma, S. 2012. Analysis of genome-wide association studies with multiple outcomes using penalization. *PLoS ONE*, **7**, e51198.

Liu, J., Huang, J., and Ma, S. 2013a. Incorporating network structure in integrative analysis of cancer prognosis data. *Genetic Epidemiology*, **37**, 173–183.

Liu, J., Huang, J., and Ma, S. 2013b. Integrative analysis of multiple cancer prognosis datasets under the heterogeneity model. *Springer Proceedings in Mathematics and Statistics*, **55**, 257–269.

Liu, J., Shi, X., Huang, J., and Ma, S. 2013c. Penalized integrative analysis of high-dimensional omics data. *http://works.bepress.com/shuangge/*.

Liu, J., Huang, J., and Ma, S. 2014a. Integrative analysis of cancer diagnosis studies with composite penalization. *Scandinavian Journal of Statistics*, **41**, 87–103.

Liu, J., Huang, J., Zhang, Y., Lan, Q., Rothman, N., Zheng, T., and Ma, S. 2014b. Integrative analysis of prognosis data on multiple cancer subtypes. *Biometrics*, **70**, 480–488.

Lorenz, A. J., Chao, S., Asoro, F. G., Heffner, E. L., Hayashi, T., Iwata, H., Smith, K. P., Sorrells, M. E., and Jannink, J. L. 2011. Genomic selection in plant breeding: knowledge and prospects. *Advances in Agronomy*, **110**, 77–123.

Ma, S., Huang, J., and Moran, M. 2009. Identification of genes associated with multiple cancers via integrative analysis. *BMC Genomics*, **10**, 535.

Ma, S., Huang, J., and Song, X. 2011. Integrative analysis and variable selection with multiple high-dimensional datasets. *Biostatistics*, **12**, 763–775.

Ma, S., Dai, Y., Huang, J., and Xie, Y. 2012. Identification of breast cancer prognosis markers via integrative analysis. *Computational Statistics and Data Analysis*, **56**, 2718–2728.

Simon, N., Friedman, J., Hastie, T., and Tinshirani, R. 2013. A sparse-group lasso. *Journal of Computational and Graphical Statistics*, **22**, 231–245.

Tseng, P. 2001. Convergence of a block coordinate descent method for nondifferentiable minimization. *Journal of Optimization Theory and Applications*, **109**, 475–494.

Valdar, W., Scolberg, L. C., Gauguier, D., Burnett, S., Klenerman, P., Cookson, W., Taylor, M., Rawlins, J., Mott, R., and Flint, J. 2006a. Genome-wide genetic association of complex traits in heterogeneous stock mice. *Nature Genetics*, **174**, 879–887.

Valdar, W., Scolberg, L., Gauguier, D., Cookson, W., Rawlins, J., Mott, R., and Flint, J. 2006b. Genetic and environmental effects on complex traits in mice. *Genetics*, **174**, 959–984.

Zhang, B., and Horvath, S. 2005. A general framework for weighted gene co-expression network analysis. *Statistical Applications in Genetics and Molecular Biology*, **4**, 45.

Zhang, C. H. 2010. Nearly unbiased variable selection under minimax concave penalty. *Annals of Statistics*, **38**, 894–942.

Zhang, C. H., and Huang, J. 2008. The sparsity and bias of the lasso selection in high-dimensional linear regression. *Annals of Statistics*, **36**, 1567–1594.

9

A Bayesian Graphical Model for Integrative Analysis of TCGA Data: BayesGraph for TCGA Integration

YANXUN XU, YITAN ZHU, AND YUAN JI

9.1 Introduction

9.1.1 A Brief Introduction of TCGA

The Cancer Genome Atlas (TCGA) is a research project supported by the National Cancer Institute and the National Human Genome Research Institute to chart the genomic changes involved in more than 20 types of cancer (Network, 2008, 2012). TCGA generates the most comprehensive cancer genomic data consisting of whole-genome measurements of multiple features (such as DNA sequence, copy number, methylation, and expressions) on thousands of matched cancer patient samples. TCGA cancer genomic data have already been widely used for cancer research during the past a few years (The Cancer Genome Atlas, https://tcga-data.nci.nih.gov/tcga/). At the end of 2012, the number of monthly unique visitors to the TCGA data portal reaches close to 1000. More than 200 grant applications cite TCGA data in 2012 and 157 papers using TCGA data were published. We expect the growth of TCGA data usage be dramatic in the next few years.

A hallmark of TCGA and TCGA data is *multimodality*. That is, *multiple* genomic characterizations, such as DNA copy number, gene expression, protein expression. are measured for the same set of biological samples across *multiple* cancers. Integrative analyses of the multimodal data provide opportunities for a systematic examination across genomic spectrum of cancer. In particular, we apply a class of Bayesian graphical models to study intragenic interactions between three genomic features, mRNA gene expression, DNA copy number variation (CNV) and DNA methylation.

Transcription is a critical genetic process in which DNA is transcribed to RNA. Perturbation of transcription directly affects mRNA expression and hence the subsequent protein production, leading to pathological states. Genetic variations such as CNVs and DNA methylations of the same gene frequently

205

contribute to disrupted gene expression. Such disruption can be detected by learning the intragenic functional interaction between the associated variations. CNVs result in an abnormal number of copies of DNA and thus change the gene expression level and associated phenotypes. For example, a deletion (loss of both DNA copies) of *PAX5* has been found to be associated with acute lymphoblastic leukemia (Shlien and Malkin, 2009). DNA methylation is a biochemical modification that adds a methyl group to the 5 position of the cytosine pyrimidine ring or the number 6 nitrogen of the adenine purine ring. There is strong evidence that abnormal hypermethylation at the gene promoter region results in transcriptional silencing of tumor suppressor genes. Also, aberrant DNA methylation patterns have been associated with a large number of human malignancies such as cancer, lupus, and a range of birth defects (Robertson, 2005). Therefore, elucidating tumor-specific methylation changes will shed light on potential clinical applications in cancer diagnosis, prognosis and therapeutics (Das and Singal, 2004).

Most literature focused on pair-wise integrations, between CNV and mRNA or between methylation and mRNA (Bussey et al., 2006; Waaijenborg et al., 2008; Menezes et al., 2009; Choi et al., 2010; Amandine et al., 2010; Trentini et al., 2013). To investigate the relationship between CNV and mRNA expression, early studies used simple statistical methods such as permutation tests. Bussey et al. (2006) computed the Pearson's correlation coefficients and tested the significance of correlations using false discovery rate (FDR) control. Waaijenborg et al. (2008) proposed a penalized canonical correlation analysis to study genome-wide association between DNA copy number and mRNA expression. Menezes et al. (2009) modeled the relationship of DNA copy number and mRNA expression by a linear model based on a modified correlation coefficient and an explorative Wilcoxon test. Choi et al. (2010) described a Bayesian double-layered mixture model that directly modeled the stochastic nature of CNV and identified abnormally expressed genes due to aberrant copy number. Amandine et al. (2010) and Trentini et al. (2013) investigated the effect of methylation on mRNA expression in glioblastoma, and CNV on mRNA expression in breast cancer, respectively.

Both CNV and DNA methylation play important roles in transcription, so an integrated analysis that models all three features together is most appropriate for understanding the interactions among them. However, inference on intragenic interactions requires all three features, copy number, methylation, and gene expression, measured across matched biological samples. TCGA data are perfectly suited for such an integrative analysis that involves three different features. Denoting with C, M, and E the three genomic features of CNV, methylation, and mRNA expression, we apply a Bayesian graphical model (Mitra et al., 2013) to learn the dependence structure of the three features through a graph. The vertices of the graph represent the features, and the presence or absence of edges indicates the conditional dependence or independence between the features, respectively. For example, an edge between M and E and a lack of edge between C and E implies methylation-controlled transcription, which is robust to copy number changes. In other words, the mRNA expression is sensitive to methylational variation but not copy number variation.

Transcriptional regulation is a complex biological process involving factors beyond CNV and DNA methylation. For example, transcription factor genes affect the expression of their targeted genes, and post-transcriptional modifications such as miRNA-mRNA interaction also play an important role on the final gene expression level. These factors can be easily incorporated into the framework of our Bayesian graphical models, each treated as a distinct node in the graph. In this sense, the work presented in the chapter is essential to future work of integration of other factors.

9.1.2 Graphical Models

Exploring conditional independence among a set of random variables is a classical statistical inference problem. For a set of continuous variables following a multivariate normal distribution, the problem reduces to inference for the inverse covariance matrix. However inference becomes challenging when the number of variables is large, as covariance estimates become unstable. One solution is to impose certain constraints on the covariance matrix, acting as latent layers in the model framework, and decided based on a scoring criterion. Combining this general idea with Gaussian modeling of the data one could apply the well-known Gaussian graphical models (GGMs). As the name implies, GGMs use a graph treating the variables as vertices and imposing a multivariate Gaussian distribution on the variables. The graph pictorially represents the conditional dependence between the variables. More importantly it implies the form of the joint distribution. Zeroes in the inverse covariance matrix, conditional independence, correspond to the absence of edges between the vertices. The problem of finding an optimal graph is thus equivalent to finding optimal constraints on the inverse covariance matrix. Yuan and Lin (2007) formulated this as a covariance optimization problem with a linear constraint on the number of positive entries in the covariance matrix. Lasso techniques were used to obtain a solution. However, the solution is sensitive to the optimization path, and the proposed method is not very effective in controlling the number of false positives.

Alternatively, Bayesian inference enables a natural stochastic exploration of the graphical space by introducing priors on the graph itself to regularize the otherwise unstable estimation and performing Markov chain Monte Carlo (MCMC) simulations. A commonly used prior is the inverse Wishart (IW) prior on the covariance matrix. The advantages of this prior are the ensured positive definiteness, ease of posterior computations using Gibbs sampling, and shrinkage of covariance estimates toward a particular structure. For details see, for example, Daniels and Pourahmadi (2002). These models, however, cannot be applied universally to all possible graphical structures and must be regulated depending on the type of graphs of interest. See Dawid and Lauritzen (1993) and Dobra et al. (2004) for a discussion.

For the application to the TCGA data, we consider Markov random fields (MRF) models, and introduce binary latent indicators of presence of genomics variations. We propose to perform full Bayesian inference on this more general graphical structure.

9.1.3 Markov Random Fields

We define an MRF as a pair $G = (V, \mathcal{E})$, where V is a set of vertices and $\mathcal{E}$ is a set of undirected edges. The vertices correspond to the variables, in our case genomic features, C, M, and E for a single gene. The edges in $\mathcal{E}$ are a subset of $\{\{i, j\}, i \neq j \in V\}$. A path is defined as an ordered set of vertices $(i_0, i_1, \ldots i_n)$ such that $\{i_{k-1}, i_k\} \in \mathcal{E}$ for $k = 1, \ldots, n$. To understand how an MRF encodes conditional independence, we first define the Markov property for an MRF.

Definition: Markov Property of an MRF. Let V_a, V_b, $V_c \subseteq V$ denote three subsets of vertices. We define V_a as conditionally independent of V_b given V_c if every path from a vertex $i_1 \in V_a$ to a vertex $i_2 \in V_b$ passes through a vertex $i_3 \in V_c$. This specification of conditional independence is called the global Markov property.

The Markov property of an MRF can also be defined alternatively. For example, we can restate the Markov property as a local condition in which each variable is conditionally independent of others given its neighbors. This is known as the local Markov property. The local Markov property is equivalent to the global Markov property under the positivity condition, which simply states that the probability of all points in the discrete sample space is strictly greater than zero. The Hammersley Clifford theorem assures us that any given conditional independence structure can be represented by an MRF, see Besag (1974). For our TCGA inference, there are only three vertices, local and global

Markov properties are the same. The absence of an edge between any two vertices implies the conditional independence of the corresponding two variables. For example, if there is no edge between M and E, but an edge between C and E, the mRNA expression is dependent on copy number but not methylation. This could be explained by methylation localized to only a single copy of the gene but not other copies.

Note that we do not consider directed edges. For TCGA applications, MRFs that do not consider directionality are suitable for two reasons. First, there are not time-course data in TCGA thus virtually eliminating the possibility of performing formal statistical inference based on directed graphs. Second, if needed, directionality can be easily deduced from biological knowledge. For example, an edge between C and E of the same gene implies that the CNV of that gene affects the mRNA expression of the gene, $C \to E$.

9.2 Methods

9.2.1 Probability Model

For a single gene, data are arranged in an $S \times T$ matrix $Y = [y_{it}]$ with rows i representing the genomic features of the gene, columns t representing different samples, and each element y_{it} representing the measurement of each feature for each sample, $i = 1, \ldots, S$ and $t = 1, 2, \ldots, T$. The proposed model introduces latent trinary indicators $z_{it} \in \{-1, 0, 1\}$ with interpretation as under-, regular and over-expression of the corresponding measurement as follows:

$$z_{it} = \begin{cases} -1 & \text{abnormally low measurement} \\ 0 & \text{normal measurement} \\ 1 & \text{abnormally high measurement} \end{cases}$$

Using z_{it} we apply the mixture model proposed by Parmigiani et al. (2002) for y_{it} given by

$$(y_{it} - \mu_i) \mid z_{it}, \boldsymbol{\theta}_i \sim I[z_{it} = -1]U(y_{it} \mid -k_{i-}, 0) + I[z_{it} = 0]N(y_{it} \mid 0, \sigma_i^2)$$

$$+I[z_{it} = 1]U(y_{it} \mid 0, k_{i+}), \tag{9.1}$$

where $I[\cdot]$ is the indicator function, μ_i is the random effect of feature i, $U(A)$ denotes a uniform distribution over the set A, and $N(\cdot \mid \mu, \sigma^2)$ denotes a normal distribution with mean μ and variance σ^2. In words, we assume a mixture model with uniform, normal and uniform components corresponding to under-, regular and over-expression. The vector $\boldsymbol{\theta}_i = (\mu_i, \sigma_i^2, k_{i-}, k_{i+})$ collects all the other parameters.

We subsequently convert the trinary variable z_{it} to a binary variable e_{it} with $p(z_{it}|e_{it} = 0) = \delta_{-1}(z_{it})$, and

$$p(z_{it} = 0|\pi_i, e_{it} = 1) = \pi_i, \quad p(z_{it} = 1|\pi_i, e_{it} = 1) = 1 - \pi_i.$$

This conversion is devised to set up the following graphical model.

Denote $V = \{1, \ldots, S\}$ the set of S vertices representing S genomic features. We use a graph on these vertices to characterize the dependence structure across different features. Recall that a graph is a pair $G = \{V, \mathcal{E}\}$ where $\mathcal{E}$ is a set of undirected edges $\{i, j\}, i, j \in V$. A graph G is used to describe the conditional independence structure of a set of variables indexed by V, for example the binary indicators $\{e_{it}, \; i \in V\}$ in the case of our application. The absence of an edge $\{i, j\}$ indicates conditional independence of e_{it} and e_{jt} given the remaining variables $e_{kt}, \; k \neq i, k \neq j$. Any joint probability model $p(e_{1t}, \ldots, e_{St})$ that respects the dependence structure G can be written as (Besag, 1974)

$$p(e_t \mid \boldsymbol{\beta}, G) = p(0 \mid \boldsymbol{\beta}, G) \times \exp\left\{\sum_{i=1}^{S} \beta_i e_{it} + \sum_{\{i,j\} \in V; i < j} \beta_{ij} e_{it} e_{jt}\right\}, \tag{9.2}$$

where $e_t = (e_{1t}, \ldots, e_{St})$ and $\boldsymbol{\beta} = (\beta_1, \ldots, \beta_S, \beta_{12}, \ldots, \beta_{S-1,S})$. For instance, we have $\boldsymbol{\beta} = (\beta_1, \beta_2, \beta_3, \beta_{12}, \beta_{23}, \beta_{13})$ when $S = 3$. Coefficients β_{ij} are non-zero only when the corresponding edge is included in the graph. Model (9.2) is known as the autologistic model.

Caragea and Kaiser (2009) and Hughes et al. (2011) proposed an alternative MCMC scheme called centered parametrization, which improves mixing of posterior simulation and simplifies prior specification. The centered version is used in the form of

$$p(e_t \mid \boldsymbol{\beta}, G) = p(0 \mid \boldsymbol{\beta}, G) \exp\left\{\sum_{i=1}^{S} \beta_i e_{it} + \sum_{\{i,j\} \in V; i < j} \beta_{ij}(e_{it} - v_i)(e_{jt} - v_j)\right\}, \tag{9.3}$$

where $v_i = \exp(\beta_i)/\{1 + \exp(\beta_i)\}$.

The joint model factors as

$$p(Y, z, e, \pi, \theta, \boldsymbol{\beta}, G) = p(Y \mid z, \theta)p(z \mid e, \pi)p(e \mid \boldsymbol{\beta}, G)p(\theta)p(\boldsymbol{\beta} \mid G)p(G). \tag{9.4}$$

We introduce the priors $p(\boldsymbol{\theta})$, $p(\boldsymbol{\beta} \mid G)$, and $p(G)$ next. Let $\text{Ga}(a, b)$ denote a gamma distribution with mean a/b. We assume conditionally conjugate priors

$$\mu_i \sim N(0, \tau_\mu), \quad \frac{1}{\sigma_i^2} \sim \text{Ga}(\gamma_\sigma, \lambda_\sigma),$$

$$\frac{1}{k_{i-}} \sim \text{Ga}(\gamma_{k_{i-}}, \lambda_{k_{i-}}), \quad \frac{1}{k_{i+}} \sim \text{Ga}(\gamma_{k_{i+}}, \lambda_{k_{i+}}),$$

$$\beta_\star \sim N(0, \sigma_\beta^2), \quad \pi_i \sim U(0, 1),$$

where $\beta_\star$ stands for the coefficients β_i, β_{ij} in (9.3).

Last, we define a model $p(G)$ with two general options, and use one for TCGA analysis. Let $G_0 = (V, \mathcal{E}_0)$ be a prior guess of the dependence structure. For genomic inference, G_0 can be often easily elicited. For example, one could connect the edge between C and E, since CNV is biologically known to be positively related to gene expression. Therefore, G_0 woud be a graph with three vertices C, M, and E and an edge set $\mathcal{E}_0$ that has only one edge connecting C and E. Knowing G_0, the first option of the prior of G is based on the number of changes to G_0 by assuming a geometric kernel

$$p(G) \propto \rho^{d(G, G_0)}, \tag{9.5}$$

where $d(G, G_0) = |\mathcal{E} \cap \mathcal{E}_0^c| + |\mathcal{E}^c \cap \mathcal{E}_0|$ and $\rho \in (0, 1)$. This prior setting imposes less weight on graphs that are more distant from G_0 and the weights decrease exponentially when d increases. The prior (9.5) works well for large graphs with say, 10 vertices. In real data applications for TCGA, we suggest an alternative uniform distribution over all possible graphs since there are only three vertices in the graph G. For instance, we only need to consider up to 8 possible graphs with three available features. Therefore, we define a uniform distribution $p(G)$ over all possible graphs. Each possible graph is given a prior probability of 1/8.

9.2.2 Markov Chain Monte Carlo (MCMC) Simulations

We carry out posterior inference for model (9.4) using MCMC simulations. Each iteration of the MCMC scheme includes the following transition probabilities

$$p(e \mid Y, \pi, \alpha, \theta, \beta, G), \ p(z \mid Y, \alpha, e), \ p(\pi \mid z),$$

$$p(\theta \mid Y, z, \alpha), \ p(\beta \mid e, G), \ p(G \mid e, \beta).$$

We start by generating e_{it} from its complete conditional posterior. Following the update of e, we generate values for z from complete conditional posterior

$p(z \mid Y, \alpha, e)$. If $e_{it} = 0$, the update is deterministic, $z_{it} = -1$. If $e_{it} = 1$, the update requires a Bernoulli draw for $z_{it} = 0$ versus $z_{it} = 1$. The update of parameters θ is straightforward. Resampling G and the regression coefficients β could be challenging in large graphs, essentially because of the difficult evaluation of the normalization constant $p(0 \mid \beta, G)$ in (9.3) (see, e.g. Mitra et al., 2013). However, here the graph has only three features, $p(G)$ is only supported over $2^3 = 8$ possible graphs, making the evaluation of the normalization constant straightforward. Thus, resampling G and β reduces to straightforward trans-dimensional MCMC as in Green (1995).

9.2.3 Posterior Inference using False Discovery Rates

Statistically, owing to our fully model-based inference using posterior probabilities, we can easily assess the noise associated with the genomic data. This is a major advantage of our proposed Bayesian modeling approach over other algorithm-driven methods. Denoting λ a generic symbol for a probability of interest, and adopting the methods introduced by Newton et al. (2004) and Müller et al. (2004), we compute the posterior expected false discovery rate ($p\overline{\text{FDR}}$) for a given cutoff λ_0, given by

$$
p\overline{\text{FDR}}(\lambda_0) = \frac{\sum_{i=1}^{S} \sum_{j>i} (1 - \hat{\lambda}_{ij}) I(\hat{\lambda}_{ij} \leq \lambda_0)}{\sum_{i=1}^{S} \sum_{j>i} (\hat{\lambda}_{ij} \leq \lambda_0)},
$$

where $I(\cdot)$ is the indicator function and $\hat{\lambda}_{ij}$ is a posterior estimate of λ_{ij}. Different cutoff values λ_0 can be used for FDR control such that $p\overline{\text{FDR}} < p_0$ for a desirable rate p_0.

9.3 Simulation Study

9.3.1 Simulation Setup

We examined the performance of our model with three simulated data sets, each with $T = 350$ samples and $S = 3, 4, 5$ features, respectively. For each simulation, a true graph G was first generated. For each pair of vertices $\{i, j\}$, we generated the edge with probability 0.5. For each imputed edge $\{i, j\}$, we generated values of β_{ij} from $N(\mu_1, 0.5^2)$, with $\mu_1 \sim U(-4, 4)$. We generated β_i, the autologistic intercept in (9.3) from $N(\mu_2, 0.5^2)$, and $\mu_2 \sim U(-0.4, 0.4)$. Then, we generated e for $T = 350$ samples. Since $p(z_{it} \mid e_{it} = 0) = \delta_{-1}(z_{it})$, $p(z_{it} = 0 \mid \pi_i, e_{it} = 1) = \pi_i$, and $p(z_{it} = 1 \mid \pi_i, e_{it} = 1) = 1 - \pi_i$, we first generated $\pi_i \sim U(0.2, 0.8)$ and then generate z. Furthermore, we set

$\mu_i = 0$, $\sigma_i = 0.316$, $k_{i-} = 4$, and $k_{i+} = 4$ for each feature i. The hyper-parameters were $\tau_\mu = 1$, $\gamma_\sigma = 2$, $\lambda_\sigma = 0.1$, $\gamma_{k_+} = 11$, $\lambda_{k_+} = 40$, $\gamma_{k_-} = 11$, $\lambda_{k_-} = 40$, and $\sigma_\beta = \sqrt{10}$.

In Figure 9.1, the plots (left) present the three simulated true graphs. Black edges and red edges represent positive and negative relationships, respectively. For instance, in data set 1, features 1 and 2 are positively related, and features 1 and 3 are negatively related. Features 2 and 3 are conditionally independent given feature 1, $\beta_{23} = 0$ in the autologistic model (9.3).

9.3.2 Simulation Results

We implemented the proposed graphical model to compute the posterior summaries for each simulated data set. The posterior estimates were obtained by MCMC posterior simulation with 10,000 iterations, of which the first 5,000 were discarded as burn-in. We calculated the posterior inclusion probability q_{ij} for each possible edge $\{i, j\}$, defined as

$$q_{ij} = \frac{1}{5000} \sum I(\{i, j\} \in \mathcal{E})$$

substituting the edge set $\mathcal{E}$ of the imputed graph for each iteration of the MCMC. We obtained the posterior estimated graph $\hat{G}$ by thresholding, based on criterion $\{q_{ij} > q_0\}$, using the posterior inclusion probability q_{ij} for each edge. The threshold q_0 was chosen so that the posterior expected false discovery rate $\overline{pFDR}(q_0) \leq 0.1$. We also reported parameter estimates of regression coefficients $\bar{\boldsymbol{\beta}} = E(\boldsymbol{\beta} \mid Y)$, the posterior mean for the autologistic coefficients. Figure 9.1 plots the posterior estimated graph for the simulated data. The number next to each edge represents either the true value (left) or the posterior mean β_{ij}'s (right). We can see that the estimated graph match the simulation truth for all three datasets, with similar estimated values of the β's.

Since graph G is modeled as a random variable, we also reported the inference $r = P(G = G_0 | data)$, here G_0 is the simulation truth. For the three data sets $r = 0.534, 0.52$, and 0.236, respectively. The last r value is smaller since the true graph in that last simulation had more edges, increasing the complexity in the estimation. In other words, the less sparse the true graph is, the less powerful the inference.

The sign of β_{ij} has an intuitively appealing interpretation related to the effect of the j-th feature on the probability of presence of i-th feature, keeping the other feature fixed. Let $e_{-ij} = \boldsymbol{e} \setminus \{e_{it}, e_{jt}\}$. It can be easily shown that β_{ij} is the log odds ratio of e_{it} and e_{jt} through simple algebra, where $\beta_{ij} > 0$ implies that $p(e_{it} = 1 \mid e_{jt} = 1, e_{-ij}) > p(e_{it} = 1 \mid e_{jt} = 0, e_{-ij})$.

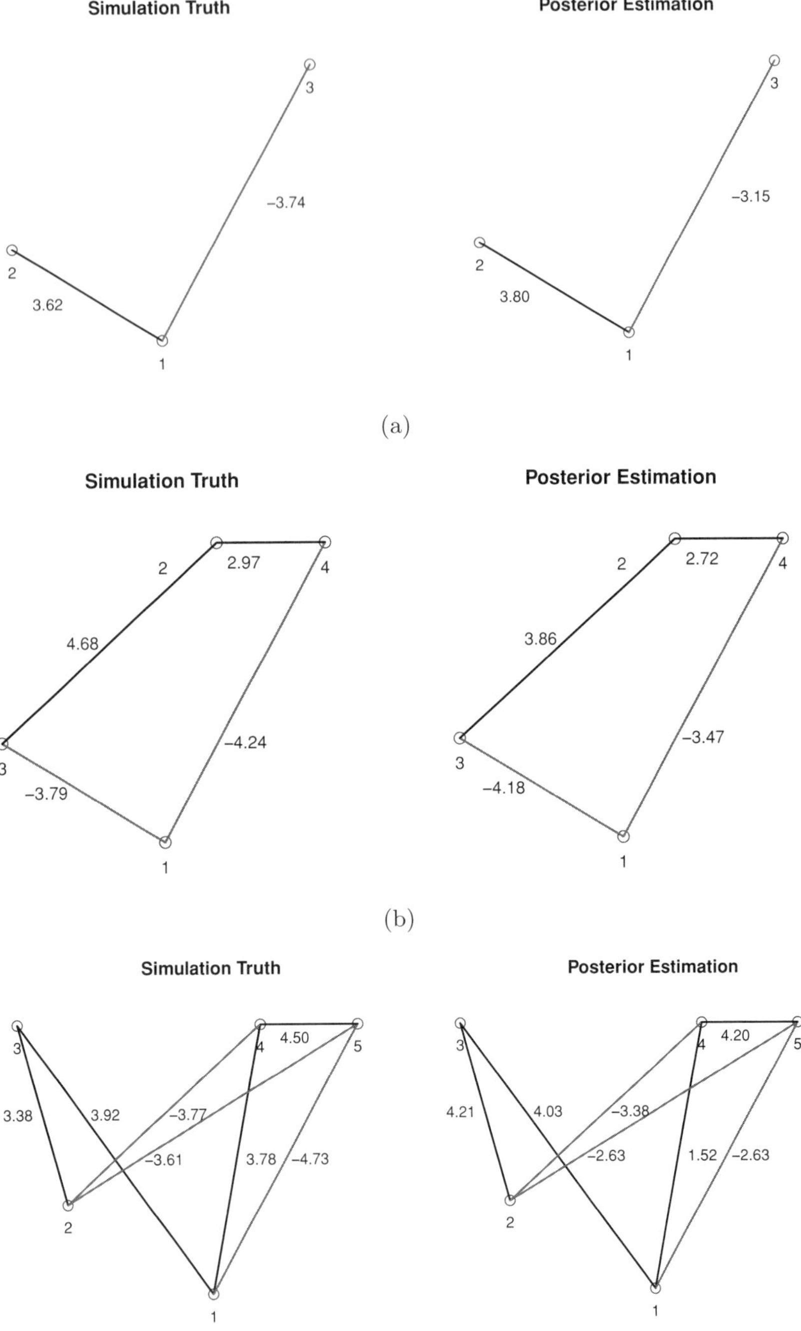

Figure 9.1 The simulation truth versus the estimated graph for three simulated data sets. Edge colors black and red represent positive and negative relationships. The solid line represents that the edge exists. The number next to each edge represents either the true value or the posterior mean of the autologistic coefficients β_{ij}'s. The estimated graph based on posterior inference is identical to the simulation truth.

9.3.3 Comparison to Partial Correlation

We show that the proposed Bayesian graphical model is different and arguably more powerful than the simple partial correlation, which computes the association of two random variables in the presence of other random variables. We considered a special case in which we had four features and the true graph was a rhombus, as in data set 2 in the simulation: feature 1 connected with 2, 2 connected with 3, 3 connected with 4, 4 connected with 1. There were no edges between 1 and 3, between 2 and 4 as shown in Figure 9.1 (data set 2). After implementing the proposed graphical model, the posterior estimated graph was the same as the true graph, and the posterior estimates of β were close to the truth.

To compare with inference based on partial correlation, we used R function PCOR.TEST to infer conditional independence between two variables. According to the simulation truth, features 1 and 3 are conditionally independent given both 2 and 4, but dependent conditional on either 2 or 4, but not both. Our graphical model obtained the right inference, providing the identical graphical presentation (Figure 9.1, data set 2). However, using partial correlation we would conclude that features 1 and 3 were conditionally independent given feature 2 ($p < 0.01$), and also conditionally independent given feature 4 ($p < 0.01$). Therefore, partial correlation cannot capture the conditional dependence in this case. two variables (2 and 4).

9.4 TCGA Integrative Analysis

We implemented the proposed graphical model and performed integrative inference based on TCGA data. For each of the 19,304 genes, we performed parallelized analyses separately and summarized results. We aimed to discover the unknown dependence structure among the three features for each gene and display it as a three-vertice graph. We included patient tumor samples from multiple cancer types for analysis. Table 9.1 shows the number of samples selected from each cancer type available from TCGA for our analyses. Specifically, the tumor samples were profiled using genomic assay platforms, including RNA-Seq for mRNA gene expression, genome-wide human SNP array 6.0 for DNA copy number, and HumanMethylation450 BeadChip for DNA methylation. For each gene, all the measurements corresponding to the same genomic feature were standardized across samples within each cancer type to have mean 0 and standard deviation 1. And samples across different cancer types were combined. In this study, we were looking for dependency of genomic features that are common and conserved across different cancer conditions. Namely, the different types of cancer were considered

Table 9.1 *TCGA cancer types and the number of samples included for the integrative analysis*

Cancer type	Samples
Bladder urothelial carcinoma [BLCA]	50
Breast invasive carcinoma [BRCA]	151
Colon adenocarcinoma [COAD]	20
Glioblastoma multiforme [GBM]	26
Head and neck squamous cell carcinoma [HNSC]	202
Kidney renal clear cell carcinoma [KIRC]	252
Lung adenocarcinoma [LUAD]	184
Lung squamous cell carcinoma [LUSC]	103
Skin cutaneous melanoma [SKCM]	162
Thyroid carcinoma [THCA]	204
Uterine corpus endometrioid carcinoma [UCEC]	94
Total	1448

collectively as one meta disease and those shared patterns with strong signals across multiple cancer types can be identified during this process, such as the KEGG Pathways in Cancer that is a cancer molecular regulation map consisting of general and shared cancer signaling mechanisms (http://www.genome.jp/kegg-bin/show_pathway?hsa05200) Genes having missing values in more than 25% of samples for any genomic feature were removed. After filtering, 19,304 genes had complete data including gene expression, DNA copy number, and DNA methylation. We obtained inference results for each of the genes by

Table 9.2 *Summary of the posterior inference for an arbitrarily selected set of genes. A column with name C-E, for example, represents the edge that connects C (CNV) and E (gene expression). PMB: posterior mean of beta $\hat{\beta}_{ij}$ for the edge; PP: posterior selection probability q_{ij}; TGP: top graph probability $Pr(G = \hat{G} \mid data)$ where $\hat{G}$ is the posterior mode of G*

Gene Symbol	C-E(PMB)	C-E(PP)	M-E(PMB)	E-E(PP)	C-M(PMB)	C-M(PP)	TGP
LRRC69	9.3751	1	−6.9453	1	10.1702	1	1
PIGZ	4.7111	0.9375	4.3072	1	−4.2246	0.9475	0.915
SOX3	0.0571	0.1625	−3.3121	1	−0.3238	0.24	0.6225
SETBP1	4.0906	1	4.4236	1	−1.4236	0.6475	0.6475
INF2	4.0893	1	2.1539	0.855	−0.2056	0.2925	0.605
FOXI3	−1.8923	0.7025	−4.0463	1	0.0639	0.1975	0.5725

Posterior Estimation for Gene BCR   Posterior Estimation for Gene GBP4

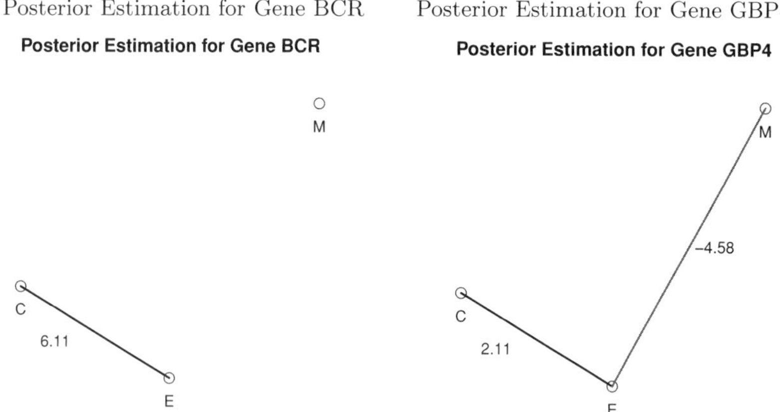

Figure 9.2 Posterior estimated graphs for genes *BCR* and *GBP4*. Black edges represent positive relationships and red edges represent negative relationships. The number next to each edge is the posterior mean of β_{ij}.

running parallel computing jobs on a computer cluster, each job producing the described MCMC posterior samples based on a run of 10,000 iterations with 5,000 burn-in.

We provide an Excel table in supplementary material that summarizes the gene name, posterior mean for the autologistic coefficients, the posterior inclusion probability and FDR for each edge, and the posterior probability for the most frequent graph for all 19,304 genes. The supplementary file can be found at www.ma.utexas.edu/users/yxu/. Genes are listed in descending order according to the posterior probability $Pr(G = \hat{G} \mid data)$, where $\hat{G}$ is the posterior mode. There are 1,006 genes whose $Pr(G = \hat{G} \mid data) > 0.6$. When the cutoff is 0.8, there are 230 genes. When the cutoff is 0.9, there are 141 genes. Table 9.2 shows inferences of several randomly selected genes for demonstration.

Figure 9.2 presents the posterior estimated graphs by including the edges with the largest q_{ij}'s such that $p\overline{\text{FDR}} < 0.1$. Figure 9.3 displays the smooth scatter plots between any two features for genes *BCR* and *GBP4*. From these two figures, we can see that the empirical trends exhibited in the scatter plots are consistent with our model estimation. For instance, there is a strong positive correlation between mRNA expression and CNV for *BCR* in Figure 9.3, and the posterior mean obtained by the proposed graphical model for the mRNA expression-CNV edge in Figure 9.2 is 6.11, indicating a strong positive dependency between the two features.

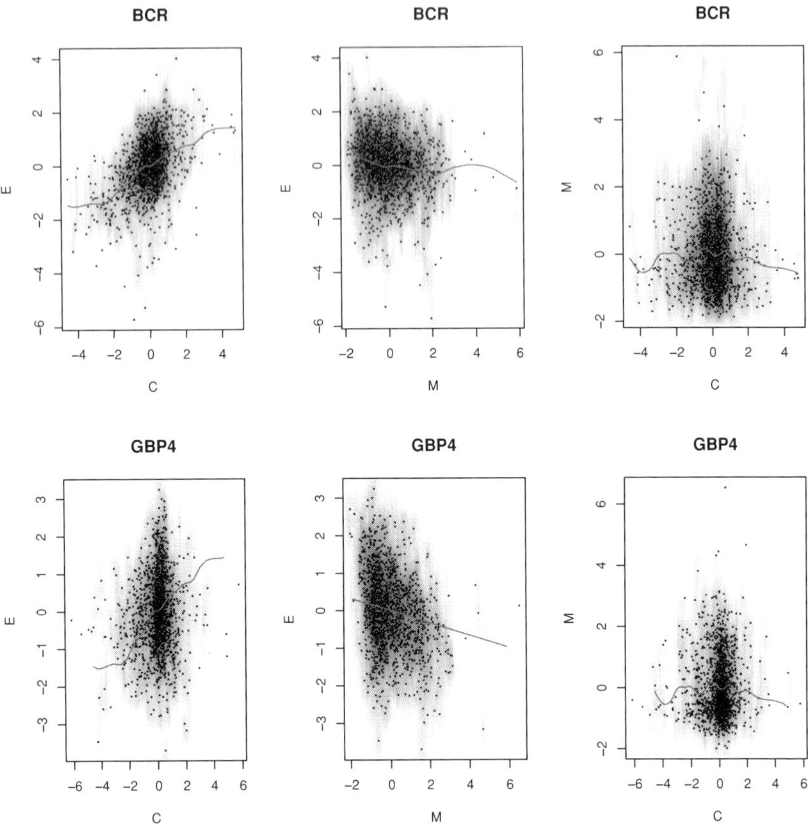

Figure 9.3 Smooth scatter plots of pairwise relationship among features C, M and E for genes *BCR* and *GBP4*. The red line in each smooth scatter plot is the lowess smoother. Dots correspond to the raw expression measurements from the TCGA data with 1,448 samples in Table 9.2.

9.5 Discussion and Conclusion

To study dependence structure of three genetic characterizations, CNV, DNA methylation, and mRNA expression, we propose a Bayesian graphical model. The inferred graph gives a clear representation of the regulatory relationships involving the three genetic features. For example, the mRNA expression of gene *BCR* is sensitive to copy number changes but robust to DNA methylation, while the mRNA expression of gene *GBP4* is sensitive to both copy number change and DNA methylation. Currently, we summarize mRNA, CNV and ME quantifications at the gene level, which means each type of measurement is represented by one single value for a gene. For more complicated considerations such as multiple mRNA isoforms for a gene, we can easily extend our original model by including each isoform as a node. The model and inference mechanism stay the same. Our model can also be easily applied to more features to learn the regulation relationship between features of different genes. For example, gene A could be a transcription factor for gene B if we detect strong relationship between protein expression of gene A and mRNA expression of gene B. As the size of graph increases, scalable and informative priors on large graphs for practical posterior inference need to be taken (Mitra et al., 2013). We are making a comprehensive list of these relationships using the entire TCGA data, expanding the effort to include more cancer types and more features such as microRNA and protein expression.

References

Amandine, E., Marc, A., de Tayrac Marie, V.E., Frederique, G., Stephan, S., Abderrahmane, H., Laurent, R., Philippe, M., Veronique, Q., and Jean, M. 2010. DNA methylation in glioblastoma: impact on gene expression and clinical outcome. *BMC Genomics*, **11**.

Besag, J. 1974. Spatial interaction and the statistical analysis of lattice systems. *Journal of the Royal Statistical Society: Series B)*, **36**(2), 192–236.

Bussey, K.J., Chin, K., Lababidi, S., Reimers, M., Reinhold, W.C., Kuo, W.L., Gwadry, F., et al. 2006. Integrating data on DNA copy number with gene expression levels and drug sensitivities in the NCI-60 cell line panel. *Molecular Cancer Therapeutics*, **5**(4), 853.

Caragea, P.C., and Kaiser, M.S. 2009. Autologistic models with interpretable parameters. *Journal of Agricultural, Biological, and Environmental Statistics*, **14**(3), 281–300.

Choi, Hyungwon, Qin, Zhaohui S, and Ghosh, Debashis. 2010. A double-layered mixture model for the joint analysis of DNA copy number and gene expression data. *Journal of Computational Biology*, **17**(2), 121–137.

Daniels, Michael J, and Pourahmadi, Mohsen. 2002. Bayesian analysis of covariance matrices and dynamic models for longitudinal data. *Biometrika*, **89**(3), 553–566.

Das, P.M., and Singal, R. 2004. DNA methylation and cancer. *Journal of Clinical Oncology*, **22**(22), 4632–4642.

Dawid, A Philip, and Lauritzen, Steffen L. 1993. Hyper Markov laws in the statistical analysis of decomposable graphical models. *The Annals of Statistics*, **21**(3), 1272–1317.

Dobra, Adrian, Hans, Chris, Jones, Beatrix, Nevins, Joseph R, Yao, Guang, and West, Mike. 2004. Sparse graphical models for exploring gene expression data. *Journal of Multivariate Analysis*, **90**(1), 196–212.

Green, P.J. 1995. Reversible jump Markov chain Monte Carlo computation and Bayesian model determination. *Biometrika*, **82**(4), 711–732.

Hughes, John, Haran, Murali, and Caragea, Petruţa C. 2011. Autologistic models for binary data on a lattice. *Environmetrics*, **22**(7), 857–871.

Menezes, R.X., Boetzer, M., Sieswerda, M., Van Ommen, G.J.B., and Boer, J.M. 2009. Integrated analysis of DNA copy number and gene expression microarray data using gene sets. *BMC bioinformatics*, **10**(1), 203.

Mitra, Riten, Müller, Peter, Liang, Shoudan, Yue, Lu, and Ji, Yuan. 2013. A bayesian graphical model for ChIP-seq data on histone modifications. *Journal of the American Statistical Association*, **108**(501), 69–80.

Müller, P., Parmigiani, G., Robert, C., and Rousseau, J. 2004. Optimal sample size for multiple testing. *Journal of the American Statistical Association*, **99**(468), 990–1001.

Network, The Cancer Genome Atlas. 2008. Comprehensive genomic characterization defines human glioblastoma genes and core pathways. *Nature*, **455**(7216), 1061–1068.

Network, The Cancer Genome Atlas. 2012. Comprehensive molecular portraits of human breast tumours. *Nature*, **490**(7418), 61–70.

Newton, Michael A, Noueiry, Amine, Sarkar, Deepayan, and Ahlquist, Paul. 2004. Detecting differential gene expression with a semiparametric hierarchical mixture method. *Biostatistics*, **5**(2), 155–176.

Parmigiani, G., Garrett, E.S., Anbazhagan, R., and Gabrielson, E. 2002. A Statistical Framework for Expression-Based Molecular Classification in Cancer. *Journal of the Royal Statistical Society: Series B*, **64**(4), 717–736.

Robertson, K.D. 2005. DNA methylation and human disease. *Nature Reviews Genetics*, **6**(8), 597–610.

Shlien, Adam, and Malkin, David. 2009. Copy number variations and cancer. *Genome Med*, **1**(6), 62.

Trentini, Filippo, Ji, Yuan, Iwamoto, Takayuki, Qi, Yuan, Pusztai, Lajos, and Müller, Peter. 2013. Bayesian mixture models for assessment of gene differential behaviour and prediction of pCR through the Integration of copy number and gene expression data. *PloS One*, **8**(7), e68071.

Waaijenborg, S., de Witt Hamer, V., Philip, C., and Zwinderman, A. 2008. Quantifying the association between gene expressions and DNA-markers by penalized canonical correlation analysis. *Statistical Applications in Genetics and Molecular Biology*, **7**(1), 1–29.

Yuan, Ming, and Lin, Yi. 2007. Model selection and estimation in the Gaussian graphical model. *Biometrika*, **94**(1), 19–35.

10

Bayesian Models for Flexible Integrative Analysis of Multi-Platform Genomics Data

ELIZABETH J McGUFFEY, JEFFREY S MORRIS, GANIRAJU C MANYAM, RAYMOND J CARROLL, AND VEERABHADRAN BALADANDAYUTHAPANI

Abstract

We present hierarchical Bayesian models to integrate an arbitrary number of genomic data platforms incorporating known biological relationships between platforms, with the goal of identifying biomarkers significantly related to a clinical phenotype. Our integrative approach offers increased power and lower false discovery rates, and our model structure allows us to not only identify which gene(s) is (are) significantly related to the outcome, but also to understand which upstream platform(s) is (are) modulating the effect(s). We present both a linear and a more flexible non-linear formulation of our model, with the latter allowing for detection of non-linear dependencies between the platform-specific features. We illustrate our method using both formulations on a multi-platform brain tumor dataset. We identify several important genes related to cancer progression, along with the corresponding mechanistic information, and discuss and compare the results obtained from the two formulations.

10.1 Introduction

Traditional cancer treatments include surgery, chemotherapy, and radiation. Although these treatments can be lifesaving, after a tumor is removed it is well known that the cancer can relapse [8, 11, 13], and both chemotherapy and radiation treatments have terrible, sometimes permanent, side effects including nausea and vomiting, hair loss, mouth sores, nerve damage, peeling skin, tinnitus, infertility, organ damage, and secondary cancer [16, 1]. The allure of a treatment with lower chances of relapse (thus increasing patient survival) and with reduced side effects (thus improving patient quality of life) has motivated researchers to investigate the development of therapeutic strategies that target the specific genetic causes of a particular type of cancer. In particular, the availability of such therapies facilitates the practice of personalized medicine; a patent's genetic profile can be assayed, and his/her treatment and dosages can be chosen to address the genetic abnormalities specific to the observed profile.

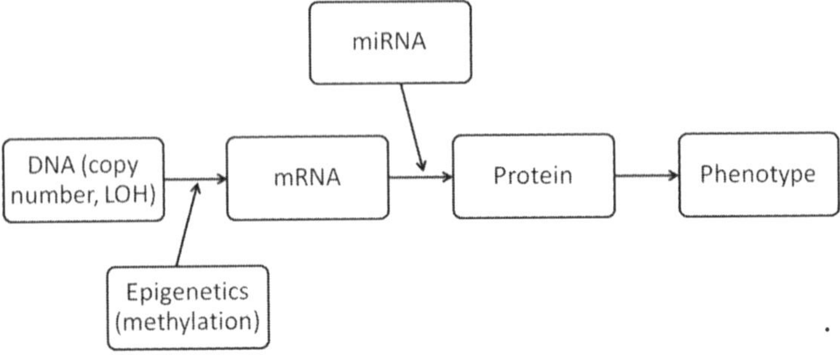

Figure 10.1 Schematic representation of the multiple molecular platforms and their biological relationships.

This approach offers the potential to increase treatment efficacy and decrease incidence of negative side effects on the level of the individual patient.

In order to develop such a targeted treatment, it is first necessary to understand the mechanics of cancer development and progression on a molecular level. In general, within a cell DNA is transcribed to messenger RNA (mRNA); mRNA is then translated into a protein; and a specific action is carried out by the protein. The segments of mRNA that code for different proteins are known as genes, and cancer is believed to involve a complex interaction of these genes. In addition to the direct DNA to mRNA to protein process, there are many different genetic alterations and interferences, such as methylation, copy number, loss of heterozygosity (LOH), etc., that have the potential to affect gene expression (the abundance of mRNA) and thus eventually impact clinical outcomes that manifest as symptoms of disease development (see Fig. 10.1). The discovery of which genes are significantly associated with patient specific outcomes as well as understanding the biological mechanisms associated with such genes' expression is a critical step in the development of targeted therapies.

Due to rapid technological advances combined with decreasing costs, multi-platform datasets are becoming increasingly available on matched patients/tumor samples. Such datasets provide measurements from multiple platforms (mRNA, protein, methylation, copy number, genotype, etc.) on each patient involved in the data collection/study. (We note that when we use the term "platform" in this chapter, we are referring to a biological entity or the molecular characteristic such as methylation, copy number, expression, etc.) While previous analyses generally studied the effect of a single platform on a clinical outcome, these comprehensive datasets have motivated the development of statistical models that integrate data from several platforms, facilitating a deeper and more complete understanding of the genetic causes of cancer development and progression and offering the potential to increase power and lower false discovery rates [23]. Statistical methods attempting to achieve this goal face several analytic challenges: high-dimensionality, complex correlation structures, and unclear (biological) interpretations.

1. *High-dimensionality* arises when the number of predictors/variables (usually on the order of thousands) is greater than the number of patients/samples (usually on the order of a few hundred); this is common when modeling genomic data. It is well-known that only a few of these variables play an active role in disease modulation. Thus, effective strategies need to be developed to address these challenges, especially by inducing shrinkage/sparsity.
2. The issue of *complex correlation structures* refers to the correlations between genes (within a platform), as well as the correlations between platforms that arise due to their complex biological relationships. A model that accounts for these biological relationships should produce more accurate and efficient effect estimates.
3. Lastly, *interpretations* can become complex, especially depending on the technique employed to handle high-dimensionality. Even if the method provides clear interpretations, it is important that there is a straightforward scientific translation. If the results are to be useful in developing targeted therapies, estimates from an integrative model should offer direct insight into the mechanics and the genes related to the clinical outcome.

Many models have been developed recently that face these challenges and integrate data from multiple platforms. The Cancer Genome Atlas (TCGA) has performed large-scale studies on ovarian cancer and Glioblastoma Multiforme (GBM), a brain tumor, in which they analyzed data from multiple platforms such as gene mutations, microRNA expression, and mRNA expression. TCGA researchers began by analyzing each platform separately and then combined the platform results to draw integrated conclusions. The ovarian cancer study identified influential gene pathways and made new discoveries regarding gene interactions, while the GBM study discovered a previously unknown link between MGMT methylation and the mutation spectra of mismatch repair genes [5, 14]. By combining the information gained from multiple platforms, instead of focusing on only one type of data, researchers were able to make novel discoveries and gain important insights into the genetic factors of these cancers.

Other methods integrate multiple platforms directly by incorporating them all into a single model. Tyekucheva et al. proposed integrating data from multiple platforms as predictors in a single logistic model. Their method identifies significant gene sets, and they show that their integrative approach has more power than approaches using a single platform of data [21]. Lanckriet et al.'s integrative method, in which data from each genomic platform is first represented as a kernel function and then all the kernels are included in a classification model, was also shown to have more power than an analogous single-platform approach [12]. Another approach that integrates multiple genomic platforms in a single model was proposed by Shen et al. Their approach, iCluster, uses joint latent variable models to cluster samples into tumor subtypes. Through applications to breast and lung cancer data, iCluster identified potential novel tumor subtypes [18]. Verhaak et al. performed an analysis on the GBM data from TCGA in which they combined gene expression data from multiple types of microarray assays in order to classify tumors into four distinct subtypes, as well as to identify which genes have significant influence on the classification. Since the subtypes respond differently to treatments, this classification information is valuable in making treatment decisions [22]. These and other integrative methods improve our understanding of the genetic causes and mechanics of cancer, and they allow us to make increasingly informed prognosis and treatment decisions.

The integrative Bayesian analysis of genomics data (iBAG) model we present in this chapter was originally proposed by Wang, et al. [23], and was later generalized by Jennings, et al [9]. It is a two-step Bayesian hierarchical model that integrates data from an arbitrary number of platforms while taking into account the biological relationships between DNA characteristics (such as methylation and copy number) and RNA-level entities (such as gene expression). In Section 10.2 we present the model details (both for the linear case and with a novel non-linear extension), and we explain how each of the challenges described above is overcome. In Section 10.3 we illustrate the method on a publicly available Glioblastoma Multiforme (GBM) dataset, and we offer a discussion in Section 10.4.

10.2 iBAG Models

The basic construction of the iBAG model consists of two components: a *mechanistic model* that attempts to capture mechanistic information by partitioning the gene expression into components explained by different upstream platforms, and a *clinical model* that subsequently incorporates these components to model the effects on a clinical outcome of interest. Through the joint estimation of these components, we not only identify the genes significantly related to the clinical outcome, but we also gain insight into the biological mechanisms modulating these effects based on the information from the upstream platforms. (Note that throughout this chapter when we refer to "modulation" by one or more platforms, we mean regulation by that platform or by the interaction across platforms that regulate at different levels. This is not to be confused with the term "modulation" as it is used regarding dynamic conditions in network data.) We present the linear construction first for ease of exposition (Section 10.2.1) and subsequently propose a more flexible non-linear extension in Section 10.2.2.

First we present the notation common to both the linear and non-linear formulations. Let $n =$ number of patients, $J =$ number of platforms being integrated, and $p_j =$ number of genes from platform j, and then define the following terms:

- $mRNA_g$ is the level of gene expression for gene g (where $g = 1, \ldots, \max(p_j);\ j = 1, \ldots, J$) and is of dimension $(n \times 1)$.
- X_{jg} is the part of gene g expression that is attributed to upstream platform j and is of dimension $(n \times 1)$. Specifically, X_{jg} is the product of some platform-specific jth predictor and a fitted coefficient. Details are expounded below.
- O_g represents the "other" (remaining) part of the gene expression that is explained by something other than the $J - 1$ upstream platforms, and is of dimension $(n \times 1)$.
- The covariates X_j and O are the vectorized gene expression effects attributed to platform j $(j = 1, \ldots, J - 1)$ and other sources respectively, and are estimated from the mechanistic model. Each X_j is of dimension $(n \times p_j)$, and O is of dimension $(n \times \max(p_j))$.
- Y denotes the clinical outcome and is of dimension $(n \times 1)$; assumed continuous for now.
- $\boldsymbol{\beta}_j$ is the vector of effects of platform j on Y and is of dimension $(p_j \times 1)$.
- ϵ is the error term and is of dimension $(n \times 1)$.

10.2.1 Linear Case

The linear iBAG model is as follows:

$$mRNA_g = \sum_{j=1}^{J-1} X_{jg} + O_g \tag{10.1}$$

$$Y = \sum_{j=1}^{J-1} X_j \boldsymbol{\beta}_j + O\boldsymbol{\beta}_J + \epsilon \tag{10.2}$$

where Equation 10.1 is the mechanistic model, and Equation 10.2 is the clinical model.

Mechanistic Model

The mechanistic model takes into account the biological relationships between platforms by modeling gene expression as it is affected by each of an arbitrary number of upstream platforms known to influence gene expression. Each of the terms in the mechanistic model are defined as follows:

In the linear case, estimation of the components X_{jg} and O_g is done via least squares regression, and a separate regression is fit for each gene. The mRNA expression is typically summarized for each gene, but the number of raw data values from the other platforms associated with each gene can vary greatly, due to multiple measurements within each gene, i.e., different probes mapped to a specific genomic locations within the gene. For example, in the data we use in Section 10.3, the number of methylation values per gene ranges from 1 to 8, and the number of copy number values per gene ranges from 1 to 16. Thus, to prevent complications due to high-dimensionality, and to match to specific genes, for each gene we begin by performing separate principal component analyses (PCA) on each of the upstream platforms. For each upstream platform, we retain the principal components that account for at least 90% of the variation, and then regress $mRNA_g$ on the corresponding PC scores. For gene g our estimate of X_{jg} is the linear combination of PC scores from platform j and the corresponding coefficient estimates; specifically

$$X_{jg} = \sum_{k=1}^{K} R_{jgk} B_{jgk} \tag{10.3}$$

where R_{jgk} is the raw data value from platform j for gene g with $K = 1$ if there is only a single raw data value, or the kth PC score from platform j for gene g if there are multiple markers, and B_{jgk} is the vector of regression coefficients. O_{jg} is then estimated as the residuals from the least squares regression.

After these steps have been carried out for each gene, we have partitioned each gene into J pieces: $j - 1$ X_{jg}'s that represent the part of gene g expression explained by upstream platform j, and 1 O_g that represents the part of gene g expression not explained by any of the included upstream platforms. We carry forward the X_{jg}'s and O_g's into the clinical model.

Clinical Model

This component models the effect that each of the pieces estimated in the mechanistic model has on a (continuous) clinical outcome. Each of the terms in the clinical model (Eq. 10.2) are defined as follows:

Our main goal is accurate and efficient estimation of the effects $\boldsymbol{\beta}_j$. If a β_{jg}, the coefficient associated with platform j of gene g, is flagged as "significant", then this reveals several aspects: (1) gene i is related to the clinical outcome, (2) platform j is modulating the expression, and (3) a unit increase of gene g expression attributed to platform j is associated with a β_{jg} change in the clinical outcome. To achieve the expected sparsity in the β's (and combat high-dimensionality) we employ a Bayesian hierarchical setup, including a sparsity-inducing prior on the β_j's. As stated previously, it is also important to preserving the estimates of the truly large/important β's, so we must choose a prior that allows for flexible shrinkage. Two popular choices of shrinkage priors are the spike and slab prior and the Laplace prior. The spike and slab is a mixture of a "spike" at 0 and a Gaussian "slab" distribution. It does facilitate shrinkage, but the shrinkage asymptotes to a constant, resulting in large effects being shrunk just as much as small effects. In our scenario we want to avoid this because we believe the large effects are the ones that are truly important, while the effects close to zero are not of interest. The Laplace prior (or the Normal-Exponential prior) results in the Bayesian lasso formulation [17]. The shrinkage offered by the Bayesian lasso is more flexible than that offered by many other priors, including spike and slab; however the single hyperparameter in the Laplace prior limits its flexibility, and the estimates of large effects are shrunk toward zero along with the smaller effects [6].

To induce sparsity but still allow for accurate estimation of truly large, non-zero effects, we choose to employ the Normal-Gamma prior, which provides more adaptive shrinkage that the Lasso prior through its two hyperparameters [6]. Our complete hierarchy (based on a formulation by Griffin and Brown [6]) is as follows:

$$\mathbf{Y} = \text{Normal}(X\boldsymbol{\beta}, \sigma^2 \mathbf{I}_n);$$

$$\boldsymbol{\beta} = \text{Normal}(\mathbf{0}_{\tilde{p}}, D_\psi) \text{ where } D_\psi = \text{diag}(\psi_{11}, \ldots, \psi_{1p_1}, \ldots, \psi_{J1}, \ldots \psi_{Jp_J});$$

$$\psi_{jg} = \text{Gamma}(\lambda_j, 1/(2\gamma_j^2));$$

$$\sigma^2 = \text{InverseGamma}(a, b);$$

$$\lambda_j = \text{Exponential}(c);$$

$$\gamma_j^{-2} = \text{Gamma}(\tilde{a}, \tilde{b}/(2\lambda_j)),$$

where $\tilde{p} = \sum_{j=1}^{J} p_j$ is the total number of predictors in the model. All of the complete conditional distributions are in closed form except for that of λ_j, so we estimate the parameters via Gibbs sampling, with a random walk Metropolis-Hastings update step for the λ_j's. (See Appendix A for complete conditionals.)

Marker Selection

After we obtain MCMC samples, we can obtain point estimates of the parameters in the clinical model (in particular, β) by simply taking the mean of our posterior samples. We also need to identify which markers (gene/platform combinations) to flag as significantly related to the clinical outcome. We use a method based on the median probability model [4], and we flag important markers through these steps:

1. Based on practical considerations, define a minimum effect size δ that is of interest. Further define (δ_-, δ_+) as the region such that you want to flag markers whose effects are outside that region, i.e., less than δ_- or greater than δ_+.

2. Given S MCMC samples and $\beta_{jg}^{(s)}$ is the β_{jg} sample from iteration s, calculate the following posterior probabilities: $p_+(x_{jg}) = \sum_{s=1}^{S} \mathbf{I}(\beta_{jg}^{(s)} > \delta_+)/S$, the posterior probability that β_{jg} is greater than the practical cutoff δ_+, and $p_-(x_{jg}) = \sum_{s=1}^{S} \mathbf{I}(\beta_{jg}^{(s)} < \delta_-)/S$, the posterior probability that β_{jg} is smaller than the practical cutoff δ_-.

3. Flag a marker as significant if either $p_+(\beta_{jg})$ or $p_-(\beta_{jg})$ is greater that 0.5.

10.2.2 Non-Linear Extensions

Previous publications [23, 9] have been limited to linear iBAG models, while here we introduce a nonlinear version of the model. In the linear iBAG case, we assume that the predictors in the mechanistic model (Eq. 10.1) are linearly related to the gene expression. Depending on the upstream platforms included in the analysis and the available information (or lack thereof) regarding how those platforms affect gene expression, such an assumption may not be reasonable. In that case, for each gene we still begin by performing PCA on each platform's raw values, but a more flexible modeling technique for estimating the pieces of the mechanistic model might be necessary to accommodate the additional flexibility.

To achieve this flexibility in the mechanistic model, we propose using generalized additive models (GAM), a class of models proposed by Hastie and Tibshirani in 1986 that replaces the linear $X\beta$ in the generalized linear model formulation for exponential family responses with a sum of smooth functions [7]. The specific formulation for our non-linear mechanistic model is as follows:

$$g\{E(\mathrm{mRNA}_g)\} = b_0 + \sum_{j=1}^{J} \sum_{k=1}^{K_j} f_{jgk}(R_{jgk}) \tag{10.4}$$

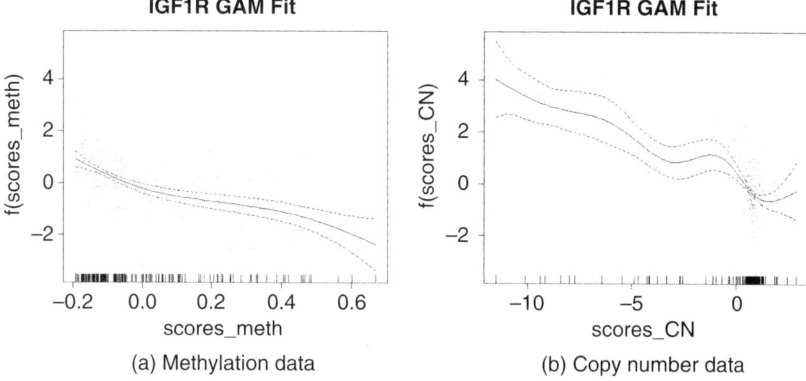

Figure 10.2 For gene *IGF1R* the fitted smooth curves for the methylation data ($f(R_{1,18,1})$) and for the copy number data ($f(R_{1,18,1})$) are plotted. The dots are the partial residuals, i.e., the residuals that would have arisen from not including the predictor of interest (methylation for panel (a) and copy number for panel (b)) but keeping the other estimates fixed. The hash marks on the x-asis are the data values, and the error bounds extend 2 standard deviations above and below the smooth estimate.

where $g(\cdot)$ is a specified link function, $f_{jgk}(\cdot)$ is a smooth function, and R_{jgk} is the kth PC score for platform j of gene g (or the raw values for platform j of gene g if $k = 1$).

The choices we must make are what to use as the $g(\cdot)$ and $f(\cdot)$ functions. The $g(\cdot)$ function is a link function chosen based on the exponential family chosen for the outcome; in our case, we have continuous mRNA values as the response, so we model them as Normal with $g(\cdot)$ as the identity funcion. As for the smooth function $f(\cdot)$, we require a good fit to the data, but one that does not *overfit*. We propose using penalized regression splines, which fit the data closely but include a penalty for too much "wiggliness" as the fit becomes closer to an interpolation. We will implement this model using Wood's R package mgcv and take advantage of the option of automatic smoothness selection for the penalty parameter using GCV (generalized cross validation) [25]. Figure 10.2 illustrates the fit from GAM for IGF1R, one of the genes we will use in our analysis in Section 10.3. This particular gene only has one value for each of two platforms (methylation and copy number), so there were only two smooth functions needed ($f(R_{1,18,1})$ and $f(R_{2,18,1})$), both of which are shown.

After fitting the nonlinear mechanistic model using GAM, we estimate each X_{jg} as $\sum_{k=1}^{K_j} \widehat{f_{jgk}}(R_{jgk})$ and the O_i's as the residuals. We carry these forth into the clinical model, which is the same hierarchical Bayesian model presented in the linear case. The estimation of parameters in the clinical model progresses as previously discussed in the linear case, with the Normal-Gamma prior facilitating adaptive shrinkage. In general, two potential disadvantages of using nonparametric smooth functions as the $f(\cdot)$ functions are (1) a lack of parsimony, and (2) possible difficulties in obtaining a straightforward scientific interpretation of the parameter estimates. For some scenarios this can be problematic, but our goal in this step is simply to partition the gene expression (as accurately as possible) into pieces explained by the different upstream platforms; we are not concerned about parsimony or precise interpretations until the clinical model (Eq. 10.2). Since the estimates from the smooth functions give us the partitioned pieces we need, while the lack of parametric assumptions provides the flexibility for producing a closer fit to the data, we can use them without reservation.

10.3 Illustrations

We apply our method to a publicly available Glioblastoma Multiforme (GBM) dataset from The Cancer Genome Atlas (TCGA). TCGA is an organization that began in 2006 with the goal of compiling and analyzing comprehensive genomic datasets for different types of cancer. Thus far, they provide data on over 20 types of cancer, with GBM being one of the first that they studied [19]. The American Cancer Society estimates that in 2014, there will be 23,380 new cases of brain and nervous system cancers, with 14,320 American fatalities from such cancers [3]. Gliobastoma Multiforme (GBM) is one of the most common and lethal brain tumors, primarily affecting people between 45 and 70 years old [2]. Without radiation treatment, a person with GBM usually lives less than three months; even with radiation treatment, the typical survival time is under fifteen months [10]. Our goal is to apply our method to identify prognostic biomarkers related to GBM development and patient survival, which could potentially be used in the development of a targeted therapy.

10.3.1 Data Description

We consider a subset of the available TCGA GBM data, consisting of mRNA expression, two upstream platforms known to affect gene expression (methylation and copy number), and uncensored survival time (in days) for 163 patients. Our copy number data is level 2 data from the HG_CGH_244A platform; it is the normalized signal for copy number alterations of aggregated regions per probe. Our methylation data is level 3 data from the HumanMethylation27K arrays; it is the methylated sites along a gene (probe level data). Our expression data is level 3 data (summarized per gene) from the Affymetrix profiled HT_HG_U133A platform [20]. Since our clinical response is survival time, we use an accelerated failure time (AFT) model, taking Y (in Eq. 10.2) to be log(survival) [24]. We focus our analysis on 49 genes from 3 signaling pathways important to GBM: RTK/PI3K, P53, and RB [15]. One gene has no methylation data, so we remove its corresponding column in the X matrix; any effect due to methylation would then be captured by the "other" predictor in the clinical model.

After standardizing the predictors and imputing the few missing values, we apply our method and obtain 10,000 MCMC samples, with 500 used as burn-in. In determining whether to flag a marker as "significant" we choose our practical minimum effect size to correspond to a 5% change in survival time, resulting in $(\delta_-, \delta_+) = (\log(0.95), \log(1.05))$. As discussed in Section 10.2, we flag a marker as significant if $p_+(\beta_{jg}) > 0.5$ or if $p_-(\beta_{jg}) > 0.5$.

10.3.2 Results

Using the linear formulation of our method, we identify 21 significant prognostic markers, 12 of them positive markers (more gene expression explained by that platform, better prognosis) and 9 of them negative markers (more gene expression explained by that platform, worse prognosis). Figures 10.3A and 10.4A show the posterior probabilities of β_{jg} being less than δ_- and greater than δ_+, respectively. Figure 10.5A shows the posterior means of each β_{jg} as well as which markers were flagged as significant. The genes with the 12 positive markers were PDGFRB, FGFR1, CCND2, PIK3R2, IRS1, CDKN2C, TP53, PIK3CA and PDGFRA. In the following results details, the percentage given after the gene name is the percent of that gene expression explained by the flagged platform under the linear formulation (see Appendix B for calculation details). Genes PDGFRB (0.2%), FGFR1 (2.0%), and CCND2 (6.0%) were determined to be related to patient survival through methylation effects, while expression of PIK3R2 (7.6%), IRS1 (4.8%), CDKN2C (19.7%), and TP53 (16.0%) were related to patient survival through copy number. For PIK3CA (70.0%), PDGFRA (44.5%), PDGFRB (98.5%), CCND2 (81.2%), and TP53 (83.1%), gene expression was related to patient survival through some other unspecified mechanism. The genes with the 9 negative markers were IGF1R, FGFR2, ARAF, GRB2, FGFR1, and MDM2. IGF1R (0.3%), FGFR2 (6.1%), ARAF (0.1%), and GRB2 (0.4%), were related to clinical response through methylation, while FGFR1 (8.8%), ARAF (2.8%), and MDM2 (81.4%) were related through copy number, and IGF1R (84.9%) and MDM2 (18.4%) were related through some mechanism other than methylation or copy number. Note that seven genes (IGF1R, PDGFRB, FGFR1, ARAF, CCND2, MDM2, and TP53) are found to be significant on two different platforms.

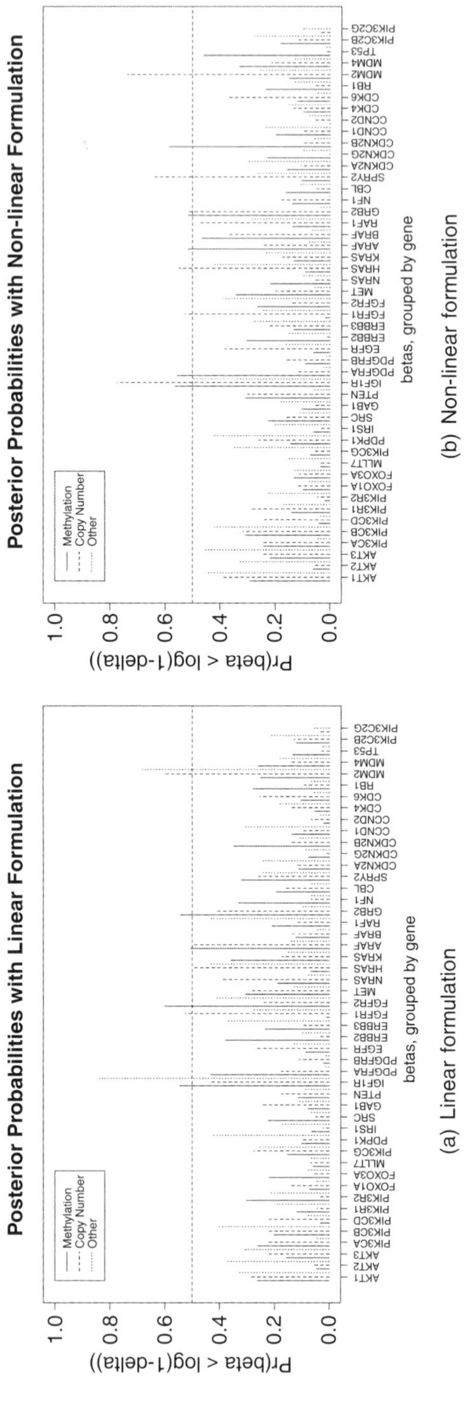

Figure 10.3 The posterior probabilities (based on MCMC samples) that $\beta_{jg} < \delta_-$.

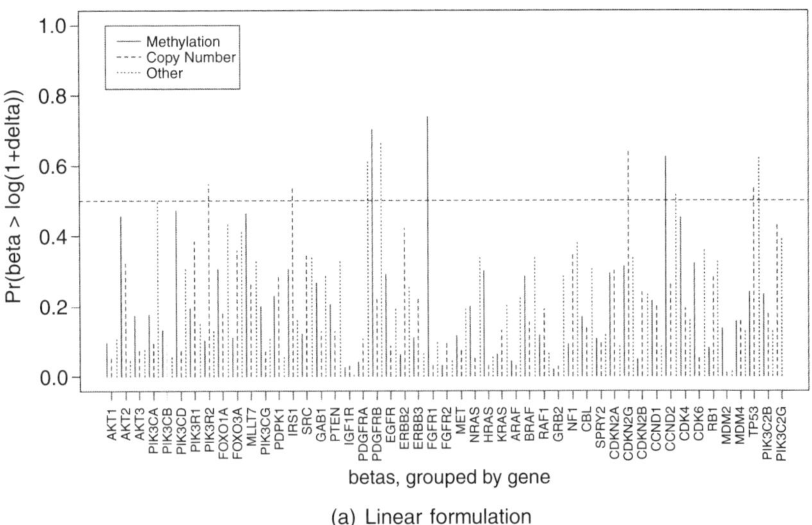

(a) Linear formulation

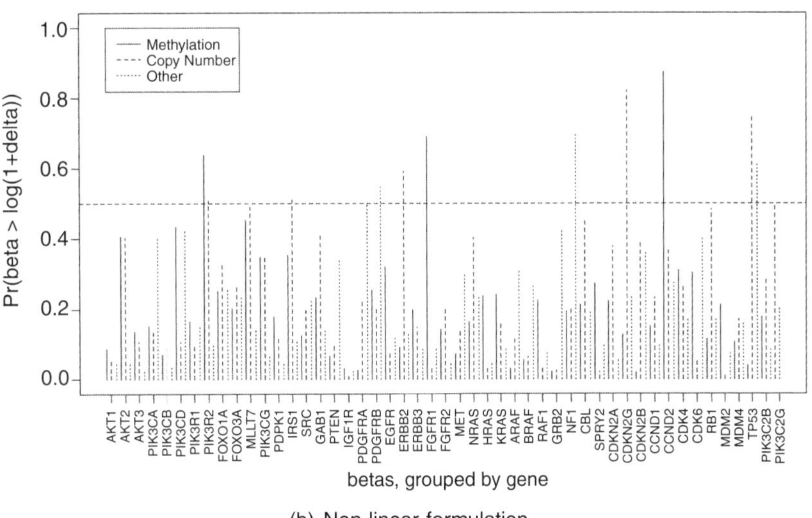

(b) Non-linear formulation

Figure 10.4 The posterior probabilities (based on MCMC samples) that $\beta_{jg} > \delta_+$.

Our method has not only identified 14 genes as having a significant effect on survival, but it has also determined which platform(s) of those genes is (are) modulating the effect. In addition, we can use the posterior means of the effects to gain more specific insight; a one-unit increase in gene g expression

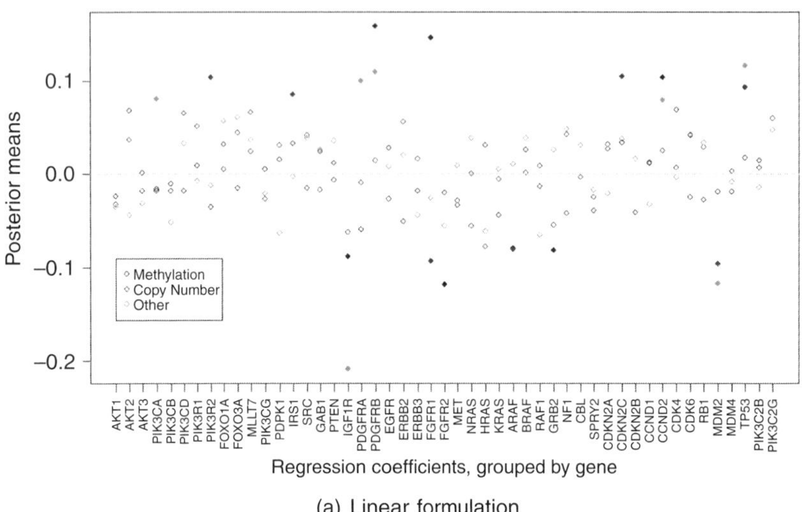

(a) Linear formulation

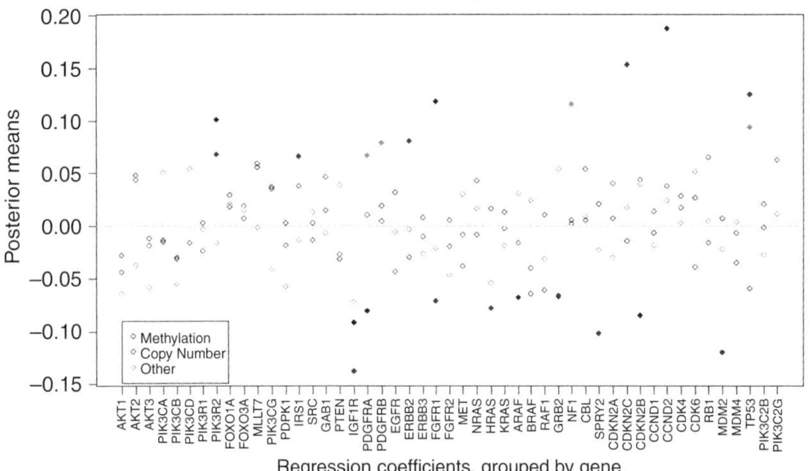

(b) Non-linear formulation

Figure 10.5 The estimates (posterior means) of the regression coefficients in the clinical model (β_{jg}'s) are shown, with the multiple platforms for each gene labeled by color. Solid plot markers indicate that the effect was found to be significant.

attributed to platform j is estimated to result in a $\{\exp(\widehat{\beta}_{jg}) - 1\} \times 100\%$ change in survival time.

Using the non-linear formulation of our method, we see some of the same markers flagged, but there are also some differences: We identify 23 significant prognostic markers, 12 positive markers and 11 negative markers. Figures 10.3B and 10.4B show the posterior probabilities of β_{jg} being less than δ_- and greater than δ_+, respectively. Figure 10.5B shows the posterior means of each β_{jg} as well as which markers were flagged as significant. The genes with the 12 positive markers were PIK3R2, FGFR1, CCND2, IRS1, ERBB2, CDKN2C, TP53, PDGFRA, PDGFRB, and NF1. In the following results details, the percentage given after the gene name is the percent of that gene expression explained by the flagged platform under the non-linear formulation (see Appendix B for calculation details). Genes PIK3R2 (2.9%), FGFR1 (3.7%), and CCND2 (8.9%) were determined to be related to clinical outcome through methylation effects, while expression of PIK3R2 (9.8%), IRS1 (4.8%), ERBB2 (8.0%), CDKN2C (46.1%), and TP53 (20.3%) were related to clinical outcome through copy number. For PDGFRA (44.5%), PDGFRB (96.3%), NF1 (40.0%), and TP53 (70.7%), gene expression was related to clinical outcome through some other unspecified mechanism. The genes with the 11 negative markers were IGF1R, PDGFRA, ARAF, GRB2, CDKN2B, FGFR1, HRAS, SPRY2, and MDM2. IGF1R ($<$0.1%), PDGFRA (19.6%), ARAF (0.2%), GRB2 (0.3%), and CDKN2B (6.7%) were related to clinical response through methylation, while IGFR1 (23.8%), FGFR1 (19.3%), HRAS (13.9%), GRB2 (23.7%), SPRY2 (16.1%), and MDM2 (86.9%) were related through copy number, Note that six genes (PIK3R2, FGFR1, TP53, PDGFRA, IGR1R, and GRB2) are found to be significant on two different platforms.

There are 14 markers flagged identically (same gene, same platform, same sign) by both the linear and non-linear model formulations. Differences between the results in moving from the linear formulation to the non-linear formulation include the following. First, PIK3CA and FGFR2 are no longer flagged, while ERBB2, NF1, CDKN2B, HRAS, and SPRY2 appear as significant. There are also four instances where a gene goes from being flagged on 2 platforms to only being flagged on 1: (1) Instead of being identified on methylation and "other" platforms, PDGFRB is only identified on other platforms. (2) Instead of being flagged on methylation and other platforms, CCND2 is only flagged on methylation effects. (3) ARAF goes from being flagged on both methylation and copy number to only methylation. (4) MDM2 goes from being flagged on both copy number and other platforms to only copy number. In addition there are three genes that go from being flagged on one platform to being identified as significant on two: (1) PIK3R2 is still flagged on copy number, but it is also

Table 10.1 *Results: Negative markers*

AKT1 (M,CN,O)	**IGF1R** (<u>*M*</u>,<u>CN</u>,*O*)	CBL (M,CN,O)
AKT2 (M,CN,O)	**PDGFRA** (<u>M</u>,CN,O)	**SPRY2** (M,<u>CN</u>,O)
AKT3 (M,CN,O)	PDGFRB (M,CN,O)	CDKN2A (M,CN,O)
PIK3CA (M,CN,O)	EGFR (M,CN,O)	CDKN2C (M,CN,O)
PIK3CB (M,CN,O)	ERBB2 (M,CN,O)	**CDKN2B** (<u>M</u>,CN,O)
PIK3CD (M,CN,O)	ERBB3 (M,CN,O)	CCND1 (M,CN,O)
PIK3R1 (M,CN,O)	**FGFR1** (M,<u>*CN*</u>,O)	CCND2 (M,CN,O)
PIK3R2 (M,CN,O)	**FGFR2** (*M*,CN,O)	CDK4 (M,CN,O)
FOXO1A (M,CN,O)	MET (M,CN,O)	CDK6 (M,CN,O)
FOXO3A (M,CN,O)	NRAS (M,CN,O)	RB1 (M,CN,O)
MLLT7 (M,CN,O)	**HRAS** (M,<u>CN</u>,O)	**MDM2** (M,<u>*CN*</u>,*O*)
PIK3CG (M,CN,O)	KRAS (M,CN,O)	MDM4 (M,CN,O)
PDPK1 (M,CN,O)	**ARAF** (<u>*M*</u>,*CN*,O)	TP53 (M,CN,O)
IRS1 (M,CN,O)	BRAF (M,CN,O)	PIK3C2B (M,CN,O)
SRC (M,CN,O)	RAF1 (M,CN,O)	PIK3C2G (CN,O)
GAB1 (M,CN,O)	**GRB2** (<u>*M*</u>,<u>CN</u>,O)	
PTEN (M,CN,O)	NF1 (M,CN,O)	

Note: All 49 genes appearing in the data are listed, along with the three platforms (M,CN,O). Genes are bolded if they were found to have a significant negative prognostic marker on any platform by either the linear or non-linear formulation. Italic platforms indicate that the platform was flagged by the linear formulation, and underlined platforms indicate that the platform was flagged by the non-linear formulation.

flagged on methylation effects. (2) PDGFRA is still flagged as a positive marker modulated by other platforms, but it becomes flagged as a negative marker on methylation in addition. (2) GRB2 is flagged on copy number effects, in addition to maintaining significance on methylation effects. Finally, we see IGF1R still flagged on 2 platforms, but one of them changes; instead of being flagged on both methylation and other effects, IGF1R is flagged on methylation and copy number effects. Tables 10.1 and 10.2 display the markers flagged as important negative and positive prognostic markers, respectively, and they also provide a comparison of results from the linear versus non-linear formulations.

Since in the case of our data we have no compelling reason to believe that the relationship between methylation, copy number, and gene expression is truly linear, we consider the results from the non-linear formulation to be more accurate. The fact that there are differences in the results at all speaks to the fact that we must be careful not to make unfounded assumptions on the specific structure of the biological relationships modeled in the mechanistic model. With a more flexible partitioning of gene expression, we see 2 genes no longer found to be significant, while 5 previously unidentified genes were flagged. Also, we see differences in results on one platform for 8 other genes; the

Table 10.2 *Results: Positive markers*

AKT1 (M,CN,O)	IGF1R (M,CN,O)	CBL (M,CN,O)
AKT2 (M,CN,O)	**PDGFRA** (M,CN,$\underline{O}$)	SPRY2 (M,CN,O)
AKT3 (M,CN,O)	**PDGFRB** (*M*,CN,$\underline{O}$)	CDKN2A (M,CN,O)
PIK3CA (M,CN,*O*)	EGFR (M,CN,O)	**CDKN2C** (M,$\underline{CN}$,O)
PIK3CB (M,CN,O)	**ERBB2** (M,$\underline{CN}$,O)	CDKN2B (M,CN,O)
PIK3CD (M,CN,O)	ERBB3 (M,CN,O)	CCND1 (M,CN,O)
PIK3R1 (M,CN,O)	**FGFR1** (*M*,CN,O)	**CCND2** (*M*,CN,*O*)
PIK3R2 ($\underline{M,CN}$,O)	FGFR2 (M,CN,O)	CDK4 (M,CN,O)
FOXO1A (M,CN,O)	MET (M,CN,O)	CDK6 (M,CN,O)
FOXO3A (M,CN,O)	NRAS (M,CN,O)	RB1 (M,CN,O)
MLLT7 (M,CN,O)	HRAS (M,CN,O)	MDM2 (M,CN,O)
PIK3CG (M,CN,O)	KRAS (M,CN,O)	MDM4 (M,CN,O)
PDPK1 (M,CN,O)	ARAF (M,CN,O)	**TP53** (M,$\underline{CN}$,*O*)
IRS1 (M,$\underline{CN}$,O)	BRAF (M,CN,O)	PIK3C2B (M,CN,O)
SRC (M,CN,O)	RAF1 (M,CN,O)	PIK3C2G (CN,O)
GAB1 (M,CN,O)	GRB2 (M,CN,O)	
PTEN (M,CN,O)	**NF1** (M,CN,$\underline{O}$)	

Note: All 49 genes appearing in the data are listed, along with the three platforms (M,CN,O). Genes are bolded if they were found to have a significant positive prognostic marker on any platform by either the linear or non-linear formulation. Italic platforms indicate that the platform was flagged by the linear formulation, and underlined platforms indicate that the platform was flagged by the non-linear formulation.

difference in partitioning in the mechanistic model appears to have had a direct effect on which platform(s) was (were) identified as significant, reinforcing the importance of accuracy in that first step.

10.4 Discussion

We have proposed a two-step, hierarchical Bayesian model to integrate different types of genomic data in a single model with the goal of identifying genetic markers significant to a clinical outcome. In addition we have presented two different formulations of the first step of the model (the mechanistic model) that facilitate more or less flexibility of gene expression partitioning, depending on the amount and nature of prior knowledge regarding the structure of biological relationships among data platforms. After applying our method to a TCGA brain tumor dataset, we identify 21 significant prognostic markers on 14 genes using the linear formulation, and 23 significant prognostic markers on 17 genes using the non-linear formulation.

Advantages of our method include: (1) Our model flexibly incorporates multiple types of genetic data in a single model. (2) We utilize principal components and a Normal-Gamma prior on the effects to effectively induce flexible shrinkage and combat high-dimensionality. (3) Our model accounts for the known biological relationships between DNA characteristics and RNA-level entities, which allows us to not only identify which genes are significant to the clinical outcome, but to also obtain the mechanistic information of which platform(s) is (are) modulating the effect. This information is critical to the development of targeted cancer therapies. (4) Our effect estimates have a direct interpretation. (5) Our method can be applied to gain insight regarding any type of cancer, as long as an appropriate dataset is available and the biological relationships among the platforms is understood.

Note that we currently use two steps (mechanistic and clinical models), but additional layers could be constructed to incorporate other platforms such as protein expression data. For example, protein expression would be regressed on mRNA expression, with an additional step of mRNA regressed on platforms such as copy number, miRNA, mutation status, etc. When the data is available, incorporating this additional step would give the potential of obtaining even more insight into the mechanics of what exactly is driving the clinical phenotype expression.

Appendix A: Complete Conditionals

$$\boldsymbol{\beta}|\text{rest} \sim \text{Normal}\{(X^{\mathrm{T}}X + \sigma^2 D_\psi^{-1})^{-1} X^{\mathrm{T}}\mathbf{Y}, \sigma^2 (X^{\mathrm{T}}X + \sigma^2 D_\psi^{-1})^{-1}\}$$

$$\sigma^2|\text{rest} \sim \text{Inv.Gamma}(a = a + n/2, b = b + \{(\mathbf{Y} - X\boldsymbol{\beta})^{\mathrm{T}}(\mathbf{Y} - X\boldsymbol{\beta})\}/2)$$

$$\psi_{jg}|\text{rest} \sim \text{Gen.Inv.Gaussian}(a = \gamma_j^{-2}, \quad b = \beta_{jg}^2, p = \lambda_j - 1/2), \text{ where}$$

$$V = \text{Gen.Inv.Gaussian}(a, b, p) \text{ has density}$$

$$(a/b)^{p/2} v^{p-1} \exp\{-(av + b/v)/2\}/\{2K_p(\sqrt{ab})\}, \text{ where } K_p(\cdot)$$

is a modified Bessel function of the second kind.

$$\lambda_j|\text{rest} \sim (1/\lambda_j)^{\tilde{a}} \exp\{-\tilde{b}\gamma_j^{-2}/(2\lambda_j) - c\lambda_j\} \left(\prod_{g=1}^{p_j} \psi_{jg}^{\lambda_j}\right) / [\{\Gamma(\lambda_j)\}^{p_j} (2\gamma_j^2)^{p_j\lambda_j}]$$

$$\gamma_j^{-2}|\text{rest} \sim \text{Gamma}(a = p_j\lambda_j + \tilde{a}, b = (\tilde{b}/\lambda_j + \sum_{i=1}^{p_j} \psi_{jg})/2)$$

In the Metropolis-Hastings update step, the proposed value is $\lambda_j^* = \exp(\sigma_\lambda^2 z)\lambda_j$ where $z \sim \text{Normal}(0, 1)$ and the tuning parameter σ_λ^2 is chosen to result in an acceptance rate between 20% and 30%. The acceptance probability is then $\min\left\{1, \frac{\pi(\lambda_j^*)}{\pi(\lambda_j)}\left(\frac{\Gamma(\lambda_j)}{\Gamma(\lambda_j^*)}\right)^{P_j}\left((2\gamma_j^2)^{-P_j}\prod_{g=1}^{P_j}\psi_{jg}\right)^{\lambda_j^*-\lambda_j}\left(\frac{\lambda_j^*}{\lambda_j}\right)\right\}$ where $\pi(\lambda_j) = (1/\lambda_j)^{\tilde{a}}\exp\{-\tilde{b}\gamma_j^{-2}/(2\lambda_j) - c\lambda_j\}$, the prior for λ_j.

Appendix B: Partitioning Explained Variation

Regardless of linear or non-linear formulation, after estimation of the mechanistic model (for one gene) we have

$$y = \widehat{b_0} + M + CN + O \tag{10.5}$$

where y is the gene expression. We carry forward M, CN, O and disregarding $\widehat{b_0}$ as a kind of mean-centering. The sums of squares that we use to calculate proportions of explained variances are:

$$SST = \sum_{i=1}^{n}(y_i - \bar{y})^2 \tag{10.6}$$

$$SSM = \sum_{i=1}^{n}(\widehat{y}_i^{(M)} - \bar{y})^2 \tag{10.7}$$

$$SSCN = \sum_{i=1}^{n}(\widehat{y}_i^{(CN)} - \bar{y})^2 \tag{10.8}$$

$$SSE = SST - SSM - SSCN \tag{10.9}$$

where $\widehat{y}_i^{(M)} = \widehat{b_0} + M$ and $\widehat{y}_i^{(CN)} = \widehat{b_0} + CN$. Then the proportion of variance explained by methylation is SSM/SST; the proportion explained by copy number is $SSCN/SST$; and the proportion explained by something other than methylation or copy number is SSE/SST.

Acknowledgments

JSM and VB's research is partially supported by NIH grant R01 CA160736 and the Cancer Center Support Grant (CCSG) (P30 CA016672). The work of EJ and RJC was supported by a grant from the National Cancer Institute (R37-CA057030). The content is solely the responsibility of the authors and does not necessarily represent the official views of the National Cancer Institute or the National Institutes of Health. We would also like to thank the TCGA consortium for making the data publicly available.

References

1 Side effects of radiation therapy. http://www.cancer.net/navigating-cancer-care/ how-cancer-treated/radiation-therapy/side-effects-radiation-therapy, January 2012. Cancer.net.

2 Glioblastoma multiforme. http://www.aans.org/Patient\%20Information/ Conditions\%20and\%20Treatments/Glioblastoma\%20Multiforme.aspx, June 2012. American Association of Neurological Surgeons.

3 American Cancer Society, Atlanta, GA. *American Cancer Society: Cancer Facts and Figures 2014*, 2014.

4 Maria Maddalena Barbieri and James O. Berger. Optimal predictive model selection. *The Annals of Statistics*, 2004.

5 D. Bell, A. Berchuck, M. Birrer, J. Chien, D. Cramer, F. Dao, R. Dhir, P. DiSaia, H. Gabra, P. Glenn, and *et al*. Integrated genomic analyses of ovarian carcinoma. *Nature*, 474(7353):609–615, Jun 2011.

6 Jim E. Griffin and Philip J. Brown. Inference with normal-gamma prior distributions in regression problems. *Baysian Analysis*, 5(1):171–188, 2010.

7 Trevor Hastie and Robert Tibshirani. Generalized additive models. *Statistical Science*, 1:297–310, 1986.

8 Kenji Ikeda, Satoshi Saitoh, Akihito Tsubota, Yasuji Arase, Kazuaki Chayama, Hiromitsu Kumada, Goro Watanabe, and Masahiko Tsurumaru. Risk factors for tumor recurrence and prognosis after curative resection of hepatocellular carcinoma. *Cancer*, 71(1):19–25, 1993.

9 E. M. Jennings, J. S. Morris, R. J. Carroll, G. C. Manyam, and V. Baladandayuthapani. Bayesian methods for expression-based integration of various types of genomics data. *EURASIP J Bioinform Syst Biol*, 2013(1):13, 2013.

10 D. R. Johnson and B. P. O'Neill. Glioblastoma survival in the United States before and during the temozolomide era. *J. Neurooncol.*, 107(2):359–364, Apr 2012.

11 F. R. Khuri, E. S. Kim, J. J. Lee, R. J. Winn, S. E. Benner, S. M. Lippman, K. K. Fu, J. S. Cooper, E. E. Vokes, R. M. Chamberlain, B. Williams, T. F. Pajak, H. Goepfert, and W. K. Hong. The impact of smoking status, disease stage, and index tumor site on second primary tumor incidence and tumor recurrence in the head and neck retinoid chemoprevention trial. *Cancer Epidemiol. Biomarkers Prev.*, 10(8):823–829, Aug 2001.

12 G. R. Lanckriet, T. De Bie, N. Cristianini, M. I. Jordan, and W. S. Noble. A statistical framework for genomic data fusion. *Bioinformatics*, 20(16):2626–2635, Nov 2004.

13 N. Martini, M. S. Bains, M. E. Burt, M. F. Zakowski, P. McCormack, V. W. Rusch, and R. J. Ginsberg. Incidence of local recurrence and second primary tumors in resected stage I lung cancer. *J. Thorac. Cardiovasc. Surg.*, 109(1):120–129, Jan 1995.

14 R. McLendon, A. Friedman, D. Bigner, E. G. Van Meir, D. J. Brat, G. M. Mastrogianakis, J. J. Olson, T. Mikkelsen, N. Lehman, K. Aldape, and *et al*. Comprehensive genomic characterization defines human glioblastoma genes and core pathways. *Nature*, 455(7216):1061–1068, Oct 2008.

15 Pathway analysis of genetic alterations in glioblastoma (TCGA). http://cbio. mskcc.org/cancergenomics/gbm/pathways/, 2012. Memorial Sloan-Kettering Cancer Center.

16 M Dror Michaelson and William K Oh. Treatment-related toxicity in men with testicular germ cell tumors. www.uptodate.com, October 2013.

17 Trevor Park and George Casella. The bayesian lasso. *Journal of the American Statistical Association*, 103(482):681–686, 2008.

18 R. Shen, A. B. Olshen, and M. Ladanyi. Integrative clustering of multiple genomic data types using a joint latent variable model with application to breast and lung cancer subtype analysis. *Bioinformatics*, 25(22):2906–2912, Nov 2009.

19 Program overview. http://cancergenome.nih.gov/abouttcga/overview, 2012. The Cancer Genome Atlas.

20 Data levels and data types. https://tcga-data.nci.nih.gov/tcga/tcgaDataType.jsp. TCGA.

21 Svitlana Tyekucheva, Luigi Marchionni, Rachel Karchin, and Giovanni Parmigiani. Integrating diverse genomic data using gene sets. *Genome Biology*, 2011.

22 R. G. Verhaak, K. A. Hoadley, E. Purdom, V. Wang, Y. Qi, M. D. Wilkerson, C. R. Miller, L. Ding, T. Golub, J. P. Mesirov, G. Alexe, M. Lawrence, M. O'Kelly, P. Tamayo, B. A. Weir, S. Gabriel, W. Winckler, S. Gupta, L. Jakkula, H. S. Feiler, J. G. Hodgson, C. D. James, J. N. Sarkaria, C. Brennan, A. Kahn, P. T. Spellman, R. K. Wilson, T. P. Speed, J. W. Gray, M. Meyerson, G. Getz, C. M. Perou, and D. N. Hayes. Integrated genomic analysis identifies clinically relevant subtypes of glioblastoma characterized by abnormalities in PDGFRA, IDH1, EGFR, and NF1. *Cancer Cell*, 17(1):98–110, Jan 2010.

23 W. Wang, V. Baladandayuthapani, J. S. Morris, B. M. Broom, G. Manyam, and K. A. Do. iBAG: integrative Bayesian analysis of high-dimensional multiplatform genomics data. *Bioinformatics*, 29(2):149–159, Jan 2013.

24 L. J. Wei. The accelerated failure time model: A useful alternative to the cox regression model in survival analysis. *Statistics in Medicine*, 11(14–15):1871–1879, 1992.

25 Simon Wood. Package 'mgcv'. http://cran.r-project.org/web/packages/mgcv/mgcv.pdf, January 2014.

11

Exploratory Methods to Integrate Multisource Data

ERIC F. LOCK AND ANDREW B. NOBEL

Abstract

In biomedical research, a growing number of measurement platforms and technologies are being used to assess diverse but related information on a set of common samples. This motivates integrative methods for *multisource* data, in which multiple data sets are derived from a common set of objects. This chapter addresses exploratory methods for multisource data. We discuss the need for flexible approaches that simultaneously model the dependence and the heterogeneity of the sources and describe two such methods in detail: joint and individual variation explained (JIVE) and Bayesian consensus clustering (BCC). JIVE is a general decomposition of variation for multisource data. The decomposition consists of three terms: a low-rank approximation capturing joint variation across sources, low-rank approximations for structured variation individual to each source, and residual noise. JIVE quantifies the amount of joint variation between sources, reduces dimensionality, and allows for visual exploration of joint and individual structure. BCC is a tool to cluster a set of objects based on multisource data. The model permits a separate clustering of the objects for each source that adhere loosely to an overall clustering. We describe a Bayesian framework for simultaneous estimation of the source-specific and overall clusterings. We illustrate JIVE and BCC with applications to publicly available data from the Cancer Genome Atlas.

11.1 Introduction

The field of *exploratory data analysis* encompasses a wide range of statistical methods that are used to identify, summarize, and display the main characteristics of a data set. Exploratory methods are not meant to test prespecified hypotheses or make out-of-sample predictions. Rather, they are meant to aid our overall understanding of the data and to guide further analyses. Exploratory approaches play an important role in the analysis of high-dimensional omics data, where scientific hypotheses can be innumerable.

Well-established multivariate techniques are used to explore a single high-dimensional data set. We list three such approaches:

- *Association mining methods.* These measure hundreds to millions of pairwise variable associations, which can be visualized as networks or correlation heatmaps.
- *Factorization methods like principal components analysis (PCA) and probabilistic factor models.* These help to reduce the dimensionality of the data and identify the primary modes of variation in a data set.
- *Clustering methods, such as hierarchical or model-based clustering.* These simplify subsequent analyses and interpretation by grouping similar samples and allow for more insightful visualization of the data (e.g., in ordered heatmaps).

In this chapter, we discuss the extension of the preceding approaches to *multisource data*, in which multiple heterogeneous data sets describe a common set of objects. Each data set represents a distinct mode of measurement or domain. Multisource data are common in biomedical research, where a growing number of platforms and technologies are used to measure diverse but related information.

Well-established multivariate techniques such as those described earlier can be used to explore each data source separately. However, separate analyses may lack power and will not capture intersource dependencies, which are often of scientific interest. At the other extreme, a joint analysis that ignores the heterogeneity of the data may not capture important features that are specific to each data source. Furthermore, methods that solely try to capture intersource dependencies are often not robust to structured variation specific to each data source (see Sections 11.4.3 and 11.5.3 for illustrations). Exploratory methods that simultaneously model shared features and features that are specific to each data source have recently been developed as flexible alternatives to fully separate or fully joint analyses. The demand for such flexibly integrative methods motivates a very dynamic area of statistics and bioinformatics research.

The Cancer Genome Atlas (TCGA) is a large-scale collaborative effort to collect and catalog data from several genomic technologies (TCGA Research Network, 2008). The integrative analysis of data from these disparate sources provides a more comprehensive understanding of cancer genetics and molecular biology. Throughout this chapter, we illustrate methodology with applications to data from TCGA.

The rest of this chapter is organized as follows. In Section 11.2, we introduce a mathematical framework and notation for multisource data. In Section 11.3, we describe the present landscape of exploratory approaches for multisource data, with a focus on factorization and clustering methods. In Section 11.4, we describe joint and individual explained (JIVE), a factorization method for multisource data. In Section 11.5, we describe Bayesian consensus clustering (BCC), a clustering method for multisource data. Both JIVE and BCC are flexibly integrative in the sense that they simultaneously model the dependence and the heterogeneity of the sources.

11.2 Formal Framework

Here we introduce a framework and related notation for multisource data. Let $\mathbb{X}_m$ represent the data for source m, $m = 1, \ldots, M$. For the methods discussed in this chapter, $\mathbb{X}_m$ is a data matrix. Columns of $\mathbb{X}_m$ represent cases or objects (which are common for $m = 1, \ldots, M$), and rows represent variables for measurement technology m. Thus each data matrix has the same number of columns N but a potentially different number of rows $p_1, p_2, \ldots, p_M$:

$$\mathbb{X}_1 : p_1 \times N$$

$$\mathbb{X}_2 : p_2 \times N$$

$$\vdots$$

$$\mathbb{X}_M : p_M \times N$$

These M matrices may be combined vertically:

$$\mathbb{X} = \begin{bmatrix} \mathbb{X}_1 \\ \mathbb{X}_2 \\ \vdots \\ \mathbb{X}_M \end{bmatrix} : p \times N$$

where $p = p_1 + p_2 + \cdots + p_M$.

To take a simple example, suppose we have N biological samples. Then $\mathbb{X}_1$ may represent gene expression measurements (of dimension p_1), $\mathbb{X}_2$ may represent genotype information (of dimension p_2), and $\mathbb{X}_3$ may represent metabolite concentrations (of dimension p_3). Variables are likely dependent across these three sources, but they contain different types of information.

Direct analysis of the joint data X can be problematic as the size and scale of the constituent data sources often differ. To remove baseline differences, it helps to center the data by subtracting the mean within each row. Sources may also have different dimension (p_m) or variability. To circumvent cases where "the largest data set wins," it helps to scale each data source by its total sum of squares. In particular, for each m, define $\mathbb{X}_m^{\text{scaled}} = \frac{\mathbb{X}_m}{||\mathbb{X}_m||}$, where $|| \cdot ||$ defines the Frobenius norm $||A||^2 = \sum_{i,j} a_{ij}^2$. Then, $||\mathbb{X}_i^{\text{scaled}}|| = 1$ for each i, and each source contributes equally to the sum of squares of the concatenated matrix:

$$\mathbb{X}^{\text{scaled}} = \begin{bmatrix} \mathbb{X}_1^{\text{scaled}} \\ \vdots \\ \mathbb{X}_M^{\text{scaled}} \end{bmatrix} \tag{11.1}$$

11.3 Related Methods

Here we focus on two categories of exploratory methods for multisource data: *factorization methods* and *clustering methods*.

11.3.1 Factorization Methods

PCA is a common factorization method for high-dimensional data. The first r principal components yield a rank r approximation of $\mathbb{X}$:

$$\mathbb{X} \approx U S$$

where U is a $p \times r$ matrix of variable *loadings* and S is an $r \times N$ matrix of sample *scores*. PCA can identify important low-dimensional structure in the samples (given by the scores) and the variables (given by the loadings). For a more complete discussion of PCA and its application to high-dimensional omics data, see Wall et al. (2003). In the multisource context, PCA of the block-scaled matrix $\mathbb{X}^{\text{scaled}}$ in Eq. (11.1) has been called *consensus PCA* (Wold et al., 1996; Westerhuis et al., 1998). Consensus PCA synthesizes information from multiple sources but does not distinguish between joint and individual effects.

Canonical correlation analysis (CCA) (Hotelling, 1936) is a popular method to globally examine the relation between two data sources. If $\mathbb{X}_1$ and $\mathbb{X}_2$ are two data matrices on a common set of samples, the first pair of CCA loadings u_1 and u_2 are unit vectors maximizing $\text{Corr}(u_1^T \mathbb{X}_1, u_2^T \mathbb{X}_2)$. Geometrically, u_1 and u_2 give the pair of directions that maximize the correlation between $\mathbb{X}_1$ and $\mathbb{X}_2$. Sample projections on the loadings vectors, $u_1^T \mathbb{X}_1$ and $u_2^T \mathbb{X}_2$, give the sample scores for $\mathbb{X}_1$ and $\mathbb{X}_2$. Subsequent CCA directions are found by enforcing orthogonality with previous directions. For data sets with $p_1 > n$ or $p_2 > n$, the CCA directions are not well defined, and overfitting is often a problem even when $p_1, p_2 < n$. Hence, standard CCA is typically not appropriate for high-dimensional data. Alternatively, *partial least squares* (PLS) (Wold, 1985) directions are defined similarly to CCA but maximize covariance rather than correlation.

Multiple canonical correlation analysis (mCCA) (Witten and Tibshirani, 2009) is an extension of CCA and PLS to more than two sources. For $\mathbb{X}_1, \mathbb{X}_2, \dots, \mathbb{X}_M$, as in Section 11.2, standardized so that each row has mean 0 and standard deviation 1, the standard mCCA loading vectors $u_1, u_2, \dots, u_M$ satisfy

$$\underset{||u_1||=\cdots=||u_M||=1}{\operatorname{argmax}} \quad \sum_{i<j} u_i^T \mathbb{X}_i \mathbb{X}_j^T u_j = \sum_{i<j} \operatorname{Cov}(u_i^T \mathbb{X}_i, u_j^T \mathbb{X}_j)$$

The preceding methods are designed to explore associations and identify common structure across multiple data sources. This does not provide a full view, as structured variation individual to a data source is also often of scientific interest. Moreover, failure to account for individual structure can obscure joint structure. Trygg and Wold (2003) examine how structured variation in $\mathbb{X}_1$ not associated with $\mathbb{X}_2$ (and vice versa) can drastically alter PLS scores, making interpretation problematic. Their solution, called O2-PLS, removes structured variation in $\mathbb{X}_1$ not linearly related to $\mathbb{X}_2$ (and vice versa) from the PLS components. As such, O2-PLS components are often more representative of the true joint structure between two data sources. More recently, general factorization methods have been developed to capture structure that is jointly present across multiple data sources and structure that is individual to each data source. These methods include *joint and individual variation explained* (JIVE) (Lock et al., 2013), *common and individual feature analysis* (CIFA) (Zhou et al., 2013), and *OnPLS* (Lofstedt, 2011). We describe JIVE in detail in Section 11.4.

11.3.2 Clustering Methods

Clustering is a very widely used exploratory tool to identify similar groups of objects (e.g., clinically relevant disease subtypes). Most applications of clustering multisource data follow one of two general approaches:

1. clustering of each data source separately, potentially followed by a post hoc integration of these separate clusterings
2. combining all data sources to determine a single "joint" clustering

Under approach 1, the level of agreement between the separate clusterings may be measured by the *adjusted Rand index* (Hubert and Arabie, 1985) or a similar statistic. Furthermore, *consensus clustering* (also called *ensemble clustering*) can be used to determine an overall partition of the objects that agrees the most with the source-specific clusterings. Several objective functions and algorithms to perform consensus clustering have been proposed (for a survey, see Nguyen and Caruana, 2007). Consensus clustering is most commonly used to combine multiple clustering algorithms, or multiple realizations of the same clustering algorithm, on a single data set. Consensus clustering has also been used to integrate multisource biomedical data (Filkov and Skiena, 2004; Cancer Genome Atlas Network, 2012). Such an approach is attractive in that it models source-specific features yet still determines an overall clustering, which is often of practical interest. However, the two-stage process of performing entirely separate clusterings followed by post hoc integration limits the power to identify and exploit shared structure (see Section 11.5.3 for an illustration of this phenomenon).

Approach 2 effectively exploits shared structure, at the expense of failing to recognize features that are specific to each data source. Within a model-based statistical framework, one can find the clustering that maximizes a joint likelihood. Assuming that each source is conditionally independent given the clustering, the joint likelihood is the product of the likelihood functions for each data source. This approach is used by Kormaksson et al. (2012) in the context of integrating gene expression and DNA methylation data. The *iCluster* method (Shen et al., 2009; Mo et al., 2013), described in Chapter 7, performs clustering by first fitting a Gaussian latent factor model to the joint likelihood; clusters are then determined by K-means clustering of the factor scores. Rey and Roth (2012) propose a dependency-seeking model in which the goal is to find a clustering that accounts for associations across the data sources.

A more flexible alternative to approaches 1 and 2 is to allow for separate but dependent source clusterings. Dependent models have been used to simultaneously cluster gene expression and proteomic data (Rogers et al., 2008), gene expression and transcription factor binding data (Savage et al., 2010), and gene expression and copy number data (Yuan et al., 2011). Kirk et al. (2012) describe a more general dependence model for two or more data sources. Their approach, called *multiple data set integration* (MDI), uses a statistical framework to cluster each data source while simultaneously modeling the pairwise dependence between clusterings. Savage et al. (2013) use MDI to integrate gene expression, methylation, microRNA and copy number data for glioblastoma tumor samples from TCGA. The pairwise dependence model does not explicitly model adherence to an overall clustering, which is often of practical interest. *Bayesian consensus clustering* (BCC) (Lock and Dunson, 2013) uses a different model of dependence based on the assumption that source-specific clusterings adhere loosely to an overall clustering. We describe BCC in detail in Section 11.5.

11.4 Joint and Individual Variation Explained

A challenge in the analysis of multisource data is that structured individual variation can interfere with finding important joint signal, just as joint structure can obscure important signal that is individual to a data source. JIVE is an exploratory method that decomposes a data set into a sum of three terms: a low-rank approximation capturing joint structure between data sources, low-rank approximations capturing structure individual to each data source, and residual noise. Analysis of individual structure provides a way to identify potentially useful information that exists in one data source but not in others. Accounting for individual structure also allows for more accurate estimation of what is common between data sources, as illustrated in Section 11.4.3. JIVE is robust to the dimensionality of the data: it may be used regardless of whether the dimension of a data set exceeds the sample size. Furthermore, JIVE is applicable to data sets with more than two data sources and has a simple algebraic interpretation. Software to perform JIVE is available at https:// genome.unc.edu/jive/.

11.4.1 Model

Let $\mathbb{X}_1, \mathbb{X}_2, \ldots, \mathbb{X}_M$ be appropriately scaled data sources on the same set of samples (as in Section 11.2). Variation that is consistent across data sources in the concatenated matrix $\mathbb{X}$ is represented by a single $p \times N$ matrix of rank $r < \text{rank}(\mathbb{X})$. Call this the *joint structure* of $\mathbb{X}$. For each $\mathbb{X}_m$, structured variation in $\mathbb{X}_m$ unrelated to the other data sources is represented by a $p_m \times N$ matrix of rank $r_m < \text{rank}(\mathbb{X}_m)$. Call these the *individual structures* for each $\mathbb{X}_m$. The sum of joint and individual structure gives a low-rank decomposition approximating the data $\mathbb{X}$.

More formally, let J_m represent the joint structure of $\mathbb{X}_m$ and A_m the individual structure of $\mathbb{X}_m$. The unified model is

$$X_1 = J_1 + A_1 + \epsilon_1$$

$$\vdots \qquad\qquad (11.2)$$

$$\mathbb{X}_M = J_M + A_M + \epsilon_M$$

where ϵ_m are $p_m \times N$ error matrices of independent entries with $\text{E}(\epsilon_m) = 0_{p_m \times N}$. Let

$$J = \begin{bmatrix} J_1 \\ J_2 \\ \vdots \\ J_M \end{bmatrix}$$

then $\text{rank}(J) = r$ and $\text{rank}(A_m) = r_m$ for $m = 1, \ldots, M$. Furthermore, we require that the rows of joint and individual structure are orthogonal, that is, $J A_m^T = 0_{p \times p_m}$ for $i = 1, \ldots, M$. Intuitively, this means that sample patterns responsible for joint structure between sources are unrelated to sample patterns responsible for individual structure. This requirement does not constrain the model in that any matrix in the form (11.2) can be written equivalently with orthogonality between joint and individual structure. Furthermore, the orthogonality constraint assures that the joint and individual components in (11.2) are uniquely determined (see Lock et al., 2012, for more details).

The JIVE model can be factorized as in PCA. The rank r joint structure matrix J can be written as US, where U is a $p \times r$ loading matrix and S is an $r \times N$ score matrix. Let

$$
U = \begin{bmatrix} U_1 \\ \vdots \\ U_M \end{bmatrix}
$$

where U_m gives the loadings of the joint structure corresponding to the rows of $\mathbb{X}_m$. The rank r_m individual structure matrix A_m for $\mathbb{X}_m$ can be written as $W_m S_m$, where W_m is a $p_m \times r_m$ loading matrix and S_m is an $r_m \times N$ score matrix. Then, the low-rank decomposition of $\mathbb{X}_m$ into joint and individual structure is given by $\mathbb{X}_m \approx U_m S + W_m S_m$. This gives the factorized model

$$
\mathbb{X}_1 = U_1 S + W_1 S_1 + R_1
$$

$$
\vdots \tag{11.3}
$$

$$
\mathbb{X}_M = U_M S + W_M S_M + R_M
$$

Joint structure is represented by the common score matrix S. These scores summarize patterns in the samples that explain variability across multiple data types. The loading matrices U_m indicate how these joint scores are expressed in the rows (variables) of data source m. The score matrices S_m summarize sample patterns individual to data source m, with variable loadings W_m.

11.4.2 Estimation

Here we discuss estimation of the model for fixed ranks $r, r_1, \ldots, r_M$. The choice of these ranks is important to accurately quantify the amount of joint and individual structure. Lock et al. (2012) describes a permutation testing approach to rank selection.

As in PCA, the JIVE model is estimated by minimizing the sum of squared error. Let R be the $p \times N$ matrix of residuals after accounting for joint and individual structure:

$$R = \begin{bmatrix} R_1 \\ \vdots \\ R_M \end{bmatrix} = \begin{bmatrix} \mathbb{X}_1 - J_1 - A_1 \\ \vdots \\ \mathbb{X}_M - J_M - A_M \end{bmatrix}$$

We estimate the matrices J and $A_1, \ldots, A_M$ by minimizing $||R||^2$ under the given ranks. This is accomplished by iteratively estimating joint and individual structure:

- given J, find $A_1, \ldots, A_M$ to minimize $||R||$
- given $A_1, \ldots, A_M$, find J to minimize $||R||$
- repeat until convergence

The joint structure J minimizing $||R||$ is equal to the first r terms in the singular value decomposition (SVD) of $\mathbb{X}$ with individual structure removed. The estimated individual structure for $\mathbb{X}_m$ is equal to the first r_m terms of the SVD of $\mathbb{X}_m$ with the joint structure removed. The estimate of individual structure for $\mathbb{X}_i$ will not change those for $\mathbb{X}_j$, $j \neq i$, and hence the M individual approximations minimize $||R||$ for fixed joint structure. Computing time can be improved by reducing the dimensionality of $X_1, \ldots, X_M$ at the outset via their SVD. Pseudocode for this iterative algorithm is given in Lock et al. (2012).

The iterative method is monotone in the sense that $||R||$ decreases at each step. Thus $||R||$ converges to a coordinate-wise minimum, which can't be improved by changing the estimated joint or individual structure.

In many practical applications, important structure between samples or objects is present on only a small subset of the measured variables. Extensions of the JIVE algorithm allow for sparsity in the variables, so that some variable loadings for joint and individual structure are exactly 0. These are discussed in Lock et al. (2013).

11.4.3 Illustrative Example

As a basic illustration of JIVE, we generate two matrices, $\mathbb{X}_1$ and $\mathbb{X}_2$, with simple patterns corresponding to joint and individual structure. The simulated data are depicted in Figure 11.1A. Both $\mathbb{X}_1$ and $\mathbb{X}_2$ are of dimension 50×100, that is, each has 50 variables measured for the same 100 objects. A common pattern of 100 independent standard normal variables is added to half of the rows in $\mathbb{X}_1$ and half of the rows in $\mathbb{X}_2$. This represents the joint structure between the

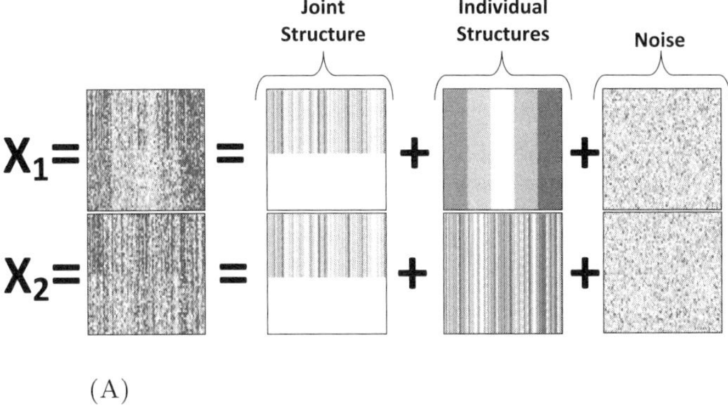

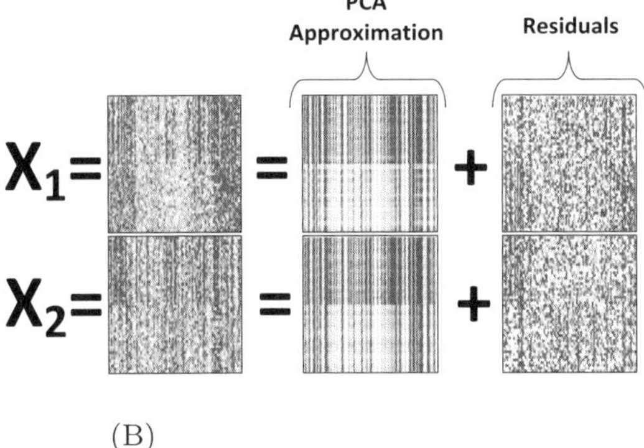

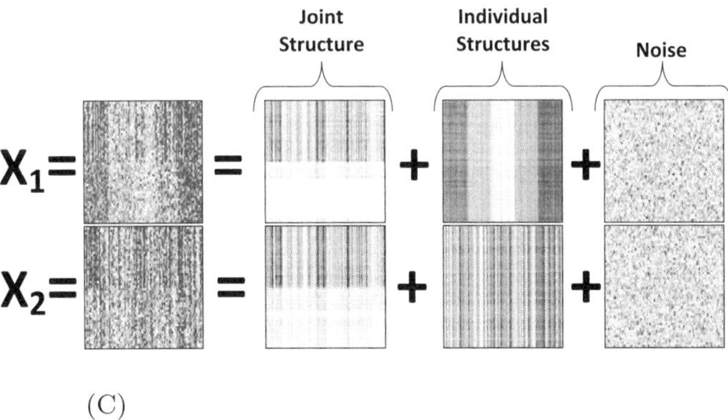

Figure 11.1 (A) Simulated data. $\mathbb{X}_1$ and $\mathbb{X}_2$ as the sum of simulated joint structure, individual structure, and noise. (B) PCA estimate. Approximation resulting from the first consensus principal component. (C) JIVE estimates. Blue corresponds to negative values, red to positive values.

two data sets. Structure individual to $\mathbb{X}_1$ is generated by partitioning the objects into five groups, each of size 20. Those columns corresponding to group 1, 2, 3, 4, or 5 have $-2, -1, 0, 1, 2$ added to each row of $\mathbb{X}_1$, respectively. Structure individual to $\mathbb{X}_2$ is generated similarly, but the groups are randomly determined and are therefore independent of the groups in $\mathbb{X}_1$. Finally, independent $N(0, 1)$ noise is added to both $\mathbb{X}_1$ and $\mathbb{X}_2$. Note that the important joint structure is visually obscured.

The joint structure represents an underlying phenomenon that contributes to several variables in both $\mathbb{X}_1$ and $\mathbb{X}_2$. Practically, the individual structure in $\mathbb{X}_1$ may correspond to an experimental grouping of the measured variables in $\mathbb{X}_1$ not present in $\mathbb{X}_2$, for example, batch effects in microarray data. Our goal is to identify both the common underlying phenomenon and individual group effects.

Consensus PCA of the concatenated matrix $[\mathbb{X}_1' \ \mathbb{X}_2']'$ does a poor job of finding the joint structure. Figure 11.1B shows a weak association between the approximation derived from the first principal component and the common signal that represents joint structure. Consensus PCA indiscriminately maximizes variation explained, and here each principal component is a mix of joint and individual structure.

Like consensus PCA, PLS and CCA also fail to identify the underlying joint signal. This is because PLS is prone to interference from individual structure and CCA is prone to overfitting. See Lock et al. (2013) for an illustration of these effects.

Next we consider the JIVE analysis of $\mathbb{X}_1$ and $\mathbb{X}_2$. Figure 11.1C shows a heatmap of the JIVE estimate for joint structure, JIVE estimates for individual structure, and the fit given by the sum of joint and indivual structure. Estimates closely resemble the true signal in Figure 11.1A. Importantly, JIVE is able to find the true joint signal between the two data sets. Furthermore, estimates for individual structure do a good job of distinguishing the groups specific to $\mathbb{X}_1$ and $\mathbb{X}_2$.

11.4.4 Application to TCGA Data

Lock et al. (2013) use JIVE to integrate microRNA (miRNA) and gene expression data for glioblastoma multiforme brain tumor samples. Here we present a similar application of JIVE to miRNA and gene expression data for 348 breast cancer tumor samples, from TCGA. Breast cancers are common and can be fatal. However, cases are very heterogeneous, and a deeper understanding of systematic distinctions between the tumor samples may lead to more targeted therapies. The Cancer Genome Atlas Network (2012) classified these 348 breast tumor samples into four subtypes: basal-like, HER2-enriched, luminal

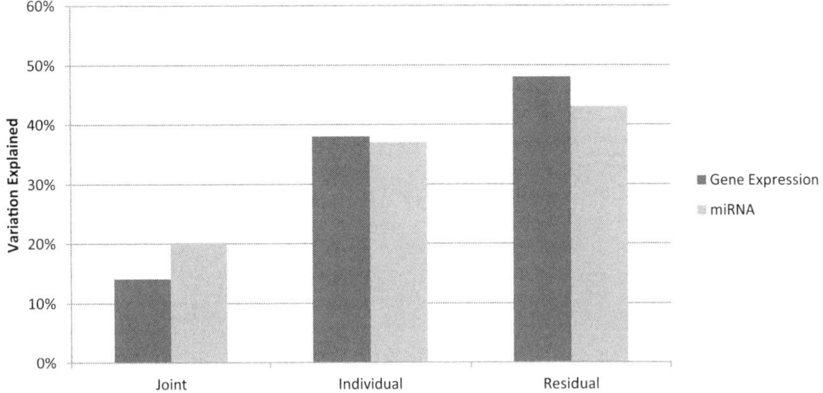

Figure 11.2 Percentage of variation (sum of squares) explained by estimated joint structure, individual structure, and residual noise for miRNA and gene expression data.

A, and luminal B. These well-known subtypes have distinct gene expression characteristics and inform more targeted therapies. Current biological ideas suggest that miRNA function primarily as posttranscriptional regulators of gene expression, and so we are interested in how miRNA-gene interactions relate to subtype distinctions.

For each tumor sample, there are measures of intensity for $p_1 = 655$ miRNAs and $p_2 = 17,814$ genes. These data are publicly available from the TCGA data portal at https://tcga-data.nci.nih.gov/docs/publications/brca_2012/.

As the gene expression and miRNA data differ in dimension and variability, they were scaled as in Eq. (11.1). Permutation testing was used to determine the ranks of estimated joint and individual structure. The test (using $\alpha = 0.01$, and 1000 permutations) identified

- rank 4 joint structure
- rank 22 structure individual to gene expression
- rank 9 structure individual to miRNA

The percentage of variation (sum of squares) explained in each data set by joint structure, individual structure, and residual noise is shown in Figure 11.2. This illustrates how the JIVE decomposition can be used to quantify and compare the amount of shared and individual variation between sources. As shown in Figure 11.2, joint structure is responsible for more variation in miRNA than in gene expression (20% and 13%, respectively), and the gene expression data have a considerable amount of structured variation (38%) that is unrelated to miRNA. This is consistent with current biological understanding, as miRNAs are just one of several factors that can influence gene expression.

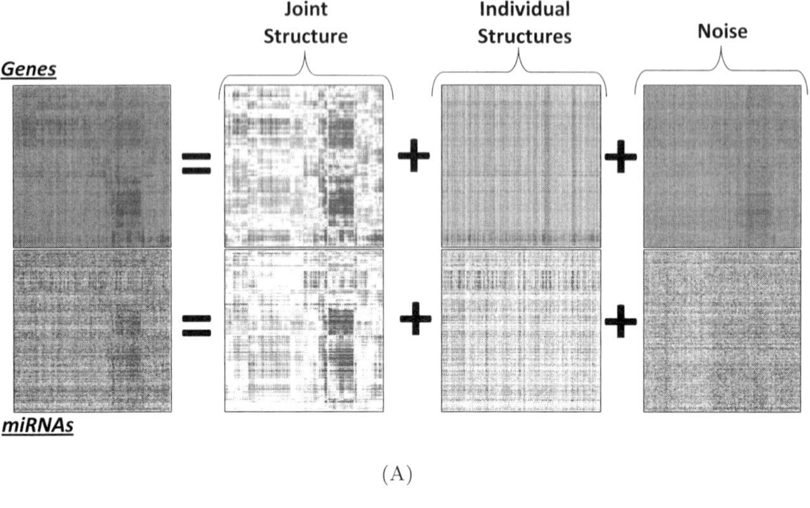

(A)

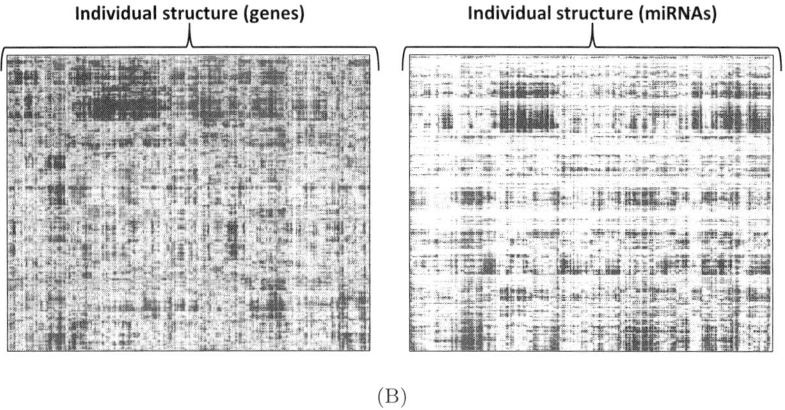

(B)

Figure 11.3 (A) Heatmaps of low-rank estimates for joint structure, individual structure, and residual noise in the gene expression and miRNA data; columns have the same order in all heatmaps. (B) Another view of the low-rank estimates for individual structure, in which columns are ordered separately for gene expression and miRNA data. Blue corresponds to negative values, red to positive values.

Heatmaps of the low-rank estimates for joint structure, individual structure, and residual noise are shown in Figure 11.3A. Rows and columns are ordered consistently in all heatmaps. Ordering was determined by hierarchical clustering of the joint structure, and this reveals sample (column) groups that are clearly distinguished in both data sources. Figure 11.3B provides an alternative view of the estimates for individual structure, in which rows and columns are ordered based on separate hierarchical clustering of each individual

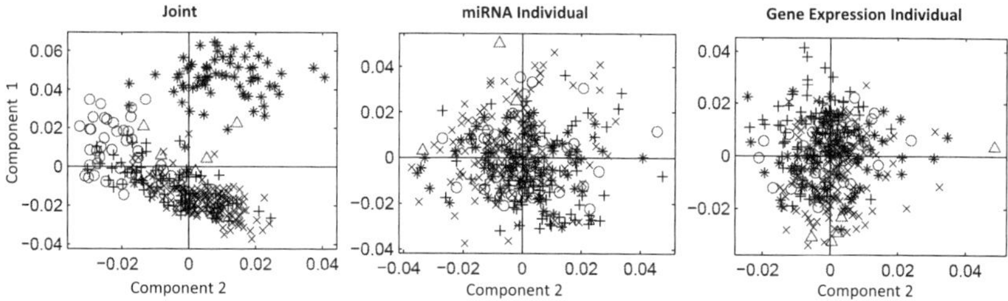

Figure 11.4 Scatter plots of sample scores for the first two joint components, first two individual miRNA components, and first two individual gene expression components. Subtypes are given by symbols: HER2 (O), basal (*), luminal A (X), and luminal B (+).

structure matrix; these reveal structured individual variation that is not apparent in Figure 11.3.

Sample scores for joint structure, matrix S in Eq. (11.3), reveal sample patterns that are present across the miRNA and gene expression data. Sample scores for individual structure, matrices S_1 and S_2 in Eq. (11.3), reveal sample patterns that are individual to each data source. Figure 11.4 shows separate scatter plots of the sample scores for the first two principal components of estimated joint structure, the first two components individual to miRNA, and the first two components individual to gene expression. Subtype distinctions are clearly present in the scatter plot of joint scores, and basal-like samples are particularly well distinguished. Subtype effects are much less visually apparent in either of the individual scatter plots, although luminal A samples are slightly distinguished in the gene expression plot.

As the subtypes are defined by gene expression clustering, their appearance in Figure 11.4 is not surprising. However, the clustering apparent in the joint plot shows involvement of miRNA in the differentiation of these subtypes. It is interesting that subtype effects are not obvious in either scatter plot for individual structure, suggesting that this variation is driven by other biological components. This is remarkable, as the fraction of gene expression variation explained by joint structure is small (see Figure 11.2). The JIVE analysis suggests that miRNA may play a substantial role in the clinical heterogeneity of breast tumors.

11.5 Bayesian Consensus Clustering

BCC is a multisource extension of model-based clustering methods. The model permits a separate clustering of the objects for each data source that adhere loosely to an overall clustering. We have developed a general and computationally scalable Bayesian framework for simultaneous estimation of the source-specific and overall clustering. BCC learns the level of clustering agreement between sources and therefore acts as a flexible bridge between separate clustering and joint clustering.

BCC is related to consensus clustering (introduced in Section 11.3), as an overall clustering is derived from multiple separate clusterings. However, BCC differs from traditional consensus clustering methods in three key aspects:

1. Both the source-specific clusterings and the consensus clustering are modeled in a statistical way that allows for uncertainty in all parameters.
2. The source-specific clusterings and the consensus clustering are estimated simultaneously rather than in two stages. This permits borrowing of information across sources for more accurate cluster assignments.
3. The strength of association to the consensus clustering for each data source is learned from the data and accounted for in the model.

Software to perform BCC is available at http://www.tc.umn.edu/~elock/Software.html.

11.5.1 Model

The probabalistic framework for BCC uses Dirichlet mixtures to model the source-specific and overall clusterings. We introduce the Dirichlet mixture model for clustering a single data set before describing the multisource model.

Dirichlet Mixture Model

Given data X_n for N objects ($n = 1, \ldots, N$), the goal is to partition these objects into at most K clusters. Typically X_n is a multidimensional vector, but we present the model in sufficient generality to allow for more complex data structures. Let $f(X_n|\theta)$ define a probability model for X_n given parameter(s) θ. For example, f may be a Gaussian density defined by the mean and variance $\theta = (\mu, \sigma^2)$. Each X_n is drawn independently from a mixture distribution with K components, specified by the parameters $\theta_1, \ldots, \theta_K$. Let $C_n \in \{1, \ldots, K\}$ represent the component corresponding to X_n and π_k be the probability that an arbitrary object belongs to cluster k:

$$\pi_k = P(C_n = k)$$

Then, the generative model is

$$X_n \sim f(\cdot|\theta_k) \text{ with probability } \pi_k$$

Under a Bayesian framework, one can put a prior distribution on $\Pi = (\pi_1, \ldots, \pi_K)$ and the parameter set $\Theta = (\theta_1, \ldots, \theta_K)$. It is natural to use a

Dirichlet prior distribution for Π. Standard computational methods such as Gibbs sampling can then be used to approximate the posterior distribution for Π, Θ, and $\mathbb{C} = (C_1, \ldots, C_N)$. The Dirichlet prior is characterized by a K-dimensional concentration parameter β of positive reals. Low prior concentration (e.g., $\beta_k \leq 1$) will allow some of the estimated π_k to be small, and therefore N objects may not represent all K clusters.

Multisource Model

We extend the Dirichlet mixture model to accommodate data from M sources $\mathbb{X}_1, \ldots, \mathbb{X}_M$, as intoduced in Section 11.2; X_{mn} represents data m for object n. Each data source requires a probability model $f_m(X_{mn}|\theta_m)$ parametrized by θ_m. Under the general framework presented here, each $\mathbb{X}_m$ may have disparate structure. For example, X_{1n} may give an image where f_1 defines the spectral density for a Gaussian random field, while X_{2n} may give a categorical vector where f_2 defines a multivariate probability mass function.

We assume there is a separate clustering of the objects for each data source but that these adhere loosely to an overall clustering. Formally, each X_{mn} $n = 1, \ldots, N$ is drawn independently from a K-component mixture distribution specified by the parameters $\theta_{m1}, \ldots, \theta_{mK}$. Let $L_{mn} \in \{1, \ldots, K\}$ represent the component corresponding to X_{mn}. Furthermore, let $C_n \in \{1, \ldots, K\}$ represent the overall mixture component for object n. The source-specific clusterings $\mathbb{L}_m = (L_{m1}, \ldots, L_{mN})$ are dependent on the overall clustering $\mathbb{C} = (C_1, \ldots, C_N)$:

$$P(L_{mn} = k|C_n) = v(k, C_n, \alpha_m)$$

The dependence function v has the simple form

$$v(L_{mn}, C_n, \alpha_m) = \begin{cases} \alpha_m \text{ if } C_n = L_{mn} \\ \frac{1 - \alpha_m}{K - 1} \text{ otherwise} \end{cases} \tag{11.4}$$

where $\alpha_m \in [\frac{1}{K}, 1]$ controls the adherence of data source m to the overall clustering. More simply, α_m is the probability that $L_{mn} = C_n$. So, if $\alpha_m = 1$, then $\mathbb{L}_m = \mathbb{C}$. The α_m are estimated from the data together with $\mathbb{C}$ and $\mathbb{L}_1, \ldots, \mathbb{L}_m$. In practice, we estimate each α_m separately or assume that $\alpha_1 = \ldots = \alpha_M$, and hence each data source adheres equally to the overall clustering. The latter is favored when $M = 2$ for identifiability reasons. More complex models that permit dependence of the α_ms are also potentially useful.

The data $\mathbb{X}_m$ are independent of $\mathbb{C}$ conditional on the source-specific clustering $\mathbb{L}_m$. Hence, $\mathbb{C}$ serves only to unify $\mathbb{L}_1, \ldots, \mathbb{L}_M$. The conditional model is

$$P(L_{mn} = k | X_{mn}, C_n, \theta_{mk}) \propto v(k, C_n, \alpha_m) f_m(X_{mn} | \theta_{mk})$$

Let π_k be the probability that an object belongs to the overall cluster k:

$$\pi_k = P(C_n = k)$$

We assume a Dirichlet(β) prior distribution for $\Pi = (\pi_1, \ldots, \pi_K)$. The probability that an object belongs to a given source-specific cluster follows directly:

$$P(L_{mn} = k | \Pi) = \pi_k \alpha_m + (1 - \pi_k) \frac{1 - \alpha_m}{K - 1} \tag{11.5}$$

Moreover, a simple application of Bayes rule gives the conditional distribution of $\mathbb{C}$:

$$P(C_n = k | \mathbb{L}, \Pi, \alpha) \propto \pi_k \prod_{m=1}^{M} v(L_{mn}, k, \alpha_m)$$

where v is defined as in (11.4).

The number of possible clusters K is the same for $\mathbb{L}_1, \ldots, \mathbb{L}_M$ and $\mathbb{C}$. The link function v naturally aligns the cluster labels, as cases in which the clusterings are not well aligned (a permutation of the labels would give better agreement) will have low posterior probability. The number of clusters actually represented may vary, and generally the source-specific clusterings $\mathbb{L}_m$ will represent more clusters than $\mathbb{C}$, rather than vice versa. This follows from Eq. (11.5) and is illustrated in Lock and Dunson (2013, Supplemental Section 2). Intuitively, if object n is not allocated to any overall cluster in data source m (i.e., $L_{mn} \notin \mathbb{C}$), then X_{mn} does not conform well to any overall pattern in the data.

Integrating over the overall clustering C gives the joint marginal distribution of $\mathbb{L}_1, \ldots, \mathbb{L}_M$:

$$P(\{L_{mn} = k_m\}_{m=1}^{M} | \Pi, \alpha) \propto \sum_{k=1}^{K} \pi_k \prod_{m=1}^{M} v(k_m, k, \alpha_m) \tag{11.6}$$

11.5.2 Estimation

Here we present a general Bayesian framework for estimating the multisource clustering model. We employ a Gibbs sampling procedure to approximate the posterior distribution for the parameters introduced in Section 11.5.1. The algorithm is general in that we do not assume any specific form for the f_m

and the parameters θ_{mk}. We use conjugate prior distributions for α_m, Π, and (if possible) θ_{mk}. Markov chain Monte Carlo (MCMC) proceeds by iteratively sampling from the conditional posterior distributions of $\{\Theta_m\}_{m=1}^M$, $\{\mathbb{L}_m\}_{m=1}^M$, $\{\alpha_m\}_{m=1}^M$, $\mathbb{C}$, and Π. For full computational details, including the prior and derived posterior distribution for each parameter, see Lock and Dunson (2013).

Each sampling iteration produces a different realization of the clusterings $\mathbb{C}, \mathbb{L}_1, \ldots, \mathbb{L}_m$, and together these samples approximate the posterior distribution for the overall and source-specific clusterings. However, a point estimate may be desired for each of $\mathbb{C}, \mathbb{L}_1, \ldots, \mathbb{L}_m$ to facilitate interpretation of the clusters. In this respect, methods that aggregate over the MCMC iterations to produce a single clustering, such as that described in Dahl (2006), can be used.

It is possible to derive a similar sampling procedure using only the marginal form for the source-specific clusterings given in Eq. (11.6). However, the overall clustering C is also of interest in most applications. Furthermore, incorporating C into the algorithm can actually improve computational efficiency dramatically, especially if M is large. As presented, each MCMC iteration can be completed in $O(MNK)$ operations. If the full joint marginal distribution of $L_1, \ldots, L_M$ is used, the computational burden increases exponentially with M.

For each iteration, C_n is determined randomly from a distribution that gives higher probability to clusters that are prevalent in $\{L_{1n}, \ldots, L_{mn}\}$. In this sense, $\mathbb{C}$ is determined by a random consensus clustering of the source-specific clusterings. For this reason, we refer to the algorithm as Bayesian consensus clustering.

The BCC software has been designed for multivariate continuous data using a normal-gamma conjugate prior distribution for cluster-specific means and variances. Full computational details for this implementation are given in Lock and Dunson (2013, Supplemental Section 1.1). The software is open source and may be modified for use with alternative likelihood models (e.g., for categorical or functional data).

One can infer the number of clusters in the model by specifying a large value for the maximum number of clusters K, for example, $K = N$. The number of clusters realized in $\mathbb{C}$ and the $\mathbb{L}_m$ may still be small. However, we find that this is not the case for high-dimensional structured data such as those used for the genomics application in Section 11.5.4. The model tends to select a large number of clusters even if the Dirichlet prior concentration parameters β_0 are very small. The number of clusters realized using a Dirichlet process increases with the sample size; hence, if the number of mixture components is indeed finite, the estimated number of clusters is inconsistent as $N \to \infty$ (Miller and Harrison, 2013). This is undesirable for exploratory applications in which the goal is to identify a small number of interpretable clusters.

Alternatively, we consider a heuristic approach that selects the value of K that gives maximum adherence to an overall clustering. For each K, the estimated adherence parameters $\alpha_m \in [\frac{1}{K}, 1]$ are mapped to the unit interval by the linear transformation

$$\alpha_m^* = \frac{K\alpha_m - 1}{K - 1}$$

We then select the value of K that results in the highest mean adjusted adherence:

$$\bar{\alpha}^* = \frac{1}{M} \sum_{m=1}^{M} \alpha_m^*$$

This approach will generally select a small number of clusters that reveal shared structure across the data sources.

11.5.3 Illustrative Example

To illustrate the flexibility and advantages of BCC in terms of clustering accuracy, we generate simulated data sources $\mathbb{X}_1$ and $\mathbb{X}_2$ as follows:

1. Let $\mathbb{C}$ define two clusters, where $C_n = 1$ for $n \in \{1, \ldots, 100\}$ and $C_n = 2$ for $n \in \{101, \ldots, 200\}$.
2. Draw α from a Uniform(0.5, 1) distribution.
3. For $m = 1, 2$ and $n = 1, \ldots, 200$, generate $L_{mn} \in \{1, 2\}$ with probabilities $P(L_{mn} = C_n) = \alpha$ and $P(L_{mn} \neq C_n) = 1 - \alpha$.
4. For $m = 1, 2$, draw values X_{mn} from a Normal(1, 1) distribution if $L_{mn} = 1$ and from a Normal(-1, 1) distribution if $L_{mn} = 2$.

The signal distinguishing the two clusters is weak enough so that there is substantial overlap within each simulated data source. We generate 100 simulations and compare the results for four model-based clustering approaches:

1. separate clustering, in which a finite Dirichlet mixture model is used to determine a clustering separately for $\mathbb{X}_1$ and $\mathbb{X}_2$
2. joint clustering, in which a finite Dirichlet mixture model is used to determine a single clustering for the concatenated data $[\mathbb{X}_1' \ \mathbb{X}_2']'$

3. dependent clustering, in which we model the pairwise dependence between each data source
4. Bayesian consensus clustering

The full implementation details for each method are given in Lock and Dunson (2013, Supplemental Section 5). We also perform an analogous simulation with $M = 3$ sources.

We consider the relative error for each model in terms of the average number of incorrect cluster assignments:

$$\text{Source error} = \frac{\sum_{m=1}^{M} \sum_{n=1}^{N} \mathbb{1}\{\hat{L}_{mn} \neq L_{mn}\}}{MN}$$

$$\text{Overall error} = \frac{\sum_{n=1}^{N} \mathbb{1}\{\hat{C}_n \neq C_n\}}{N}$$

where $\mathbb{1}$ is the indicator function. For joint clustering, the source clusters $\hat{\mathbb{L}}_m$ are identical. For separate and dependent clustering, we determine an overall clustering by maximizing the posterior expected adjusted Rand index (Fritsch and Ickstadt, 2009) of the source clusterings.

The relative error for each clustering method with $M = 2$ and $M = 3$ sources is shown in Figure 11.5. Smooth curves are fit to the results for each method using LOESS (Cleveland, 1979) and display the relative clustering error for each method as a function of α. Not surprisingly, joint clustering performs well for $\alpha \approx 1$ (perfect agreement) and separate clustering performs well when $\alpha \approx 0.5$ (no relationship). BCC and dependent clustering learn the level of cluster agreement and hence serve as a flexible bridge between these two extremes. Dependent clustering does not perform as well with $M = 3$ sources, as the pairwise dependence model does not assume an overall clustering and therefore has less power to learn the underlying structure for $M > 2$.

11.5.4 Application to TCGA Data

We apply BCC to multisource genomic data on breast cancer tumor samples from TCGA. For a common set of 348 samples (introduced in Section 11.4.4), our processed data set includes

- RNA expression (GE) data for 645 genes
- DNA methylation (ME) data for 574 probes
- miRNA expression (miRNA) data for 423 miRNAs
- reverse phase protein array (RPPA) data for 171 proteins

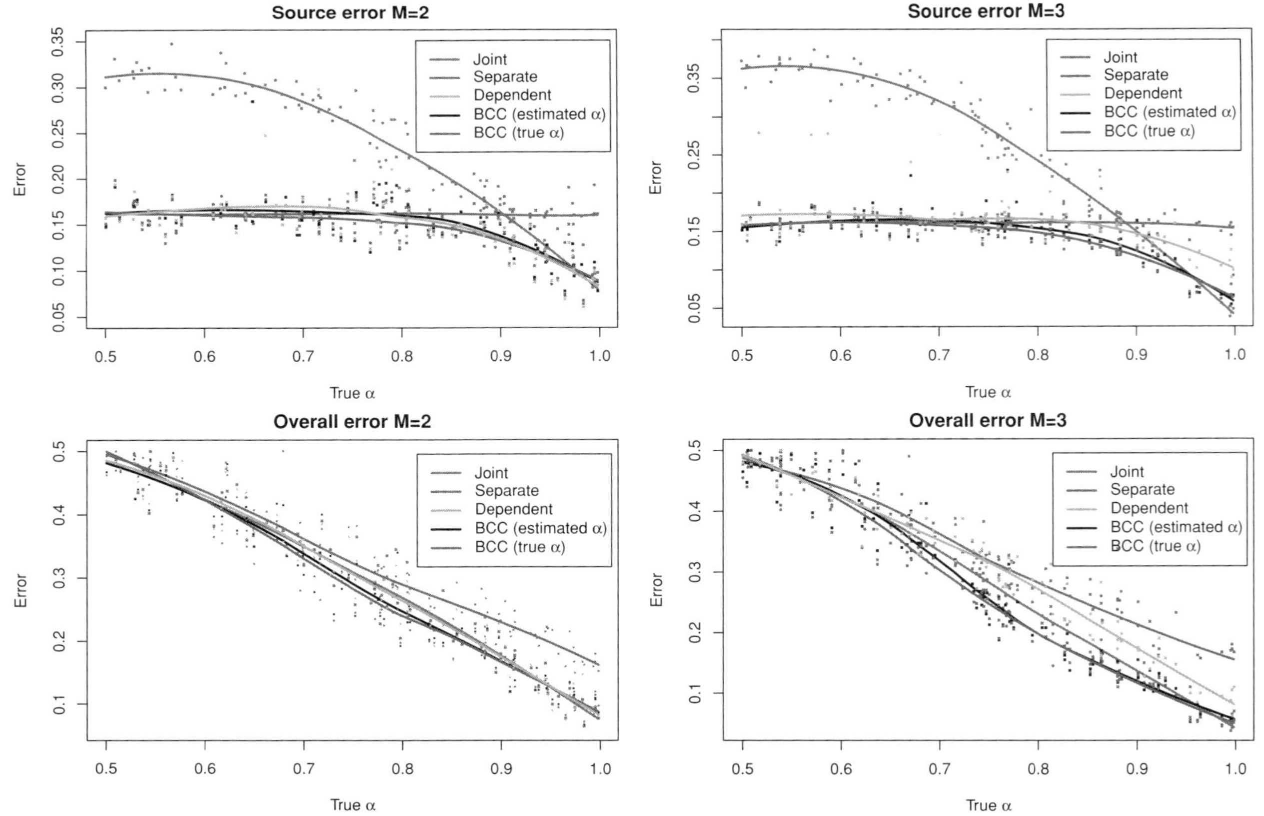

Figure 11.5 Source-specific and overall clustering error for 100 simulations with $M = 2$ and $M = 3$ data sources, shown for joint clustering, separate clustering, dependent clustering, BCC, and BCC using the true α. A LOESS curve displays clustering error as a function of α for each method.

These four data sources are measured on different platforms and represent different biological components. However, they all represent genomic data for the same sample set, and it is reasonable to expect some shared structure. These data are publicly available from the TCGA Data Portal. See http:// www.tc.umn.edu/~elock/Software.html for R code to completely reproduce the following analysis, including instructions on how to download and process these data.

Breast cancer is a very heterogeneous disease and is therefore a natural candidate for clustering. Previous studies have found anywhere from 2 (Duan, 2013) to 10 (Curtis et al., 2012) distinct clusters based on a variety of characteristics. In particular, four comprehensive sample subtypes were previously identified based on a multisource consensus clustering of the TCGA data (Cancer Genome Atlas Network, 2012). These correspond closely to the well-known molecular subtypes basal, luminal A, luminal B, and HER2. These subtypes were shown to be clinically relevant, as they may be used for more targeted therapies and prognoses.

We use the heuristic described in Section 11.5.2 to select the number of clusters for BCC, with intent to determine a clustering that is well represented across the four genomic data sources. We select $K = 3$ clusters, and posterior probability estimates were converted to hard clusterings via the approach described in Dahl (2006) to facilitate comparison and visualization.

Figure 11.6 provides a point-cloud view of each data set given by a scatter plot of the first two principal components. The overall and source-specific cluster index is shown for each sample, as well as a point estimate and approximate 95% credible interval for the adherence parameter α. The GE data have the highest adherence to the overall clustering ($\alpha = 0.91$); this makes biological sense, as RNA expression is thought to have a direct causal relationship with each of the other three data sources. The four data sources show different sample structure, and the source-specific clusters are more well distinguished than the overall clusters in each plot. However, the overall clusters are clearly represented to some degree in all four plots. Hence, the flexible yet integrative approach of BCC seems justified for these data.

Further details regarding the preceding analysis are given in Lock and Dunson (2013, Section 3.3 and Supplemental Section 6). These include an in-depth scientific interpretation of the clusters, the prior specifications for the model, charts that illustrate mixing over the MCMC draws, a comparison of the source-specific clusterings L_{mn} to source-specific subtypes defined by TCGA, clustering heatmaps for each data source, and short-term survival curves for each overall cluster.

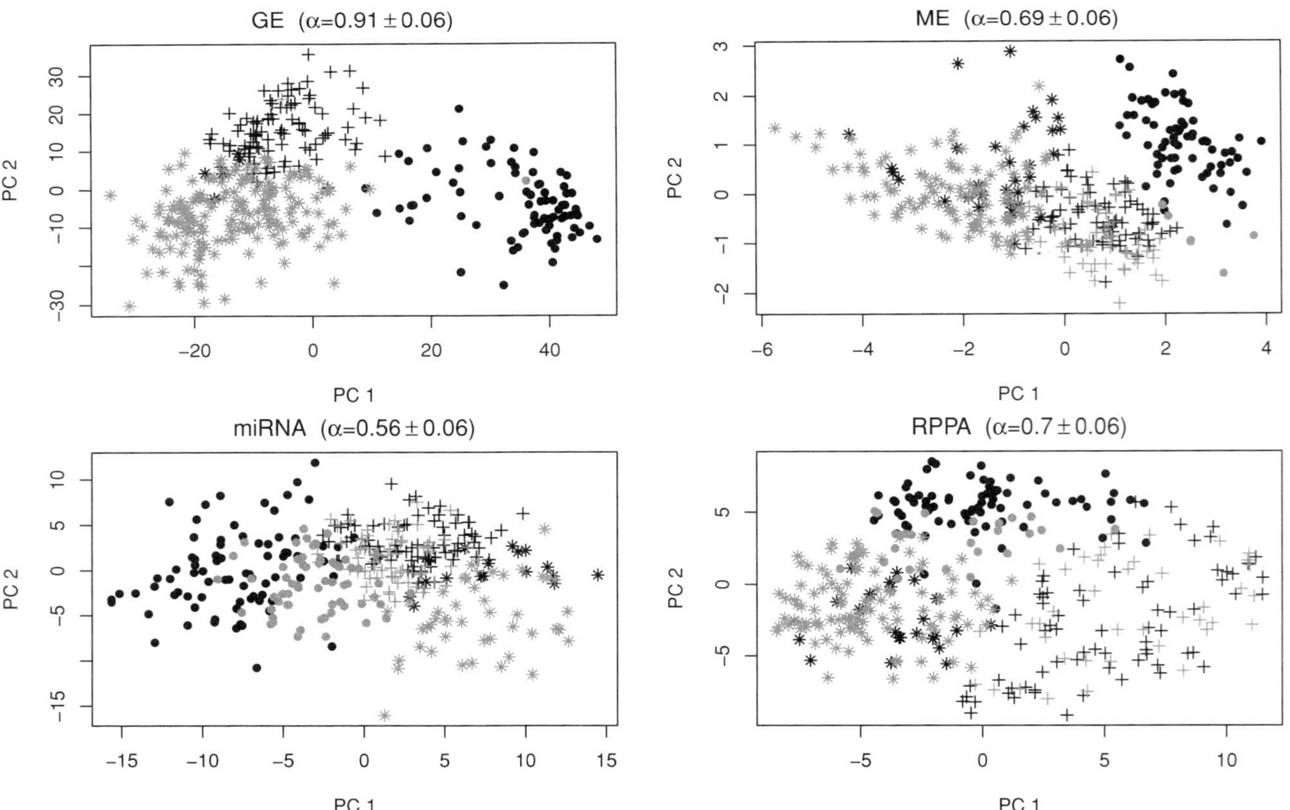

Figure 11.6 PCA plots for each data source. Sample points are shaded by overall cluster; cluster 1 is black, cluster 2 is dark gray, and cluster 3 is light gray. Symbols indicate source-specific cluster; cluster 1 is circles, cluster 2 is crosses, and cluster 3 is stars.

11.6 Conclusion and Discussion

In this chapter, we have presented exploratory methods for multisource data, with a focus on factorization and clustering. We have described the advantages of methods that simultaneously explore the dependence and heterogeneity of the data sources, and we have illustrated these advantages using JIVE for factorization and BCC for clustering. Some other considerations for a given method include its computational burden, model complexity, and incorporation of statistical uncertainty; we conclude with a brief discussion of these related issues.

Computational burden is often a concern when integrating large omics data sources. Use of JIVE or BCC to integrate large ($p > 100,000$) genomic data sets is computationally feasible but not trivial. The JIVE application presented in Section 11.4.4 took approximately 20 minutes of computing time on a 2.3 GHz laptop with 4 GB ram. The BCC application presented in Section 11.5.4 takes approximately 25 minutes of computing time to run 10,000 MCMC iterations, and draws appear to converge to a stationary distribution after approximately 500 iterations. Computing time for the BCC algorithm scales linearly with the number of sources (M), the sample size (N), the number of clusters (K), and the combined dimension of all data sources (p) for each MCMC iteration.

There is often a trade-off between computational burden and the complexity of a given method. Moreover, more complex methods often require more effort to interpret, which is an important consideration for exploratory analysis. The methods described in this chapter are more complex than fully separate or fully joint analyses of multisource data. They convey more information but are also generally more computationally intensive. However, both JIVE and BCC also make simplifying assumptions regarding the dependence between data sources. The implicit model for JIVE assumes that dependencies are characterized by a linear low-dimensional subspace, and the model for BCC assumes dependencies are characterized by a single overall clustering. Alternative factorization methods that model nonlinear dependence (Lawrence, 2005), or methods that model clustering dependence between each pair of data sources (Kirk et al., 2012), are more computationally intensive and require more effort to interpret.

BCC estimates a full probability model, accounting for statistical uncertainty in all parameters. This is an attractive property, but in the integration of large omics data, the use of a such a (Bayesan or frequentist) model is often not computationally or analytically feasible. For example, the K-means method followed by post hoc integration (Cancer Genome Atlas Network, 2012) does not model uncertainty but is computationally faster, and this conveys an advantage for exceptionally large data sets. The JIVE method does not have a probability model for uncertainty, and other multisource factorization methods that do model uncertainty are more computationally intensive (Ray et al., 2014). In general, the creation of methods that measure statistical uncertainty and scale well for big data problems is an important challenge.

Acknowledgments

This work was supported in part by grant R01-ES017436 from the National Institute of Environmental Health Sciences (NIEHS), grant DMS-1310002 from the National Science Foundation (NSF), and grant R01-MH101819-01 from the National Institutes of Health (NIH).

References

Cancer Genome Atlas Network. 2012. Comprehensive molecular portraits of human breast tumours. *Nature*, **490**(7418), 61–70.

Cleveland, W. S. 1979. Robust locally weighted regression and smoothing scatterplots. *Journal of the American Statistical Association*, **74**, 829–836.

Curtis, Christina, Shah, Sohrab P, Chin, Suet-Feung, Turashvili, Gulisa, Rueda, Oscar M, Dunning, Mark J, Speed, Doug, et al. 2012. The genomic and transcriptomic architecture of 2,000 breast tumours reveals novel subgroups. *Nature*, **486**(7403), 346–352.

Dahl, DB. 2006. *Model-Based Clustering for Expression Data via a Dirichlet Process Mixture Model*. Cambridge University Press.

Duan, Q, Kou, Y, Clark, NR, Gordonov, S, and Maayan, A. 2013. Metasignatures identify two major subtypes of breast cancer. *CPT: Pharmacometrics and Systems Pharmacology*, e35.

Filkov, Vladimir, and Skiena, Steven. 2004. Heterogeneous data integration with the consensus clustering formalism. Pages 110–123 of: *Data Integration in the Life Sciences*. Springer.

Fritsch, Arno, and Ickstadt, Katja. 2009. Improved criteria for clustering based on the posterior similarity matrix. *Bayesian Analysis*, **4**(2), 367–391.

Hotelling, H. 1936. Relations between two sets of variants. *Biometrika*, **28**, 321–377.

Hubert, Lawrence, and Arabie, Phipps. 1985. Comparing partitions. *Journal of Classification*, **2**, 193–218.

Kirk, Paul, Griffin, Jim E, Savage, Richard S, Ghahramani, Zoubin, and Wild, David L. 2012. Bayesian correlated clustering to integrate multiple datasets. *Bioinformatics*, **28**(24), 3290–3297.

Kormaksson, M, Booth, JG, Figueroa, ME, and Melnick, A. 2012. Integrative model-based clustering of microarray methylation and expression data. *Annals of Applied Statistics*, **6**(3), 1327–1347.

Lawrence, Neil. 2005. Probabilistic non-linear principal component analysis with Gaussian process latent variable models. *Journal of Machine Learning Research*, **6**, 1783–1816.

Lock, EF, and Dunson, DB. 2013. Bayesian consensus clustering. *Bioinformatics*, **29**(20), 2610–2616.

Lock, Eric F, Hoadley, Katherine A, Marron, JS, and Nobel, Andrew B. 2012. Supplement to "Joint and individual variation explained (JIVE) for integrated analysis of multiple data types." doi:10.1214/12-AOAS597SUPP.

Lock, Eric F, Hoadley, Katherine A, Marron, JS, and Nobel, Andrew B. 2013. Joint and individual variation explained (JIVE) for integrated analysis of multiple data types. *Annals of Applied Statistics*, **7**(1), 523.

Miller, Jeffrey W, and Harrison, Matthew T. 2013. A simple example of Dirichlet process mixture inconsistency for the number of components. *arXiv:1301.2708*.

Mo, Qianxing, Wang, Sijian, Seshan, Venkatraman E, Olshen, Adam B, Schultz, Nikolaus, Sander, Chris, Powers, R Scott, Ladanyi, Marc, and Shen, Ronglai. 2013. Pattern discovery and cancer gene identification in integrated cancer genomic data. *Proceedings of the National Academy of Sciences of the United States of America*.

Nguyen, Nam, and Caruana, Rich. 2007. Consensus clusterings. Pages 607–612 of: *Proceedings of the 7th IEEE International Conference on Data Mining (ICDM 2007), October 28–31, 2007, Omaha, Nebraska, USA*. IEEE Computer Society.

Ray, Priyadip, Zheng, Lingling, Lucas, Joseph, and Carin, Lawrence. 2014. Bayesian joint analysis of heterogeneous genomics data. *Bioinformatics*, **30**(10), 1370–1376.

Rey, Melanie, and Roth, Volker. 2012. Copula mixture model for dependency-seeking clustering. Pages 927–934 of: Langford, John, and Pineau, Joelle (eds.), *Proceedings of the 29th International Conference on Machine Learning (ICML-12)*. ICML '12. New York: Omnipress.

Rogers, Simon, Girolami, Mark, Kolch, Walter, Waters, Katrina M, Liu, Tao, Thrall, Brian, and Wiley, H Steven. 2008. Investigating the correspondence between transcriptomic and proteomic expression profiles using coupled cluster models. *Bioinformatics*, **24**(24), 2894–2900.

Savage, Richard S, Ghahramani, Zoubin, Griffin, Jim E, Bernard, J, and Wild, David L. 2010. Discovering transcriptional modules by Bayesian data integration. *Bioinformatics*, **26**(12), i158–i167.

Savage, Richard S, Ghahramani, Zoubin, Griffin, Jim E, Kirk, Paul, and Wild, David L. 2013. Identifying cancer subtypes in glioblastoma by combining genomic, transcriptomic and epigenomic data. arXiv:1304.3577.

Shen, R, Olshen, AB, and Ladanyi, M. 2009. Integrative clustering of multiple genomic data types using a joint latent variable model with application to breast and lung cancer subtype analysis. *Bioinformatics*, **25**(22), 2906–2912.

TCGA Research Network. 2008. Comprehensive genomic characterization defines human glioblastoma genes and core pathways. *Nature*, **455**, 1061–1068.

Trygg, J, and Wold, S. 2003. O2-PLS, a two-block (XCY) latent variable regression (LVR) method with an integral OSC filter. *Journal of Chemometrics*, **17**(1), 53–64.

Wall, M, Rechtstiener, A, and Rocha, L. 2003. Singular value decomposition and principal component analysis. Pages 91–109 of: Berrar, DP, Dubitzky, W, and Granzow, M (eds.), *A Practical Approach to Microarray Data Analysis*. Kluwer: Norwell, MA.

Westerhuis, JA, Kourti, T, and MacGregor, JF. 1998. Analysis of multiblock and hierarchical PCA and PLS models. *Journal of Chemometrics*, **12**(5), 301–321.

Witten, DM, and Tibshirani, RJ. 2009. Extensions of sparse canonical correlation analysis with applications to genomic data. *Statistical Applications in Genetics and Molecular Biology*, **8**(1).

Wold, H. 1985. Partial least squares. Pages 581–591 of: Kotz, S, and Johnson, NL (eds.), *Encyclopedia of Statistical Sciences* (Vol. 6). Wiley: New York.

Wold, S, Kettaneh, N, and Tjessem, K. 1996. Hierarchical multiblock PLS and PC models for easier model interpretation and as an alternative to variable selection. *Journal of Chemometrics*, **10**(5–6), 463–482.

Yuan, Yinyin, Savage, Richard S, and Markowetz, Florian. 2011. Patient-specific data fusion defines prognostic cancer subtypes. *PLoS Computational Biology*, **7**(10), e1002227.

Zhou, Guoxu, Cichocki, Andrzej, and Xie, Shengli. 2013. Common and individual features analysis: Beyond canonical correlation analysis. arXiv:1212.3913.

VERTICAL INTEGRATIVE ANALYSIS (METHODS SPECIALIZED TO PARTICULAR DATA TYPES)

12

eQTL and Directed Graphical Model

WEI SUN AND MIN JIN HA

Abstract

Gene expression quantitative trait loci (eQTL) are genetic loci that are associated with gene expression traits. The study of the eQTL, or the genetic basis of gene expression variation, not only improves our understanding of gene expression regulation but also brings insights on the functional roles of genetic variations that influence phenotypic outcomes, such as complex human diseases. In contrast to genome-wide association studies, where the signal-to-noise ratio is often low, the eQTLs often have stronger influence on gene expression variation, and hundreds or thousands of eQTLs may be recovered. We conjecture that one of the major applications of eQTL findings is to construct directed graphical models of gene expression data. In this chapter, we review the methods for eQTL mapping, constructing directed graphical models, and the approaches to construct directed graphical models using eQTL data.

12.1 Introduction

The expression of a gene may be associated with the genotype of one or more genetic loci, and such loci are often referred to as gene expression quantitative trait loci (eQTLs). An eQTL study is an integrated study of genetic variants and gene expression across a group of samples. In many eQTL studies, phenotype data (e.g., disease status or drug response) are also collected, and it is of great interest to use eQTL results to inform or guide the phenotype study. A promising approach toward this goal is to construct a directed gene-gene network using eQTL data. In this chapter, we provide reviews and discussions on constructing directed graphical models using eQTL data.

It has been well appreciated that a gene network perspective is crucial to understanding the molecular basis of complex traits, such as many human diseases (Barabási et al., 2011; Marbach et al., 2012). Gene networks can be studied by undirected or directed graphs. For example, a protein-protein interaction graph, where two proteins are connected if they interact with each

271

other, is an undirected graph. A biological pathway often corresponds to a directed graph. The meaning of a directed edge within a pathway depends on the nature of the pathway. In a gene regulation pathway, an edge $A \rightarrow B$ indicates A regulates B. In a signaling pathway, an edge $A \rightarrow B$ indicates signal is transmitted from A to B. Pathway-level analysis is a crucial step to understanding the molecular basis of complex traits, including many human diseases. The complex traits, by definition, are due to accumulative effects of many perturbations of a biological system. Although such perturbations may vary across individuals, they may converge at pathway level. In other words, the patients of a certain disease may carry different genetic mutations and be subject to different environmental exposures, however, such genetic and environmental factors may lead to the malfunction of the same pathway and then cause the clinical phenotypes of the disease. Therefore, it is of great importance to study pathways, and to this end, we need to characterize the causal relations of the genes and proteins within each pathway. For example, many cancer drugs have been developed to target particular proteins in signaling pathways. If we understand the underlying directed graphical models of the signaling pathways, then we may predict the consequences of certain cancer drugs and recommend optimal drug combinations.

Although we have briefly described the importance of directed graphical models in genomic studies, it is well known that interventions or perturbations are needed to infer causal relations. A directed graphical model for gene expression data can include up to tens of thousands of genes, and thus a huge number of interventions/perturbations are needed to infer causal relations. Building directed graphical models using systematic experimental interventions, such as gene knockout or RNA interference, is not feasible yet. The eQTLs of gene expression provide natural perturbations to the expression of a large number of genes and thus enable promising approaches to construct directed graphical models using eQTL data.

In the remainder of this chapter, we first introduce the genetic architecture of gene expression, which justifies the feasibility of using eQTLs to construct directed gene-gene networks. Directed graphical models include both directed cyclic graphs (DCGs) and directed acyclic graphs (DAGs), whereas the former allow cycles within the graphs and the latter do not. We mainly focus on DAGs in this chapter. DAG has been well studied in computer science and has only recently raised attention in statistical society. We give an introduction to the existing methods for DAG construction and then review the statistical methods of DAG/DCG estimation using eQTL data. Then we conclude the chapter with discussions and potential future directions.

12.2 The Genetic Architecture of Gene Expression

12.2.1 Local eQTL versus Distant eQTL

The first genome-wide eQTL study was conducted using 40 yeast sergeants that are offspring of two parental yeast stains (Brem et al., 2002). In that study, Brem et al. (2002) identified several eQTL hot spots, which are genetic loci that are associated with the expression of more genes than expected by chance. These findings have stimulated a series of studies to identify eQTL modules, where each module includes a group of genes genetically regulated by a single locus (Sun et al., 2007).

Several eQTL studies in mouse have also identified eQTL hot spots (Bystrykh et al., 2005; van Nas et al., 2010). However, recent large-scale eQTL studies have demonstrated that such eQTL hot spots are rare in human populations. In fact, even distant eQTLs (i.e., the eQTLs that are distant from their targeted genes, e.g., > 1 Mb) are rare. In one of the largest human eQTL studies, Wright et al. (2014) systematically assessed heritability of gene expression and eQTLs using 2752 twin individuals, including both members of 1308 twin pairs and one member of 136 twin pairs. Figure 12.1 shows the eQTL results using about half of these samples (1263 unrelated individuals). The eQTLs of 18,392 genes were associated with more than 8 million single nucleotide polymorphisms (SNPs). The genotypes of these SNPs were imputed based on 1000 Genome references, and SNPs with low imputation quality or low minor allele frequency (MAF) were removed. From this figure, it is apparent that the vast majority of the eQTLs are local eQTLs. More specifically, Wright et al. (2014) reported 9640 genes (out of 18,392 genes) with local eQTL ($q < 0.01$), among which 9148 (94.9%) were replicated ($q < 0.1$) in an independent study with 1895 independent subjects. In contrast, 348 distant eQTLs ($q < 0.001$, plus several careful QC steps) were reported, among which 165 (47.4%) were replicated ($q < 0.1$) in the independent study.

In this chapter, we emphasize understanding *human* diseases using eQTL data, and thus this type of genetic architecture of gene expression is expected, that is, abundant local eQTLs and few distant eQTLs.

12.2.2 eQTL Mapping

The term *eQTL mapping* refers the process of conducting genome-wide association analysis for all the gene expression traits. In this subsection, we briefly introduce existing methods for eQTL mapping, starting with clarification of the concepts of *cis-* and *trans*-eQTL.

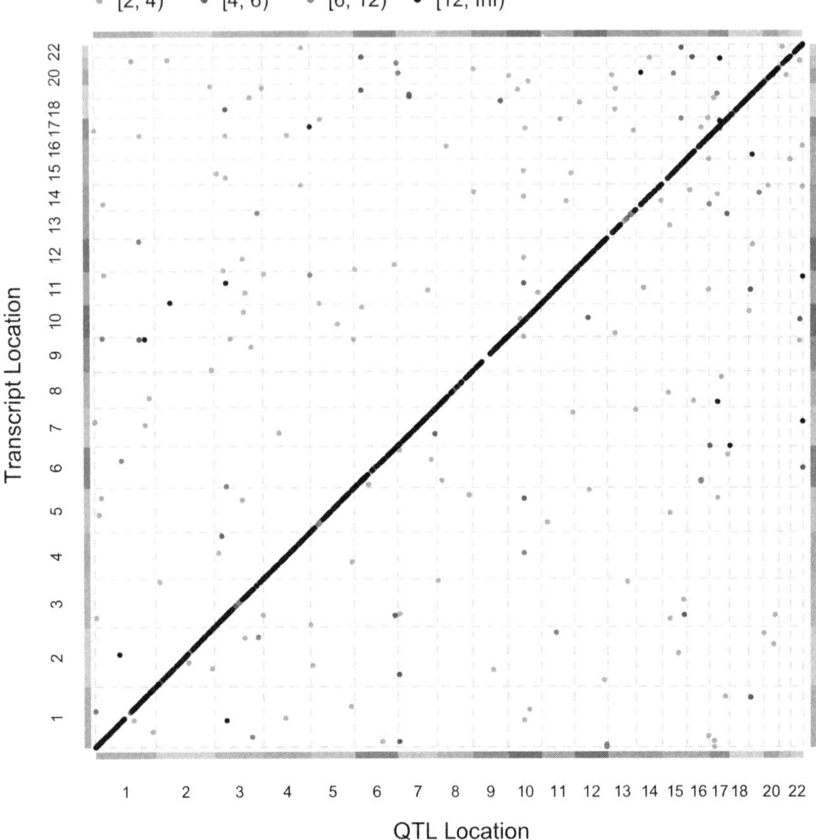

Figure 12.1 eQTL results of gene expression and SNP genotypes measured from whole blood of 1263 unrelated individuals. Each point indicates a significant association between one gene and one SNP. The color of a point reflects the range of $-\log_{10}(q\text{-value})$, which is labeled at the top of the figure.

The terms *cis-* and *trans-*eQTL have been abused to refer to local and distant eQTL, respectively. Here we precisely define them as follows. The Latin words *cis* and *trans* mean "on the same side" and "across", respectively. A *cis-*eQTL is located on the same chromosome as its target gene, but it is not necessarily "local" to the gene. A *cis-*eQTL modifies gene expression in an allele-specific manner. For example, a mutation in a maternal allele can only affect gene expression from maternal alleles. By contrast, a *trans-*eQTL of a gene can be located anywhere in the genome, and it influences the expression of both alleles of its target gene to the same extent. *Cis-* and *trans-*eQTLs can be distinguished

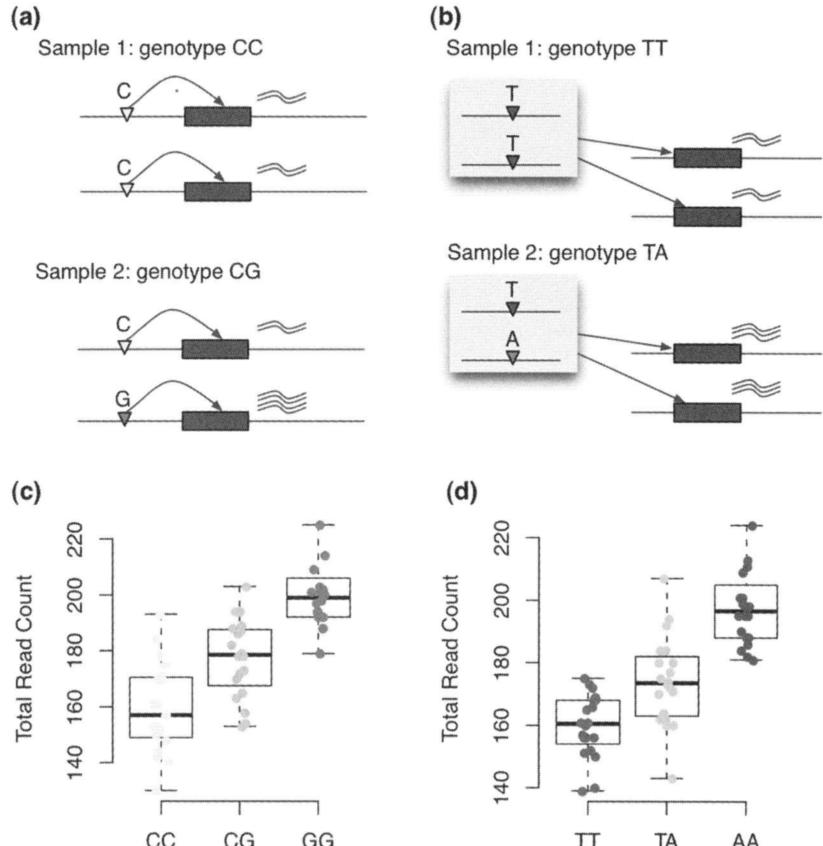

Figure 12.2 (A) An example of a *cis*-eQTL in two samples. In sample 2, where the candidate eQTL (the SNP for which we test association) has a heterozygous genotype CG, the expressions of the two alleles are different. (B) An example of a *trans*-eQTL in two samples. In sample 2, where the candidate eQTL has a heterozygous genotype TA, the expressions of the two alleles are the same. (C) Simulated data for a *cis*-eQTL across 60 samples with 20 samples within each genotype class. (D) Simulated data for a *trans*-eQTL across 60 samples with 20 samples within each genotype class. This figure is adapted from Sun and Hu (2013, Figure 12.1).

by allele-specific gene expression (Figures 12.2A and 12.2B), which can be measured by RNA-seq, but they cannot be distinguished by the total expression (Figures 12.2C and 12.2D).

The eQTL mapping methods/software can be grouped by two criteria. One is to use allele-specific gene expression (ASE) or not, and the other is to jointly analyze multiple gene expression traits or not. Without considering

ASE or jointly analyzing multiple gene expression traits, the problem of eQTL mapping reduces to a large number of genome-wide association studies, where the number of gene expression traits may range from 10,000 to 40,000. The major concern is, then, the computational time. Currently the computationally most efficient software for eQTL mapping is Matrix eQTL (Shabalin, 2012). By calculating test statistics using matrix operations, matrix eQTL can be hundreds of times faster than other existing software.

With RNA-seq data, one can assess the ASE of each sample by counting the number of RNA-seq reads that overlap with any heterozygous SNPs of this sample. As illustrated in Figure 12.2, total expression and ASE are both informative for mapping *cis*-eQTL, and it is desirable to distinguish *cis*- and *trans*-eQTL and to jointly model both types of data for *cis*-eQTL mapping. Currently there is only one method available for this purpose, a so-called TReCASE model (Sun, 2012). This method requires information on the haplotypes connecting the candidate eQTL and its target gene. Usually the haplotypes are imputed from genotype data and are accurate within a relatively short distance (e.g., a few million base pairs), and thus this software is mainly useful for local eQTL mapping.

The gene expression traits presented in an eQTL study are not independent of each other. Their dependence may be revealed by their coexpression pattern or shared functionality. Several methods have been developed to exploit such information for eQTL mapping. For example, Pan (2009) proposed a penalized regression method for eQTL mapping while accounting for shared functions of the gene expression traits, and those shared functions are specified *a priori* by a gene pathway or network. In another study, Cai et al. (2013a) jointly estimated the correlation structure of all gene expression traits and the eQTLs.

12.2.3 eQTLs and Causal Inference

We have mentioned that interventions or perturbations are needed for DAG estimation. Now we would like to clarify the difference of between these two concepts. Intervention often implies setting one or more random variables of the DAG to particular values or states. For example, in a DAG of binary random variables taking values 0 or 1, an intervention may set one or more variables to 0. A more general form of intervention is stochastic intervention (Korb et al., 2004), which sets one or more random variables to the values drawn from particular distribution(s). For example, in a DAG that models continuous random variables, a stochastic intervention may set one variable by a value drawn from a normal distribution with mean 0 and standard deviation 1. An intervention on a random variable effectively removes its connections to its parents. For example, in a DAG of $Y_1 \to Y_2 \to Y_3$, an intervention on Y_2 reduces the DAG to be $Y_1 \quad Y_2 \to Y_3$.

Given the more precise definition of intervention, it is clear that *the eQTL effects are not interventions on gene expression*. They are perturbations in the sense that an eQTL modifies the expression of its target but does not cut its connections with its parents. We may include the DNA genotypes as random variables in the DAG; then the interventions are applied to DNA genotypes, and the consequences are passed to gene expression through eQTL effects. This type of indirect intervention has been referred to as surrogate experiments (Bareinboim and Pearl, 2012).

Next we justify that the variation of DNA genotype across individuals can be considered as consequence of randomized experiment. In an individual with diploid genome, each gene has two alleles, one from father and one from mother. During meiosis, the two alleles of a gene are randomly assigned to each of its daughter cells, which is called Mendelian randomization (Smith, 2007; Sheehan et al., 2008). This constitutes a randomized experiment. In addition, in eQTL studies, the design of experiment (i.e., interventions on DNA genotype rather than gene expression) is also consistent with our intuition that DNA genotype affects gene expression, rather than vice versa.

To use eQTL to derive a causal gene expression network, we also need to separate direct and indirect eQTL effects. For example, given a SNP, denoted by X, and two genes, denoted by Y_1 and Y_2, if the causal relation is $X \to Y_1 \to Y_2$, then X may appear to be an eQTL for both Y_1 and Y_2. We need to know that X directly affects Y_1 but indirectly affects Y_2 for the purpose of DAG estimation. Such information can be obtained by separating *cis*-eQTL and *trans*-eQTL using RNA-seq data (Sun, 2012; Sun and Hu, 2013). All the *cis*-eQTLs directly influence their target genes and a *trans*-eQTL may influence its target's expression directly or indirectly. Therefore, it is desirable to use only *cis*-eQTLs for DAG construction.

12.3 DAG Estimation

12.3.1 An Overview

A directed acyclic graph (DAG) is a useful tool to study causal relations among a set of random variables, where each vertex represents a random variable and each directed edge represents a causal relation (Pearl, 2009). We consider a DAG $\mathcal{G} = (V, E)$, where $V = \{1, \ldots, p\}$ denotes a set of vertices corresponding to random variables $Y_1, \ldots, Y_p$, and E is a collection of directed edges. Denote the parent vertices of vertex i by pa_i, and the corresponding random variables by Y_{pa_i}; then the likelihood of $Y_1, \ldots, Y_p$ can be decomposed based

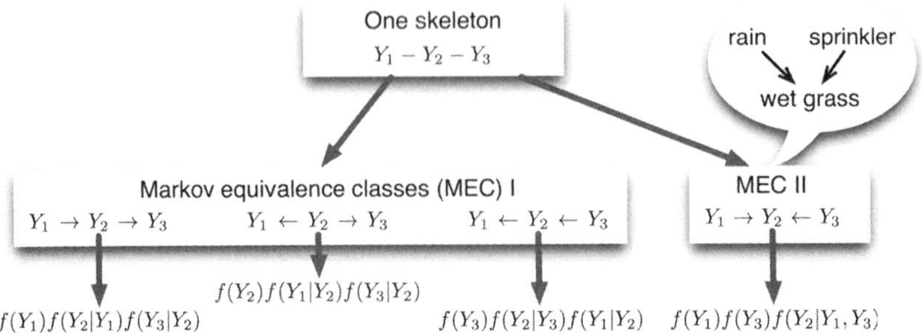

Figure 12.3 An illustration of four DAGs that have the same skeleton. These four DAGs form two Markov equivalence classes (MECs). The likelihood of each DAG is written out based on Eq. (12.1).

on the Markov property that Y_i is independent with all the remaining variables given its parents Y_{pa_i}:

$$f(Y_1, \ldots, Y_p) = \prod_{i=1}^{p} f(Y_i | Y_{\mathrm{pa}_i}) \tag{12.1}$$

Observational data provide a set of conditional independence relations among the random variables, which may be compatible with multiple DAGs. The collection of all the DAGs corresponding to the same set of conditional independence restrictions constitutes a *Markov equivalence* class. More specifically, all the DAGs of one Markov equivalence class have the same *skeleton* and the same set of *v-structures*, and these two concepts are defined as follows:

- If we remove the directions of all the edges in a DAG, the resulting undirected graph is the *skeleton* of the DAG.
- A v-structure is a structure of $Y_1 \rightarrow Y_2 \leftarrow Y_3$, where Y_1 and Y_3 are not directly connected.

Observational data can only identify a Markov equivalence class but cannot distinguish the DAGs within a Markov equivalence class. For example, the DAGs $Y_1 \rightarrow Y_2 \rightarrow Y_3$, $Y_1 \leftarrow Y_2 \leftarrow Y_3$, and $Y_1 \leftarrow Y_2 \rightarrow Y_3$ all represent the (conditional) dependence or independence that $Y_1 \sim Y_2$, $Y_2 \sim Y_3$, $Y_1 \sim Y_3$, and $Y_1 \perp Y_3|Y_2$, where $A \sim B$ indicates A and B are dependent, and $A \perp B|C$ indicates that A is independent of B given C. Therefore these three DAGs belong to the same Markov equivalence class and cannot be distinguished by observational data (Figure 12.3). In contrast, the DAG $Y_1 \rightarrow Y_2 \leftarrow Y_3$, which have the same skeleton but different v-structures from the other three

DAGs, represent different conditional independent assumptions that $Y_1 \perp Y_3$ and $Y_1 \sim Y_3|Y_2$ (Figure 12.3). The relation $Y_1 \sim Y_3|Y_2$ might not be intuitive at first glance. It states that given their shared child vertex, the two parent vertices are not independent. A classical example of this relation is that given wet grass (the child vertex), the events of rain and "sprinkler being on" are not independent. Given a DAG, the likelihood of the three random variables Y_1, Y_2, and Y_3 can be written out based on Eq. (12.1). It is easy to show, using the formula of conditional probability, that the likelihoods of the three DAGs belonging to Markov Equivalence Class I are the same.

Extensive efforts have been devoted to DAG (or Markov equivalence class) estimation. The existing methods can be classified into three categories. The first category includes the search-and-score methods that search for the DAG that maximizes or minimizes a predefined score, such as the Bayesian information criterion (BIC) or minimum description length (MDL). The second category includes the constraint-based methods that construct DAGs by assessing conditional independence of random variables. The third category includes the hybrid methods that combine more than one method of the first two categories. Next we give details of a few representative methods.

12.3.2 The Search-and-Score Methods

The greedy equivalence search (GES) algorithm (Chickering, 2003) is an example of a search-and-score method. GES searches for the DAG that maximizes BIC, given observational, instead of interventional, data. Instead of searching the space of all the DAGs, GES searches across the Markov equivalence classes. This approach significantly improves computational efficiency but does not lose any accuracy because, as mentioned earlier in this chapter, the DAGs within a MEC cannot be distinguished by observational data. The BIC can be considered as an l_0-penalized likelihood, and a recent work discusses the theoretical properties of l_0-penalized maximum likelihood estimates of DAG skeletons (van de Geer and Bühlmann, 2013).

Hauser and Bühlmann (2012) extended the GES by introducing *interventional Markov equivalence classes*, which are equivalence classes of DAGs under multiple interventions. These interventional Markov equivalence classes define a finer partition of DAGs than (observational) Markov equivalence classes, and thus the underlying DAG can be identified with less uncertainty. For example, three DAGs $Y_1 \to Y_2 \to Y_3, Y_1 \leftarrow Y_2 \to Y_3$, and $Y_1 \leftarrow Y_2 \leftarrow Y_3$ form an (observational) Markov equivalence class. Given intervention on Y_2, Y_2 is separated from its parent, and thus the DAGs become $Y_1 \; Y_2 \to Y_3$,

$Y_1 \leftarrow Y_2 \rightarrow Y_3$, and $Y_1 \leftarrow Y_2 \; Y_3$. These three DAGs have distinct likelihoods, and they form three interventional Markov equivalence classes given the intervention on Y_2.

Another example of the search-and-score methods is the order-search method (Teyssier and Koller, 2005), which searches across the ordering of all the p vertices. This strategy is supported by the fact that the number of orderings ($p!$) is much smaller than the number of DAGs ($p!2^{\binom{p}{2}}$) and that DAG estimation given the order of the variables is a much easier task. For example, when the partial ordering of the variables is known (i.e., for each vertex j, all of its ancestors, but none of its descendants, belong to the vertex set $\{1, \ldots, j-1\}$), estimation of DAG under multivariate Guassian assumption is equivalent to $p-1$ penalized regression using Y_j as response and $Y_1, Y_2, \ldots, Y_{j-1}$ as covariates, for $j = 2, \ldots, p$ (Shojaie and Michailidis, 2010).

12.3.3 The PC Algorithm and Related Methods

The PC algorithm, which is named after the first names of its authors Peter Spirtes and Clark Glymour (Spirtes et al., 2000), is representative of the constraint-based methods. The PC algorithm starts with a complete graph (a graph where any two vertices are connected), and then it "thins" the graph by testing whether any two vertices are independent or conditionally independent given one or more other vertices. This delivers the skeleton of the DAG, and finally a set of deterministic rules is applied to orient part of the edges in the skeleton. Kalisch and Bühlmann (2007) proved estimation consistency of the PC algorithm when $p = O(n^a)$ for $a > 0$, where p is the number of vertices and n is sample size. Specifically, each test of conditional independence has a certain probability of making a mistake, and they showed that under some regularity and sparsity conditions, the summation of these mistaken probabilities goes to 0. The results of the PC algorithm depend on the order of the edges to be assessed. Colombo and Maathuis (2012) proposed a modification of the PC algorithm that overcomes such order dependency. This new method, named as PC stable, can substantially improve the performance of the PC algorithm.

12.3.4 Hybrid Methods

The max-min hill-climbing (MMHC) algorithm (Tsamardinos et al., 2006) first estimates the skeleton of the DAG using a constraint-based method (the max-min part of the algorithm) and then orients the edges using a search-and-score technique (the hill-climbing part of the algorithm). The max-min step conceptually resembles the forward-backward regression. More specifically, the parents or children of a vertex Y_j are identified by sequentially selecting vertices associated with Y_j, given all the subsets of the current parents-children set. In the hill-climbing step, the best DAG search starts with an empty graph, and then operations including edge addition, deletion, or direction reversion are employed to improve the score of the graph, while the edge addition is limited to the edges identified in the max-min step.

Another hybrid algorithm (Schmidt et al., 2007) uses a penalized regression with l_1 penalty to replace the max-min step of the MMHC algorithm. An l_1 penalized regression, which uses one variable Y_j as response and all the other variables as covariates, aims to identify the parents, children, and coparents of Y_j, which constitute the so-called Markov blanket of Y_j (Aliferis et al., 2010). The false positive edges due to coparent relations can be removed in the later hill-climbing step of the algorithm. This penalized estimation is also referred to as neighborhood selection for estimating a Gaussian graphical model (GGM) (Maathuis et al., 2009). Fu and Zhou (2013) proposed a penalized likelihood–based method for DAG estimation using both observation and interventional data and an adaptive lasso (l_1) penalty.

We have developed a hybrid algorithm named the PenPC algorithm (Ha et al., 2014) to estimate the DAG skeleton. The PenPC algorithm is a two-step algorithm that combines penalized estimation and the PC algorithm. It first adapts a neighborhood selection method to estimate the Markov blanket of each vertex and then applies a modified PC algorithm to remove false positive edges between coparents. We employ the log penalty (Mazumder et al., 2011) for neighborhood selection. The log penalty can be written as $p_{\lambda,\tau}(|b|) = \lambda \log(|b| + \tau)$, where b is the coefficient to be penalized and λ and τ are two tuning parameters. Comparing with the l_1 penalty, the log penalty significantly improves the accuracy of neighborhood selection. In fact, it has been shown that the l_1 penalty cannot achieve variable selection consistency (i.e., accurate selection of the Markov blanket in this case) unless the correlations between associated and unassociated covariates are weak (Zou, 2006). This assumption may be too strong for a DAG estimation problem. For example, assume the underlying DAG is $Y_1 \rightarrow Y_2 \rightarrow Y_3$. To select the neighborhood of Y_1, we can apply a penalized regression of Y_1 versus Y_2 and Y_3. Then the l_1 penalized estimate picks up Y_2 but not Y_3 if the correlation between Y_2 and Y_3 is weak. This contradicts the graphical structure that Y_2 and Y_3 are directly connected. In the second step of the PenPC algorithm, we remove false positives due to coparent relations. After adding up the uncertainty of neighborhood selection in the first step and the cumulative mistaken probabilities in the second step, we can still obtain a consistent estimate of the DAG skeleton while allowing $p = O(\exp\{n^a\})$.

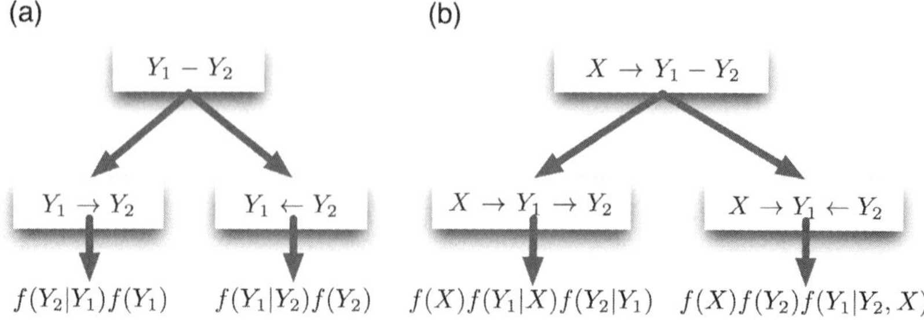

Figure 12.4 An illustration that adding an eQTL can distinguish two directions $Y_1 \rightarrow Y_2$ and $Y_1 \leftarrow Y_2$.

12.4 Directed Graphical Model Estimation Using eQTL Data

The intuition that eQTL can help estimation of edge direction can be obtained from the following simple example. Consider a network of two genes Y_1 and Y_2. If Y_1 and Y_2 are coexpressed, there is an undirected edge $Y_1 - Y_2$ in the graph. We cannot distinguish the two directions $Y_1 \rightarrow Y_2$ and $Y_1 \leftarrow Y_2$ because two DAGs encode the same dependence assumption $Y_1 \sim Y_2$ and thus have the same likelihood (Figure 12.4A):

$$f(Y_2|Y_1)f(Y_1) = f(Y_1, Y_2) = f(Y_1|Y_2)f(Y_2)$$

If we know that Y_1 has an eQTL, denoted by X, then the partially directed graph is $X \rightarrow Y_1 - Y_2$, and the possible DAG is $X \rightarrow Y_1 \rightarrow Y_2$ or $X \rightarrow Y_1 \leftarrow Y_2$. These two graphs can be distinguished because they encode different conditional independence assumptions. $X \rightarrow Y_1 \rightarrow Y_2$ implies $X \perp Y_2|Y_1$ and $X \rightarrow Y_1 \leftarrow Y_2$ implies $X \sim Y_2|Y_1$, and thus they have different likelihoods (Figure 12.4B):

$$L(X \rightarrow Y_1 \rightarrow Y_2) = f(X)f(Y_1|X)f(Y_2|Y_1)$$

$$L(X \rightarrow Y_1 \leftarrow Y_2) = f(X)f(Y_2)f(Y_1|Y_2, X)$$

The problem becomes more challenging when we have multiple or even tens of thousands of genes in the network. We first give a brief introduction to directed graphical model estimation using eQTL data, and then we give more details on a few representative approaches. While a previous section focuses on the DAG estimation method, this section covers methods for either DAG estimation or directed cyclic graph (DCG) estimation.

Previous studies have used eQTL data to dissect the causal relations among three variables, including an eQTL, a gene expression trait, and a third variable, which could be a clinical phenotype (Schadt et al., 2005), another gene expression trait (Kulp and Jagalur, 2006; Chen et al., 2007), or the activity of a transcription factor (Sun et al., 2007). Neto et al. (2008) employed eQTL for directed graphical model estimation in two steps. They first estimated an undirected graph (i.e., the skeleton) using the PC algorithm, which does not require eQTL data, and then used eQTLs to orient the edges in the undirected graph. Later they extended this method to jointly estimate causal networks of gene expression traits and the underlying genetic architecture using Bayesian model averaging and a modified Metropolis-Hastings algorithm (Neto et al., 2010). Hageman et al. (2011) jointly estimated a gene-gene network and eQTLs using a Bayesian method, while placing constraints on the network through a structural prior.

Another type of approach for graphical model estimation is structural equation models (SEM), which permit both cyclic and acyclic graphs. Li et al. (2006) employed a score-based model selection method. Logsdon and Mezey (2010) estimated a network skeleton by applying a penalized regression with an adaptive lasso penalty for each gene expression trait and then transformed the skeleton into a DAG or a directed cyclic graph (DCG) based on eQTL perturbations. Cai et al. (2013b) extended the work of Logsdon and Mezey (2010) by providing the adaptive lasso the initial parameter estimates from penalized regressions using the lasso penalty.

Next we discuss in depth a few representative approaches.

12.4.1 QTL Directed Dependency Graph (QDG)

Neto et al. (2008) developed the QTL directed dependency graph (QDG) method and implemented it in R package `qtlnet`. The QDG method was originally designed to study the relations of multiple phenotypes given their QTLs, though it can be applied to eQTL studies as well. The QDG method assumes "multiple QTL associated with these traits had previously been determined page 1090." It has the following steps:

1. Construct a network skeleton from the PC algorithm.
2. Distinguish QTLs with direct and indirect effects. They first identified QTLs that affect two connected phenotypes and then checked whether the QTL is independent with one phenotype given the other.
3. Orient each edge by LOD score, which is the $\log_{10}$ likelihood ratio for the edge $Y_i \rightarrow Y_j$ versus $Y_j \rightarrow Y_i$ given all the vertices (either phenotype or DNA genotype) connected to Y_i or Y_j.
4. Randomly choose an order of all the edges, and then following this order, sequentially update the directions of the edges using the LOD score

conditioning on the vertices that are parents of Y_i or Y_j. Continue such updating until no more edges change direction. Note that the LOD score of this step is different from that of step 3, where the conditioning set includes all the neighboring vertices.

5. Repeat step 4 1000 times, and choose the graph with the highest score, which could be a likelihood-based measure of fit.

In a later paper, Neto et al. (2010) developed a new method named QTL-net, which jointly estimates the graphical structure of the phenotypes and the underlying genetic architecture. This method would be computationally too demanding to study genome-wide eQTL data with tens of thousands genes and millions of SNPs. In addition, the genetic architecture of human gene expression is relatively simple, with the vast majority of the eQTLs being local eQTLs. Therefore it may be a reasonable approximation to assume the genetic architectures only involve local eQTLs and thus skip (or greatly simplify) the estimation of the underlying genetic architecture. In contrast to QDG, which reports the most likely graph, the QTLnet reports graph structure based on Bayesian model averaging. In other words, the posterior probability of edge $Y_i \rightarrow Y_j$ is the summation of the posterior probabilities of the graphs that have the edge $Y_i \rightarrow Y_j$.

12.4.2 *A Bayesian Framework for DAG Construction using eQTL Data*

Hageman et al. (2011) proposed a Bayesian framework for DAG inference with gene expression as continuous variables and genotype at genetic markers as discrete variables. The causal relation between genotype and gene expression is constrained so that genotype affects gene expression, but not vice versa. The prior distribution of the graph was assigned as

$$P(G) \propto e^{-\tau e(G)}$$

where

$$e(G) = \sum_{i,j=1}^{N} |B_{i,j} - G_{i,j}|$$

and $0 \leq B_{i,j} \leq 1$ expresses the prior belief of a directed edge from Y_i to Y_j. Parameter τ can be considered as a tuning parameter. Hageman et al. (2011) chose to set $\tau = 0.1$ and $B_{i,j} = 0$ for all i and j.

The likelihood function is parameterized such as each vertex is a linear function of a number of verities plus possible eQTLs. They assume each regression coefficient of the linear models follows a normal distribution and the residual variance of the linear model follows an inverse gamma distribution. A Metroplis-Hastings strategy was developed to sample networks from posterior distribution. In addition to single-edge proposals of adding, deleting, and reversing an edge, Hageman et al. (2011) also introduced a reversible edge (REV) proposal that reverses an edge and then samples a new parent set to maintain the acyclic constraint of the graph.

12.4.3 Structure Equation Models

Structure equation models (SEMs) have been widely used in sociology, psychology, and other areas (Bollen, 1989). Recently, SEMs have been employed to estimate directed graphical models using eQTL or QTL data (Li et al., 2006, 2008). A few methods have been developed to add penalization terms into the SEM (Logsdon and Mezey, 2010; Cai et al., 2013b) to address the challenge of high dimensionality of gene expression data.

Let n, m, and p be the sample size, the number of genes, and the number of SNPs, respectively. Denote the gene expression matrix by $\mathbf{Y}_{n \times m}$ and denote the genotype data matrix by $\mathbf{X}_{n \times p}$. Then the SEM can be written as

$$\mathbf{Y} = \mathbf{YB} + \mathbf{XF} + \mathbf{E} \tag{12.2}$$

$\mathbf{B}$ is an $m \times m$ matrix that defines the directed graph among the m genes. Denote the (j, k)th element of $\mathbf{B}$ by b_{jk}, which is the "causal" effect from gene j to gene k. If both b_{jk} and b_{kj} are not 0, then there is a loop between Y_j and Y_k. The diagonal elements of $\mathbf{B}$ are usually set to be 0, reflecting the commonly used assumption of no self-loop. $\mathbf{F}$ is a $p \times m$ matrix indicating eQTL effects. $\mathbf{E}$ is the matrix of residual error.

Li et al. (2006) applied SEM to study the relations of a few phenotypes that have overlapping genetic architecture. They first constructed an initial graphical model using QTL data and then refined this model. Specifically, they first identified QTLs such as $X_1 \to Y_1$. Then they identified causal relations such as $X_1 \to Y_1 \to Y_2$ by comparing model $Y_2 \sim X_1$ and $Y_2 \sim X_1 + Y_1$. If the LOD score of Y_2 versus X_1 has a relatively large change, then it supports the relation $X_1 \to Y_1 \to Y_2$, that is, Y_1 mediates the relation between X_1 and Y_2. Relations like $X_1 \to Y_1$ and $X_1 \to Y_1 \to Y_2$ constitute the initial graphical model. Then they assessed the fit of this model and refined it by adding, removing, or reversing the direction of an edge.

Liu et al. (2008) proposed a more sophisticated strategy to build an initial graphical model that is tailored to eQTL data. For example, they distinguished direct and indirect eQTL effects, for example, $X_1 \to Y_2$ (X_1 is a direct eQTL effect of Y_2) versus $X_1 \rightarrow Y_1 \rightarrow Y_2$ (X_1 is an indirect eQTL effect of Y_2), and they also developed strategies to select a regulator around an eQTL region. Given the graph topology, Liu et al. (2008) obtained the MLE of the parameters by factoring the likelihood while collapsing the vertices of a cyclic component into one vertex. Within a cyclic component, MLE is obtained by a genetic algorithm (GA). Then a backward-forward algorithm was employed to iteratively drop and add an edge.

Logsdon and Mezey (2010) proposed a SEM estimation method using eQTL data, while restricting their analysis on a subset of genes such that each gene has a strong local eQTL, and the local eQTLs of any two genes have weak or no correlation. Logsdon and Mezey's (2010) method has two steps: estimation of an undirected graph by neighborhood selection, followed by edge orientation. Specifically, the neighborhood selection step involves p penalized regressions with an adaptive lasso penalty, where p is the number of genes. In each penalized regression, the response variable is the expression of one gene, and the predictors are the expression of all the other genes plus the preselected local eQTL of this gene. The initial weight of adaptive lasso was obtained from an initial round of lasso regression. This neighborhood selection may induce some false positive edges, such as an edge between X_1 and Y_2, while the true relation is $X_1 \rightarrow Y_1 \leftarrow Y_2$. Logsdon and Mezey (2010) removed such false positives by assessing marginal correlation between X_1 and Y_2. As pointed out by the authors, this method relies on very restrictive assumptions that each gene has a strong local eQTL and any two local eQTLs are weakly correlated. In fact, in their real data analysis on a yeast eQTL data set, only 35 of 5727 genes satisfied these restrictions.

Cai et al. (2013b) also proposed a method of penalized estimation of SEM using the adaptive lasso penalty, though they estimated the initial weights of the adaptive lasso by ridge regression instead of lasso as Logsdon and Mezey (2010) did. Cai et al. (2013b) also proposed a coordinate ascending algorithm to directly solve the MLE, instead of using a neighborhood selection approach.

12.5 Discussion

In this chapter, we have reviewed statistical methods for eQTL mapping and for estimation of directed graphical models with or without eQTL data. We have listed some software and their web addresses in Table 12.1.

Because eQTL data often provide much-needed perturbations for direction estimation and directed graphical models can have a profound impact on disease

Table 12.1 *Methods and software*

Type	Method	Reference
eQTL mapping	matrix eQTL	Shabalin (2012)
		http://www.bios.unc.edu/research/genomic_software/Matrix_eQTL/
	TReCASE	Sun (2012)
		http://www.bios.unc.edu/~weisun/software/asSeq.htm
DAG construction	PC algorithm	Kalisch and Bühlmann (2007)
		http://cran.r-project.org/web/packages/pcalg/index.html
	PenPC algorithm	Ha et al. (2014)
		http://www.bios.unc.edu/~weisun/software/PenPC.htm
DAG construction using eQTL data	QDG and QTLnet	Neto et al. (2008, 2010)
		http://cran.r-project.org/web/packages/qtlnet/index.html

treatment, we expect this application of eQTL data will become increasingly popular in the near future. In fact, this approach can be applied to other types of data as well. For example, one may use the same methods to study the relation between genetic markers and protein abundance, that is, the protein QTL (pQTL) data.

The acyclic assumption of a DAG may appear restrictive for a gene-gene network because there may be feedback loops in gene expression regulation. The SEM reviewed in this chapter is one approach to allowing cyclic relations. Allowing cyclic relations does require a larger number of data (e.g., interventions) to fully identify the underlying model. An alternative solution is to construct a dynamic Bayesian network using time course data (Husmeier, 2003). This is a situation where the natural ordering of the variables is available through time information and thus penalized regression itself is able to identify the DAG skeleton by estimating conditional autoregressive correlations. The main challenge would be that the time course data usually have a limited number of time points and thus augmenting data from other sources would be useful.

Once graphic models are estimated, it would be interesting to compare such models across different populations (e.g., cancer patients of different subtypes). For example, differential coexpression analysis (DCA) has been developed to uncover causative gene-regulatory mechanisms (Rhinn et al., 2013). We expected DCA, or in combination with differential expression analysis, would uncover novel findings that are otherwise undetected by network analysis in a single condition or differential expression analysis across conditions.

References

Aliferis, Constantin F., Statnikov, Alexander, Tsamardinos, Ioannis, Mani, Subramani, and Koutsoukos, Xenofon D. 2010. Local causal and Markov blanket induction for causal discovery and feature selection for classification part I: algorithms and empirical evaluation. *Journal of Machine Learning Research*, **11**, 171–234.

Barabási, Albert-László, Gulbahce, Natali, and Loscalzo, Joseph. 2011. Network medicine: a network-based approach to human disease. *Nature Reviews Genetics*, **12**(1), 56–68.

Bareinboim, Elias, and Pearl, Judea. 2012. Causal inference by surrogate experiments: z-identifiability. arXiv:1210.4842.

Bollen, K. A. 1989. *Structure Equations with Latent Variables*. Wiley-Interscience.

Brem, Rachel B., Yvert, Gaël, Clinton, Rebecca, and Kruglyak, Leonid. 2002. Genetic dissection of transcriptional regulation in budding yeast. *Science*, **296**(5568), 752–755.

Bystrykh, Leonid, Weersing, Ellen, Dontje, Bert, Sutton, Sue, Pletcher, Mathew T., Wiltshire, Tim, Su, Andrew I., et al. 2005. Uncovering regulatory pathways that affect hematopoietic stem cell function using "genetical genomics." *Nature Genetics*, **37**(3), 225–232.

Cai, T. Tony, Li, Hongzhe, Liu, Weidong, and Xie, Jichun. 2013a. Covariate-adjusted precision matrix estimation with an application in genetical genomics. *Biometrika*, **100**(1), 139–156.

Cai, Xiaodong, Bazerque, Juan Andres, and Giannakis, Georgios B. 2013b. Inference of gene regulatory networks with sparse structural equation models exploiting genetic perturbations. *PLoS Computational Biology*, **9**(5), e1003068.

Chen, Lin S., Emmert-Streib, Frank, Storey, John D., et al. 2007. Harnessing naturally randomized transcription to infer regulatory relationships among genes. *Genome Biology*, **8**(10), R219.

Chickering, David Maxwell. 2003. Optimal structure identification with greedy search. *Journal of Machine Learning Research*, **3**, 507–554.

Colombo, D., and Maathuis, M. H. 2012. A modification of the PC algorithm yielding order-independent skeletons. arXiv:1211.3295.

Fu, Fei, and Zhou, Qing. 2013. Learning sparse causal Gaussian networks with experimental intervention: regularization and coordinate descent. *Journal of the American Statistical Association*, **108**(501), 288–300.

Ha, M. J., Sun, W., and Xie, J. 2014. PenPC: a two-step approach to estimate the skeletons of high dimensional directed acyclic graphs. ArXiv e-prints, May.

Hageman, Rachael S., Leduc, Magalie S., Korstanje, Ron, Paigen, Beverly, and Churchill, Gary A. 2011. A Bayesian framework for inference of the genotype–phenotype map for segregating populations. *Genetics*, **187**(4), 1163–1170.

Hauser, Alain, and Bühlmann, Peter. 2012. Characterization and greedy learning of interventional Markov equivalence classes of directed acyclic graphs. *Journal of Machine Learning Research*, **13**, 2409–2464.

Husmeier, Dirk. 2003. Sensitivity and specificity of inferring genetic regulatory interactions from microarray experiments with dynamic Bayesian networks. *Bioinformatics*, **19**(17), 2271–2282.

Kalisch, Markus, and Bühlmann, Peter. 2007. Estimating high-dimensional directed acyclic graphs with the PC-algorithm. *Journal of Machine Learning Research*, **8**, 613–636.

Korb, Kevin B., Hope, Lucas R., Nicholson, Ann E., and Axnick, Karl. 2004. Varieties of causal intervention. Pages 322–331 of: *PRICAI 2004: Trends in Artificial Intelligence*. Springer.

Kulp, David C., and Jagalur, Manjunatha. 2006. Causal inference of regulator-target pairs by gene mapping of expression phenotypes. *BMC Genomics*, **7**(1), 125.

Li, Renhua, Tsaih, Shirng-Wern, Shockley, Keith, Stylianou, Ioannis M., Wergedal, Jon, Paigen, Beverly, and Churchill, Gary A. 2006. Structural model analysis of multiple quantitative traits. *PLoS Genetics*, **2**(7), e114.

Liu, Bing, de la Fuente, Alberto, and Hoeschele, Ina. 2008. Gene network inference via structural equation modeling in genetical genomics experiments. *Genetics*, **178**(3), 1763–1776.

Logsdon, Benjamin A., and Mezey, Jason. 2010. Gene expression network reconstruction by convex feature selection when incorporating genetic perturbations. *PLoS Computational Biology*, **6**(12), e1001014.

Maathuis, M. H., Kalisch, M., and Bühlmann, P. 2009. Estimating high-dimensional intervention effects from observational data. *Annals of Statistics*, **37**(6A), 3133–3164.

Marbach, Daniel, Costello, James C., Küffner, Robert, Vega, Nicole M., Prill, Robert J., Camacho, Diogo M., Allison, Kyle R., et al. 2012. Wisdom of crowds for robust gene network inference. *Nature Methods*, **9**(8), 796–804.

Mazumder, Rahul, Friedman, Jerome H., and Hastie, Trevor. 2011. SparseNet: Coordinate descent with nonconvex penalties. *Journal of the American Statistical Association*, **106**(495), 1125–1138.

Neto, Elias Chaibub, Ferrara, Christine T., Attie, Alan D., and Yandell, Brian S. 2008. Inferring causal phenotype networks from segregating populations. *Genetics*, **179**(2), 1089–1100.

Neto, Elias Chaibub, Keller, Mark P., Attie, Alan D., and Yandell, Brian S. 2010. Causal graphical models in systems genetics: a unified framework for joint inference of causal network and genetic architecture for correlated phenotypes. *Annals of Applied Statistics*, **4**(1), 320–339.

Pan, Wei. 2009. Network-based multiple locus linkage analysis of expression traits. *Bioinformatics*, **25**(11), 1390–1396.

Pearl, J. 2009. *Causality: Models, Reasoning and Inference*. Cambridge University Press.

Rhinn, Herve, Fujita, Ryousuke, Qiang, Liang, Cheng, Rong, Lee, Joseph H., and Abeliovich, Asa. 2013. Integrative genomics identifies APOE ε4 effectors in Alzheimer's disease. *Nature*, **500**(7460), 45–50.

Schadt, Eric E., Lamb, John, Yang, Xia, Zhu, Jun, Edwards, Steve, GuhaThakurta, Debraj, Sieberts, Solveig K., et al. 2005. An integrative genomics approach to infer causal associations between gene expression and disease. *Nature Genetics*, **37**(7), 710–717.

Schmidt, Mark, Niculescu-Mizil, Alexandru, and Murphy, Kevin. 2007. Learning graphical model structure using L1-regularization paths. *AAAI*, **7**, 1278–1283.

Shabalin, Andrey A. 2012. Matrix eQTL: ultra fast eQTL analysis via large matrix operations. *Bioinformatics*, **28**(10), 1353–1358.

Sheehan, N. A., Didelez, V., Burton, P. R., and Tobin, M. D. 2008. Mendelian randomisation and causal inference in observational epidemiology. *PLoS Medicine*, **5**(8), e177.

Shojaie, Ali, and Michailidis, George. 2010. Penalized likelihood methods for estimation of sparse high-dimensional directed acyclic graphs. *Biometrika*, **97**(3), 519–538.

Smith, George Davey. 2007. Capitalizing on Mendelian randomization to assess the effects of treatments. *Journal of the Royal Society of Medicine*, **100**(9), 432–435.

Spirtes, P., Glymour, C. N., and Scheines, R. 2000. *Causation, Prediction and Search*. Vol. 81. MIT Press.

Sun, Wei. 2012. A statistical framework for eQTL mapping using RNA-seq data. *Biometrics*, **68**(1), 1–11.

Sun, Wei, and Hu, Yijuan. 2013. eQTL mapping using RNA-seq data. *Statistics in Biosciences*, **5**(1), 198–219.

Sun, Wei, Yu, Tianwei, and Li, Ker-Chau. 2007. Detection of eQTL modules mediated by activity levels of transcription factors. *Bioinformatics*, **23**(17), 2290–2297.

Teyssier, Marc, and Koller, Daphne. 2005. Ordering-based search: A simple and effective algorithm for learning Bayesian networks. Pages 584–590 of: *Proceedings of the Twenty-First Conference on Uncertainty in Artificial Intelligence*.

Tsamardinos, Ioannis, Brown, Laura E., and Aliferis, Constantin F. 2006. The max-min hill-climbing Bayesian network structure learning algorithm. *Machine Learning*, **65**(1), 31–78.

van de Geer, Sara, and Bühlmann, Peter. 2013. l0-penalized maximum likelihood for sparse directed acyclic graphs. *Annals of Statistics*, **41**(2), 536–567.

van Nas, Atila, Ingram-Drake, Leslie, Sinsheimer, Janet S., Wang, Susanna S., Schadt, Eric E., Drake, Thomas, and Lusis, Aldons J. 2010. Expression quantitative trait loci: replication, tissue-and sex-specificity in mice. *Genetics*, **185**(3), 1059–1068.

Wright, Fred A., Sullivan, Patrick F., Brooks, Andrew I., Zou, Fei, Sun, Wei, Xia, Kai, Madar, Vered, et al. 2014. Heritability and genomics of gene expression in peripheral blood. *Nature Genetics*, **46**(5), 430–437.

Zou, Hui. 2006. The adaptive lasso and its oracle properties. *Journal of the American Statistical Association*, **101**(476), 1418–1429.

13

MicroRNAs: Target Prediction and Involvement in Gene Regulatory Networks

PANAYIOTIS V. BENOS

Abstract

MicroRNAs (miRNAs), the small, noncoding RNA molecules, were first discovered by Victor Ambros in 1993 as negative regulators of *C. elegans* gene *lin-4*. They were later identified as key regulators of gene expression in plants and animals. Their role in disease mechanisms and disease prognosis and diagnosis as well as in the precise regulation of the developmental program of many animals has been undisputable. In this chapter, we discuss two key issues related to their function: (1) miRNA:mRNA targeting and (2) miRNA involvement in gene regulatory networks. We present the fundamental principles on which the major computational algorithms are based for predicting miRNA targets and inferring miRNA-involving regulatory networks.

13.1 Introduction

MicroRNAs (miRNAs) are small, noncoding RNAs that act as regulators of gene expression. Their central role in development and disease [1] has been well established [1–3]. As biomarkers, miRNAs can be more accurate than mRNAs because they constitute the final, fully functional product and not some intermediate state. With the discovery that miRNAs are circulating in the blood plasma [4–6], there was an explosion of research in this area, the results of which tie miRNAs to cell-cell communication [7, 8] as well as diseases like myocardial injury [9] and cardiovascular diseases [10–12], pulmonary hypertension [13], interstitial fibrosis [14], and cancer [15–21], to name a few.

In the genome, they are found in the introns of protein coding genes or in the intergenic regions. They are transcribed by their own promoter or by the promoter of the host gene. Interestingly, as many as 25% of the intronic miRNAs have their own promoter too [22, 23]. After transcription, they undergo a number of processing steps that result in a single-stranded, mature RNA of 20–22 nts, which is loaded to the Argonaute protein, an RNAse H enzyme. Functional miRNAs, located in small introns, have been also reported that bypass the

291

initial step of miRNA processing [24–26]. In complex with the Argonaute protein miRNAs form duplexes with their targets on mRNAs via Watson-Crick RNA base pairing. The way the miRNAs are attached to Argonaute implies that their $5'$-end sequence (especially nucleotides 2–7 or 2–8, also named "seed sequence") initiates the binding, and near-perfect matching is required for this region [27]. If perfect matching in this region is not achieved, more extensive complementarity in the rest of the target will help stabilize the molecule complex. In terms of target location, the most studied and best understood miRNA targets are those located in the $3'$-UTR of the genes, and this is where all currently available algorithms for predicting miRNA targets are focusing. Targets have also been found in the $5'$-UTR of the genes [28] and in their open reading frames, although their functional role is less clear.

In general, miRNAs are considered to be gene expression inhibitors, where the formation of the miRNA:mRNA duplexes results in inhibition of protein translation, acceleration of mRNA degradation, or both. In some cases, however, miRNA targets have shown to prolong mRNA half-life, for example, by interfering with RNAse binding. Notably, the algorithms for inferring miRNA-mRNA networks typically identify many mRNA:miRNA targeting pairs whose expression is positively correlated. Finally, the role of miRNAs in chromatin remodeling has also been reported.

13.2 Popular microRNA Target Prediction Algorithms

There are a number of available computational methods for miRNA target prediction, and they all have one thing in common: low prediction overlap. The study of Hua et al. [29] showed that three popular algorithms overlap 26% to 28% when analyzing data from rat neural tissue. Hammell et al. [30] found even smaller overlap between another four algorithms. Part of the problem stems from the variety of the target characteristics each algorithm considers and the assumptions they make. Another reason is the difference in training data sets that were available at the time each algorithm was developed. Betel et al. [31] have a relatively recent, unbiased comparison of the different features most algorithms use. In that study, the context score seems to be the strongest predictor of a binding site, followed by the alignment score and the energy score. A very nice review of target prediction algorithms and target site characteristics was published by David Bartel in 2009 [27]. In the following, we briefly describe some of the most commonly used algorithms. The basic characteristics of various miRNA target prediction algorithms are also summarized in Table 13.1.

Table 13.1 *Basic characteristics of the main miRNA target prediction algorithms*

Algorithm	Features utilized	Web server	Ref
TargetScan	Seed pairing (very stringent), context score	targetscan.org	[34]
miRanda	Pairing energy	microrna.org (scores)	[32]
PITA	Pairing energy/stability, site accessibility	http://genie.weizmann.ac.il/pubs/mir07/mir07_prediction.html	[37]
ComiR	Site occupancy used to improve four prediction algorithms; improved scores combined with SVM	www.benoslab.pitt.edu/comir	[40]
rna22	Pairing to islands of conserved n-mers	https://cm.jefferson.edu/rna22/Interactive/	[41]
PicTar	Seed pairing for perfect seeds; otherwise, overall pairing stability	http://pictar.mdc-berlin.de/	[38]
mirSVR	Structural, context, global features	microrna.org (scores)	[31]
mirWIP	Seed pairing, overall pairing stability, site accessibility	http://146.189.76.171/query.php	[30]

miRanda [32] is one of the first algorithms developed and it is still popular today. miRanda first searches for sequences that are similar to the (whole) miRNA by using standard sequence alignment methods (i.e., dynamic programming). Then it calculates the energy of the miRNA:target duplex, filters out those targets that are not consistently conserved across species, and finally ranks the remaining targets. miRanda was first developed for *Drosophila* targets for which a number of closely and more distantly related species were sequenced. It was later adapted to work with vertebrate sequences [33] by comparing human targets to mouse and zebrafish to fugu fish.

TargetScan [34], one of the most popular algorithms, takes advantage of the structure of the miRNA:mRNA duplex to simplify the target search. TargetScan initially searches for perfect matches of the 7-nt-long seed sequence of a given miRNA on the 3′-UTR of a potential target gene and then extends each seed to find the miRNA:target duplex with the best energy. Two scores are assigned to this mRNA:miRNA pair. The first is a Z-score, which represents the two binding strength of all targets of the given miRNA on the 3′-UTR of the given mRNA. The second is an *R*-score, which represents the evolutionary conservation of the targets. For mammalian genomes, TargetScan also provides a *context score*, which incorporates five general sequence and structural features of the targets.

By doing so, it helps predict site efficacy without the need for evolutionary conservation [35]. We should note, however, that recent studies have challenged the notion of perfect complementarity of the seed sequence as a prerequisite for binding [36].

PITA [37], like miRanda, uses thermodynamics to determine a miRNA target. However, it also includes information on target accessibility. To put it simply, if a miRNA target is located in an RNA stemloop, then the total energy of the miRNA:mRNA target pairing will be the energy of the miRNA:target duplex minus the energy required to "open" the stemloop structure and make the target accessible. Although the algorithm represented very well the variability the authors observed in mutation experiments (where specific mutations diminished target accessibility), in an independent comparison, the PITA energy score did not yield as good results [31].

PicTar [38] is the first attempt to predict targets of multiple miRNAs. Similar to TargetScan, PicTar gives emphasis to the seed sequence, in which perfect or nearly perfect matching is required. Targets with imperfect matching of the seed sequence are also considered, but they have to pass an additional overall hybridization filter. PicTar, a hidden Markov model algorithm, makes extensive use of multispecies alignments. Given all these, it's perhaps not surprising that the predictions between PicTar and TargetScan show considerable overlap [39].

ComiR [40] is the first algorithm that incorporates the miRNA expression level in the binding model. The basic idea is that the decision of whether a target sequence is occupied by a miRNA is not determined only by the sequence characteristics (i.e., complementarity, binding energy, etc.) but by the expression level (concentration) of the miRNA in the cell. Intuitively, one will expect that when a miRNA is expressed at low levels, it will bind only on its available high-affinity targets. However, when its expression increases, then as the high-affinity targets are already occupied, the miRNA will start targeting the suboptimal targets.

Rna22 [41] is one of the few algorithms that does not require the knowledge of the organism's miRNA-ome for searching for miRNA targets nor directly uses evolutionary conservation for determining a target. Rna22 works in the following way. First, a pattern finding algorithm (TEIRESIAS) runs on all mature miRNA sequences in a miRNA database (RFAM). Second, significance is assigned to each pattern by comparing it to a second-order Markov chain sequence of the genome. Finally, the 3′-UTRs of the organism are searched for "islands" containing more than 30 statistically significant patterns. If miRNA sequences are known for this species, then the islands of significant targets can be matched to them.

mirSVR [31] is a support vector regression algorithm that uses various features to predict miRNA targets. The features used are structural (miRanda-derived miRNA:target duplex features), local context features (e.g., flanking AU context score), and global context features (e.g., position of the target in the 3′-UTR). In this publication, the context score seems to be the dominant performance factor for the mirSVR.

mirWIP [30] uses seed pairing, overall hybridization energy, and target accessibility to infer targets. These parameters are derived from chromatin immunoprecipitation experiments in *C. elegans*.

13.3 Fundamentals of miRNA Binding

13.3.1 The Thermodynamics of Binding

miRNAs, loaded on AGO proteins, recognize their targets on mRNAs via Watson-Crick base complementarity. The more extensive the complementarity is, the stronger is the binding. For predicting miRNA targets, current algorithms simplify the problem by ignoring the AGO contribution and treat it as an RNA:RNA hybridization problem. Another simplification that most algorithms developed before 2012 do is that they partition the problem into single miRNA-to-single target. Thermodynamic energies are calculated as if only the miRNA and a single target were in the mix and they are allowed to reach equilibrium. In reality, the miRNA binding to a given target occurs in a cell where many targets with different hybridization energies are present for any given miRNA. In such an environment, a lowly expressed miRNA is expected to bind almost solely to its high-affinity targets for the time determined by the thermodynamic kinetics of the interaction. However, as the miRNA expression increases, the high-affinity targets become bound practically 100% of the time and are thus essentially "removed" from the target mixture. Subsequently, the rest of the miRNA molecules will start bind to targets of suboptimal energy. This phenomenon is known as "target exclusivity" (a given target cannot be bound by more than one miRNA molecule at a time) and the Fermi-Dirac distribution describes it well (Eq. (13.1)):

$$P(mi\,R_i{:}T_j) = \frac{1}{1 + e^{(E_{ij} - \mu_i)/RT}} \tag{13.1}$$

where E_{ij} is the binding energy of the miRNA to the specific target and μ_i is the total binding potential of the miRNA-i. This can be derived as follows. Let's assume that for a given miRNA-i($mi\,R_i$) and a target-j (T_j), the $[mi\,R_i]$ and $[T_j]$ denote the concentrations of the free miRNA and target molecules, respectively, and $[mi R_i{:}T_j]$ is the concentration of the bound miRNA:target molecules. Then the equilibrium binding constant is defined as

$$K_{ij} = \frac{[mi R_i{:}T_j]}{[mi R_i][T_j]}$$

If we define $\mu_i := \text{RT} \bullet \ln([mi R_i])$ as an approximation to the total binding potential of the miRNA, and given that $E_{ij} = \text{RT} \bullet \ln([K_{ij}])$, then the probability of binding will be

$$P(mi R_i{:}T_j) = \frac{[mi R_i{:}T_j]}{[mi R_i{:}T_j] + [T_j]} = \frac{1}{1 + \frac{1}{K_{ij}[mi R_i]}} = \frac{1}{1 + e^{(E_{ij} - \mu_i)/RT}}$$

If the miRNA is in excess, then $[mi R_i]$ can be approximated by the total expression of this miRNA, and in this case, Eq. (13.1) will provide a measure of the probability of binding of $mi R_i$ to T_j.

13.3.2 *Effect of Single and Multiple miRNA Targets*

Another issue with modeling the effect of a set of miRNAs on the expression level of a gene is the effect of the single targets. The magnitude of the effect might depend on the target itself and the expression level of the miRNAs in the set. The current view is that each individual target causes a relatively mild reduction of the mRNA expression, which can be used to "buffer" noisy signals [42]. Under this model, the big effects that are sometimes observed will come from the combinatorial effect of multiple miRNA targets on individual mRNAs. For example, silencing of let-7d was shown to cause on its own epithelial-to-mesenchymal transition (EMT) of lung cells through its HMGA2 targeting [43]. But HMGA2 has six predicted let-7 targets on its 3′-UTR. Regarding modeling the combinatorial effect of multiple miRNA targets, additivity appears to be a valid assumption [40]. In this case, one can sum up the probabilities of all targets of all miRNAs that bind to a given mRNA to get a comprehensive "regulation potential" score for the set of miRNAs with the given expression levels. It is interesting to note that under this model, the same set of miRNAs but with different miRNA expression levels will give different regulation scores, which is not the case for most target prediction algorithms today.

13.3.3 *Fermi-Dirac Combination of Targets*

But how does the combination of the Fermi-Dirac probabilities fare against a naive target combination method? As we mentioned before, with the exception of TargetScan and PicTar, the overlap between target prediction algorithms is

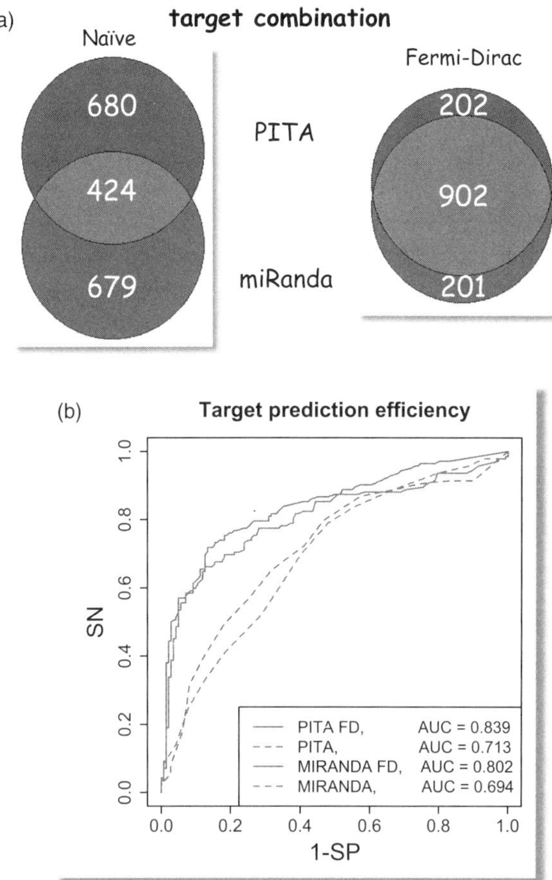

Figure 13.1 Using Fermi-Dirac to improve multiple target prediction. (A) Target overlap between miRanda and PITA using (left) the naive target combination method and (right) Fermi-Dirac. (B) ROC curves of miRanda and PITA on predicting mRNA targeted by multiple miRNAs. Dotted lines are naive combination of targets; solid lines are Fermi-Dirac combination of targets.

poor. This extends when one tries to predict the "targeting potential" of a given mRNA by a set of miRNAs. Figure 13.1A (left) shows the overlap between miRanda and PITA on the top ∼1100 mRNAs (data from [44]) when target scores are combined using the naive method (i.e., an mRNA is targeted by a set of miRNAs if it is a target of at least one of them). However, when the targets are ranked using Eq. (13.1) over all targets of all miRNAs present in the cell, then the overlap on the top 1100 targets increases considerably (from 38% to 82%) (Figure 13.1A, right).

This shows that combining targets of multiple miRNAs quantitatively results in a better overlap between two major target prediction algorithms. We also found that it leads to overall better performance of these algorithms in identifying true and false targets. Using miRNA targets obtained through Argonaute immunoprecipitation (Ago-IP) in *Drosophila* cells and confirmed by Argonaute silencing, we found that combining targets of multiple miRNAs with Eq. (13.1) improves the overall performance of miRanda and PITA (Figure 13.1B).

13.4 ComiR: Quantitative Modeling of Combinatorial miRNA Targeting

ComiR is a relatively new miRNA targeting method [40]. Unlike other methods, ComiR predicts whether a given mRNA will be targeted by *a set* of miRNAs *with known concentrations*. ComiR improves the scoring potential of four popular miRNA target finding methods (miRanda, PITA, TargetScan, mirSVR) by incorporating miRNA expression into the calculation of the corresponding targeting potential scores. (1) ComiR uses the energies calculated by miRanda in the Fermi-Dirac equation to calculate a combinatorial score, CS_{miRanda}, of all targets of all miRNAs on a given mRNA (Eq. (13.2)):

$$CS_{\mathrm{miRanda}} = \sum_{i=1}^{NMIR} \sum_{j=1}^{NTrg_i} P(miR_i{:}T_j) = \sum_{i=1}^{NMIR} \sum_{j=1}^{NTrgt_i} \frac{1}{1 + e^{(E_{ij}-\mu_i)/RT}} \quad (13.2)$$

In Eq. (13.2), *NMIR* is the number of miRNAs in the data set with a predicted target on the given mRNA, and $NTrgt_i$ is the number of targets of miRNA-i on this mRNA. (2) In a similar way, Eq. (13.2) is used for the calculation of CS_{PITA} score, with energies E_{ij} now obtained from the PITA algorithm. (3) TargetScan predictions are not primarily based on thermodynamic energies. Thus targeting potential score $CS_{\mathrm{TargetScan}}$ is simply a weighted sum of the TargetScan scores of all miRNA targets on a given mRNA (Eq. (13.3)):

$$CS_{\mathrm{TargetScan}} = \sum_{i=1}^{NMIR} \sum_{j=1}^{NTrgt_i} \mu_i \cdot TS\_score(miR_i, T_j) \quad (13.3)$$

In this case, *TS_score* is the TargetScan score for target the miRNA miR_i – target T_j pair. In Eq. (13.3), the miRNA expression level μ_i de facto restricts miRNA only to those with observable expression. (4) In a similar way, ComiR calculates CS_{mirSVR}, the targeting potential score for mirSVR.

Combining targets quantitatively by weighting the scores with the miRNA expression levels improves the performance of miRanda and PITA (Figure 13.1B) as well as TargetScan and mirSVR (unpublished data). ComiR takes advantage of this and integrates the four targeting potential scores

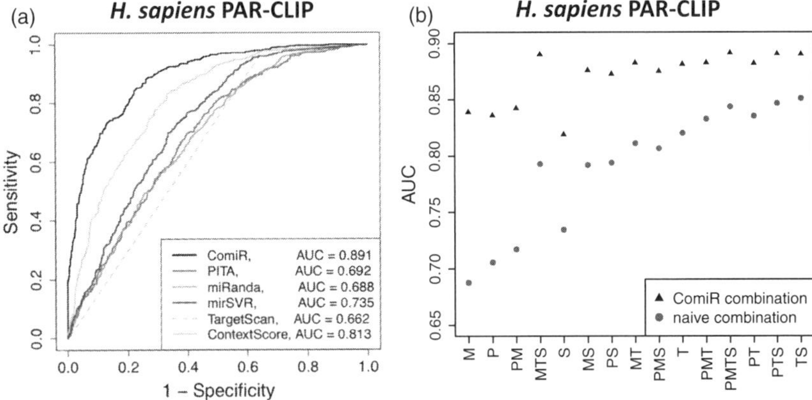

Figure 13.2 ComiR shows significantly improved performance in predicting targets of multiple miRNAs. (A) ROC curves of ComiR (black line) compared to TargetScan (dotted orange line, without conservation score; solid orange line, context score), miRanda (green line), PITA (blue line), and mirSVR (red line). All differences in area under the curve (AUC) are statistically significant. (B) Comparison of SVMs integrating different tools. Regardless of the tools used, the ComiR combination of scores (black triangles) outperforms the naive score combination (red circles). The largest improvement of ComiR scoring is for miRanda and PITA. M, miRanda; P, PITA; T, TargetScan; S, mirSVR.

($CS_{miRanda}$, CS_{PITA}, $CS_{TargetScan}$, CS_{mirSVR}) into one using a support vector machine (SVM). The training data set was obtained by combining data from two publicly available *Drosophila* data sets. One data set contained the miRNA targets identified from an AGO1-IP in *Drosophila* S2 cells [44]. The other data set contained the mRNAs that were increased in expression after AGO1 depletion [45]. Interestingly, only 13% of the AGO1-bound mRNA from the AGO1-IP experiment was found to be upregulated after the AGO1 depletion. One factor might be the noisiness of the IP method. Another might be the fact that miRNAs may have only a moderate effect on most mRNAs [46]. In any case, this is a factor one needs to consider when studying miRNA effects on genes and disease networks. This 13% was the positive data set used in ComiR training, whereas the negative data set was a balanced set of mRNA who were not AGO1-bound and the AGO1 depletion did not change their expression. We showed that the *Drosophila*-trained ComiR model outperforms the four basic prediction algorithms even when applied to human PAR-CLIP data [47] (Figure 13.2A). Actually, improved performance of the ComiR *Drosophila*-trained models on the human data set is observed regardless of the combination of methods used to train the SVM (Figure 13.2B), showing that quantitative modeling of miRNA targeting offers an advantage for identifying the genes that are affected by *a set* of miRNAs over decisions based on naive target combination.

13.5 Comparison of Different Algorithms for miRNA Target Prediction

In the presentation of their mirSVR algorithm, Betel et al. [31] also tested the ability of different features at predicting the true targets. They reported that the context score, incorporated in TargetScan, was the best predictor for many miRNAs. The alignment score, used by miRanda to evaluate the miRNA:site alignment, was the second best performing; and the energy score, used by PITA, was last, and in some cases the energy score and target expression difference were negatively correlated. Remember that PITA takes into consideration the mRNA secondary structure to determine target accessibility. So, its underperformance in that test may be explained by the fact that mRNAs and miRNAs are constantly bound to proteins, which might alter the natural fold of the mRNA in different ways. So, the theoretical mRNA fold might not reflect what happens in vivo. Notably, in our comparison of single target predictor algorithms, for scores normalized by miRNA expression, we also observed TargetScan performing the best, followed by miRanda, and PITA, while mirSVR had the worse performance (Figure 13.2B, black triangles). However, the question asked there is somewhat different, namely, is gene *A* predicted to be regulated by a *set* of miRNAs with known expression levels? In the Betel et al. [31] comparison, the question asked was, is gene *A* predicted to be regulated by miRNA *m*? In conclusion, if one wants to know what are the possible targets of a given miRNA, TargetScan and mirSVR seem to perform the best. But if one wants to know whether a given mRNA is regulated by a set of miRNAs, whose expression levels are known, then ComiR is probably better suited for the job. Rna22, conversely, is the algorithm one would use when insufficient information about the miRNAs exists (e.g., in a newly sequenced organism).

13.6 miRNAs in Gene Regulatory Networks and Disease

13.6.1 miRNAs as Network Regulators

Besides acting as biomarkers, miRNAs have been implicated into various diseases through mechanistic modeling. This stems from their role in the regular function of regulatory networks and their corresponding deregulation in disease. miRNAs are frequently found to participate in a type of fundamental network module called feed-forward loop (FFL), which has been studied extensively, at least in the case of transcription factors (for a very good review, see [48]). FFL is the network module in which a master regulator (transcription factor, miRNA)

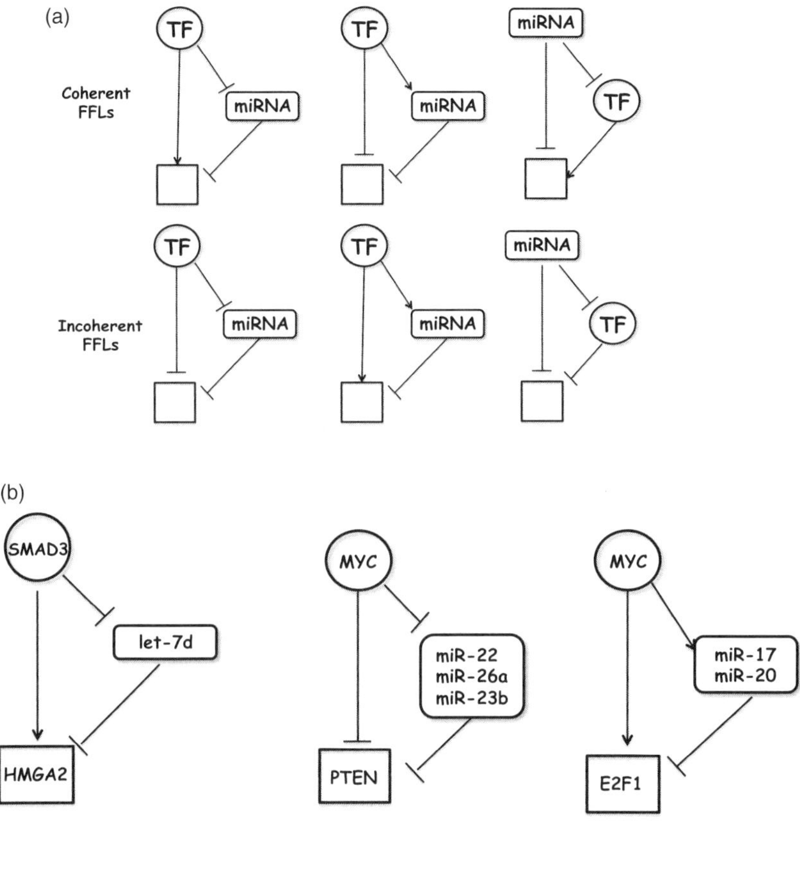

Figure 13.3 (A) Types and (B) examples of feed-forward loops (FFLs), a key regulatory network module in many biological systems. TF, transcription factor.

regulates a gene directly and through another regulator. FFLs can be broadly classified as *coherent* or *incoherent* depending on whether the net direct effect is the same as or opposite to the indirect effect. Figure 13.3A presents all the coherent and incoherent FFLs involving miRNAs. In this figure we assume that the miRNA effect is always negative. Figure 13.3B shows some known FFLs involving miRNAs [43, 49, 50]. When the action on the target gene requires both the direct and the indirect signals to be concurrently present ("AND" gate), FFLs act as "noise buffers," thus enhancing the stability of the network (homeostasis). Transient spikes in the expression of the master regulator will be absorbed (or delayed) in the indirect path and thus the target gene will not be affected. By the same token, persistent changes in expression of the master regulator will be propagated, and then the coherent FFLs will act as a switch between two cell states. This was the case when in lung epithelial cells TGF-β1 causes SMAD3 upregulation, which in turn upregulated HMGA2 directly and indirectly through silencing of let-7d (Figure 13.3B). Then this FFL switch causes the cells to undergo epithelial-to-mesenchymal transition (EMT), as Pandit et al. have shown [43]. Recently databases dedicated to FFLs for certain important genes have started appearing in the literature [50].

13.6.2 Methods for Inferring miRNA Regulatory Networks

The measurement of mRNA and miRNA expression profiles in the same individuals or samples enhances our ability to infer regulatory networks that involve both types of regulators: transcription factors and miRNAs. Most of those algorithms derived from principles that were used in the past for reverse-engineering the gene networks from mRNA expression data alone. They can be broadly divided into two categories. In one category, there are the *relevance networks*, which build networks from pairwise gene correlations. These algorithms are usually fast, can work with relatively few data points, and hence are very popular. Their main drawback is that they tend to generate many false positive edges. Examples of relevance network algorithms include ARACNe [51], one of the first such algorithms developed and widely used today, MMIA [52], MAGIA [53], and mirConnX [54], which constrains the relevance network with prior information to reduce the number of false positive edges. All these algorithms use association metrics such as correlation and mutual information to infer pairwise associations. Both parametric and nonparametric tests can be used. For linearly dependent variables, Pearson correlation coefficient is frequently used. For detecting nonlinear, monotic associations, Spearman's rho or Kendall's tau ranks correlation coefficients are more appropriate. Mutual Information, another nonparametric test, has also been used, and with good results [51], despite its drawbacks (see later).

One problem with pairwise correlations is that they are not directed. As such, when two variables, A and B, are correlated, they cannot tell whether A regulates B or vice versa. This problem can be simplified by using external information. For example, if only one of the two variables is a regulator (transcription factor, miRNA), we can infer the direction of interaction since we do not expect a nonregulator to *directly* affect the expression of another gene. This brings us to the second problem of relevance networks: the problem of conditional dependencies. The aforementioned correlated variables A and B can appear to be correlated due to a common regulator.

These problems are addressed by the second category of algorithms, the *graphical* or *regression networks*, which learn a network of interactions from regression coefficients or conditional independencies. These include regression methods like GenMir++ [55], Bayesian networks [56, 57], and random forests [58]. Although such methods are able to capture higher order interactions, they tend to become computationally inefficient when large numbers of genes are considered. In addition, the sample size they require is larger. To overcome these problems, some methods use external information to filter or cluster variables to reduce the search space. In the following, we present three characteristic methods for regulatory network inference.

ARACNe [51] is one of the first methods to infer regulatory mRNA/miRNA networks from expression data. ARACNe uses mutual information to measure the association of two variables, mRNAs or miRNAs. In addition to the lack of directionality information, which is common to most relevance network algorithms, mutual information is also lacking information about the sign of interaction because it is a nonnegative metric. In other words, when genes *A* and *B* are found to be associated, we do not know whether this is a positive or negative association. Furthermore, it requires calculation of marginal and joint probabilities of the variables, which makes it both computationally intensive and sample size sensitive [54]. Having said that, ARACNe has generated some very important results [59, 60]. Later, ARACNe was incorporated into the MINDy algorithm [61], which includes information about protein turnover, transcription complex formation, and selective enzyme recruitment.

mirConnX [54] is a constraint relevance network algorithm. First, it calculates a network from pairwise correlations. Then it overlays this network to a precomputed network of prior information derived from computational predictions of TF $\rightarrow$ gene/miRNA and miRNA $\rightarrow$ gene/TF interactions and the literature. The overlaying is done by the weighted sum of the *p*-values of the pairwise correlations with the prior probability of interaction. The correlations provide the sign of the interaction (positive/negative) and the prior information the direction of the interaction. Although any correlation measure can be used, for computational efficiency reasons, the web interface allows the choice of three correlation measures: Pearson, Spearman's rho, and Kendall's tau.

GenMiR++ [62] is a Bayesian data analysis algorithm that is used on combined mRNA/miRNA data sets. GenMiR++ maximizes the likelihood of the mRNA expression given the expression of miRNAs and the potential targets. It is a supervised learning method, which uses relatively few known targets for training. This is also one of the major criticisms of GenMiR++, given the large existing variety and complexity of the miRNAs and their targets. We should also say that technically, GenMiR++ is not a mRNA/miRNA network inference algorithm, because the associations it assigns are always from miR-NAs to genes.

PARADIGM [63] is another popular but different algorithm. PARADIGM is unique in that it tries to predict the activity of a protein given the available omics data (mRNA expression, copy number variation, DNA methylation, etc.) and a number of other prior information, such as known protein-protein interactions, pathway information, and other posttranslational regulatory events (e.g., phosphorylation). The drawback is that it currently does not incorporate miRNA targeting information.

13.7 Comparison of Different Network Inferrence Algorithms

Direct comparison of the network inference algorithms is difficult to be made in real data sets, because the known regulatory edges are frequently only a subset of the true regulatory edges and true negative edges (e.g., gene *A* definitely does not regulate gene *B*) are rarely reported in the literature. However, currently the most popular algorithm seems to be ARACNe for general network inference, whereas PARADIGM is used when specific protein activities need to be predicted.

13.8 Future Work

miRNAs are small but very important molecules that are implicated in many developmental processes and phenotypes. miRNA-mediated misregulation of gene networks frequently leads to disease. In the last decades, we have learned a lot about their biogenesis and the regulation of the target genes, but the modeling of the miRNA processes is far from perfect. There are many reasons for that. One is the lack of understanding of the mechanism of the miRNA:mRNA interactions. So far all models have ignored the fact that when miRNA contacts the mRNA, it is already in complex with proteins, which may affect the binding. Recent work has shown that many of the assumptions we have made about miRNA:mRNA binding might not be accurate [36]. Second is, the lack of appropriate quantitative models to describe the complex dynamics of miRNA interactions with multiple targets. This is known as the "sponge effect" [64], according to which change in expression of a single mRNA will indirectly affect the expression of other mRNAs who share common miRNA targets with it. For example, if gene *HMGA2*, which has six let-7 miRNA targets on its 3′-UTR, is silenced, then the "freed" let-7 molecules will target more strongly other genes with suitable targets on their 3′-UTRs. The opposite is also true. Overexpressing a gene will affect the expression of other genes by "absorbing" some of the miRNA molecules for whom it will be a new target. Third, miRNA overexpression experiments, which are frequently used as training or testing sets in miRNA target prediction algorithms, can be misleading. When a miRNA is overexpressed, it will compete with the endogenous miRNAs for the available AGO proteins and thus it might reduce the overall effect of all other miRNAs [65]. Fourth, most data sets do not include protein levels of the target genes. Thus the various analysis algorithms make the assumption that miRNA binding will lead to mRNA degradation. But mRNA degradation may be only one of the mechanisms of action for the miRNAs. Finally, miRNA binding is a transient phenomenon, which is not easy to be captured with available data, unless tight time course data become available.

All these phenomena constitute major obstacles in understanding and modeling miRNA targeting in a cell. However, the always evolving new technologies for data collection (e.g., CLIP-based methods [47, 66]), combined with high-throughput proteomics, will help resolve this complex and fascinating puzzle.

References

1 Ruepp A, Kowarsch A, Schmidl D, Buggenthin F, Brauner B, Dunger I, Fobo G, Frishman G, Montrone C, Theis FJ: PhenomiR: a knowledgebase for microRNA expression in diseases and biological processes. *Genome Biol* 2010, **11**:R6.
2 Ambros V: The functions of animal microRNAs. *Nature* 2004, **431**:350–355.
3 O'Rourke JR, Swanson MS, Harfe BD: MicroRNAs in mammalian development and tumorigenesis. *Birth Defects Res C Embryo Today* 2006, **78**:172–179.
4 Mitchell PS, Parkin RK, Kroh EM, Fritz BR, Wyman SK, Pogosova-Agadjanyan EL, Peterson A, Noteboom J, O'Briant KC, Allen A, et al: Circulating microRNAs as stable blood-based markers for cancer detection. *Proc Natl Acad Sci U S A* 2008, **105**:10513–10518.
5 Chen X, Ba Y, Ma L, Cai X, Yin Y, Wang K, Guo J, Zhang Y, Chen J, Guo X, et al: Characterization of microRNAs in serum: a novel class of biomarkers for diagnosis of cancer and other diseases. *Cell Res* 2008, **18**:997–1006.
6 Lawrie CH, Gal S, Dunlop HM, Pushkaran B, Liggins AP, Pulford K, Banham AH, Pezzella F, Boultwood J, Wainscoat JS, et al: Detection of elevated levels of tumour-associated microRNAs in serum of patients with diffuse large B-cell lymphoma. *Br J Haematol* 2008, **141**:672–675.
7 Zhang Y, Liu D, Chen X, Li J, Li L, Bian Z, Sun F, Lu J, Yin Y, Cai X, et al: Secreted monocytic miR-150 enhances targeted endothelial cell migration. *Mol Cell* 2010, **39**:133–144.
8 Turchinovich A, Samatov TR, Tonevitsky AG, Burwinkel B: Circulating miRNAs: cell-cell communication function? *Front Genet* 2013, **4**:119.
9 Wang E, Nie Y, Zhao Q, Wang W, Huang J, Liao Z, Zhang H, Hu S, Zheng Z: Circulating miRNAs reflect early myocardial injury and recovery after heart transplantation. *J Cardiothorac Surg* 2013, **8**:165.
10 Pleister A, Selemon H, Elton SM, Elton TS: Circulating miRNAs: novel biomarkers of acute coronary syndrome? *Biomark Med* 2013, **7**:287–305.

11 van Empel VP, De Windt LJ, da Costa Martins PA: Circulating miRNAs: reflecting or affecting cardiovascular disease? *Curr Hypertens Rep* 2012, **14**:498–509.
12 Egea V, Schober A, Weber C: Circulating miRNAs: messengers on the move in cardiovascular disease. *Thromb Haemost* 2012, **108**:590–591.
13 Wei C, Henderson H, Spradley C, Li L, Kim IK, Kumar S, Hong N, Arroliga AC, Gupta S: Circulating miRNAs as potential marker for pulmonary hypertension. *PLoS ONE* 2013, **8**:e64396.
14 Yuchuan H, Ya D, Jie Z, Jingqiu C, Yanrong L, Dongliang L, Changguo W, Kuoyan M, Guangneng L, Fang X, et al: Circulating miRNAs might be promising biomarkers to reflect the dynamic pathological changes in smoking-related interstitial fibrosis. *Toxicol Ind Health* 2014, **30**:182–191.
15 Ramshankar V, Krishnamurthy A: Lung cancer detection by screening – presenting circulating miRNAs as a promising next generation biomarker breakthrough. *Asian Pac J Cancer Prev* 2013, **14**:2167–2172.
16 Qu H, Xu W, Huang Y, Yang S: Circulating miRNAs: promising biomarkers of human cancer. *Asian Pac J Cancer Prev* 2011, **12**:1117–1125.
17 Madhavan D, Zucknick M, Wallwiener M, Cuk K, Modugno C, Scharpff M, Schott S, Heil J, Turchinovich A, Yang R, et al: Circulating miRNAs as surrogate markers for circulating tumor cells and prognostic markers in metastatic breast cancer. *Clin Cancer Res* 2012, **18**:5972–5982.
18 Mostert B, Sieuwerts AM, Martens JW, Sleijfer S: Diagnostic applications of cell-free and circulating tumor cell-associated miRNAs in cancer patients. *Expert Rev Mol Diagn* 2011, **11**:259–275.
19 Brase JC, Johannes M, Schlomm T, Falth M, Haese A, Steuber T, Beissbarth T, Kuner R, Sultmann H: Circulating miRNAs are correlated with tumor progression in prostate cancer. *Int J Cancer* 2011, **128**:608–616.
20 Zhao H, Shen J, Medico L, Wang D, Ambrosone CB, Liu S: A pilot study of circulating miRNAs as potential biomarkers of early stage breast cancer. *PLoS ONE* 2010, **5**:e13735.
21 Ross JS: Measuring circulating miRNAs: the new "PSA" for Breast Cancer? *Oncologist* 2010, **15**:656.
22 Corcoran DL, Pandit KV, Gordon B, Bhattacharjee A, Kaminski N, Benos PV: Features of mammalian microRNA promoters emerge from polymerase II chromatin immunoprecipitation data. *PLoS ONE* 2009, **4**:e5279.
23 Marson A, Levine SS, Cole MF, Frampton GM, Brambrink T, Johnstone S, Guenther MG, Johnston WK, Wernig M, Newman J, et al: Connecting microRNA genes to the core transcriptional regulatory circuitry of embryonic stem cells. *Cell* 2008, **134**:521–533.
24 Berezikov E, Chung WJ, Willis J, Cuppen E, Lai EC: Mammalian mirtron genes. *Mol Cell* 2007, **28**:328–336.
25 Okamura K, Hagen JW, Duan H, Tyler DM, Lai EC: The mirtron pathway generates microRNA-class regulatory RNAs in *Drosophila*. *Cell* 2007, **130**:89–100.
26 Ruby JG, Jan CH, Bartel DP: Intronic microRNA precursors that bypass Drosha processing. *Nature* 2007, **448**:83–86.
27 Bartel DP: MicroRNAs: target recognition and regulatory functions. *Cell* 2009, **136**:215–233.
28 Lee I, Ajay SS, Yook JI, Kim HS, Hong SH, Kim NH, Dhanasekaran SM, Chinnaiyan AM, Athey BD: New class of microRNA targets containing simultaneous 5′-UTR and 3′-UTR interaction sites. *Genome Res* 2009, **19**:1175–1183.

29 Hua YJ, Tang ZY, Tu K, Zhu L, Li YX, Xie L, Xiao HS: Identification and target prediction of miRNAs specifically expressed in rat neural tissue. *BMC Genomics* 2009, **10**:214.

30 Hammell M, Long D, Zhang L, Lee A, Carmack CS, Han M, Ding Y, Ambros V: mirWIP: microRNA target prediction based on microRNA-containing ribonucleoprotein-enriched transcripts. *Nat Methods* 2008, **5**:813–819.

31 Betel D, Koppal A, Agius P, Sander C, Leslie C: Comprehensive modeling of microRNA targets predicts functional non-conserved and non-canonical sites. *Genome Biol* 2010, **11**:R90.

32 Enright AJ, John B, Gaul U, Tuschl T, Sander C, Marks DS: MicroRNA targets in Drosophila. *Genome Biol* 2003, **5**:R1.

33 John B, Enright AJ, Aravin A, Tuschl T, Sander C, Marks DS: Human MicroRNA targets. *PLoS Biol* 2004, **2**:e363.

34 Lewis BP, Burge CB, Bartel DP: Conserved seed pairing, often flanked by adenosines, indicates that thousands of human genes are microRNA targets. *Cell* 2005, **120**:15–20.

35 Grimson A, Farh KK, Johnston WK, Garrett-Engele P, Lim LP, Bartel DP: MicroRNA targeting specificity in mammals: determinants beyond seed pairing. *Mol Cell* 2007, **27**:91–105.

36 Xia Z, Clark P, Huynh T, Loher P, Zhao Y, Chen HW, Ren P, Rigoutsos I, Zhou R: Molecular dynamics simulations of Ago silencing complexes reveal a large repertoire of admissible "seed-less" targets. *Sci Rep* 2012, **2**:569.

37 Kertesz M, Iovino N, Unnerstall U, Gaul U, Segal E: The role of site accessibility in microRNA target recognition. *Nat Genet* 2007, **39**:1278–1284.

38 Krek A, Grun D, Poy MN, Wolf R, Rosenberg L, Epstein EJ, MacMenamin P, da Piedade I, Gunsalus KC, Stoffel M, Rajewsky N: Combinatorial microRNA target predictions. *Nat Genet* 2005, **37**:495–500.

39 Rajewsky N: microRNA target predictions in animals. *Nat Genet* 2006, **38 Suppl:**S8–13.

40 Coronnello C, Hartmaier R, Arora A, Huleihel L, Pandit KV, Bais AS, Butterworth M, Kaminski N, Stormo GD, Oesterreich S, Benos PV: Novel modeling of combinatorial miRNA targeting identifies SNP with potential role in bone density. *PLoS Comput Biol* 2012, **8**:e1002830.

41 Miranda KC, Huynh T, Tay Y, Ang YS, Tam WL, Thomson AM, Lim B, Rigoutsos I: A pattern-based method for the identification of MicroRNA binding sites and their corresponding heteroduplexes. *Cell* 2006, **126**:1203–1217.

42 Mendell JT, Olson EN: MicroRNAs in stress signaling and human disease. *Cell* 2012, **148**:1172–1187.

43 Pandit KV, Corcoran D, Yousef H, Yarlagadda M, Tzouvelekis A, Gibson KF, Konishi K, Yousem SA, Singh M, Handley D, et al: Inhibition and role of let-7d in idiopathic pulmonary fibrosis. *Am J Respir Crit Care Med* 2010, **182**:220–229.

44 Hong X, Hammell M, Ambros V, Cohen SM: Immunopurification of Ago1 miRNPs selects for a distinct class of microRNA targets. *Proc Natl Acad Sci U S A* 2009, **106**:15085–15090.

45 Eulalio A, Rehwinkel J, Stricker M, Huntzinger E, Yang SF, Doerks T, Dorner S, Bork P, Boutros M, Izaurralde E: Target-specific requirements for enhancers of decapping in miRNA-mediated gene silencing. *Genes and Development* 2007, **21**:2558–2570.

46 Bartel DP, Chen CZ: Micromanagers of gene expression: the potentially widespread influence of metazoan microRNAs. *Nat Rev Genet* 2004, **5**:396–400.

47	Hafner M, Landthaler M, Burger L, Khorshid M, Hausser J, Berninger P, Rothballer A, Ascano M Jr., Jungkamp AC, Munschauer M, et al: Transcriptome-wide identification of RNA-binding protein and microRNA target sites by PAR-CLIP. *Cell* 2010, **141:**129–141.

48	Alon U: Network motifs: theory and experimental approaches. *Nat Rev Genet* 2007, **8:**450–461.

49	Herranz H, Cohen SM: MicroRNAs and gene regulatory networks: managing the impact of noise in biological systems. *Genes Dev* 2010, **24:**1339–1344.

50	El Baroudi M, Cora D, Bosia C, Osella M, Caselle M: A curated database of miRNA mediated feed-forward loops involving MYC as master regulator. *PLoS ONE* 2011, **6:**e14742.

51	Margolin AA, Nemenman I, Basso K, Wiggins C, Stolovitzky G, Dalla Favera R, Califano A: ARACNE: an algorithm for the reconstruction of gene regulatory networks in a mammalian cellular context. *BMC Bioinformatics* 2006, **7 Suppl 1:**S7.

52	Nam S, Li M, Choi K, Balch C, Kim S, Nephew KP: MicroRNA and mRNA integrated analysis (MMIA): a web tool for examining biological functions of microRNA expression. *Nucleic Acids Res* 2009, **37:**W356–362.

53	Wang LL, Li Y, Zhou SF: A bioinformatics approach for the phenotype prediction of nonsynonymous single nucleotide polymorphisms in human cytochromes P450. *Drug Metab Dispos* 2009, **37:**977–991.

54	Huang GT, Athanassiou C, Benos PV: mirConnX: condition-specific mRNA-microRNA network integrator. *Nucleic Acids Res* 2011, **39:**W416–423.

55	Pritchard JK, Stephens M, Rosenberg NA, Donnelly P: Association mapping in structured populations. *Am J Hum Genet* 2000, **67:**170–181.

56	Liu B, Li J, Tsykin A, Liu L, Gaur AB, Goodall GJ: Exploring complex miRNA-mRNA interactions with Bayesian networks by splitting-averaging strategy. *BMC Bioinformatics* 2009, **10:**408.

57	Le TD, Liu L, Liu B, Tsykin A, Goodall GJ, Satou K, Li J: Inferring microRNA and transcription factor regulatory networks in heterogeneous data. *BMC Bioinformatics* 2013, **14:**92.

58	Roqueiro D, Huang L, Dai Y: Identifying transcription factors and microRNAs as key regulators of pathways using Bayesian inference on known pathway structures. *Proteome Sci* 2012, **10 Suppl 1:**S15.

59	Basso K, Margolin AA, Stolovitzky G, Klein U, Dalla-Favera R, Califano A: Reverse engineering of regulatory networks in human B cells. *Nat Genet* 2005, **37:**382–390.

60	Margolin AA, Wang K, Lim WK, Kustagi M, Nemenman I, Califano A: Reverse engineering cellular networks. *Nat Protoc* 2006, **1:**662–671.

61	Wang K, Saito M, Bisikirska BC, Alvarez MJ, Lim WK, Rajbhandari P, Shen Q, Nemenman I, Basso K, Margolin AA, et al: Genome-wide identification of post-translational modulators of transcription factor activity in human B cells. *Nat Biotechnol* 2009, **27:**829–839.

62	Huang JC, Babak T, Corson TW, Chua G, Khan S, Gallie BL, Hughes TR, Blencowe BJ, Frey BJ, Morris QD: Using expression profiling data to identify human microRNA targets. *Nat Methods* 2007, **4:**1045–1049.

63	Vaske CJ, Benz SC, Sanborn JZ, Earl D, Szeto C, Zhu J, Haussler D, Stuart JM: Inference of patient-specific pathway activities from multi-dimensional cancer genomics data using PARADIGM. *Bioinformatics* 2010, **26:**i237–245.

64	Ebert MS, Sharp PA: MicroRNA sponges: progress and possibilities. *Rna* 2010, **16:**2043–2050.

65 Khan AA, Betel D, Miller ML, Sander C, Leslie CS, Marks DS: Transfection of
small RNAs globally perturbs gene regulation by endogenous microRNAs. *Nat
Biotechnol* 2009, **27**:549–555.
66 Chi SW, Zang JB, Mele A, Darnell RB: Argonaute HITS-CLIP decodes
microRNA-mRNA interaction maps. *Nature* 2009, **460**:479–486.

14

Integration of Cancer Omics Data into a Whole-Cell Pathway Model for Patient-Specific Interpretation

CHARLES VASKE, SAM NG, EVAN PAULL,
AND JOSHUA STUART

Abstract

Recent surveys of multiple types of cancer reveal that most tumors are composed of populations of different cells with distinct sets of mutations. The variety of such sub-clones are thought to fuel drug resistance – targeting any single gene or pathway nearly always selects for those sub-clones lucky enough to have evolved backup mechanisms. The picture for treating cancer is thus becoming more clear: treating one cancer means keeping an entire population of cells in check. One angle, based on the principles of combinatorics that has worked in treating HIV for example, is to target multiple pathways simultaneously to better the chances of eliminating every malignant cell in the population. However, this approach requires gene regulatory network models that accurately describe the causes underlying tumor cell fitness.

This chapter overviews computational representations of genetic pathways for captur-ing salient aspects of tumor biology such as the activation of key genes and processes, how mutations confer gain- or loss-of-function in their cognate genes, and whether explanatory subnetworks can be identified for specific subtypes of disease. Pathway modeling provides a much-needed theoretical foundation to postulate tumor vulnerabil-ities that can scale to the consideration of multiple targets. Advances in the development of such mathematical frameworks should aid the clinical application of patient-specific treatments that minimize or eliminate malignant cell populations.

14.1 Introduction

Cancer genomics utilizes a wide variety of data types on individual tumors. Microarrays have been used to profile genome copy number, gene expression, DNA methylation, and more. Profiling of DNA mutations, DNA breakpoints, gene fusions, RNA expression, and RNA editing are now becoming routine. These techniques have established that even though cancer is a genomically driven disease, any one type of genomic alteration is not enough to explain the altered behaviors and transcriptional states of various tumors. Drivers genes

310

cluster into groups of biochemically interacting genes, and individual tumors often have a mutation in just one of the genes in the cluster. For example, PIK3CA or its suppressor *PTEN* is mutated in tumor samples far more frequently than expected, but samples with both mutations are far less frequent than expected. A full understanding of the diversity and pattern of mutations and abnormal regulation in cancer therefore requires integration of data types and the cellular network.

Advances in measurement technology for DNA, RNA, and protein states have made it possible for many types of data to be collected on individual tumor samples in research cohorts and soon such data will be able to be collected, in common patient care in the clinic. Already, the interaction network is crucial in choosing proper treatments for cancer patients (Ellis et al., 2012). For example, to determine the suitability of drugs that target the cell surface receptor *EGFR*, tumors will be tested for activating mutations in the gene *KRAS*, whose protein product is a kinase downstream of *EGFR*, because its signaling renders the drug ineffectual.

With the great diversity of genomic mutations and expression states seen in nominally similar cancers, pharmaceutical treatment of tumors could be greatly benefited by a system for predicting the response of particular combinations of molecular interventions on the cancer cell. Any such system will need to model portions of the biochemical functions in the cell. What is our current state of knowledge of such biochemical interactions in human cells? Is it sufficient to build a predictive system? Do we have enough data to learn the statistical parameters of such systems, and can we find a balance between a model complex enough to capture the biology, while remaining computationally tractable?

In this chapter we present a few early attempts at such whole-cell models of biochemical function. PARADIGM (Vaske et al., 2010; Sedgewick et al., 2013) is a system that models the genomic, mRNA, protein, and activity state of genes as well as the activity states of multimolecule complexes with a probabilistic graphical model. PARADIGM-SHIFT uses PARADIGM's model to predict the functional impact of a mutation in a tumor. Finally, TieDIE uses a completely different mathematical model, heat diffusion, to isolate important subnetworks while integrating many types of data, including inferences made by PARADIGM.

14.1.1 Pathway Databases

Many bioinformatics databases curate directed interactions between biomolecules. Some databases intend to be comprehensive collections of all interactions in the literature, whereas others have a more specialized focus. One of the most comprehensive pathway databases, Pathway Commons (Cerami et al., 2011), is a collection of other pathway databases. It attempts to standardize and integrate more specialized databases and provide a common method of accessing data from other databases. It also provides helpful web-based tools for viewing and navigating pathway information.

The Pathway Interaction Database (Schaefer et al., 2009) is a collaboration between the National Cancer Institute and the Nature Publishing Group that offers high-quality curations of pathways associated with cancer. Its primary focus is on signaling pathways instead of the more commonly curated metabolic pathways and therefore is of particular interest for methods that incorporate transcriptomics and proteomics. The website for the database also has a helpful user interface for finding the connections between arbitrary molecules in the database.

Cytoscape (Shannon et al., 2003) is a biological network viewer and defines its own format for networks: Simple Interchange Format (SIF) allows the definition of edges between nodes with a single edge attribute, such as transcription or protein-protein interaction. Further annotation about the edges can be contained in a separate file, and there is an extensive user interface for customizing the display of node and edge attributes. More complex relationships than simple pairs can be represented in SIF by defining a new node for a particular reaction. Representation of nonphysical elements as first-class nodes in a graph is a common theme in pathway formats.

The BioPAX pathway format (Demir et al., 2010) is widely supported by pathway databases. It is a format based on the technologies of the semantic web, in particular, the Web Ontology Language (OWL), which is built on top of the Resource Description Framework (RDF). BioPAX supports the definition of reactions in terms of the inputs and outputs of the interactions as well as the catalysis and inhibition of reactions. The constituents of these reactions are "referenced," in that there is an intermediate object that, for example, describes the specific state of a protein, which then in turn refers to a particular protein. BioPAX supports flexible cross-referencing between databases to the protein identifier in the BioPAX to an external and canonical naming scheme. BioPAX defines pathways by grouping and optionally ordering reactions. As a technical detail, pathway databases often expose their BioPAX data in XML form, and for large and comprehensive databases, the size of the XML documents can strain traditional XML tools, requiring stream-based processing.

These pathway databases provide the basic background data for building a whole-cell pathway model.

14.1.2 Pathway Methods

There are two general classes of pathway methods for analysis of gene expression: those based on sets of (unconnected) genes and those based on structured interactions between genes. Gene set methods have been used for many years and use the fact that genes of common function will often show coordinated gene expression changes, either within individual patients or across patients. The Gene Set Enrichment Analysis (GSEA) (Mootha et al., 2003; Subramanian et al., 2005) method is an extremely popular method for analysis of gene expression. GSEA compares two classes of samples such as those sensitive and resistant to treatment, and produces a score per gene set. It is therefore less suited to patient-specific analyses since multiple examples in each of the two classes are required to form the dichotomous statistical test. The recently developed Pathifier (Drier et al., 2013) method uses principal curve analysis to model the coexpression of genes in a set among individual samples. Using this model of expression, it is then possible to determine if a new sample significantly deviates from the model of normal expression, for example, in cancer settings. The advantage of set-based methods over structured methods is that for some cellular function, we do not yet have a model of the interactions, but we may have knowledge that certain genes are coexpressed in certain conditions.

Structured pathway methods use information about gene regulatory and signaling interactions. For some diseases, such as cancer, where we have some knowledge of gene circuitry such methods can provide greater insights by ranking the pathways. One of the first methods to take advantage of structured information is now available as PathOlogist (Efroni et al., 2007; Greenblum et al., 2011). PathOlogist estimates both pathway activity and consistency on approximately 500 pathways, each consisting of on the order of a hundred entities. PathOlogist infers both interaction and gene activity and learns a probabilistic model using one type of data, typically expression. These scores are created per sample, allowing classification of samples based on pathway activities.

Signaling Pathway Impact Analysis (SPIA) uses both data on differentially expressed genes and a pathway structure to determine the importance of each gene's differential expression. SPIA uses a matrix-based representation of the pathway edge weights, which necessarily assumes linearity when combining the effects of interactions. SPIA produces a single score for the pathway across two different classes of samples, so it is somewhat difficult to apply for individual samples. The clipper method (Martini et al., 2013) attempts to extract the parts of pathways that are most perturbed. All these pathway methods attempt to integrate gene expression with a pathway model, whereas PARADIGM attempts to integrate multiple types of omics data with the pathway model.

14.1.3 Pathway-Based Mutation Assessment

The genomic landscape undergoes a plethora of alterations on the path of carcinogenesis from copy number gains and losses to mutations. To make sense of the molecular mechanisms of cancer, it is essential to distinguish the 'driver' events from a sea of 'passenger' events. Because the impact of missense mutations is less obvious to predict, many approaches have been developed to distinguish 'driver' or 'passenger' mutations. Because of positive selection pressure, we can identify 'driver' mutations as occurring at a higher frequency than expected by chance or based on biological information such as predicted impact.

As we are able to observe more cancer samples in a specific cohort, it becomes clear that certain mutations occur at a higher frequency than we would expect. This positive selection for a mutation is evidence for a driver mutation. Methods such as MutSig (Lawrence et al., 2014) were developed to detect such genes. Many 'passengers' are often called significant because of various biases toward mutation, such as the accumulation of mutations due to a gene's late replication timing, whereas more rare 'driver' mutations are missed because of a lack of statistical power in recurrence methods. Genes are scored by comparing the frequency of mutations to a 'passenger' mutation rate estimated by excluding known 'driver' mutations. The significantly mutated genes are determined by setting a threshold based on the false discovery rate.

SIFT (Ng and Henikoff, 2003) and MutationAssessor (Reva et al., 2011) classify the functional impact of mutations based on sequence conservation at the positions in which these missense mutations occur. If a particular position was highly preserved across many normal genomes, a mutation in that relatively invariant position would be predicted to be impactful. PolyPhen-2 (Adzhubei et al., 2010) also predicts the functional impact of individual events but includes additional biological factors in weighing its predictions compared to SIFT. By taking into consideration features related to the local sequence of the mutation and structural information trained against a library of known damaging alleles of human disease, Polyphen-2 is able to improve performance over SIFT.

By combining the advantages of recurrence-based methods and functional impact–based methods, Oncodrive FM (Gonzalez-Perez and Lopez-Bigas, 2012) filters out some of the more recurrent 'passenger' mutations that have low predicted functional impact and retains lowly recurrent 'driver' mutations with a high functional impact. Oncodrive FM weights each mutation based on its functional impact score to create a combined score based on functional impact and recurrence.

Pathway-based assessments can be helpful in adding an orthogonal layer of information to identify networks most impacted by 'driver mutations.' If we can determine how pathways are perturbed by mutation, we can better employ our knowledge to design interventions to treat patients.

14.1.4 Active Subnetwork Search and Discovery

Given a set of functionally disrupted genes in a cohort of patient samples, network-based methods can be used to identify subsets with either direct or indirect regulatory connections between them, increasing the chance that these altered genes are functionally related. Genes in these connected subnetworks may be more likely to be 'drivers' of the cancer phenotype, or in some cases, they may be genes that connect driver mutations but have a function that is essential for tumor survival. In either case, pathway information is critical in selecting these important genes with a low rate of false positives.

MEMo (Ciriello et al., 2012) and Dendrix (Vandin et al., 2012a) both identify pathway connections between genes with nonsynonymous exonic mutations and altered copy number profiles, enabling the discovery of certain low-frequency driver events that would have otherwise missed detection with traditional statistical methods (Network et al., 2013). Both find subnetworks that tend to have only a single functionally altered gene in any given patient then generate a background model to select only those subnetworks with genomic aberrations that are more mutually exclusive than expected by chance. This is similar in approach to jActiveModules (Shannon et al., 2003) – a cytoscape plugin that finds expression-activated subnetworks – but applied to mutation and copy number data that can be treated as binary input for a given gene and sample.

Additional methods, such as the prize-collecting Steiner tree (PCST) (Dittrich et al., 2008) framework, treat subnetwork finding as a linear optimization problem and implement methods for either an exact solution or (in the case of more complex networks and larger gene sets) an approximation to the optimal, which is computationally intractable to predict. This builds on previous work using the PCST formulation (Tuncbag et al., 2013), where continuous scores can be used for genes and putative edges in the PCST model, allowing for more flexible integration of different data types (mutation calls, expression). However, this flexibility also leads to greater complexity in the required optimization task, but recent work in applied mathematics (Ljubić et al., 2006) has improved the tractability of an exact solution to the PCST problem.

Many of these methods suffer from curation bias from the input network, resulting in the same well-connected 'hub' nodes appearing repeatedly in results, while less studied but potentially important drivers are missed. Hot-Net (Vandin et al., 2012b) corrects some of this bias through a "heat kernel" model where gene perturbations are propagated over an undirected regulatory network with the same dynamics as heat moves over a conductance network. Because greater connectivity leads to greater heat loss (as well as gain), the model mitigates the bias toward well-studied hub nodes when finding mutated subnetworks.

14.2 The PARADIGM Pathway Method

Modern omics data sets, such as copy number, mRNA expression, and protein expression, give semiquantitative estimates of the abundances of various molecules in a cell. PARADIGM integrates these omics data sets with pathway information to estimate each molecule's activity level by sharing information on the activity levels of physically interacting molecules. For example, instead of a single variable representing TP53, PARADIGM will not only have a variable for the number of copies of TP53 in the genome but also variables for the amount of TP53 mRNA and the amount of p53 protein. Additionally, PARADIGM's model will have a variable for the p53 tetramer complex, which is the complex that typically performs the biochemical function associated with TP53, transcriptional regulation. A factor graph, a type of probabilistic graphical model, ties all these variables together into a probabilistic equation.

Probabilistic graphical models have been very successfully used in a wide variety of bioinformatics applications that have large numbers of variables, starting with the large gene-gene Bayes nets (Friedman et al., 2000). With discrete variables, probabilistic graphical models can capture nonlinear dependencies between variables, whereas a continuous model, such as a Gaussian graphical model, cannot. A probabilistic graphical model decomposes a joint probability distribution over a set of variables $\mathbb{Y}$ into a set of local functions such as $f_i(y_i, y_j, y_k) : Y_i \times Y_j \times Y_k \to [0, \infty)$. Using the notation that local function i has the domain $\mathbb{Y}_i \subseteq \mathbb{Y}$,

$$P(\mathbb{Y}) = \frac{1}{Z} \prod_i f_i(\mathbb{Y}_i) \tag{14.1}$$

where Z is a normalization constant to make a proper probability space from the unnormalized local functions f_i.

PARADIGM uses parameter sharing, a common graphical modeling technique, to improve accuracy and interpretability. With parameter sharing, some factors f_i are identical functions but have different input variables. This goes further than sharing parameters between samples, as in PARADIGM every tumor is modeled with identical factors. Within a single sample, similar biological functions (e.g., transcription) are modeled with the same factor parameters. This can be represented by adding a layer of "templates" for the factors, borrowing the terminology from FACTORIE (McCallum et al., 2009). For each of n templates T_i, there are a set of m_i variable tuples, $\{\mathbb{Y}_{ij} \mid 1 \leq j \leq m_i\}$, that allows the joint probability to be represented as a product over the templates:

$$P(\mathbb{Y}) = \frac{1}{Z} \prod_{i=1}^{n} \prod_{j=1}^{m_i} f_i(\mathbb{Y}_{ij}) \tag{14.2}$$

Not all of the elements in the pathway are observable. In fact many, for example, the binding status of all known compounds, are quite difficult to measure with current technology. We therefore divide the model's set of variables into those that we can observe, $\mathbb{X}$, and those that are hidden and we wish to estimate, $\mathbb{Y}$. Each factor will take a possibly empty subset of both $\mathbb{Y}$ and $\mathbb{X}$:

$$P(\mathbb{Y} \mid \mathbb{X}) = \frac{1}{Z(\mathbb{X})} \prod_{i=1}^{n} \prod_{j=1}^{m_i} f_i(\mathbb{Y}_{ij}, \mathbb{X}_{ij}) \tag{14.3}$$

Despite the exponential number of variable settings, there are efficient approximations to solve Eq. (14.3) for the marginal distribution of $P(Y)$ for each $Y \in \mathbb{Y}$. Additionally, given sufficient data, there are methods to learn the functions f_i for various optimization goals. PARADIGM uses loopy belief propagation to estimate $P(\mathbb{Y} \mid \mathbb{X})$. Expectation maximization (EM) is used to learn the parameters for f_i that maximize the likelihood with hidden variables. This computational framework has scaled to modeling nearly a hundred thousand variables per sample and requires a few hours of computation per sample.

PARADIGM builds a specific factor graph by combining three elements: (1) a generic pathway representation (Figure 14.1A), (2) modeling details (Figure 14.1D), and (3) evidence in the form of genomics observations, resulting in a factor graph that can contain on the order of 100,000 variables (Figure 14.1C). The generic pathway representation consists of two parts: entity type definitions and interaction definitions. For each pathway entity, $p \in P$, there is an associated type, $T_p \in T$, such as protein or a complex of molecules. Each interaction is a labeled edge in the pathway graph, $e \in E$, and is defined as a tuple from $P \times P \times A$, where A is the set of possible action/association types in the pathway, such as "transcribes," "component of complex," or "phosphorylates." In the following sections, we define the detailed construction of a factor graph

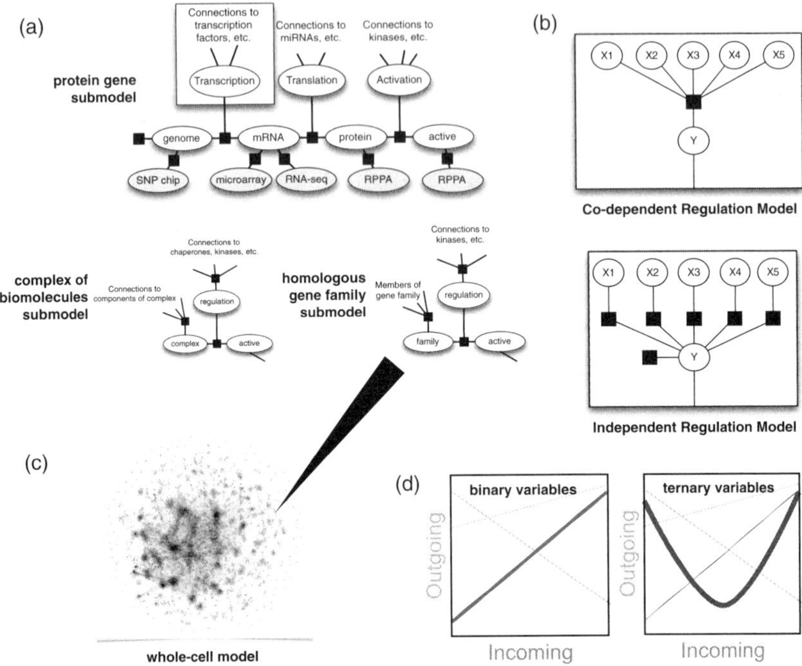

Figure 14.1 PARADIGM modeling components. (A) Default submodels in PARADIGM. Each oval represents a variable and each black box represents a factor. Lines connect factors to their constituent variables. Submodels for three different classes of biomolecules are depicted. (B) Regulation models of a child node given multiple parents. The codependent regulation model combines all possible statistical interactions between the parents at the expense of an exponential parameter space. The independent regulation model has a linear parameter space but cannot represent more complicated parent interactions. (C) The whole cell model consists of variables and factors from both the submodels in panel A as well as gene-gene interactions from panel B. (D) Binary versus ternary variables. A factor between two binary variables can represent linear transformations of signal, depicted as lines. A factor between two ternary variables can represent nonlinear transformations of signal, as depicted by the purple line.

using the modeling details supplied from molecule class submodels, interaction maps, and provided evidence.

14.2.1 Variables

In PARADIGM, each variable is a three-state summarization of a quantitative aspect of a cell. Each class of gene, protein-coding gene, RNA gene, or miRNA, is expanded out into a set of variables that correspond to the various states of that gene. In Figure 14.1A, a protein-coding gene has been expanded along the central dogma of biology into the genome state, mRNA state, protein state, and activity state, and the protein submodel encapsulates this simple representation. In addition, there are variables that represent summaries of the regulatory states, which come from *trans*-acting factors. Not shown in the figure are variables for observations of the data, which are connected to the latent variables such as the mRNA node or the genome node. The rationale for keeping a distinction between hidden and observed variables is that it allows for a much more flexible representation of the data and different data generation models. For example, for one data set, we may have gene expression measurements from a particular microarray platform, whereas on a different data set, we may have expression data from a different microarray platform as well as from RNA-seq. By separating the observation variables, we can attach multiple observations in those samples where we have multiple platforms. Additionally, we can learn platform-specific or batch-specific noise effects for each data generation method.

Each discrete variable in the model can take on three different states – up, down, or normal – relative to some control state. For some variables, such as the genome copy number state, this meaning is clear: normal means that there are two copies of the gene in the tumor genome, down means fewer copies, and up means more copies. For others, such as the mRNA expression state, the meaning can be more flexible, depending on the chosen control, that could reflect different investigational questions of interest. For example, if the control expression state is the expression of a matched normal tissue from the same patient, the PARADIGM model will be most useful for investigating the process of carcinogenesis and what is different about the tumor compared to normal tissue. If, however, the normal state is selected to be the median expression value in a cohort of similar tumors, then the model is now most useful for distinguishing between tumors in the cohort. If all the tumors have elevated expression of a gene compared to normal, and it's very consistent, then such a control will not be able to detect that the gene is important in carcinogenesis. In exchange, the median-centered control allows better differentiation between similar tumors, where the comparison to normal may hide smaller changes between the tumors.

The final outputs of PARADIGM are estimates from the "activity" variable. For protein coding genes, the activity state differs from the protein state in that it captures posttranslational modifications that allow activity. For example, even if a protein is expressed, it may not be active until it has been phosphorylated.

PARADIGM also keeps variables for groups of molecules both from complexes and gene families. Complexes are a grouping of molecules that must be bound together to act together. For example, the p53 protein acts as a tetramer, but it also acts differently when bound to additional proteins that change its transcriptional regulation behavior. Gene families are commonly annotated in pathway databases as well, when homologs share a similar function. For example, the genes in the RAS family (*KRAS*, *NRAS*, and *HRAS*) are annotated in pathway databases as sharing most of their functions, so PARADIGM models this behavior by introducing a representative state for the entire RAS family, which is active whenever any of its constituents members are active.

To maximize modeling flexibility, PARADIGM keeps a map from pathway entity types, such as protein, complex, and family, to a set of submodel variables. The submodel for each class of biomolecule is configurable by the user and is a mapping from a pathway entity type to a set of subvariables: $D_v(T_p) : T \rightarrow V$, where V is the set of subvariables. Three submodels are shown in Figure 14.1A: protein $\mapsto$ {genome, mRNA, protein, active, Transcription, Translation, Activation}, complex $\mapsto$ {complex, Activation, active}, and family $\mapsto$ {family, Activation, active}. Each of the pathway entities, p, has an entity type, T_p, and a factor graph variable is created for each subvariable type of each entity:

$$\mathbb{Y} = \{Y_{pv} \mid p \in P, v \in D_v(T_p)\} \tag{14.4}$$

A particular genomic data set, G_k, from the set of genomic data sets to be integrated, G, is a set of tuples over a sample's pathway entities' observed values: $G_k : P \times S \rightarrow$ {up, normal, down}. This results in the following observed variables in the factor graph:

$$\mathbb{X} = \{X_{pk} \mid G_k(p, s) \text{ is defined}\} \tag{14.5}$$

Note that all variables in the factor graph have two indices: the pathway entity p and a subvariable.

Submodel definitions like this allow us to integrate data across numerous omics data sets. But there is additional information that can be obtained by linking together distinct submodels. For example, TP53 may have normal genome copy number and expression levels, but a mutation in the gene may have an unknown effect. By looking at the mRNA variables of the downstream targets, we can combine the information from other genes with the TP53 data to infer whether TP53 is functioning normally or abnormally because of the mutation.

14.2.2 *Interactions and Probabilistic Factors*

Physical interactions between biomolecules offer the chance to learn much more about the state of a cell than by looking at a gene's data in isolation. In PARADIGM, the local factors of the factor graph correspond to directed physical interactions between molecules or to an observational process. The direction of the interaction corresponds to which element is changed by the interaction, or a temporally later change in the interaction. For example, when a transcription factor promotes the transcription of a gene, the transcription of the mRNA is preceded by the binding of the TF, so the mRNA node is the target of the interaction. Similarly, for a complex, the complex is created after all the constituents have bound together, so the variable for the complex is the target of all the directed interactions with the constituents.

Prior to constructing factors, a directed graph is constructed from three sources: (1) submodel edges, (2) pathway interaction edges, and (3) observation edges. Both the submodel edges and pathway edges connect elements from $\mathbb{Y}$, while all observation edges go from $\mathbb{Y}$ to $\mathbb{X}$. Submodel edges are defined per pathway element, and in the example of Figure 14.1A, the edges for the complex would be (*complex* $\to$ *active*) and (*Activation* $\to$ *active*). For every p such that $T_p = complex$, the directed graph will have edges ($Y_{p,complex} \to Y_p$, *active*) and ($Y_{p,Activation} \to Y_{active}$). An interaction map is used to translate a pathway edge (p, p', i) where typically $i \in I = \{$transcribes, component of, $\ldots\}$, to particular elements of $\mathbb{Y}$. For example, the interaction type *transcribes* maps from "active" to "Transcription," resulting in a directed edge ($Y_{p,active} \to Y_{p',Transcription}$). For each genomic data set G_k, the user specifies a specific subvariable type to attach to (such as mRNA), resulting in an edge ($Y_{pv} \to X_{pk}$) for every pathway entity p with data.

This directed graph on the variables is then converted to a proper factor graph by iteratively constructing factors at each factor graph variable. Generally, PARADIGM uses a single probabilistic factor for each variable, corresponding to

$$f_i(\mathbb{Y}_i) = P(Y_i \mid parents(Y_i)) \tag{14.6}$$

where $parents(Y_i)$ is the set of variables that have Y_i as their target. This defines $\mathbb{Y}_i = Y_i \cup parents(Y_i)$, where i is a double index with both the pathway element p and subvariable v. The process is identical for elements of $\mathbb{X}$. For those cases where $\mathbb{Y}_i$ consists entirely of Y_{pv} with identical p, that is, the connections are only between submodel elements, and parameters are shared as in Eq. (14.3).

This type of factor allows full dependencies between the parents but requires an exponential number of parameters as a function of the number of parents. Because many genes are regulated by at least 10 pathway elements, a straightforward representation of gene regulatory networks would result in computationally intractable graph structures. In early versions of PARADIGM, this was dealt with by inserting intermediate "splitting" nodes in the directed graph, prior to factor creation, such that no node had more than five parents. Recent versions use a naive Bayes–like model for the regulatory subvariables – Transcription, Translation, and Activation, as in Figure 14.1B, which creates a sparse factorization at the expense of treating all of the parents as an independent influence on the target activity.

14.2.3 Learning Interaction Parameters

The expectation maximization algorithm is used to learn the parameters of the factor graph. First, initial parameters are set on each of the factors, either uniformly or by using prior information from pathway databases to lean each edge toward activation or inhibition. Owing to the presence in the factor graph of latent variables that directly connect only to other latent variables, the particular meaning of "up" or "down" could be switched and result in a model with exactly the same likelihood. This means that randomly setting initial parameters may result in models that are less interpretable.

After initial parameters are set, an expectation step is taken by calculating for each sample the marginal distributions $P(\mathbb{Y}_i, \mid \mathbb{X})$ for every factor domain $(\mathbb{Y}_i, \mid \mathbb{X}_i)$. For discrete variables, this marginal posterior probabilities of the factors are also the sufficient statistics. Therefore the expectations can be added together across different samples, and the maximization step of EM is performed by maximum likelihood estimation of each factor's parameters used with the summed sufficient statistics. At each expectation step, the likelihood is noted, and the learning process is deemed converged when the ratio of likelihood increase to total likelihood is less than, typically, 10^{-5}.

Once learning has converged, final estimates of activity are calculated for every pathway element using the "active" subvariable's posterior and prior probabilities. This is called the *integrated pathway level* (IPL) of a pathway entity and is defined as

$$IPL(p, s) = sign(\hat{v})\log\frac{P(Y_{p,active} = \hat{v} \mid \mathbb{X}_s)}{P(Y_{p,active} = v)} \qquad (14.7)$$

where $\hat{v}$ is the value of *up, down, normal* that maximizes the log odds expression, and sign corresponds to 1, -1, or 0 for the values of v, respectively. This collapses a point on the 3-simplex to a scalar value, allowing easier interpretation by standard genomics tools such as clustering or supervised analysis.

This formulation of ternary variables, rather than the more common binary variables, has the advantage of allowing nonlinear functions. Figure 14.1D, shows plots of potential factor response as a variable's binary belief passes from false to true, or from negative IPL to positive IPL. With binary variables, a factor connected to two variables can only represent a linear function, as there are only two free parameters in a factor $P(Y \mid X)$, as the Y variable is normalized. However, with ternary variables, $P(Y \mid X)$ has six free parameters and can represent linear functions as well as nonlinear functions. Such functions are necessary to represent many known biological processes. One example is the influence of RAF on RAS inhibition. With low levels of RAF inhibition, RAS is active. Moderate amounts of RAF inhibition reduces RAS's activity, but with larger amounts of RAF inhibition, RAS regains its activity (Holderfield et al., 2013).

14.3 Applications of PARADIGM

PARADIGM transforms a set of genomic data sets with a pathway structure to create a matrix of IPL scores for every pathway element in every sample. The resulting matrix, an example of which is shown in Figure 14.2A, is amenable to the many different analysis techniques developed for gene expression matrices. Common applications of PARADIGM can be split into two broad categories of machine learning: unsupervised and supervised.

Unsupervised clustering is often used to stratify cancer samples. A typical application was performed in a cohort of breast cancer patients (Kristensen et al., 2011). Figure 14.2A shows a clustered heatmap of the IPLs from a run on copy number and expression data in the cohort. The colored bars on top indicate the clusters as discovered by unsupervised clustering by HOPACH (J van der Laan and Pollard, 2003). These clusters stratify the patients into groups with differential survival, as shown in Figure 14.2B. This same unsupervised stratification analysis was also performed for the available genomic data sets, including DNA methylation and miRNA expression. Comparing the PARADIGM stratification to the other genomic data set stratifications and the stratification by usual clinical data, the PARADIGM stratification was the most significant in terms of distinguishing patients by their outcomes (Figure 14.2C). Similar improved stratification power was found in glioblastoma multiforme (Vaske et al., 2010).

Given a stratification of patients, PARADIGM can also be used to pull out a subnetwork of differential IPLs. An example stratification is shown in which the patients of the TCGA breast cancer cohort are dichotomized based on the kind of *PIK3CA* mutations found in their tumors. Figure 14.2D shows the mutations in this cohort, which generally cluster into either the kinase domain or the helical domain. Comparing the samples with mutations in the kinase domain to samples with a mutation anywhere else, we find a differential subnet between these two classes of mutations in the same gene (Figure 14.2E).

PARADIGM has also been used as a component of cell line drug sensitivity prediction in DIRRP (Brubaker et al., 2014). DIRRP runs a standard network as well as a network with simulated rewiring from the drug intervention. The p-value of a paired T-test between the two sets of IPLs in a cell line is used as

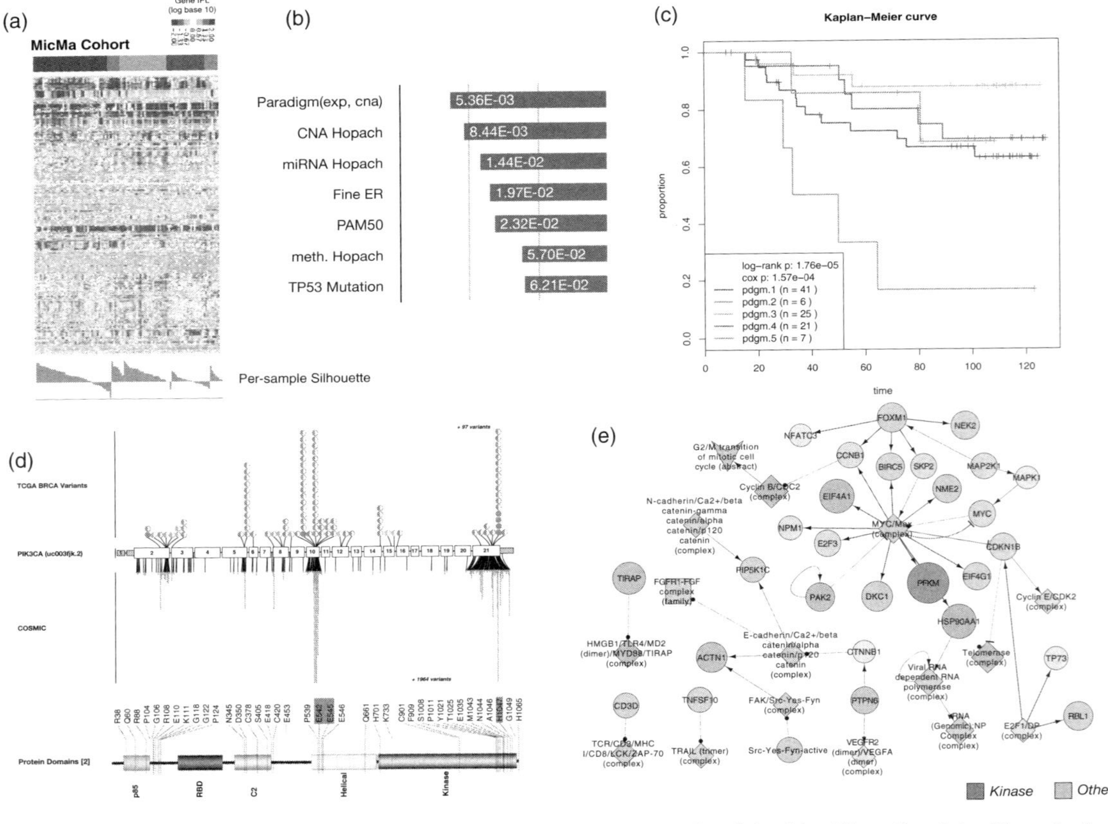

Figure 14.2 (A) Matrix of PARADIGM IPL activities. Red indicates "up" activity, blue "down" activity. The color bar shows clusters derived from HOPACH. Silhouette plots are shown at the bottom. (B) Log *p*-values of Wilcoxon tests of HOPACH unsupervised clustering of various data sets and of clinical markers. (C) Kaplan-Meier plot of HOPACH clusters from panel A. (D) Mutational landscape and domain structure of *PIK3CA*. (E) Differential network of TCGA breast cancer samples with *PIK3CA* kinase mutations versus nonkinase mutations.

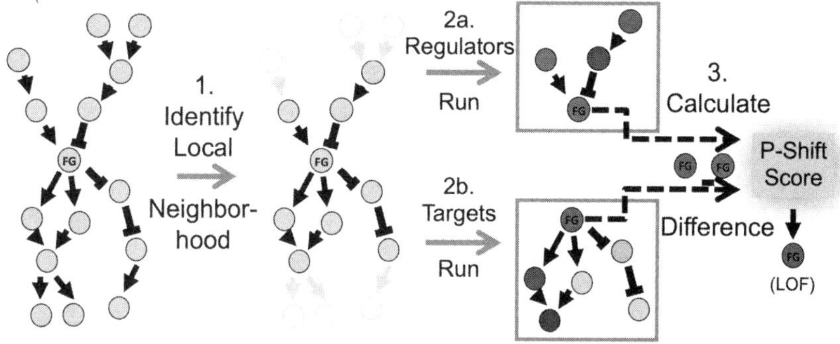

Figure 14.3 To calculate shift scores, PARADIGM-SHIFT goes through the following steps: it (1) undergoes feature selection to determine which features to include in the upstream (regulators) and downstream (targets) models, (2) calculates inferred activities for upstream and downstream networks using PARADIGM, and (3) computes the shift score as the difference between the downstream and upstream PARADIGM runs.

the prediction of sensitivity to the drug, with more perturbed networks being more predicted to be more sensitive to the drug. Accuracy and precision of these predictors over all cancers and drugs in the compendiums was 78% and 73%, respectively.

14.4 PARADIGM-SHIFT

14.4.1 PARADIGM-SHIFT Method

The PARADIGM-SHIFT (Ng et al., 2012) method predicts not only driver from passenger mutations but also whether mutations are likely to increase the function, gain-of-function (GOF), or decrease the function, loss-of-function (LOF), of the gene. This is achieved by estimating the functional impact of like several other methods a mutation on the pathway by using PARADIGM (Vaske et al., 2010). In cases of LOF or GOF mutation, there would be conflicting signal from the genomic data upstream compared to downstream of the mutated gene. For example, in the case of a LOF event the downstream targets would not be active because the protein is rendered nonfunctional and feedback circuitry could lead to an increase in activity of the upstream regulators in an attempt to rescue the function of the gene. Shift scores are computed for each sample and capture any differences in upstream and downstream signal inferred from PARADIGM. If these shift scores are significantly lower or higher in the mutant samples versus the nonmutant samples then the collection of mutations in the gene across the cohort are classified as LOF or GOF, respectively.

Figure 14.3 illustrates the steps PARADIGM-SHIFT takes to calculate the shift scores for each sample. First, feature selection is employed to determine the set of neighboring upstream regulators and downstream targets to include.

Typically, any neighbor with an absolute t-statistic greater than one standard deviation computed by contrasting its expression levels in mutant versus non-mutant samples. Each neighborhood network for a mutated gene m is split up into the upstream regulators and the downstream targets, based on the directionality of the edges in the pathway, using a maximum path length L, Fisher's score for each gene s, and score threshold T:

$$regulators_m = \{path(g, m) \mid s(g) \geq T, |path(m, g)| \leq L\}$$
$$targets_m = \{path(g, m) \mid s(g) \geq T, |path(m, g)| \leq L\}$$

where $path(a, b)$ is the set of genes included in the shortest path from a to b in the pathway network. A PARADIGM run is performed using $regulators_m$ as the network, resulting in an IPL for the target gene m of IPL_r, and a similar PARADIGM run is performed with $targets_t$ as the network, resulting in IPL_t. Finally, the shift score for m is calculated as the difference $IPL_t - IPL_r$.

The accuracy of the trained model can be assessed by using the absolute shift score as a classifier to predict the presence of a mutation in a cross-validation setting. In the case in which a mutation has a functional impact on the pathway, we would expect to observe high absolute shift scores for mutant samples, higher discrepancy between upstream and downstream signal, compared to nonmutant samples. If the model is predictive of a functional impact for a mutation, then PARADIGM-SHIFT classifies the mutation as either LOF or GOF based on the distribution of shift scores for the mutated and nonmutated samples, with negative shifts resulting in a LOF prediction and positive shifts resulting in a GOF prediction. The strength of the predicted impact (mutant separation) is determined by the t-statistic of the distribution of shift scores for the mutant against the nonmutant samples. The significance of these shifts is computed by comparing the observed mutant-separation against a background model of mutant-separation values determined using the same fixed network model, but permuting the gene labels associated with the input genomic data.

14.4.2 PARADIGM-SHIFT Application

Pathway-based methods for inferring the functional impact of mutations can not only leverage gene-level data to predict impact but also deepen our understanding of how mutations affect the surrounding network as a whole. The consequence of a mutation on the pathway is left ambiguous by other methods that leverage frequency or conservation to predict functional impact since the prediction is that the mutation is likely impactful, but does contain information about whether the mutation leads to an increase or decrease in downstream activity. For example, the Nrf2 (*NFE2L2*) oncogene is often

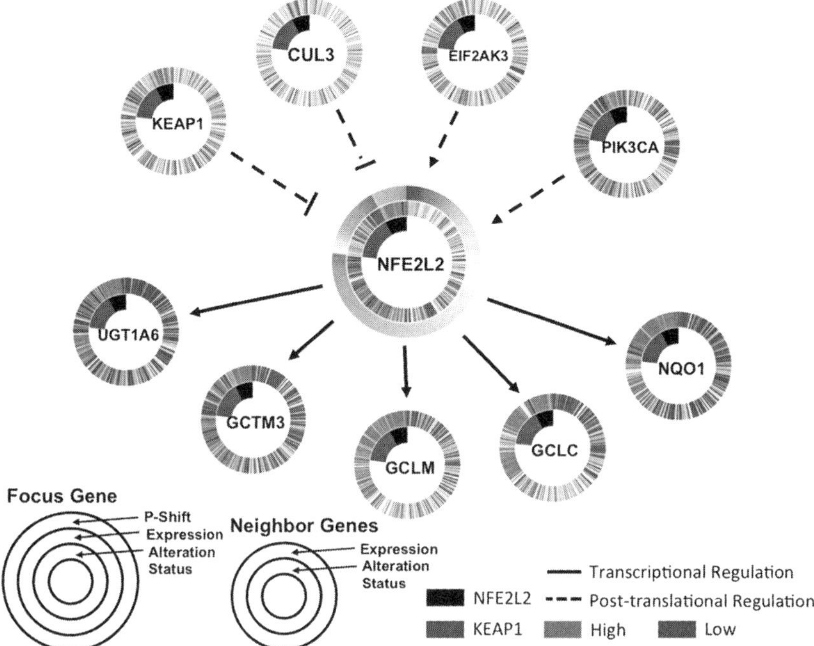

Figure 14.4 PARADIGM-SHIFT analysis of *NFE2L2* and *KEAP1* mutation on the Nrf2 signaling pathway. Circlemap display of mutation neighborhood selected around *NFE2L2*. Solid lines indicate transcriptional regulation and dashed lines indicate protein regulation. Samples were sorted first by the *NFE2L2* and *KEAP1* mutation status, then by shift score.

mutated in lung cancer; i.e. mutations are selected that increase its signaling activity that promote tumorigenesis. The PARADIGM-SHIFT algorithm was employed in a study of more than 346 samples of lung adenocarcinoma and lung squamous-cell carcinoma using genomic data and curated pathways to identify a GOF of the Nrf2 signaling pathway driven by *NFE2L2* and *KEAP1* mutation (Figure 14.4). *NFE2L2* and *KEAP1* mutation status of the samples (center ring) is shown alongside the expression data for the genes selected to infer the GOF of the Nrf2 signaling pathway. Samples were sorted first by mutation status and then by shift score, which highlights the discrepancy in the upstream and downstream signal for *NFE2L2*. There are several features upstream and downstream of *NFE2L2* that can explain the discrepancy in signal found with PARADIGM-SHIFT. *KEAP1* in *NFE2L2* mutant samples is upregulated, whereas downstream targets such as *NQO1*, *GCLC*, *GCLM*, and others appear highly expressed, indicating that the mutant *NFE2L2* is insensitive to repression by *KEAP1* and remains highly active, consistent with GOF. *KEAP1* mutation also appears to lead to a similar disruption in *NFE2L2* regulation, which accounts for higher activity of *NFE2L2* compared to nonmutant samples. While mutations in *KEAP1* and *NFE2L2* account for many of the samples with high shift score, there are also a few samples that do not harbor a mutation in *NFE2L2* or *KEAP1* that have high shift score as well. This predicts that there may be novel disruptions affecting the regulation in the pathway, such as amplifications or deletion of genes not yet known to interact with this pathway.

In this example, PARADIGM-SHIFT predicts that *NFE2L2* and *KEAP1* mutation lead to Nrf2 signaling pathway activation. An advantage of PARADIGM-SHIFT is that it can make inferences about pathway activation or deactivation regardless of whether SNV data are present or has been detected for a particular sample. Because PARADIGM-SHIFT can be run on samples without alterations in the focus gene, it can be used to look for hits to the pathway that lead to the same phenotype from copy number alterations, changes in gene expression, gene mutations and fusions, or other events. In addition, PARADIGM-SHIFT allows us to identify the genes that appear most affected by genomic alterations to these pathways, which may be helpful for identifying effective interventions for treating these cancers.

14.4.3 The TieDIE Method

Integrating multiple data sources with a pathway model can greatly improve the ability to find active subnetworks, compared with using just a single source of information such as gene expression or mutational data. The TieDIE algorithm addresses this data integration problem by applying a heat diffusion model to multiple data inputs and uses the merged output to find "linker" genes that are strongly implicated by each data input. Linker genes are then used to find subnetworks that connect genomic perturbations to transcriptional changes; these subnetworks are critical in finding genes that may be lacking in *cis*-level data but are implicated by other genes in the surrounding pathway and the logic of the corresponding interactions. Like the optimization-based PCST algorithm, TieDIE is one of a class of methods that can integrate multiple data types with pathway information to extract meaningful subnetworks. However, TieDIE is also able to generate patient-specific subnetworks by relating patient-level data to the statistically robust subnetworks generated with cohort-wide data.

14.4.4 TieDIE Applications

Building on the heat kernel model, the TieDIE method (Paull et al., 2013) takes the output of a differential gene expression analysis generated from transcriptional (RNA-Seq, microarray) data and performs heat diffusion over a supplied biological pathway to find a pathway neighborhood strongly supported by the transcriptional data. Separately, the algorithm takes a set of significantly perturbed genes found with genomic sequencing data (mutations, copy number alterations), performs heat diffusion on these inputs, and then merges the result with the diffused transcriptional data via a "linker" function, producing an overall score for each gene in a given input pathway. The algorithm uses high-scoring "linker genes" to find subnetworks connecting genomic alterations ("source" genes) to the observed transcriptional changes ("target" genes) in a patient cohort, providing a probable explanation for dysregulated transcriptional profiles (see Figure 14.5).

TieDIE is given n sets of data points for N genes that represent inferences derived from distinct data sources. These data sources may include a set of genes with functionally impacting mutations (weighted by mutation impact assessment score, using the CHASM, PARADIGM-Shift, or other methods; Carter et al., 2009; Ng et al., 2012; Dees et al., 2012; Adzhubei et al., 2010; Reva et al., 2011), significantly methylated genes, or a set of transcription factors inferred to have altered activity due to the expression of their downstream targets. In addition, a static interaction network, or graph, G, is used as a background on which to attempt to connect the data sources together into a coherent causal picture of altered gene activities. Let $X_i = [x_{i,1} \ldots x_{i,N_i}]$ represent the informal belief in a perturbation in the ith data type, for each of N_i genes in that data set. The interactions in G can be derived from curated sources such as the National Cancer Institute's Pathway Interaction Database, functional genomics predictions, or directed transcription factor to target interactions. TieDIE makes use of the adjacency matrix A of the graph G, where $A_{ij} = 1$ if node i activates node j, $A_{ij} = -1$ if node i represses or inactives node j, and 0 otherwise. The actual values of this vector depend on the data type: for instance, with transcription factors, this may represent the score from a Gene Set Enrichment Analysis (GSEA) (Mootha et al., 2003; Subramanian et al., 2005) test on gene expression data. For genomics data, these scores may be binary to reflect the presence of a mutation in each gene or weighted by mutation frequency in a patient cohort. TieDIE also requires that the total weight of X_i be normalized to some constant, c_i, so that $c_i = \sum_{j=1}^{N} x_{ij}$, and typically this constant is set to 1 to guarantee that all data types have the same influence.

A new vector of scores $\hat{X}_i$ is produced by diffusing the original values over the graph G; this process puts high scores on genes that are near the input set but also lowers the scores of input genes that are isolated (in G) from the rest of the input set. As with the HotNet method, the heat-diffusion kernel can be used as the diffusion "engine," where the diffused belief value for each gene in the graph is defined as ($\hat{X}_i = X_i * e^{-Lt}$), where L is the difference of the degree and the adjacency matrix (also called the graph Laplacian) and t is

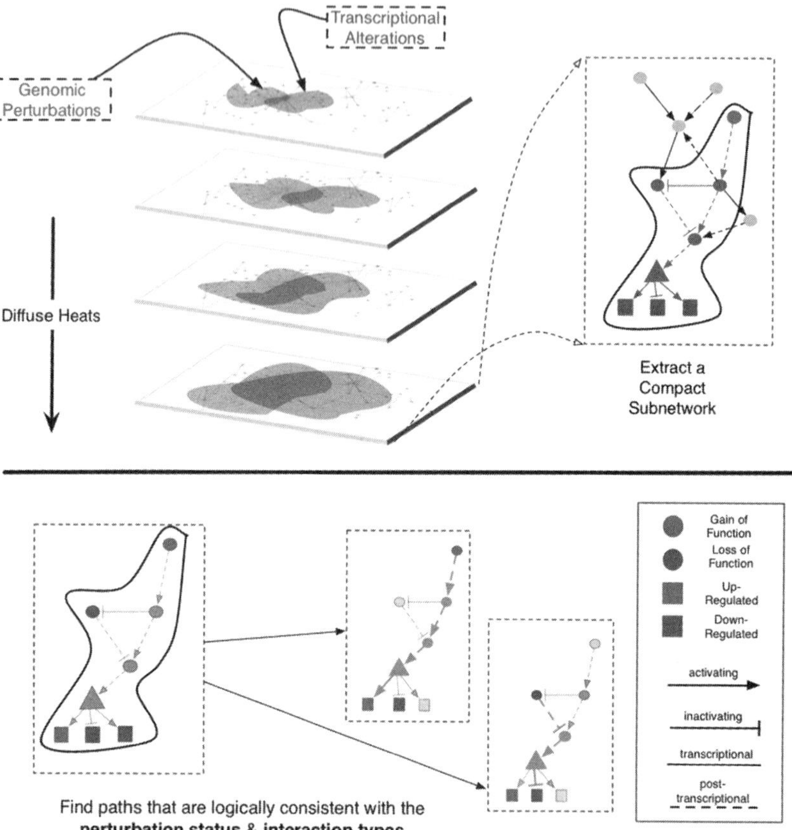

Figure 14.5 (top) Relevant genes from two distinct sets are shown by dyes diffusing on a pathway from a source set (e.g., genomically altered genes; red nodes) and a target set (e.g., transcription factors; blue nodes). "Linke" genes are shown as purple nodes, placed between the source and target sets; multiple time slices of the diffusion process are shown as stacked layers of the same network. (bottom) Subnetworks are extracted following the diffusion process. The algorithm finds all paths that connect genomic alterations to transcriptional alterations where the edge-interation logic is consistent with the sign of the source and target nodes (i.e., gain or loss of function; up- or down-regulation). After this filtering step, the union of all edges in the validated paths defines the resulting subnetwork.

a predetermined time constant. TieDIE can also use Google's PageRank, or other methods that exploit the graph topology, to refine a set of input scores. To extract genes that have intersecting evidence from all input data sets, TieDIE computes a "linker function" $f()$ of the diffused input sets $\hat{X}$ that produces a single vector of "linker scores"; this function is typically defined as the $min()$ operator. A set of linking genes is produced by thresholding the linker scores (see Figure 14.5).

To find a subnetwork that connects the genomic perturbations to transcriptional changes, the TieDIE algorithm adds an edge between each pair of nodes where both are in either the input or linker gene sets. This set of edges and nodes defines an initial subgraph g that can be further reduced by finding the subset of edges that connect source to target nodes through paths that are logically consistent with both the input data (i.e., gain or loss of function mutations; up- or down-regulation of gene expression) and the pathway interactions used. To find the consistent path subset within g (g_c), TieDIE computes a perturbation score for each source gene type tuple $e = u, v, \tau$ where $\tau \in -1, 0, 1$ is either an activating or inactivating edge. For each edge $e = u, v, \tau$ the influence score at node v is defined by the following equation:

$$I_v(e) = I_v(u, v, \tau) = \begin{cases} PS(u), \tau = +1 \\ -PS(u), \tau = -1 \end{cases}$$

TieDIE then performs a depth-first search over all directed edges from a given genomic perturbation in gene p, recursively computing the influence score for each new node. For each target node with a perturbation score that matches the directed influence score from the perturbed gene p (i.e., $I_v(p, v, \tau) = PS(v)$), the set of edges on the path ($p \rightarrow v$) is included. The algorithm is terminated at a given path depth (3, by default) and return the subgraph of all validated edges, each of which is a component of at least one logically consistent path from a source to a target gene.

The following pseudocode summarizes the steps to generate the subgraph g_c, given input source and target sets $\hat{X}_{source}, \hat{X}_{target}$, a user-supplied network size threshold S, graph G, and linker function $f()$:

procedure TIEDIE($X_{source}, X_{target}, f(), S, G$)
    $\hat{X}_{source} \leftarrow X_{source} * e^{-Lt}$
    $\hat{X}_{target} \leftarrow X_{target} * e^{-Lt}$
    $threshold \leftarrow max(f(\hat{X}_{source}, \hat{X}_{target}))$
    while $\left| f(\hat{X}_{source}, \hat{X}_{target}) \geq threshold \right| \leq S$ **do**
        $decrement(threshold)$
    end while
    $linkers \leftarrow \left\{ f(\hat{X}_{source}, \hat{X}_{target}) \geq threshold \right\}$
    $g \leftarrow \{\}$
    for $edge \in G$ **do**
        if $edge \in linkers$ **then**
            $g \leftarrow \{g \cup edge\}$
        end if
    end for
    $g_c \leftarrow \{\}$

    for all (paths $s \rightarrow t \in g$, where $\{s \in X_{source} : t \in X_{target}\}$) **do**
        if $consistent(path)$ **then**
            $g_c \leftarrow \{g_{consistent} \cup \{edge \in path\}\}$
        end if
    end for
    return g_c
end procedure

14.4.5 TieDIE Applications

In a study of 446 samples of clear cell renal carcinoma, genomic data and a protein interaction database were used with the HotNet algorithm to identify a subnetwork connecting the mutated genes *PBRM1*, *ARID1A*, and *SMARCA4*: all key genes involved in the SWI/SNF chromatin remodelling complex (Network et al., 2013). While this step found a set of mutated genes likely to effect the chromatin state and resulting transcriptional activity of key disease-driving genes, the inclusion of gene-expression data was necessary to identify the most likely transcriptional effects of these specific alterations. To identify these effects, the TieDIE algorithm was used to integrate muta-tion, expression, and pathway data; TieDIE found that the mutated genes (*PBRM1/ARID1A/SMARCA4*) were significantly close in pathway space to a set of transcription factors with altered activity in the mutated samples, pro-ducing a descriptive subnetwork linking these genes and lending support to the hypothesis that the changes in chromatin state lead to an altered expression state in this disease. These transcriptional effects were found to encompass a wide number of processes, including RAS signaling, transcriptional output (*FOS, JUN, SP1*, and *HIF1A*), immune signaling (*NFKB1A, IL6*), DNA repair, and others. The network provides a single, integrated view of multiple effects to these biological processes and may help experts select testable hypotheses and design follow-up studies.

Methods that can be used with patient data to infer specific network models are highly useful because they not only summarize the potential molecular basis for a patient's disease but can also suggest treatment options. TieDIE attempts to solve this problem by connecting the observed genomic alterations to expression changes in a tumor, revealing additional 'linking' genes that may possess drug target potential, despite. For example, a breast cancer sample of luminal A subtype is shown in Figure 14.6, where amplifications in *IGF1R* and *PAK1* and nonsynonymous mutations in *AKT1* and *TP53* are connected to expression changes downstream of *ERK1-2*-active protein, *JUN, TP53*, and *EDN1*. Intervening edges and nodes are supported by published functional

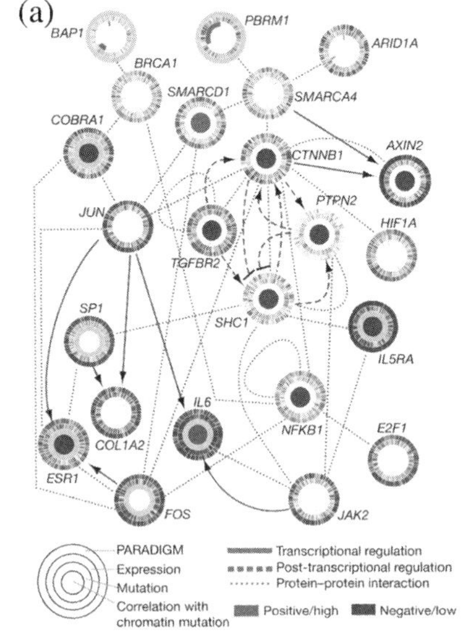

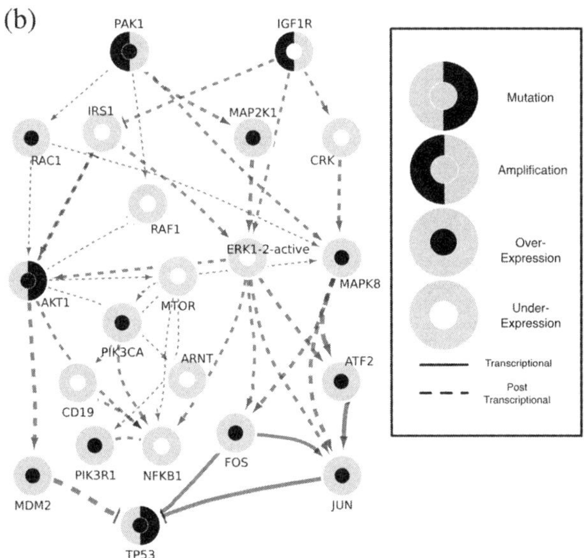

Figure 14.6 (A) TieDIE was used to assess the impact of mutations in genes known to participate in chromatin-remodeling processes (*PBRM1/ARID1A/BAP1/SETD2/ KDM5C*), and identified as significant by MutSig, in a TCGA study of clear cell renal carcinoma. TieDIE identified a significant subnetwork connecting three of these genes (*PBRM1/ARID1A/BAP1*) to active transcriptional hubs as identified by the PARADIGM method. Each gene is shown as a multiring circle with multiple levels of data, so that each "spoke" in the ring represents a single patient sample. (B) A network inferred by TieDIE for a specific TCGA luminal A breast cancer sample; the outer ring indicates the presence of amplification or mutation events, while the inner circle shows the over- or underexpression of each gene. Whereas the mutation in *AKT1* is characteristic of luminal A breast cancer subtypes, the basal-like amplification of *IGF1R* and mutation of *TP53* indicate a mixed profile of this particular sample, highlighting the need for more detailed classification in breast cancer.

interactions and were also validated by a cohort-driven TieDIE network trained on more than 400 breast cancer samples. Interestingly, this sample has events and transcriptional profiles that are characteristic of both luminal A breast cancer samples (*AKT1* mutation) and basal-like samples (*TP53* mutation, insulin-like growth factor *IGF1R* amplifications). A further investigation of the expression profiles in the surrounding network showed that *HIF1A* is active, which reflects a basal-like program of hypoxic response driving angiogenesis, which is also supported by increased *EDN1* expression. However, the basal-like expression profiles of *PIK3CA* and *IRS1* highlight the complex and mixed nature of this patient's disease possibly due to heterogeneity of subclones exhibiting a variety of luminal and basal qualities. These observations underscore the insufficiency of histological and gene-panel classification when seeking drug treatment options. The need for methods that can produce a complete molecular portrait of each patient's disease should become even more apparent with new and emerging technologies that collect data on epigenetic and structural changes to the genome; these data must be integrated with existing expression and genomics data on a per-patient basis.

References

Adzhubei, Ivan A, Schmidt, Steffen, Peshkin, Leonid, Ramensky, Vasily E, Gerasimova, Anna, Bork, Peer, Kondrashov, Alexey S, and Sunyaev, Shamil R. 2010a. A method and server for predicting damaging missense mutations. *Nature methods*, **7**(4), 248–249.

Brubaker, Douglas, Difeo, Analisa, Chen, Yanwen, Pearl, Taylor, Zhai, Kaide, Bebek, Gurkan, Chance, Mark, and Barnholtz-Sloan, Jill. 2014. Drug intervention response predictions with paradigm (dirpp) identifies drug resistant cancer cell lines and pathway mechanisms of resistance. Pages 125–135 of: *Pacific Symposium on Biocomputing*.

Cancer Genome Atlas Research Network, et al. 2013. Comprehensive molecular characterization of clear cell renal cell carcinoma. *Nature*, **499**(7456), 43–49.

Carter, Hannah, Chen, Sining, Isik, Leyla, Tyekucheva, Svitlana, Velculescu, Victor E, Kinzler, Kenneth W, Vogelstein, Bert, and Karchin, Rachel. 2009. Cancer-specific high-throughput annotation of somatic mutations: computational prediction of driver missense mutations. *Cancer research*, **69**(16), 6660–6667.

Cerami, Ethan G, Gross, Benjamin E, Demir, Emek, Rodchenkov, Igor, Babur, Özgün, Anwar, Nadia, Schultz, Nikolaus, Bader, Gary D, and Sander, Chris. 2011. Pathway Commons, a web resource for biological pathway data. *Nucleic acids research*, **39**(suppl 1), D685–D690.

Ciriello, Giovanni, Cerami, Ethan, Sander, Chris, and Schultz, Nikolaus. 2012. Mutual exclusivity analysis identifies oncogenic network modules. *Genome research*, **22**(2), 398–406.

Dees, Nathan D, Zhang, Qunyuan, Kandoth, Cyriac, Wendl, Michael C, Schierding, William, Koboldt, Daniel C, Mooney, Thomas B, et al. 2012. MuSiC: identifying mutational significance in cancer genomes. *Genome research*, **22**(8), 1589–1598.

Demir, Emek, Cary, Michael P, Paley, Suzanne, Fukuda, Ken, Lemer, Christian, Vastrik, Imre, Wu, Guanming, et al. 2010. The BioPAX community standard for pathway data sharing. *Nature biotechnology*, **28**(9), 935–942.

Dittrich, Marcus T, Klau, Gunnar W, Rosenwald, Andreas, Dandekar, Thomas, and Müller, Tobias. 2008. Identifying functional modules in protein-protein interaction networks: an integrated exact approach. *Bioinformatics*, **24**(13), i223–i231.

Drier, Yotam, Sheffer, Michal, and Domany, Eytan. 2013. Pathway-based personalized analysis of cancer. *Proceedings of the National Academy of Sciences of the United States of America*, **110**(16), 6388–6393.

Efroni, Sol, Schaefer, Carl F, and Buetow, Kenneth H. 2007. Identification of key processes underlying cancer phenotypes using biologic pathway analysis. *PloS ONE*, **2**(5), e425.

Ellis, Matthew J, Ding, Li, Shen, Dong, Luo, Jingqin, Suman, Vera J, Wallis, John W, Van Tine, Brian A, et al. 2012. Whole-genome analysis informs breast cancer response to aromatase inhibition. *Nature*, June, 1–8.

Friedman, N, Linial, M, Nachman, I, and Pe'er, D. 2000. Using Bayesian networks to analyze expression data. *Journal of computational biology*, **7**(3–4), 601–620.

Gonzalez-Perez, Abel, and Lopez-Bigas, Nuria. 2012. Functional impact bias reveals cancer drivers. *Nucleic acids research*, **40**(21), e169–e169.

Greenblum, Sharon I, Efroni, Sol, Schaefer, Carl F, and Buetow, Ken H. 2011. The PathOlogist: an automated tool for pathway-centric analysis. *BMC bioinformatics*, **12**, 133.

Holderfield, Matthew, Merritt, Hanne, Chan, John, Wallroth, Marco, Tandeske, Laura, Zhai, Huili, Tellew, John, et al. 2013. RAF inhibitors activate the MAPK pathway by relieving inhibitory autophosphorylation. *Cancer cell*, **23**(5), 594–602.

J van der Laan, Mark, and Pollard, Katherine S. 2003. A new algorithm for hybrid hierarchical clustering with visualization and the bootstrap. *Journal of statistical planning and inference*, **117**(2), 275–303.

Kristensen, Vessela N, Vaske, Charles J, Ursini-Siegel, Josie, Van Loo, Peter, Nordgard, Silje H, Sachidanandam, Ravi, Sorlie, Therese, et al. 2011. Integrated molecular profiles of invasive breast tumors and ductal carcinoma in situ (DCIS) reveal differential vascular and interleukin signaling. *Proceedings of the National Academy of Sciences of the United States of America*, **109**(8), 2802–2807.

Lawrence, Michael S, Stojanov, Petar, Polak, Paz, Kryukov, Gregory V, Cibulskis, Kristian, Sivachenko, Andrey, Carter, Scott L, et al. 2014. Mutational heterogeneity in cancer and the search for new cancer-associated genes. *Nature*, **499**(7457), 214–218.

Ljubić, Ivana, Weiskircher, René, Pferschy, Ulrich, Klau, Gunnar W, Mutzel, Petra, and Fischetti, Matteo. 2006. An algorithmic framework for the exact solution of the prize-collecting Steiner tree problem. *Mathematical programming*, **105**(2–3), 427–449.

Martini, Paolo, Sales, Gabriele, Massa, M Sofia, Chiogna, Monica, and Romualdi, Chiara. 2013. Along signal paths: an empirical gene set approach exploiting pathway topology. *Nucleic acids research*, **41**(1), e19.

McCallum, Andrew, Schultz, Karl, and Singh, Sameer. 2009. Factorie: probabilistic programming via imperatively defined factor graphs. Pages 1249–1257 of: Bengio, Y, Schuurmans, D, Lafferty, J, Williams, C K I, and Culotta, A (eds.), *Advances in Neural Information Processing Systems*, vol. 22.

Mootha, Vamsi K, Lindgren, Cecilia M, Eriksson, Karl-Fredrik, Subramanian, Aravind, Sihag, Smita, Lehar, Joseph, Puigserver, Pere, et al. 2003. PGC-1alpha-responsive genes involved in oxidative phosphorylation are coordinately downregulated in human diabetes. *Nature genetics*, **34**(3), 267–273.

Ng, Pauline C, and Henikoff, Steven. 2003. SIFT: predicting amino acid changes that affect protein function. *Nucleic acids research*, **31**(13), 3812–3814.

Ng, Sam, Collisson, Eric A, Sokolov, Artem, Goldstein, Theodore, Gonzalez-Perez, Abel, Lopez-Bigas, Nuria, Benz, Christopher, Haussler, David, and Stuart, Joshua M. 2012a. PARADIGM-SHIFT predicts the function of mutations in multiple cancers using pathway impact analysis. *Bioinformatics*, **28**(18), i640–i646.

Paull, Evan O, Carlin, Daniel E, Niepel, Mario, Sorger, Peter K, Haussler, David, and Stuart, Joshua M. 2013. Discovering causal pathways linking genomic events to transcriptional states using Tied Diffusion Through Interacting Events (TieDIE). *Bioinformatics*, **29**(21), 2757–2764.

Reva, Boris, Antipin, Yevgeniy, and Sander, Chris. 2011. Predicting the functional impact of protein mutations: application to cancer genomics. *Nucleic acids research*, **39**(17), e118–e131.

Schaefer, Carl F, Anthony, Kira, Krupa, Shiva, Buchoff, Jeffrey, Day, Matthew, Hannay, Timo, and Buetow, Kenneth H. 2009. PID: the pathway interaction database. *Nucleic acids research*, **37**(suppl 1), D674–D679.

Sedgewick, Andrew J, Benz, Stephen C, Rabizadeh, Shahrooz, Soon-Shiong, Patrick, and Vaske, Charles J. 2013. Learning subgroup-specific regulatory interactions and regulator independence with PARADIGM. *Bioinformatics (Oxford, England)*, **29**(13), i62–i70.

Shannon, Paul, Markiel, Andrew, Ozier, Owen, Baliga, Nitin S, Wang, Jonathan T, Ramage, Daniel, Amin, Nada, Schwikowski, Benno, and Ideker, Trey. 2003. Cytoscape: a software environment for integrated models of biomolecular interaction networks. *Genome research*, **13**(11), 2498–2504.

Subramanian, Aravind, Tamayo, Pablo, Mootha, Vamsi K, Mukherjee, Sayan, Ebert, Benjamin L, Gillette, Michael A, Paulovich, Amanda, et al. 2005. Gene set enrichment analysis: a knowledge-based approach for interpreting genome-wide expression profiles. *Proceedings of the National Academy of Sciences of the United States of America*, **102**(43), 15545–15550.

Tuncbag, Nurcan, Braunstein, Alfredo, Pagnani, Andrea, Huang, Shao-Shan Carol, Chayes, Jennifer, Borgs, Christian, Zecchina, Riccardo, and Fraenkel, Ernest. 2013. Simultaneous reconstruction of multiple signaling pathways via the prize-collecting Steiner forest problem. *Journal of computational biology*, **20**(2), 124–136.

Vandin, Fabio, Upfal, Eli, and Raphael, Benjamin. 2012a. Algorithms and genome sequencing: identifying driver pathways in cancer. *Computer*, **45**(3), 39–46.

Vandin, Fabio, Clay, Patrick, Upfal, Eli, and Raphael, Benjamin J. 2012b. Discovery of mutated subnetworks associated with clinical data in cancer. Pages 55–66 of: *Pacific Symposium on Biocomputing* 17. Singapore: World Scientific.

Vaske, Charles J, Benz, Stephen C, Sanborn, J Zachary, Earl, Dent, Szeto, Christopher, Zhu, Jingchun, Haussler, David, and Stuart, Joshua M. 2010. Inference of patient-specific pathway activities from multi-dimensional cancer genomics data using PARADIGM. *Bioinformatics (Oxford, England)*, **26**(12), i237–i245.

15

Analyzing Combinations of Somatic Mutations in Cancer Genomes

MARK D. M. LEISERSON AND BENJAMIN J. RAPHAEL

Abstract

In the past few years, high-throughput DNA sequencing has helped identify numerous genes that are recurrently mutated in cancer. Such recurrently mutated genes are likely to play key roles in the development of cancer. However, many other cancer genes are mutated rarely and therefore difficult to identify by their frequency of occurrence across cancer samples. Understanding the development and progression of cancer requires the identification of combinations of recurrently mutated genes in signaling and regulatory pathways. In this chapter, we discuss three approaches to identify such recurrently mutated combinations of genes: (1) evaluation of known pathways or gene sets, (2) discovery of significantly mutated subgraphs of an interaction network, and (3) identification of gene sets with mutually exclusive mutations. We demonstrate these three approaches on glioblastoma mutation data from the Cancer Genome Atlas.

15.1 Introduction

Cancer is a genetic disease caused largely by somatic mutations that accumulate in an individual's genome throughout the individual's lifetime. High-throughput DNA sequencing technologies now allow researchers to measure the somatic mutations in tumor samples from a large number of patients. A major challenge in interpreting the resulting cancer genome sequences is to distinguish the *driver* mutations responsible for cancer from the random *passenger* mutations that have no impact on the cancer. A common approach to predict driver mutations, or driver genes, is to identify recurrent mutations, or recurrently mutated genes, that are mutated in significantly more samples than expected by chance. However, most cancers exhibit extensive mutational heterogeneity, with different tumors having different combinations of driver mutations. This mutational heterogeneity implies that relatively few genes are recurrently mutated across tumors of the same cancer (sub)type, complicating the identification of driver genes. This phenomenon is visible in recent large-scale cancer

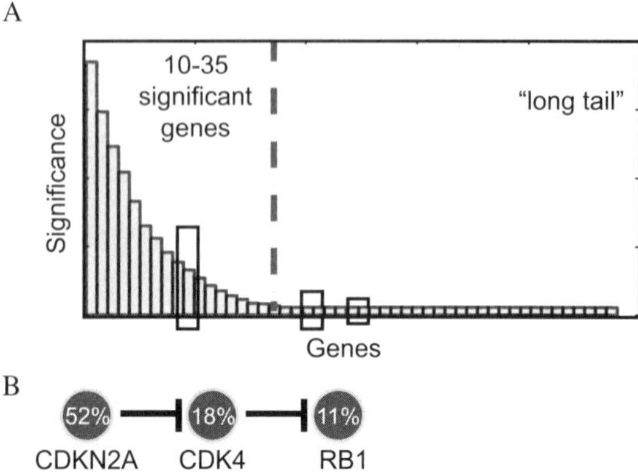

Figure 15.1 The long tail phenomenon and combinations of mutations. (A) Sorting genes by the statistical significance of the number samples with a mutation in the gene reveals the *long tail* phenomenon, whereby a small number of genes are mutated in many samples, while many genes are mutated in few samples. Recent cancer sequencing studies (The Cancer Genome Atlas Research Network, 2008, 2011, 2012a, 2012c, 2013; Kandoth et al., 2013; Creighton et al., 2013) with sample sizes of a few hundred report less than three dozen significantly mutated genes at a reasonable statistical significance level (red line). Combinations of mutated genes provide one explanation for the long tail with rare/common mutations occurring in different genes (boxed) in the same pathway. (B) Three proteins and their interactions in the Rb signaling pathway. Within each protein is the percentage of tumors in the TCGA GBM data with somatic mutations in corresponding genes, as described in Section 15.2.

sequencing studies (The Cancer Genome Atlas Research Network, 2011, 2012a, 2012b, 2012c, 2013; Creighton et al., 2013; Kandoth et al., 2013), where relatively few genes were found to be recurrently mutated, followed by a "long tail" of rarely mutated genes (Figure 15.1).

A major reason for mutational heterogeneity in cancer is that mutations target cellular regulatory and signaling pathways, each pathway composed of multiple proteins/genes. Different tumors may have mutations in different members of a given pathway, meaning that numerous combinations of mutations could lead to the same cancer (sub)type. Thus, in addition to evaluation of single genes for recurrent mutations, it is also useful to test combinations of genes, in known or novel pathways, for significant numbers of mutations.

In this chapter, we discuss three approaches to identify recurrent combinations of genes in a cohort of sequenced cancer genomes. The first approach is to test predefined gene sets for enrichment of mutations. Such gene sets are derived from prior knowledge of biological pathways or protein complexes, and we discuss the application of DAVID (Huang et al., 2009a, 2009b) and GSEA (Mootha et al., 2003; Subramanian et al., 2005) to cancer mutation data.

The second approach is to identify subnetworks of a genome-scale interaction network that are mutated more than expected. We discuss the HotNet algorithm (Vandin et al., 2011, 2012b) for this task. The third approach is to identify combinations of mutations/genes that exhibit a pattern of mutual exclusivity. We discuss the Dendrix algorithms (Vandin et al., 2012a; Leiserson et al., 2013) for this purpose. These three approaches differ in the amount of prior knowledge of gene sets that they require, ranging from full knowledge of gene sets to no prior knowledge. As the number of sequenced cancer genomes increases, approaches like HotNet and Dendrix that do not restrict the combinations of mutations to those in known pathways methods become increasingly attractive, as such approaches may be able to identify novel pathways or *cross talk* between pathways.

15.2 Sequencing Cancer Genomes and Somatic Mutations

The first step in analyzing a cancer genome is to measure the somatic mutations that are present in the genome. Three approaches are currently in use: microarray techniques, including array comparative genomic hybridization (aCGH) and single nucleotide polymorphism (SNP) arrays; whole-exome sequencing; and whole-genome sequencing. SNP arrays and aCGH are useful for detecting copy number aberrations (Alkan et al., 2011). Whole-exome sequencing approaches measure single nucleotide mutations and small indels in the protein coding regions of the genome ($\approx 1\%$ of the human genome). Whole-genome sequencing measures all classes of somatic aberrations, including single-nucleotide mutations and small indels (in coding and noncoding regions), copy number aberrations, and genome rearrangements.

Following the sequencing of a cancer genome/exome, the next step is to identify the somatic mutations that are present in the cancer genome. This is typically done by aligning sequence reads from the cancer genome/exome to the reference human genome sequence and identifying sequence variants, including single-nucleotide variants (SNVs), copy number aberrations (CNAs), and structural variants (SVs) (Meyerson et al., 2010). Usually, a matched normal sample from the same individual is also sequenced to distinguish somatic variants from inherited *germline* variants. This analysis results in a catalog of somatic mutations in the cancer genome, with some errors and missing variants due to various limitations in the sequencing technologies and variant detection algorithms.

The reported somatic mutations from the sequencing reads include a mixture of driver mutations, passenger mutations, and sequencing or variant detection errors. Nearly all cancer sequencing studies of SNVs to date have focused on nonsynonymous coding mutations that alter the sequence of proteins. Algorithms such as MutSigCV (Lawrence et al., 2013) and MuSiC (Dees et al., 2012) attempt to distinguish genes containing driver mutations from genes containing passenger mutations according to the number of samples with a mutation in the gene. These algorithms include models of the background mutation rates of genes that include factors such as gene length, gene expression, and replication timing. Different approaches are used to analyze copy number aberrations (CNAs) as the length and position of these aberrations often vary considerably across individuals. A commonly used algorithm to identify recurrent copy number aberrations shared by multiple individuals is GISTIC2 (Mermel et al., 2011). Often the recurrent aberrations identified by GISTIC2, or other similar algorithms, span multiple genes, making it difficult to determine which, if any, of the genes in the aberration are driver genes.

In the remainder of this chapter, we analyze glioblastoma (GBM) mutation data from TCGA (Chang et al., 2013).[1] to illustrate several approaches to analyze combinations of somatic mutations. The TCGA GBM mutation data set includes nonsynonymous SNVs, indels, splice-site mutations, and aCGH copy number data from 290 GBM tumor samples. We removed the SNV data from genes that were likely to be passenger genes according to MutSigCV. We identified recurrent CNAs using GISTIC2.0 (Mermel et al., 2011). The resulting data set consists of two binary (0/1) mutation event matrices. The SNV matrix records the samples that contain each measured SNV. A CNA matrix records the list of samples that contain each CNA. We refer to these two matrices as the TCGA GBM data set throughout the remainder of this chapter.

15.3 Significantly Mutated Pathways and Gene Sets

Given a list of mutated genes, there are several approaches to determine whether these genes have significant overlap with known pathways or predefined gene sets. These approaches can be divided into three categories: gene list enrichment, ranked gene list overrepresentation, and mutation overrepresentation. Popular tools that implement the first two categories are DAVID (Huang et al., 2009a,b) and GSEA (Mootha et al., 2003; Subramanian et al., 2005), respectively. These tools are agnostic to the data that were used to generate the gene list or rank the genes and thus can be applied to many types of data. The most common use of these tools is to analyze gene expression data, but in the next section we illustrate their application to mutation data from the TCGA GBM

---

[1] This data set is a superset of the data from the earlier TCGA GBM study (The Cancer Genome Atlas Research Network, 2008).

study. In contrast, the third category of approaches models the somatic mutation process in a cohort of cancer samples and tests known pathways and predefined gene sets for significant numbers of somatic mutations or significant numbers of mutated samples. Examples of such approaches include PathScan (Wendl et al., 2011) and (Boca et al., 2010). A major challenge for all of these methods is that annotated pathways and gene sets often contain many genes and overlap, which complicates the interpretation of the results. We detail this further in the next section.

15.3.1 Application of DAVID and GSEA to Cancer Mutation Data

The DAVID tool (Huang et al., 2009a, 2009b) takes a list of genes as input and then tests a collection of known pathways and gene sets for overrepresentation of genes from the input list. For cancer genome analysis, one may use a list of significantly mutated genes as input to DAVID. We ran DAVID on the 300 genes mutated in the most samples after collapsing the SNVs and CNAs from the TCGA GBM data set at the gene level. We tested against pathways and protein complexes from the KEGG (Kanehisa and Goto, 2000; Kanehisa, 2013), BBID (BBID, 2014), and BioCarta (BioCarta, 2014) databases. DAVID reported 14 enriched pathways (False Discovery Rate (FDR) < 0.05), six of which had "cancer" in their name (Table 15.1). These 14 pathways are large: each contains more than 100 genes and thus the pathways identified by DAVID are less focused than shown in Figure 15.1. In addition, only 37 of the 300 input genes overlap any of these 14 enriched pathways, and most of these 37 genes are members of many (an average of 4) of the enriched pathways. Thus, there is extensive overlap between the enriched pathways (cf. "Additional Genes" column in Table 15.1). This, combined with the large size of the enriched pathways, makes it difficult to formulate precise conclusions from the DAVID output about the ways that mutations target particular protein interactions within pathways.

The GSEA algorithm (Mootha et al., 2003; Subramanian et al., 2005) takes as input a ranked list of genes and tests whether genes from a predefined gene set are overrepresented at the top of the ranked list. The GSEA software includes MSigDB (Subramanian et al., 2005), a curated collection of gene sets that includes known pathways and other functionally related sets of genes. A straightforward way to identify significantly mutated pathways or gene sets is to apply GSEA to a ranked list of mutated genes, ranked according to a measure of statistical significance for individual genes. We applied GSEA to a list of 1375 mutated genes in the TCGA GBM data set after collapsing the SNV and CNA matrices at the gene level, ranking each gene according to the number of

Table 15.1 *Ten pathways with lowest FDR identified by DAVID on the top 300 most mutated genes in the TCGA GBM data set*

Pathway data base	Pathway	No. of overlapping genes (% pathway genes)	No. of additional genes	FDR
KEGG	Glioma	14 (4.7%)	14	2×10^{-7}
KEGG	Melanoma	14 (4.7%)	1	1×10^{-6}
KEGG	Chronic myeloid leukemia	13 (4.4%)	3	1×10^{-4}
KEGG	Non-small cell lung cancer	11 (3.7%)	0	1×10^{-4}
KEGG	Pathways in cancer	24 (8.1%)	7	2×10^{-4}
KEGG	Cell cycle	15 (5.0%)	6	2×10^{-4}
KEGG	Prostate cancer	13 (4.4%)	0	9×10^{-4}
KEGG	Small cell lung cancer	12 (4.0%)	0	3×10^{-5}
KEGG	Focal adhesion	17 (5.7%)	6	2×10^{-3}
BIOCARTA	Influence of Ras and Rho proteins on G1 to S Transition	8 (2.7%)	0	0.01

samples in which it is mutated. We tested these genes against the 1984 curated gene sets from MSigDB of size 5–200. GSEA reported 74 significant gene sets (FDR < 0.05), 13 of which had "cancer" in their name (see Table 15.2). Similar to the results from DAVID, the significant gene sets identified by GSEA tended to be large and had considerable overlap. The significant gene sets included 196 of the 1375 input genes, and each of these 196 genes was a member of an average of more than four gene sets. One of the reported gene sets is TCGA_GLIOBLASTOMA_MUTATED, which presumably was created following the first TCGA GBM publication (The Cancer Genome Atlas Research Network, 2008). This set of eight genes overlaps 7 of the 14 most mutated input genes: *EGFR, PTEN, TP53, RB1, NF1, PIK3R1*, and *PIK3CA*. This demonstrates that GSEA can make very precise predictions when the gene set database contains a small, narrowly defined gene set. However, the novelty of the predictions will always be limited by the gene sets available in the database: there are too many gene sets to test all exhaustively ($\approx 10^{19}$ sets of 5 genes in the human genome).

15.4 Significantly Mutated Subnetworks: HotNet

15.4.1 Overview

Another approach to identify mutated pathways that does not restrict attention to predefined sets of genes is to analyze protein-protein interaction (PPI)

Table 15.2 *Ten gene sets with lowest FDR identified by GSEA on the mutated genes in the TCGA GBM data set*

Gene set name	No. of overlapping genes (% gene set)	No. of additional genes	FDR
KEGG_MELANOMA	20 (28%)	20	0
KEGG_BLADDER_CANCER	9 (21%)	1	0
TCGA_GLIOBLASTOMA_MUTATED	7 (88%)	1	0
PID_RB_1PATHWAY	12 (18%)	6	0
KEGG_CHRONIC_MYELOID_LEUKEMIA	19 (26%)	4	0
BIOCARTA_CELLCYCLE_PATHWAY	8 (35%)	1	0
BIOCARTA_RACCYCD_PATHWAY	8 (31%)	0	0
BIOCARTA_ARF_PATHWAY	6 (35%)	0	0
BIOCARTA_CTCF_PATHWAY	8 (35%)	1	0
REACTOME_MITOTIC_G1_G1_S_PHASES	10 (7%)	2	0

networks. PPI networks give a whole-proteome representation of interactions between pairs of proteins, without subdividing proteins into discrete pathways. Thus, the interactions may include those between proteins in known pathways as well as cross-talk between pathways. A variety of PPI networks are available, ranging from small networks that include only well-known and well-annotated interactions to large networks that include interactions reported in high-throughput interaction arrays. The goal in using PPI networks to analyze cancer mutation data is to identify connected subnetworks that are mutated more than expected by chance. Such subnetworks may represent known pathways, novel pathways, or cross-talk between pathways. However, it is not possible to exhaustively test all possible connected subnetworks: for example, there are $\approx 10^{18}$ subnetworks of size 5 in a reasonably sized human PPI network. This creates a severe computational and statistical bottleneck, both in evaluating each subnetwork and in performing the necessary multiple hypothesis correction for the large number of statistical tests. Thus, one requires an algorithm to perform such analysis in a computationally efficient and statistically appropriate way.

In this section we present the HotNet algorithm (Vandin et al., 2011, 2012b) to find mutated subnetworks in a PPI network. The remainder of this section is organized as follows. First, we present the HotNet algorithm in Section 15.4.2. Then we demonstrate the application of HotNet to the TCGA GBM data set in Section 15.4.3. Finally, we discuss setting parameters for HotNet in Section 15.4.4. HotNet has been applied in multiple TCGA publications (The Cancer Genome Atlas Research Network, 2011, 2013; Creighton et al., 2013).

Table 15.3 *Measurements of human PPI networks (largest connected component)*

Interaction network	No. of nodes	No. of edges	Diameter	Avg. shortest path
HINT (Das and Yu, 2012)	8,269	28,497	14	4.23
HPRD (Prasad et al., 2009)	9,205	36,720	14	4.22
iRefIndex (Razick et al., 2008)	12,129	91,809	12	3.64
Multinet (Khurana et al., 2013)	14,399	109,570	9	3.39

15.4.2 HotNet Algorithm

A major challenge in identifying significantly mutated subnetworks is the topology of PPI networks. The diameters[2] of PPI networks, as well as the average shortest path between pairs of nodes in the network, are relatively low, ranging from 9–14 and 3–4 in several well-known networks (Table 15.3). This feature of the topology is largely due to the presence of nodes with many neighbors, for example, TP53 has 427 neighbors in the iRefIndex interaction network (Razick et al., 2008). Many of these high-degree nodes also tend to be highly mutated, reflecting either a genuine biological phenomenon or ascertainment bias; that is, highly mutated genes like *TP53* are likely to be well studied with many experimentally characterized interactions. These features of biological networks imply that observing that several highly mutated nodes are connected in the interaction network may not be that surprising. A method that identifies significantly mutated subnetworks must account for the global and local topology of the interaction network.

HotNet takes as input a collection of mutation scores for genes and a connected[3] PPI graph where each node corresponds to a gene/protein and each edge represents an interaction between the two genes/proteins. HotNet uses a heat diffusion process to simultaneously encode the significance score of individual gene mutations and the topology of interactions between genes in the protein-protein interaction network (Figure 15.2). Each node is assigned an initial heat according to its mutation score. This heat diffuses over the edges of the network for a fixed time t, with each node diffusing its heat to – and receiving heat from – its neighbors. As high-degree nodes with many neighbors

---

[2] The diameter of a graph $G = (V, E)$ is the longest shortest path between any pair $u, v \in V$ of nodes in the graph.

[3] In practice HotNet is often run on the largest connected component of a protein-protein interaction network, which generally includes the vast majority of nodes and interactions.

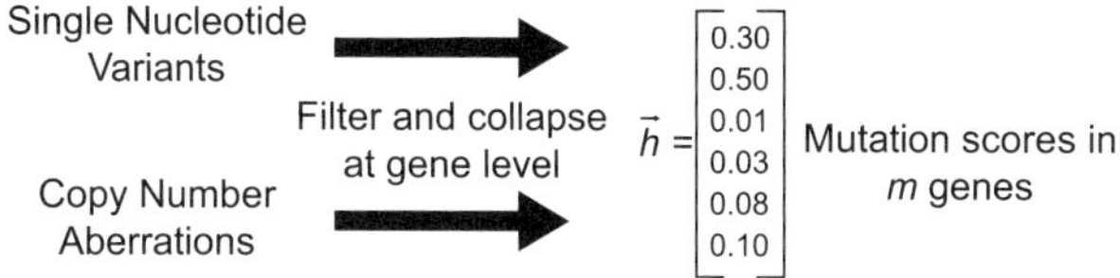

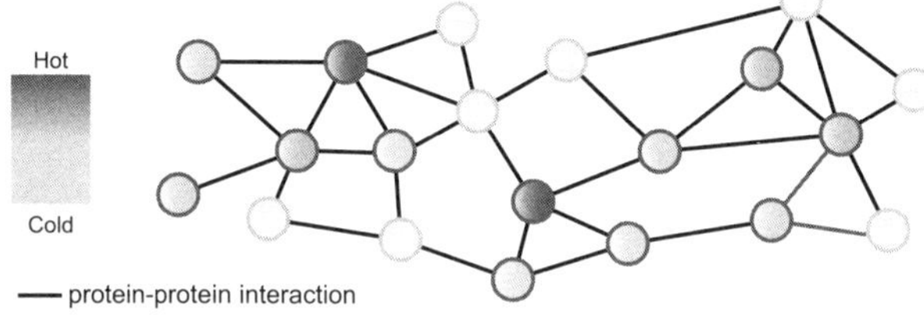

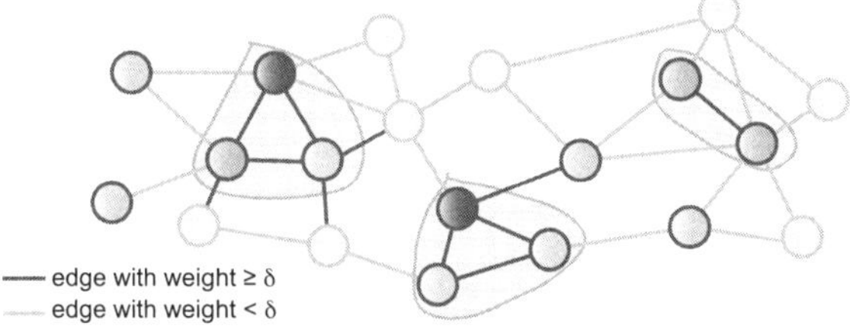

Figure 15.2 Overview of the Hotnet algorithm. (A) HotNet generates a heat score for each protein in a PPI network using single nucleotide variants and copy number aberrations from a cohort of tumors. (B) HotNet applies the heat to each protein in the network and allows it to diffuse for time t to create an edge-weighted graph. (C) HotNet partitions the graph by removing edges with weight less than a parameter δ, selected as described in the text. HotNet outputs the connected components of the partitioned graphs as significantly mutated subnetworks and evaluates their significance using a permutation test.

will diffuse heat equally to all of their neighbors, these neighbors will receive less heat than the neighbors of a lower-degree node.

We briefly describe the HotNet algorithm. Let $G = (V, E)$ be an unweighted, undirected input PPI graph with proteins as vertices and where each gene is the label for the protein it encodes. Let $\vec{h}$ be the "heat" vector of mutation scores such that $\vec{h}(i)$ is the mutation score of gene i. The HotNet algorithm consists of three steps:

1. Place heat $\vec{h}(i)$ on node i and allow it to diffuse for time t. The result is a complete edge-weighted "influence" graph $F(V, E')$, where each edge $(u, v) \in E'$ simultaneously encodes the local topology and mutation score of u and v.

2. Partition $F(V, E')$ into subnetworks (connected components) by removing edges with weight $< \delta$ (see later).
3. Assess the statistical significance of the resulting subnetworks using a two-stage statistical test. Let X_s be the number of subnetworks containing s or more genes. In the first stage of the test, Vandin et al. (2011) use a permutation test to identify the smallest s such that X_s is statistically significant. Finding that X_s is statistically significant does not imply that each of the subnetworks is significant on its own. Thus, in the second state of the test (Vandin et al., 2011), compute an upper bound to the false-discovery rate (FDR) for the X_s subnetworks.

The output of HotNet is the genes in the significantly mutated subnetworks and their statistical significance (p-value and FDR). Further details are in Vandin et al. (2011, 2012b).

15.4.3 Applying HotNet to Mutation Data

HotNet is implemented as a Python package that is run in three phases. First, HotNet generates a text file that stores the heat (mutation score) of each gene from the input mutation data (Figure 15.2A). Second, HotNet performs a permutation test to identify the minimum edge weight δ that is used to partition the weighted influence graph F (Figure 15.2C). Last, HotNet partitions F and performs the two-stage statistical test to evaluate the significance of the subnetworks.

We ran HotNet on the TCGA GBM data set and the iRefIndex PPI network (Razick et al., 2008). We first collapsed the SNV and CNA event matrices from the TCGA GBM data set at the gene level, marking a gene as mutated in a sample if it contained either an SNV or CNA in the sample. The mutation score of a gene is the number of mutated samples. We then ran HotNet using the following commands (configuration files are reproduced in the appendix):

```
# Run the entire HotNet pipeline, given
# a TSV heat file (Figure 17.1(a)) and an influence matrix
python simpleRun.py @config/simple-run.config
```

HotNet outputs each of the subnetworks it identified in both text and HTML format. Figure 15.3 shows the default web output of the subnetworks identified by HotNet on the TCGA GBM data and iRefIndex PPI network. Running HotNet on these data on a single Xeon 2.6 Ghz core with 256 Gb of RAM took 36 minutes and 26 seconds.

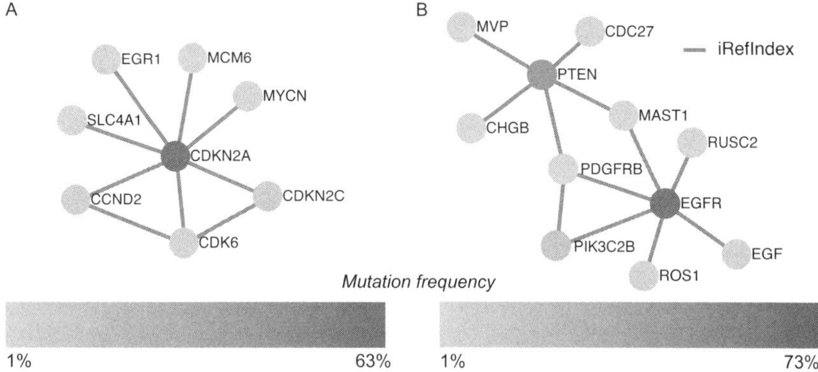

Figure 15.3 Subnetworks identified by HotNet on the TCGA GBM data set. (A) A sub-network that includes multiple members of the Rb signaling pathway. (B) A subnetwork that includes members of the RTK and PI3K signaling pathways.

HotNet also outputs the statistical significance of subnetworks for a range of minimum subnetwork sizes s. The significance is calculated using a permutation test that forms a permutational null distribution by permuting the heat scores on the nodes of the network (restricting to nodes/genes that were measured). For each s, HotNet outputs the observed number of components of size at least s, the expected number of components of size at least s according to the permutational distribution, and the permutational p-value, that is, the fraction of permuted data sets with at least the observed number of components of size at least s. Table 15.4 shows the HotNet significance table output for the TCGA GBM data set.

HotNet computes the significance for multiple minimum subnetwork sizes s because the user may be interested in identifying subnetworks of different sizes. As the permutation tests for different values of s are different hypotheses, strictly speaking the user should perform a multiple hypothesis correction, for example, the Bonferroni-corrected p-value is obtained by multiplying the reported p-value by the number of values of s that are tested. Running the statistical test with more permutations (the default is 100) will provide better estimates of the p-values.

15.4.4 HotNet Parameter Selection

The parameter δ in HotNet is the minimum edge weight used to partition the influence graph F by removing edges in E' with weight less than δ. HotNet

Table 15.4 *Statistical significance of HotNet results
on the TCGA GBM data set*

Size s	Expected	Observed	Empirical p-value
2	5.95	9	0.1
3	2.02	2	0.68
4	0.91	2	0.21
5	0.44	2	0.06
6	0.27	2	0.03
7	0.18	2	0.02
8	0.1	2	0.0
9	0.06	1	0.06
10	0.04	1	0.04

Note: For each subnetwork size s, the expected number of
components of size at least s (calculated from permuted data),
the observed number of components of size at least s, and
the empirical p-value is recorded.

sets the value of δ automatically by examining the distribution of the number of
components of size at least s on permuted data. The full procedure for setting
δ is described in Vandin et al. (2011, 2012b).

The parameter t in HotNet controls the amount of time for which the heat
diffusion process is run, and thus the influence of one gene on another. When
$t = 0$, each node i has $\vec{h}(i)$ heat, that is, no heat has diffused, whereas at
$t = \infty$, the heat diffusion process has reached equilibrium, and each node has
the same heat. Using HotNet requires setting t to some value between 0 and
∞ that captures interesting features of the underlying network's topology. One
heuristic to set t is to find a value of t such that most of the heat diffused from
any given node remains at the neighbors of the node. This intuitively makes
sense when searching for mutated subnetworks, as we are most interested in
capturing how mutations cluster in individual proteins and their immediate
neighbors.

We set t using the following procedure:

1. Choose a set N of nodes with different topological properties, for example,
   high degree and low degree.
2. Place one unit of heat on each $n \in N$, and compute the amount of heat on
   each n's neighbors for a range of values of t (e.g., $t = 0.01, 0.02, \ldots, 0.2$).
3. Choose the value of t such that most heat is on the neighbors of each $n \in N$.

Table 15.5 lists the values of t computed using this procedure for each of the
four human PPI networks described in Section 15.4.2.

Table 15.5 *Diffusion time t used by
HotNet for four human protein-protein
interaction networks*

Interaction	t
HINT (Das and Yu, 2012)	0.10
HPRD (Prasad et al., 2009)	0.10
iRefIndex (Razick et al., 2008)	0.05
Multinet (Khurana et al., 2013)	0.05

15.5 De Novo Driver Exclusivity: Dendrix and Multi-Dendrix

15.5.1 Overview

The ultimate goal for methods that analyze combinations of mutations is to find recurrently mutated combinations of mutations/mutated genes de novo, that is, without using prior knowledge to restrict the gene sets that we consider. As noted earlier, testing all combinations of mutations/genes is not feasible. Another approach is to use particular patterns of mutations that are expected to occur in pathways as a constraint on the gene sets to consider. Specifically, it is assumed that tumors contain relatively few driver mutations (Vogelstein et al., 2013) and these are distributed across the multiple pathways that are perturbed to make a cell cancerous (Hanahan and Weinberg, 2011). Thus, we expect that an individual tumor will have approximately one driver mutation per pathway. Equivalently, when looking across tumors, we expect that the genes in a pathway will exhibit a mutually exclusive pattern of mutations. At the same time, if perturbation of the pathway is important, we expect that most pathways will have a mutation in some gene in the pathway. The De Novo Driver Exclusivity (Dendrix) algorithm (Vandin et al., 2012a) and Multi-Dendrix algorithm (Leiserson et al., 2013) find one or more sets of genes that exhibit these properties of mutual exclusivity and high *coverage* (number of samples). We emphasize that Dendrix and Multi-Dendrix do *not* use any prior knowledge of pathways/gene sets to constrain the gene sets under consideration.

The remainder of this section is organized as follows. First, we present the Dendrix and Multi-Dendrix algorithms in Section 15.5.2. Then, we demonstrate the two algorithms on the example TCGA GBM data set in Section 15.5.3.

15.5.2 Dendrix and Multi-Dendrix Algorithms

The Dendrix and Multi-Dendrix algorithms both search for gene sets with mutually exclusive mutations and high coverage from cohorts of sequenced cancer genomes. The only required input data for the algorithms is a mutation matrix that encodes the status of m mutational events in a cohort of n tumor samples. A mutational event may range in scale from a single nucleotide substitution through a whole-chromosome gain/loss. Typically, we consider mutational events at the level of genes, marking a gene as mutated if it contains a mutation of a particular type; for example, we can separate SNVs/small indels from amplifications/deletions of a gene (see later). For ease of exposition, we write mutational events at the level of genes. Let A be an $m \times n$ matrix such that

$$A_{ij} = \begin{cases} 1 & \text{if gene } i \text{ is mutated in tumor } j \\ 0 & \text{otherwise} \end{cases} \tag{15.1}$$

Let $\Gamma(i) = \{j : A_{ij} = 1\}$ be the set of samples with at least one mutation in gene i. Similarly, for a set M of genes (i.e., rows of A), we define

$$\Gamma(M) = \bigcup_{g \in M} \Gamma(g) \tag{15.2}$$

The goal of both the Dendrix and Multi-Dendrix algorithms is to find sets M of k genes with high coverage and approximately exclusive mutations. To do this, Vandin et al. (2012a) define a weight $W(M)$ on a gene set that measures how strongly the gene set exhibits these properties. To describe this weight, we first define the coverage of a M to be $|\Gamma|$ and define the coverage overlap as the number of genes in M that are mutated in more than one sample:

$$\omega(M) = \sum_{g \in M} |\Gamma(g)| - |\Gamma(M)| \tag{15.3}$$

It is often possible to increase the coverage of a gene set at the expense of lower exclusivity or higher coverage overlap. A gene set M with both high coverage and approximately exclusive mutations is one that maximizes coverage and minimizes coverage overlap. To balance the trade-off between these criteria, Vandin et al. (2012a) define the weight function $W(M)$ as the difference between coverage and coverage overlap of M:

$$W(M) = |\Gamma(M)| - \omega(M) = 2|\Gamma(M)| - \sum_{g \in M} |\Gamma(g)| \tag{15.4}$$

Dendrix

The goal of Dendrix is to find gene sets M with high weight $W(M)$ (see Figure 15.4). One approach to find such gene sets is to exhaustively compute $W(M)$ for all $\binom{m}{k}$ gene sets M of k genes. However, the number of gene sets to examine increases exponentially in m, and thus, except for small values of m

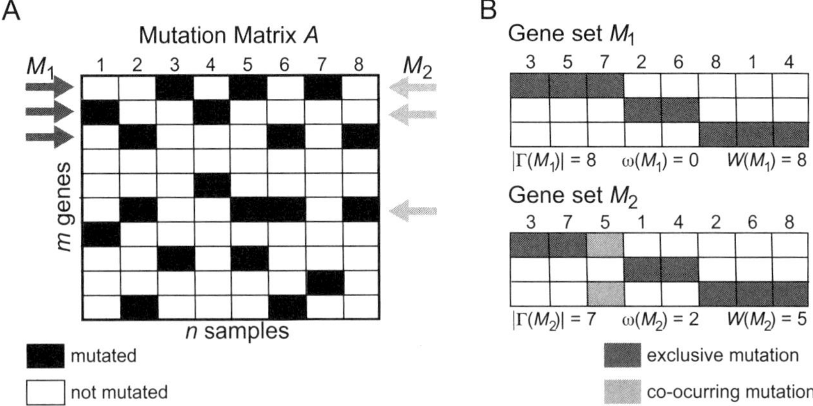

Figure 15.4 Overview of the Dendrix algorithm. (A) Dendrix takes as input a mutation matrix A (here shown at the level of individual genes) and finds gene sets with high weight W (i.e., approximately exclusive mutations and high coverage). The red and the green arrows point to the two highest-weight gene sets in A of size $k = 3$, respectively. Note that the gene sets overlap by two genes. (B) Mutation matrices for gene sets M_1 and M_2, with samples sorted independently in each set to illustrate exclusivity. Below each set are the coverage $|\Gamma|$, coverage overlap ω, and weight W.

and k, testing all gene sets is infeasible. Further compounding the challenge, Vandin et al. (2012a) showed that finding the set M of maximum weight $W(M)$ is NP-hard and thus cannot be solved efficiently for an arbitrary mutation matrix.

To overcome this challenge, Vandin et al. (2012a) introduced a Markov chain Monte Carlo (MCMC) algorithm to randomly sample gene sets in proportion to their weight. MCMC is a general method for selecting random samples from a set Ω according to the probability distribution π (Brooks, 1998). The MCMC algorithm constructs an ergodic Markov chain with stationary distribution π and then runs the chain long enough such that the emissions from the Markov chain are sampled according to π. In this case, the sample space Ω is the set of gene sets with k genes, and we set

$$\pi(M) = \frac{\exp\{W(M)\}}{\sum_{M' \in \Omega} \exp\{W(M')\}} \tag{15.5}$$

for any gene set M.

After running the MCMC algorithm for a suitably large number of iterations, Dendrix samples gene sets according to the distribution π such that higher-weight gene sets are output exponentially more often than the lower-weight gene sets. However, a single run of Dendrix will identify both optimal and suboptimal gene sets.

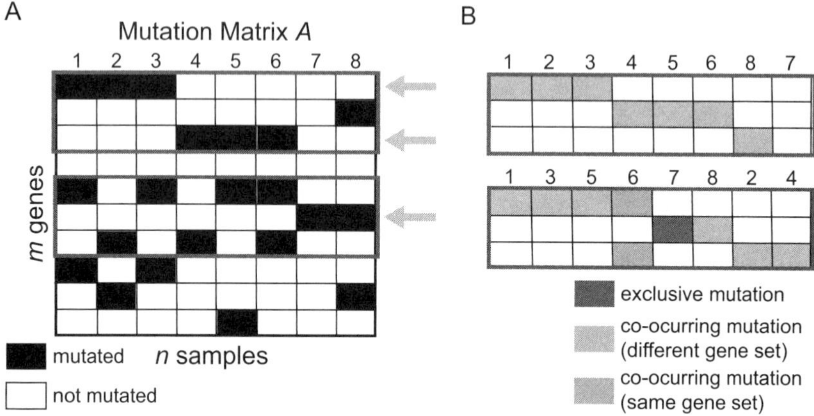

Figure 15.5 Overview of the Multi-Dendrix algorithm. (A) Multi-Dendrix identifies multiple gene sets simultaneously from a mutation matrix A. Shown in blue and red are the two gene sets of size $k = 3$ with maximum total weight. Methods that identify multiple gene sets *iteratively* would report the gene set with maximum weight indicated by the green arrows and thus would not find the optimal collection of gene sets. (B) The collection of exclusive gene sets identified by Multi-Dendrix from mutation matrix A. The samples (columns) in panel B are ordered to show the exclusivity of the mutations in the both gene sets. The mutations in each gene set largely exclusive (blue and green), with many co-occurring mutations between gene sets (green) and only one co-occurring mutation within the same gene set (orange).

Multi-Dendrix

The motivation for Multi-Dendrix is that tumors are expected to have driver mutations in *multiple* pathways. For example, Hanahan and Weinberg (2011) discuss how cancer requires mutations to signaling pathways involved in proliferation (e.g., the MAPK/ERK pathway) and the differentiation (e.g., the TGF-β pathway), among others. Because multiple pathways are important for cancer development, we expect most individuals to harbor mutations in more than one pathway. The original Dendrix algorithm found multiple gene sets in an iterative fashion. However, this is not always the best approach, as highly mutated genes from different pathways will sometimes be grouped together in a high-weight set at the first iteration (Figure 15.5). Multi-Dendrix (Leiserson et al., 2013) finds multiple high-weight sets *simultaneously*, thus better characterizing the patterns of mutual exclusivity within pathways and high coverage of each individual pathway.

Multi-Dendrix searches for a collection $\mathbf{M} = (M_1, \ldots, M_t)$ of t gene sets such that the sum of the weights $W(\mathbf{M}_i)$ is maximized. This problem is also

NP-hard, but Multi-Dendrix solves the problem by formulating it as an integer linear program (ILP) and uses highly optimized algorithms for solving ILPs that are sufficient for analysis of whole-exome data from hundreds of tumors. One issue searching for mutually exclusive gene sets is that one often does not know the number or size of gene sets to consider. Multi-Dendrix addresses this issue by running for a range parameters – varying the number t of gene sets, the minimum gene set size $k_{\min}$, and the maximum gene set size $k_{\max}$ – to find a collection of stable sets, called *core modules*. Multi-Dendrix identifies the core modules from the optimal collections found across a range of parameters using a complete edge-weighted graph where genes are nodes and the edges between genes are weighted by the number of parameter settings for which the genes were grouped into the same gene set. The core modules are the connected components of the graph after removing low-weight edges.

15.5.3 Applying Dendrix and Multi-Dendrix to Mutation Data

In this section, we give an example of running Dendrix and Multi-Dendrix on the TCGA GBM data. For this example, we collapse the SNVs at the gene level, although we note that it is possible to collapse SNVs at the level of protein domains or even analyze individual SNVs. For the CNA data, we analyze each amplification/deletion output by GISTIC2, even if the aberration spans multiple genes (see Section 15.2). That is, we do not collapse the CNA event matrix at the gene level. The ability for Dendrix and Multi-Dendrix to analyze mutation data at the event level instead of the gene level is a major difference between these algorithms and HotNet.

Running Dendrix

Dendrix is run in two phases. First, the `Dendrix.py` script performs MCMC to sample gene sets in proportion to their weight. For each sampled gene set, Dendrix reports the weight of the gene set and its *frequency*, or number of times it was sampled. Second, the `PermutationTestDendrix.py` script computes the statistical significance of the weight of any sampled gene set.

1. *Run Dendrix*. `Dendrix.py` requires seven command-line arguments to run. We run Dendrix on the TCGA GBM mutation data by using the following command-line:

```
python Dendrix.py TCGA-GBM-mutations.tsv 3 5 10000000 \
    TCGA-GBM-genes.txt 1 100
```

As Dendrix loads the mutation data and starts the MCMC, it will report its progress with the following output:

```
Load genes...
Loading mutations...
Number of genes: 302
Cleaning sample_mutatedGenes table...
```

Once Dendrix has output this message, the MCMC algorithm has begun; in our example 10^7 iterations of MCMC are performed. On a single Xeon 2.6Ghz on a machine with 256 Gb of RAM, Dendrix's runtime on the TCGA GBM data is 43 minutes and 31 seconds.

Dendrix outputs two tab-separated files for each experiment it runs, each labeled by the experiment number: `sets_frequencyOrder_experiment0.txt` and `sets_weightOrder_experiment0.txt`. Both files contain the same information for each sampled gene set: the genes in the gene set, frequency with which the gene set was sampled, and the weight of the gene set. The difference between the two files is that `frequencyOrder` sorts the gene sets by the frequency while `weightOrder` sorts the gene sets by their weight. We report the first 10 lines of the `sets_frequencyOrder_experiment0.txt` file. Note that because Dendrix is a stochastic algorithm, the sampling frequencies in the first column will vary from run to run.

```
Total visited: 339
98165   CDK4(A)   RB1       CDKN2A(D)   222
92      CDK4(A)   DGKD      CDKN2A(D)   208
69      MDM2(A)   RB1       CDKN2A(D)   207
44      CDK4(A)   SLC6A14   CDKN2A(D)   206
43      CDK4(A)   UGT2B4    CDKN2A(D)   206
38      CDK4(A)   PRDM15    CDKN2A(D)   206
34      CDK4(A)   KLK6      CDKN2A(D)   206
33      CDK4(A)   NBPF9     CDKN2A(D)   206
33      CDK4(A)   FOXR2     CDKN2A(D)   206
```

2. *Computing statistical significance with the Dendrix permutation test.* After running `Dendrix.py`, the next phase of Dendrix is to determine the statistical significance of the results using the Dendrix permutation test, which is implemented in the script `PermutationTestDendrix.py`. The Dendrix permutation test seeks to determine how surprising it is to observe that the nth highest weight gene set has weight W, conditioned on the mutation frequencies of each gene in the mutation matrix. We run the Dendrix

permutation test on the TCGA GBM data using the following command-line:

```
python PermutationTestDendrix.py TCGA-GBM-mutations.tsv 3 5 \
  1000000 TCGA-GBM-genes.txt 100 222 1
```

As the Dendrix permutation test begins running, it will report its progress with the following output:

```
Load genes...
Loading mutations...
302
Number of genes: 302
Cleaning sample_mutatedGenes table...
```

The output of the `PermutationTestDendrix.py` is one file, `p_value_dendrix.txt`, that contains one string: the p-value from the permutation test. In this case, none of the highest-weight gene sets in 100 permutations had a weight of at least 222, so the contents of `p_value_dendrix.txt` are

```
0.0
```

Because only 100 permutations were performed, the output 0.0 is interpreted as $p < 0.01$. The runtime for the Dendrix permutation test on the TCGA GBM data was longer than one day.

Running Multi-Dendrix

Multi-Dendrix is run in three phases. First, Multi-Dendrix identifies collections of gene sets with maximum weight across a range of gene set sizes and number of gene sets. Then, each of these collections is evaluated for statistical significance. Last, Multi-Dendrix reports core modules, groups of genes that are found together across the majority of parameter values.

Multi-Dendrix is implemented as a single Python script that will execute the entire Multi-Dendrix pipeline. We can run the Multi-Dendrix pipeline on the TCGA GBM data using the following command-line:

```
python multi_dendrix_pipeline.py -n TCGA-GBM -o output/ -v \
 -k_min 3 -k_max 5  -t_min 2 -t_max 4 -c 5 \
 -m TCGA-GBM-mutations.tsv  -g TCGA-GBM-genes.txt
```

Multi-Dendrix outputs each collection it identified for particular values of t, k_{min}, and k_{max}, as well as the core modules, in both text and HTML format.

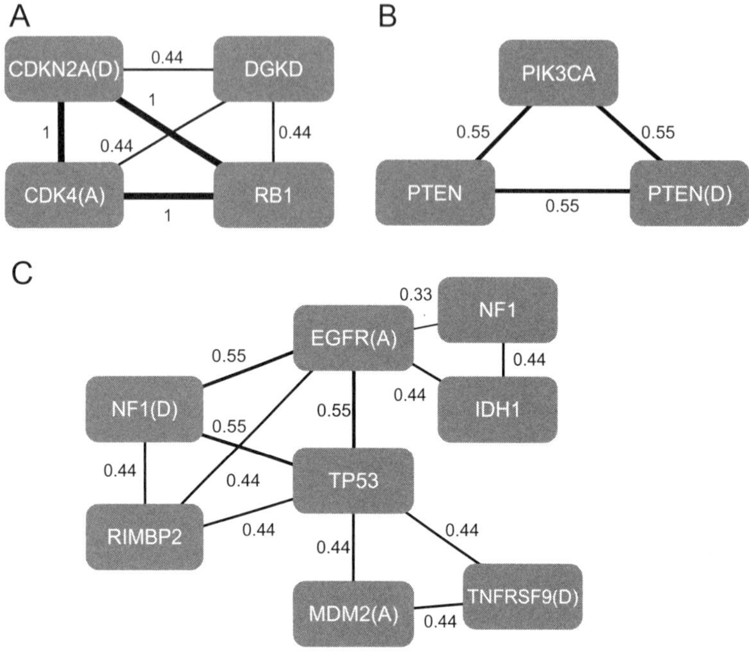

Figure 15.6 Core modules identified by Multi-Dendrix on the TCGA GBM data set. (A) indicates amplification of the gene; (D) indicates deletion of the gene; and gene name alone indicates SNVs and small indels in gene. Each pair of genes is connected by an edge weighted by the proportion of times both genes were reported by Multi-Dendrix in the same gene set.

For example, the text output of a single collection of $t = 2$ gene sets of $k_{\min} = k_{\max} = 3$ genes on the TCGA GBM data set consists of the following:

```
Weight Gene Set
222   CDK4(A),  CDKNA(D)  RB1
188   EGFR(A),  TP53,     NF1(D)
```

Here, (A) indicates amplification of the gene; (D) indicates deletion of the gene; and gene name alone indicates SNVs and small indels in gene.

The core modules identified by Multi-Dendrix on the TCGA GBM data set are shown in Figure 15.6.

Cancer (Sub)type Analysis

One of the confounding factors when searching for significantly mutated pathways by finding patterns of mutually exclusive mutations is that the samples may come from a mixture of cancer types or subtypes. Pairs of (sub)type-specific mutated genes will appear to be mutually exclusive, but not because they are members of the same biological pathway. The Multi-Dendrix Python package includes a module to analyze mutation data for (sub)type-specific mutations in the case that the (sub)types are known in advance. The module uses Fisher's exact test to test for associations between the presence or absence of a mutational event and the (sub)type of the sample. The module tests every gene against every (sub)type in the mutation data and reports the Bonferronni-corrected p-values for each such association.

We applied Multi-Dendrix's (sub)type specificity test to the TCGA GBM data set. We first downloaded the subtype classification for 248 of the 290 GBM samples from Firehose.[4] This subtype classification was generated by clustering of mRNA expression data. We then analyzed the subtype-specific mutations in the TCGA GBM data set using the following Python code:

```
# Import module
import multi_dendrix.subtypes as Sub


# File locations
MUTATION_MATRIX = "../data/pancan-gbm/mtx/gbm-pancan-f1.mtx"
PATIENT_LIST  = "../data/pancan-gbm/sample-w-cnmf-subtype.tsv"
GENE_LIST = "../data/pancan-gbm/mtx/gbm-pancan-f1-genes.lst"
OUTPUT_FILE  = "output/pancan-gbm/subtype-specific-mutations.tsv"


# Run (sub)type specificity test
args = [ "-m", MUTATION_MATRIX, "-p", PATIENT_LIST, "-v",
        "-g", GENE_LIST, "-o" OUTPUT_FILE ]
Sub.run(Sub.parse_args(args))
```

Running this analysis takes less than a second. By default, Multi-Dendrix outputs all gene-(sub)type pairs where the gene has a surprising number of mutations in the (sub)type (corrected $p < 0.05$). The top three most significant genes from the results of running the (sub)type analysis are shown in Table 15.6. The significant genes are largely consistent with the four subtypes identified by Verhaak et al. (2010): *NF1* is predominately mutated and deleted in the mesenchymal subtype (subtype 1); *IDH1* is only mutated in the proneural subtype (subtype 2), and most amplifications and mutations in *EGFR* are found in the classical subtype (subtype 3).

---

[4] Consensus nonnegative matrix factorization (NMF) subtypes were downloaded from http://gdac.broadinstitute.org/runs/analyses_2012_10_24/data/GBM/20121024/.

Table 15.6 *Genes enriched for subtype-specific mutations in the TCGA GBM mutation data set*

(Sub)type (no. of samples)	Gene	Type mutations	Nontype mutations	Corrected p-value
1 (78)	*RB1*	13	6	0.002
	NF1(D)	14	10	0.002
	NF1	14	10	0.014
2 (87)	*IDH1*	14	0	6×10^{-7}
	TP53	39	33	0.0003
	EGFR(A)	29	93	0.0009
3 (83)	*EGFR(A)*	63	59	5×10^{-9}
	EGFR	37	34	0.0005
	TP53	12	60	0.001

Note: For each gene, we show the subtype where it is enriched, the number of mutations in the gene in the subtype, the number of mutations in the gene in other samples, and the Bonferronni-corrected p-value.

15.6 Discussion and Conclusion

This chapter surveyed several algorithms for identifying the combinations of somatic mutations that are recurrent across multiple cancer genomes. Finding such combinations provides an explanation for the observed mutational heterogeneity in cancer and suggests that different members of the same pathway have driver mutations in different individuals. These algorithms analyze known pathways and gene sets (DAVID and GSEA), find subnetworks of protein-protein interaction networks (HotNet), or find combinations of mutually exclusive mutation (Dendrix/Multi-Dendrix).

There remain many challenges to be overcome in the analysis of combinations of mutations. First, most of the methods rely on mutation data derived from other algorithms, including variant calling and recurrent mutation/aberration detection algorithms. These algorithms are often tuned to achieve reasonable sensitivity/specificity for single mutation/gene analysis, but different requirements may be optimal for analysis of combinations of mutations. Second, as the cost of DNA sequencing continues to decline, methods for analyzing combinations must scale to many thousands of mutations/samples. Finally, another challenge is to develop better network models of biological processes. Most current networks represent a composite of the undirected physical protein interactions that occur in different cell types and conditions. Networks that are tissue or condition specific, as well as those that include multiple interaction types (e.g., regulatory interactions) and also provide richer annotation of interactions (e.g., gene A inhibits/activates gene B), will better represent the underlying biology. Constructing such networks and using them to analyze cancer mutation data are priorities for future studies.

Appendix: Config File Used for Running HotNet on TCGA GBM Data

```
# config/simple.config
--runname PanCan-GBM
--infmat_file /data/compbio/datasets/HeatKernels/IREFINDEX/
  9.0/iref_inf_0.05.mat
--infmat_index_file /data/compbio/datasets/HeatKernels/
  IREFINDEX/9.0/iref_index_genes
--edge_file /data/compbio/datasets/HeatKernels/IREFINDEX/
  9.0/iref_edge_list
--network_name iRefIndex
--heat_file output/iref/pancan-gbm-heat.tsv
--output_directory output/iref/pancan-gbm/
--num_permutations 100
--no-parallel
```

References

Alkan, Can, Coe, Bradley P, and Eichler, Evan E. 2011. Genome structural variation discovery and genotyping. *Nature reviews genetics*, **12**(5), 363–376.

BBID. 2014. *Biological Biochemical Image Database*. http://bbid.irp.nia.nih.gov/.

BioCarta. 2014. *BioCarta*. http://www.biocarta.com/.

Boca, Simina M, Kinzler, Kenneth W, Velculescu, Victor E, Vogelstein, Bert, and Parmigiani, Giovanni. 2010. Patient-oriented gene set analysis for cancer mutation data. *Genome biology*, **11**(11), R112.

Brooks, Stephen P. 1998. Markov chain Monte Carlo method and its application. *Journal of the Royal Statistical Society*, **47**(1), 69–100.

Chang, Kyle, Creighton, Chad J, Davis, Caleb, Donehower, Lawrence, Drummond, Jennifer, et al. 2013. The Cancer Genome Atlas Pan-Cancer analysis project. *Nature genetics*, **45**(10), 1113–1120.

Creighton, Chad J, Morgan, Margaret, Gunaratne, Preethi H, Wheeler, David A, Gibbs, Richard A, et al. 2013. Comprehensive molecular characterization of clear cell renal cell carcinoma. *Nature*, June.

Das, Jishnu, and Yu, Haiyuan. 2012. HINT: High-quality protein interactomes and their applications in understanding human disease. *BMC systems biology*, **6**, 92.

Dees, Nathan D, Zhang, Qunyuan, Kandoth, Cyriac, Wendl, Michael C, Schierding, William, et al. 2012. MuSiC: identifying mutational significance in cancer genomes. *Genome research*, **22**(8), 1589–1598.

Hanahan, Douglas, and Weinberg, Robert A. 2011. Hallmarks of cancer: the next generation. *Cell*, **144**(5), 646–674.

Huang, Da Wei, Sherman, Brad T, and Lempicki, Richard A. 2009a. Bioinformatics enrichment tools: paths toward the comprehensive functional analysis of large gene lists. *Nucleic acids research*, **37**(1), 1–13.

Huang, Da Wei, Sherman, Brad T, and Lempicki, Richard A. 2009b. Systematic and integrative analysis of large gene lists using DAVID bioinformatics resources. *Nat protoc*, **4**(1), 44–57.

Kandoth, Cyriac, Schultz, Nikolaus, Cherniack, Andrew D, Akbani, Rehan, Liu, Yuexin, et al. 2013. Integrated genomic characterization of endometrial carcinoma. *Nature*, **497**(7447), 67–73.

Kanehisa, Minoru. 2013. Molecular network analysis of diseases and drugs in KEGG. *Methods in molecular biology*, **939**, 263–275.

Kanehisa, M, and Goto, S. 2000. KEGG: Kyoto encyclopedia of genes and genomes. *Nucleic acids research*, **28**(1), 27–30.

Khurana, Ekta, Fu, Yao, Chen, Jieming, and Gerstein, Mark. 2013. Interpretation of genomic variants using a unified biological network approach. *PLoS computational biology*, **9**(3), e1002886.

Lawrence, Michael S, Stojanov, Petar, Polak, Paz, Kryukov, Gregory V, Cibulskis, Kristian, et al. 2013. Mutational heterogeneity in cancer and the search for new cancer-associated genes. *Nature*, **499**, 214–218.

Leiserson, Mark D M, Blokh, Dima, Sharan, Roded, and Raphael, Benjamin J. 2013. Simultaneous identification of multiple driver pathways in cancer. *PLoS computational biology*, **9**(5), e1003054.

Mermel, Craig H, Schumacher, Steven E, Hill, Barbara, Meyerson, Matthew L, Beroukhim, Rameen, and Getz, Gad. 2011. GISTIC2.0 facilitates sensitive and confident localization of the targets of focal somatic copy-number alteration in human cancers. *Genome biology*, **12**(4), R41.

Meyerson, Matthew, Gabriel, Stacey, and Getz, Gad. 2010. Advances in understanding cancer genomes through second-generation sequencing. *Nature reviews genetics*, **11**(10), 685–696.

Mootha, Vamsi K, Lindgren, Cecilia M, Eriksson, Karl-Fredrik, Subramanian, Aravind, Sihag, Smita, et al. 2003. PGC-1alpha-responsive genes involved in oxidative phosphorylation are coordinately downregulated in human diabetes. *Nature genetics*, **34**(3), 267–273.

Prasad, T S Keshava, Goel, Renu, Kandasamy, Kumaran, Keerthikumar, Shivakumar, Kumar, Sameer, et al. 2009. Human Protein Reference Database – 2009 update. *Nucleic acids research*, **37**(Database issue), D767–D772.

Razick, Sabry, Magklaras, George, and Donaldson, Ian M. 2008. iRefIndex: a consolidated protein interaction database with provenance. *BMC bioinformatics*, **9**(Jan.), 405.

Subramanian, Aravind, Tamayo, Pablo, Mootha, Vamsi K, Mukherjee, Sayan, Ebert, Benjamin L, et al. 2005. Gene set enrichment analysis: a knowledge-based approach for interpreting genome-wide expression profiles. *Proceedings of the National Academy of Sciences of the United States of America*, **102**(43), 15545–15550.

The Cancer Genome Atlas Research Network. 2008. Comprehensive genomic characterization defines human glioblastoma genes and core pathways. *Nature*, **455**(7216), 1061–1068.

The Cancer Genome Atlas Research Network. 2011. Integrated genomic analyses of ovarian carcinoma. *Nature*, **474**(7353), 609–615.

The Cancer Genome Atlas Research Network. 2012a. Comprehensive genomic characterization of squamous cell lung cancers. *Nature*, **489**(7417), 519–525.

The Cancer Genome Atlas Research Network. 2012b. Comprehensive molecular characterization of human colon and rectal cancer. *Nature*, **487**(7407), 330–337.

The Cancer Genome Atlas Research Network. 2012c. Comprehensive molecular portraits of human breast tumours. *Nature*, **490**(7418), 61–70.

The Cancer Genome Atlas Research Network. 2013. Genomic and epigenomic landscapes of adult de novo acute myeloid leukemia. *New England journal of medicine*, **368**(22), 2059–2074.

Vandin, Fabio, Upfal, Eli, and Raphael, Benjamin J. 2011. Algorithms for detecting significantly mutated pathways in cancer. *Journal of computational biology*, **18**(3), 507–522.

Vandin, Fabio, Upfal, Eli, and Raphael, Benjamin J. 2012a. De novo discovery of mutated driver pathways in cancer. *Genome research*, **22**(July), 375–385.

Vandin, Fabio, Clay, Patrick, Upfal, E L I, and Raphael, Benjamin J. 2012b. Discovery of mutated subnetworks associated with clinical data in cancer. In: *Pacific Symposium on Biocomputing*.

Verhaak, Roel G W, Hoadley, Katherine A, Purdom, Elizabeth, Wang, Victoria, Qi, Yuan, et al. 2010. Integrated genomic analysis identifies clinically relevant subtypes of glioblastoma characterized by abnormalities in PDGFRA, IDH1, EGFR, and NF1. *Cancer cell*, **17**(1), 98–110.

Vogelstein, Bert, Papadopoulos, Nickolas, Velculescu, Victor E, Zhou, Shibin, Diaz, Luis A, and Kinzler, Kenneth W. 2013. Cancer genome landscapes. *Science (New York, N.Y.)*, **339**(6127), 1546–1558.

Wendl, Michael C, Wallis, John W, Lin, Ling, Kandoth, Cyriac, Mardis, Elaine R, Wilson, Richard K, and Ding, Li. 2011. PathScan: a tool for discerning mutational significance in groups of putative cancer genes. *Bioinformatics (Oxford, England)*, **27**(12), 1595–1602.

16

A Mass-Action-Based Model for Gene Expression Regulation in Dynamic Systems

GUOSHOU TEO, CHRISTINE VOGEL, DEBASHIS GHOSH,
SINAE KIM, AND HYUNGWON CHOI

Abstract

Although joint analysis of multiple omics data sets is often discussed in the context of analyzing large-scale genomic data in clinical or population studies, data integration is also useful for systems biology studies that investigate biological mechanisms in model systems under controlled environment. In this chapter, a model-based method is developed to simultaneously analyze time course transcriptomic and proteomic data sets to quantitatively dissect the contribution of RNA-level and protein-level regulation to the variation in gene expression. The statistical method is based on a mass-action-based model for protein synthesis and degradation rates of individual genes, and change points in the stochastic process of the kinetic parameters are derived to identify distinct patterns of regulation of gene expression in time course profiles. A sampling-based inference procedure using Markov chain Monte Carlo is implemented, and the posterior probabilities of change points in the ratio of protein synthesis and degradation are used to control the Bayesian false discovery rate. The method is illustrated using a yeast data set monitoring mRNA and protein expression in hyperosmolarity shock, where stress response functions are immediately invoked by up-regulation at the mRNA and protein levels and translational machinery is shut down in the early time points but reactivated later in time points at the protein levels.

16.1 Introduction

The process of RNA synthesis (transcription) is closely related with protein synthesis (translation) according to the central dogma of molecular biology (Crick, 1970). Considering gene expression as an array of biochemical processes to produce gene products, regulation of gene expression is a highly complex mechanism with multiple access points through transcriptional, posttranscriptional, translational, and posttranslational regulations. For instance, when cells

Adapted with permission from Teo et al., "PECA: a novel statistical tool for deconvoluting time-dependent gene expression regulation," *J. Proteome Res.*, 2014; 13(1):29–37. Copyright 2014 American Chemical Society.

encounter environmental stress, they are challenged to reprogram the transcriptome first (all messenger RNAs) to confer increased viability and fitness in the new environment and further adjust protein expression and additional posttranslational regulations (Causton et al., 2001). However, the dynamic relationship between the transcriptome and the proteome has remained elusive due to the lack of technology to measure protein expression at a scale comparable to gene expression, and it is of great interest to investigate how much of transcriptional and translational regulation determines the fate of the final gene products (Warringer et al., 2010; Garre et al., 2012).

To achieve this aim, proteome-wide expression data sets must be generated with sufficient coverage and quantitative precision, especially in a time-resolved manner. Thanks to recent advances in large-scale high-resolution mass spectrometry (MS), comprehensive quantitative proteomics data sets are now becoming available with longitudinal designs (e.g., following a treatment of interest) (Cox et al., 2005; Soufi et al., 2009; Fournier et al., 2010; Warringer et al., 2010; Schwanhausser et al., 2011; Lee et al., 2011; Garre et al., 2012). For example, a few recent studies used time course transcriptomic and proteomic data sets to monitor stress response in yeast and described the distinct roles of regulation at the level of RNAs and proteins where the variation in protein expression was only partially explained by transcription changes (Lee et al., 2011; Vogel et al., 2011). Although these results are intriguing, statistical analysis was limited to linear correlation or the analysis of variance in these studies, separately applied to RNA and protein data sets. In other words, there is no generalizable statistical method to jointly model the two data sets to objectively extract biological signals of regulation at different molecular levels.

With the emergence of these new data sets, the time is now ripe to develop robust statistical methods to identify candidate genes that are regulated at the RNA and/or protein levels and to quantitatively dissect the different layers of gene expression control. Because the final protein concentration is the combined result of these processes, the key task is to construct a mathematical model of gene regulation, equipped with appropriate kinetic parameters for transcription, translation, and the respective degradation. In this framework, the synthesis and degradation rates can be inferred from the data and formally tested for significant changes, providing statistically rigorous interpretation of the regulation activities that resulted in the observed concentration changes for each protein. In other words, we aim to convert expression data into information on the rates of concentration changes and regulation.

In this work, we propose a statistical modeling framework called Protein Expression Control Analysis (PECA) to identify genes putatively regulated at the RNA or protein levels based on parallel time course data sets of mRNAs and proteins. Adopting the kinetic mass-action model used in the simulation exercise by Lee et al. (2011), PECA dissects the change in the protein concentration during each time interval (i.e., the period between adjacent time points) into two potential sources: the change in the concentration of mRNA transcripts and the change in the protein synthesis/degradation rate ratios. This deconvolution renders the inferred protein rate ratios specific to the regulation at the protein level. As explained later, the same model can be posited to infer the RNA rate ratios to determine RNA-level regulation, under the reasonable assumption that the DNA copy numbers do not change over time. For both analyses, PECA derives the posterior probability that the rate ratio of synthesis versus degradation changed at each time point (before and after each time point), along with the associated false discovery rates (FDR) (Müller et al., 2006). Hence this scoring framework leads to an unbiased statistical framework of regulation changes at both molecular levels.

We remark that there are a few methods for analyzing time course data sets in the current statistics and bioinformatics literature (Storey et al., 2005; Park et al., 2003; Conesa et al., 2006; Tai and Speed, 2006). However, these methods are not suitable for the multi-omics data of our interest, especially for detecting regulation at the RNA and protein levels simultaneously. First, those methods are designed to analyze single-source data sets (e.g., transcriptomics data alone), and they do not explicitly model the kinetic parameters of synthesis and degradation. Second, they are not able to account for the contribution of mRNA concentration changes when analyzing protein-level regulation. Third, they perform statistical tests whether the expression has changed anywhere in the time course, not the temporal changes of regulatory parameters at specific time points and the direction of change, which is offered by PECA.

The rest of the chapter is organized as follows. We first present the statistical model and propose a straightforward estimation procedure using a Markov chain Monte Carlo sampler. We evaluate the performance of our approach with simulation studies and report the reanalysis of the yeast data by Lee et al. (2011).

16.2 Method

16.2.1 Change-Point Model for Gene Expression Regulation

Suppose that we have parallel gene and protein expression data $X = \{x_{jit}\}$ and $Y = \{y_{jit}\}$ for protein $i = 1, \ldots, I$ in replicates $j = 1, \ldots, N$ observed over time points $(h_0, \ldots, h_T)$. Time h_0 indicates the time point before the samples are treated or the baseline of subsequent time points. We assume that the protein expression measurements follow log normal distributions,

$$y_{jit} \sim \mathcal{LN}\left(\eta_{jit}, \tau_i^2\right)$$

after proper normalization of the data. Our goal is to infer the protein synthesis rate κ_{it}^s and the degradation rate κ_{it}^d during the interval (h_t, h_{t+1}) of length $\Delta h_t = (h_{t+1} - h_t)$ for protein i. More importantly, the mean parameters are related between adjacent time points, as follows:

$$\eta_{ji,t+1} = \eta_{jit} + \Delta h_t \left(x_{jit}\kappa_{it}^s - \eta_{jit}\kappa_{it}^d\right) \tag{16.1}$$

for $t = 0, 1, \ldots, T - 1$. At time t, the mRNA abundance is x_t and the current protein abundance y_t, and we would expect that the protein abundance will increase or decrease by $x_t\kappa_{it}^s - \eta_t\kappa_{it}^d$. This is based on the mass-action kinetic action model, which underpins the simulation model of Lee et al. (2011).

Equation (16.1) is a straightforward representation of time course profile of mean parameters as a simultaneous outcome of synthesis and degradation of each molecule. If the abundance of a protein is regulated by transcriptional regulation only, then we assume that the two parameters $\{(\kappa_{it}^s, \kappa_{it}^d)\}_{t=0}^{T-1}$ do not change over time. By contrast, if the protein is regulated by altering either the synthesis or the degradation rate, we assume that $\{(\kappa_{it}^s, \kappa_{it}^d)\}_{t=0}^{T-1}$ change over time. Translational regulation is a useful mechanism to react to sudden changes because transcriptional regulation of protein expression entails a lengthy chain of cellular processes, such as transport of mRNA from nucleus to ribosome, some biological functions require an immediate response via translation control of synthesis and degradation at the protein level (Sonenberg and Hinnebusch, 2009). Thus proteomic response to such an environment shock can be delivered by altering protein synthesis and degradation directly, rather than altering the concentration level of their precursors (mRNAs).

To detect the change in these rate parameters, we formulated a change point model to describe the probability distribution of $\kappa_i^s = (\kappa_{i0}^s, \ldots, \kappa_{i,T-1}^s)$, as follows. We first note that κ_{it}^s and κ_{it}^d are always positive because they are rate parameters by definition, and thus the issue of identifiability arises. This is expected because we model the change in protein expression as the difference of two positive values, where there can be an infinite number of solutions. Hence we impose the restriction $\kappa_{it}^d = 1 - \kappa_{it}^s$ for all i. This condition does not undermine the aim of this model because our interest is ultimately in the rate ratio $\kappa_{it}^s/\kappa_{it}^d$. Under this simplex constraint, it suffices to keep track of κ_{it}^s only. For protein i, let C_i and $|C_i|$ denote the set of time points $\{t : \kappa_{i,t-1}^s \neq \kappa_{it}^s | 0, 1, \ldots, T - 1\}$ and the size of the set, respectively. If the elements of κ_i^s remained constant across time, C_i is an empty set; if some elements of κ_i^s

were distinct from others, C_i is the set of all intermediate time points from 1 to $T - 1$ with different adjacent rates. Given a specific configuration of C_i, we can reparameterize this model by $\boldsymbol{\theta}_i = (C_i, \{(\kappa'_{it})\}_{t=0}^{|C_i|})$ where $\kappa'_{it} = \kappa^s_{it}/(\kappa^s_{it} + \kappa^d_{it})$, which further reduces to $\kappa'_{it} = \kappa^s_{it}$ under the simplex constraint. We remark that this change point model resembles the well-known model of Green (1995), but our model is simpler than his because change points can occur at the observed time points only. This is a reasonable choice since there are often a few time points in dynamic expression studies (often fewer than 10 time points), but the location of change points can be easily incorporated in the model for data sets with sufficiently dense time points.

16.3 Estimation and Inference

To estimate the model parameters, we construct a MCMC sampler that combines standard Metropolis-Hastings updates and dimension switching updates in the form of reversible-jump MCMC (Green, 1995).

First, the likelihood of the entire model is

$$(\text{likelihood}) = \prod_{i=1}^{I}\prod_{j=1}^{N}\prod_{t=0}^{T} \frac{1}{y_{jit}\tau_i\sqrt{2\pi}} \exp\left[-\frac{1}{2\tau_i^2}(\ln(y_{jit}) - \ln(\eta_{jit}))^2\right]$$

where

$$\eta_{jit} = \eta_{ji0} + \sum_{\ell=0}^{t-1} \Delta h_\ell \left(x_{ji\ell}\kappa'_{i\ell} - \eta_{ji\ell}(1 - \kappa'_{i\ell})\right)$$

We specify prior distributions that are the least informative in our view:

$$\eta_{ji0} \sim \mathcal{N}(0, 100^2) \quad \text{for} \quad j = 1, \ldots, N$$

$$\kappa'_{i\ell} \sim \mathcal{U}(0, 1) \quad \text{for} \quad \ell = 0, \ldots, |C_i|$$

$$\tau_i^{-2} \sim \mathcal{G}(a_\tau, b_\tau)$$

for fixed C_i for all i, where $\mathcal{N}, \mathcal{U}, \mathcal{G}$ denote normal, uniform, and gamma distributions, respectively. We also assume that the change point configuration C_i has the following prior:

$$\pi(C_i) \propto \varphi^{|C_i|}(1 - \varphi)^{T-1-|C_i|}$$

where we set $\varphi = 0.5$, assuming that nothing is known a priori about the chance of having a change point in any of the proteins.

To elicit the hyperprior parameters (a_τ, b_τ), we first calculate the sample variance of the protein intensities across all time points in each replicate and plug in the maximum likelihood estimates for the shape parameter a and scale parameter b:

$$(a_\tau, b_\tau) = \underset{a,b}{\arg\max} \left[\left(\frac{b^a}{\Gamma(a)} \right)^{N \times I} \left(\prod_{i,j} v_{ij} \right)^{-\alpha-1} \exp\left(-b \sum_{i,j} \frac{1}{v_{ij}} \right) \right]$$

where v_{ij} is the sample variance for protein i replicate j.

In summary, the prior can be written as

$$(\text{prior}) \propto \prod_{i=1}^{I} \left\{ \frac{b_\tau^{a_\tau}}{\Gamma(a_\tau)} (\tau_i^2)^{-a_\tau-1} e^{-\frac{b_\tau}{\tau_i^2}} \cdot \prod_j \phi\left(\frac{\eta_{ji0}}{100}\right) \cdot \varphi^{|C_i|}(1-\varphi)^{T-1-|C_i|} \right\}$$

where the prior for $\{\kappa'_{it}\}$ is omitted conditional on the fact that they are all on the unit interval, and ϕ denotes standard normal density.

The model parameters are updated in the following order:

$$\{\eta_{ji0}\}_{j=1}^{N} \to \tau_i^2 \to \{\kappa'_{it}\}_{t=0}^{T-1} \to C_i$$

for all i. This whole cycle is repeated for 5000 iterations for the burn-in period and $M = 20,000$ iterations for the main iteration with thinning of 20 samples, in both the simulation and data analysis sections that follow. We use hat and tilde symbols to denote current and proposal values, respectively.

1. We first start with η_{ji0} by a Metropolis-Hastings step, with proposal value $\tilde{\eta}_{ji0}$ drawn from $\mathcal{N}(\hat{\eta}_{ji0}, 0.1^2)$, and compute the Metropolis-Hastings ratio to complete the update. Because this parameter is involved in the mean values at all time points, the likelihood has to be evaluated at all time points for updating each of these parameters.
2. Next, we draw the variance parameter τ_i^2 by Gibbs sampling from inverse gamma distribution $\mathcal{IG}(a_\tau + N(T+1)/2, b_\tau + \sum_{j,t}(y_{jit} - \eta_{jit})^2/2)$.
3. Next, we draw $\{\kappa'_{i\ell}\}$ for $\ell = 0, \ldots, |C_i|$ under the fixed C_i for each protein i. We use random walk Metropolis-Hastings steps to update them, that is, we draw a proposal value $\tilde{\kappa}'_{i\ell}$ from $\mathcal{N}(\hat{\kappa}'_{i\ell}, 0.1^2)$ and accept or reject afterward.
4. Finally, we update the change point configuration C_i. There are two different moves: birth of a new change point and removal (death) of an existing change point. Because these two moves are reversible in notation, we just describe the birth move here. Suppose that $\hat{\kappa}'_{i\ell}$ covers a time period (h_t, h_{t+m}) that contains at least one observation time(s). Then we propose a birth of a new change point $h^* \in \{h_{t+1}, \ldots, h_{t+m-1}\}$ within the interval (chosen from one of the intermediate time points) and break the current rate parameter into two daughter parameters, namely, $(\tilde{\kappa}'_{i\ell}, \tilde{\kappa}'_{i,\ell+1})$, where it is required to meet
$$(h^* - h_t) \cdot \text{logit}(\tilde{\kappa}'_{i\ell}) + (h_{t+m} - h^*) \cdot \text{logit}(\tilde{\kappa}'_{i,\ell+1}) = (h_{t+m} - h_t) \cdot \text{logit}(\hat{\kappa}'_{i\ell})$$

with a random perturbation such that

$$\frac{\tilde{\kappa}'_{i,\ell+1}}{1 - \tilde{\kappa}'_{i,\ell+1}} = \frac{1-u}{u} \frac{\tilde{\kappa}'_{i\ell}}{1 - \tilde{\kappa}'_{i\ell}}$$

with $u \sim \text{Uniform}(0, 1)$. Under this transformation, the Jacobian is $\frac{(\tilde{\kappa}'_{i\ell}(1-\tilde{\kappa}'_{i\ell})+\tilde{\kappa}'_{i,\ell+1}(1-\tilde{\kappa}'_{i,\ell+1}))^2}{\hat{\kappa}'_{i\ell}(1-\hat{\kappa}'_{i\ell})}$ for $(\hat{\kappa}'_{i\ell}, u) \to (\tilde{\kappa}'_{i\ell}, \tilde{\kappa}'_{i,\ell+1})$. Hence the Metropolis-Hastings ratio for the birth move just equals the posterior ratio times the Jacobian as the acceptance probability of this proposal is

$$\min\{1, \text{likelihood ratio} \times \text{prior ratio} \times \text{proposal ratio} \times \text{Jacobian}\}$$

where the prior and proposal ratios are the ratios of Uniform distribution over unit intervals. Then the Metropolis-Hastings ratio becomes

$$\prod_{j,t} \left[\exp\left\{ -\frac{1}{2\tau_i^2}(\ln(y_{jit}) - \ln(\eta_{jit}))^2 \right\} \right] \frac{\varphi}{1-\varphi}$$
$$\times \frac{(\tilde{\kappa}'_{i\ell}(1 - \tilde{\kappa}'_{i\ell}) + \tilde{\kappa}'_{i,\ell+1}(1 - \tilde{\kappa}'_{i,\ell+1}))^2}{\hat{\kappa}'_{i\ell}(1 - \hat{\kappa}'_{i\ell})}$$

Using the samples drawn from the posterior distributions, we perform statistical inference as follows. Our main goal is to identify the time points where the protein rate ratio shifts, that is, $p_{it} = P(\kappa^s_{it} \neq \kappa^s_{i,t+1} | X_i, Y_i)$, where X_i and Y_i denote the gene and protein expression data for protein i, respectively. This score has the nice property that it is a marginal probability computed after accounting for the data and change point configurations at all time points. Instead of seeking the *maximum a posteriori estimate* of C_i, we perform our inference based on this probability. Denote the posterior samples of $\{\kappa'_{it}\}$ by $r_{it}^{(1)}, \ldots, r_{it}^{(M)}$ for each κ'_{it}. We first compute p_{it} by $\hat{p}_{it} = \frac{1}{M} \sum_{m=1}^{M} \mathbf{1}\{r_{it}^{(m)} \neq r_{i,t+1}^{(m)}\}$. If $\hat{p}_{it} \geq p^*$ holds for at least one t, where p^* is the probability threshold, we consider protein i to be translationally regulated. To determine an optimal threshold, we compute the Bayesian false discovery rate (BFDR) as

$$BFDR(p^*) = \frac{\sum_{i,t}(1 - \hat{p}_{it})\delta_{it}(p^*)}{\sum_{i,t} \delta_{it}(p^*)} \tag{16.2}$$

where $\delta_{it}(p^*) = 1\{\hat{p}_{it} \geq p^*\}$ (Genovese and Wasserman, 2003; Müller et al., 2006). This decision rule $\delta_{it}(\cdot)$ results in the selection of specific time points where translation regulation shifts from the preceding time period. Furthermore, we can perform functional clustering (Bar-Joseph et al., 2012) using these surrogate data $\{\kappa'_{it}\}$ instead of the raw expression data Y by the agglomerative hierarchical clustering (Eisen et al., 1998) with the Euclidean distance

Table 16.1 *Mean parameters of gene expression data in the three groups*

Group	Size	μ_0	μ_1	μ_2	μ_3	μ_4	μ_5
1	500	1.00	1.25	1.20	1.10	1.00	1.00
2	500	1.00	1.25	1.20	1.10	1.10	1.10
3	500	1.00	0.75	0.80	0.90	0.90	0.90

metric on the matrix data $\{\kappa'_{it}\}$ for the selected proteins, ultimately identifying different groups of proteins with a similar translational regulation pattern.

16.4 Simulation Study

We first conducted simulation studies to evaluate the sensitivity and specificity of the method. We simulated expression data sets for K transcripts (mRNAs) and proteins in parallel in single biological replicate across six different time points. Among these, we created three groups that are different in terms of the translational control mechanism, which emulated the protein expression profiles of up- and down-regulated proteins in Lee et al. (2011). Specifically, each of the three groups represents a different combination of transcriptional and translational regulation. Protein expression in group 1 is regulated entirely by transcriptional regulation (gene up-regulation). Protein expression in group 2 is translationally up-regulated by an increased rate of translation during the first time period in addition to the transcriptional up-regulation. This pattern is expected to occur in immediate shock conditions when direct translational regulation is required. Finally, group 3 represents the case of down-regulation in both data, where down-regulation of protein expression was driven by increased degradation rates in the late time points as well as transcriptional down-regulation in the early time points.

Here we describe the data generation process in detail. We simulated gene expression data reflecting the burst of up- and down-regulation of mRNAs between the first two time points, from log normal distribution with their respective mean parameters in each group as specified in Table 16.1 along the time course, and the variance parameters fixed at $\sigma^2 = 0.1$. To simulate protein expression data according to the turnover mechanism, we set the translation and degradation rates (κ^s, κ^d) as tabulated in Table 16.2, in which the protein synthesis rate changes by a factor of r^*. We fixed κ^d at 1, and thus r^* essentially represents the "scaled" rate ratio. This leads to the time-dependent mean expression values following the relationship in Eq. (16.1). Using these means,

Table 16.2 *Protein synthesis rates in protein
expression data in the three groups with fixed
degradation rate* $\{\kappa_t^d\} = 1$ *at all time points*

Group	κ_0^s	κ_1^s	κ_2^s	κ_3^s	κ_4^s
1	1.00	1.00	1.00	1.00	1.00
2	r^*	1.00	1.00	1.00	1.00
3	1.00	1.00	r^{*-1}	r^{*-2}	r^{*-2}

Note: Essentially r^* plays the role of protein rate ratio.

we simulated protein expression data from log normal distribution, where different variance parameters τ^2 were attempted to control the signal-to-noise ratio. Based on Eq. (16.1), the ratio e^τ / r^* can be interpreted as a variant of the coefficient of variation (CV), provided that the gene and protein expression data are properly scaled. We have evaluated the performance at different CVs, where r^* ranged from 1.5 to 2.0 and τ^2 ranged from 0.01 to 0.2. In each scenario, we looked at three different probability thresholds $p^* = 0.5, 0.6, 0.7$.

Figure 16.1 shows the results. The sensitivity, specificity, and $BFDR$ estimates in the figure were computed by averaging the results over 100 simulations of each setting. The MCMC sampler converged quickly to the posterior distribution, as illustrated in Figure 16.2 (left). First, any detection in group 1 represents false positives. For group 2, r^* increased sharply during the first time period (h_0, h_1) and thus the second time point h_1 is the true change point. Likewise for group 3, r^* decreased from unit rate twice at h_2 and h_3. Hence any detection at these time points at groups 2 and 3 represents true positives. To see the range of the CVs we cover in the simulation, consider the worst case scenario with $r^* = 1.5$ and $\tau = 0.2$. This means that the protein rate ratio increases by 50%, yet the standard deviation of the error is at about 22% ($e^{0.2} \approx 1.22$). In this case, the level of translational regulation signal will be masked by the noise. By contrast, in the scenario with $r^* = 2.0$ and $\tau = 0.01$, the protein rate ratio increases by 100% and the noise is ignorable (1%).

As expected, the proposed model performed very well in the scenarios with low CV ($\tau = 0.1$ or below), achieving almost perfect specificity (>97%) and good sensitivity (>80%) with increasing r^*. Interestingly, the sensitivity for down-regulation in group 3 at h_3 (not shown) was very low compared to the sensitivity at h_2, even though the rate ratio went down by the same factor of r^*. This is possibly because the gene expression level increased at h_3 from 0.8 to 0.9 in the simulation scheme, compensating the decrease in protein turnover. Finally, the estimated $BFDR$ was not trivially small at all three thresholds,

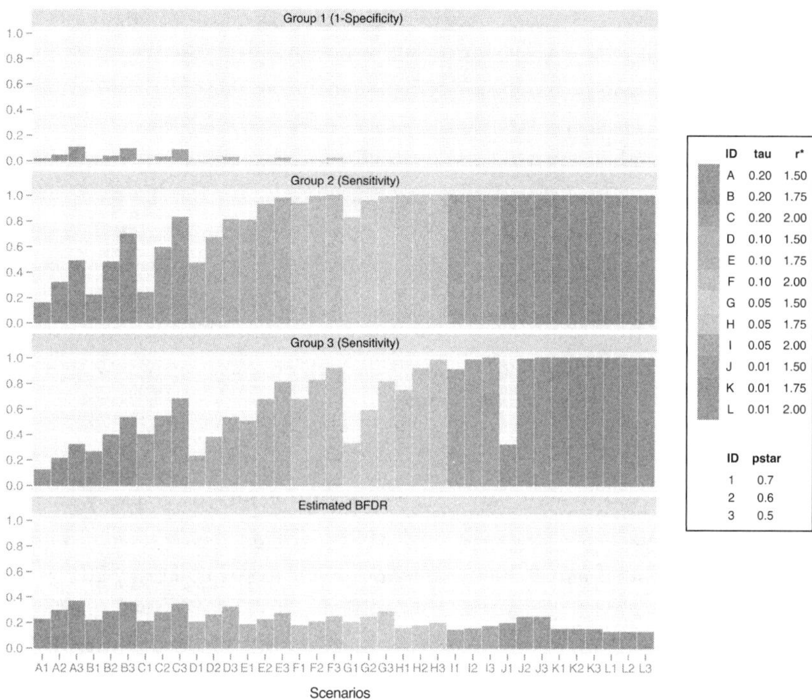

Figure 16.1   Simulation results. Proteins selected from group 1 across all time points are false positives. Proteins selected from group 2 at h_1 and group 3 at (h_2, h_3) are true positives.

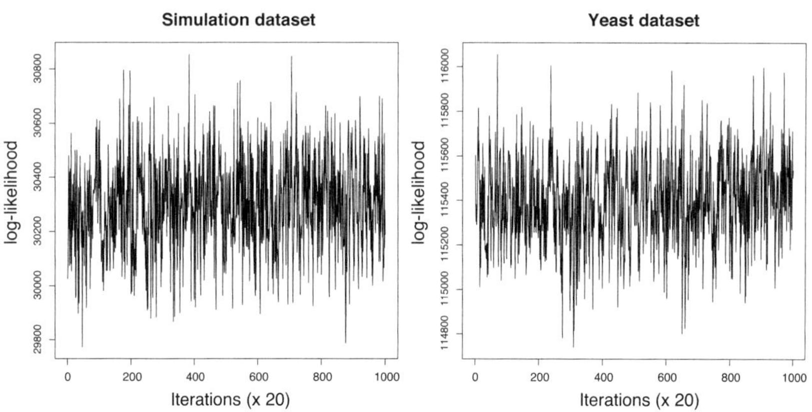

Figure 16.2   The log-likelihood trajectory of the model shows that the parameter values were drawn from the appropriate posterior distributions in the simulation data and the yeast data set.

ranging from 12% to 36% across the scenarios (last panel of Figure 16.1). However, because very few false positives were detected in group 1 in the modest signal-to-noise ratio settings (Figure 16.1, top), these estimates can be considered to be conservative.

Overall, our method showed good sensitivity and specificity for the scenarios with modest signals at all probability score thresholds. The result also suggests that the optimal threshold can be set as low as (~ 0.6) in the scenarios with a high signal-to-noise ratio, even if the associated $BFDR$ estimates may be greater than conventional FDR targets such as 5%. However, in the scenario where large variation in the mRNA abundance coexists with protein expression changes, a selection criterion that controls $BFDR$ reasonably low will be desired. We illustrate such a case in the next section.

16.5 Application: Analysis of Osmotic Shock in Yeast

Next we reanalyzed the yeast data set which profiled the cellular response to an osmotic shock using three biological replicates (Lee et al., 2011). In the experiment, 0.7M NaCl was applied to budding yeast in growth medium, where the dose of salt provides a robust physiological response but results in high viability and eventual resumption of cell growth. Samples were collected before and at 30, 60, 90, 120, and 240 minutes after treatment to capture cells acclimated to the new environment, and the samples were divided for gene and protein expression measurements using microarray and quantitative mass spectrometry respectively. For data analysis, the authors performed modified t-test to select proteins that are differentially expressed between, before, and after the treatment and used mRNA expression as a post hoc analysis to show correlated changes therein. In our analysis, we first analyzed the data for protein-level regulation inference, treating microarray data as X and mass spectrometry data as Y, as described in the methods. We also analyzed the data for RNA-level regulation, using microarray data as Y and a fictitious DNA copy number data set filled with constant element 1 as X (0 on log scale). This represents the assumption that the DNA copy number remains constant along the time course in the genome, which is a realistic assumption in normal cell populations.

Gene expression was measured using a custom Nimblegen tiled microarray platform (Gene Expression Omnibus GSE23798). Instead of quantile normalization the authors used, the data were further examined for systematic shift in expression level distribution across different samples, but it was deemed that no further normalization was necessary because such normalization may remove real signals due to the burst in transcriptional regulation upon osmotic shock (Loven et al., 2012). Protein expression was measured using an isobaric tagging and liquid chromatography and mass spectrometry on an Orbitrap Velos instrument. The mass spectrometry analysis was performed simultaneously for the samples of each biological replicate taken at different time points. This batch analysis is beneficial in time course designs because a multiplexed design can control the within-sample variation better than other designs, such as label-free protein quantitation. From this experiment, 1999 proteins were quantified consistently across the time points in at least two replicates, and among those we analyzed 1508 proteins with no missing data across all three replicate samples.

For model parameter estimation, we ran the MCMC for 20,000 iterations with thinning (every 10th sample) after 5000 iterations for the burn-in period. We elicited the same prior distributions used in the simulation studies, and the acceptance rates for the Metropolis-Hastings updates (13%) and reversible jump MCMC (21%) remained reasonably good (before thinning of the chain). We performed visual inspection of model fit by plotting the estimated level of protein expression $\{\eta_{jit}\}$ against the observed values and found that the fit was reasonably good. We also confirmed the convergence of the MCMC sampler to the posterior distribution by the trace plot of the log likelihood (Figure 16.2).

16.6 Scoring Protein-Level Regulation Changes

Using the preceding output, we extracted candidate genes subject to protein-level regulation. We selected 249 out of the 1508 proteins with the posterior probability >0.8 at any of the time points as proteins putatively regulated at the protein level, controlling the FDR at 10%. To see gene function enrichment in the proteins regulated at the protein level, we selected two clusters of 68 and 131 proteins that were up- and down-regulated respectively at 30 minutes. All 1508 proteins in the protein/RNA data set served as the background list for hypergeometric test. Similar to the transcriptome analysis, up-regulated proteins showed enrichment of stress-related functions ($p < 0.001$). By contrast, down-regulated proteins showed enrichment of the terms related to RNA processing and regulation of translation, indicating immediate shutdown of translation activities under high osmolarity ($p < 0.001$).

Similar to the RNA data, we found that the major change in the protein-level rate ratios also occurred immediately after the treatment (0~30 min). In addition, most proteins regulated at the protein level (161/199) were also significantly regulated at the RNA level, implying that the regulation of gene expression during osmotic stress response was highly coordinated at both levels, particularly at early time points. However, as only 249 genes (17%) were regulated at the protein level while 722 genes (49%) were at the RNA level at much more stringent FDR, one may hypothesize that transcriptional reprogramming is the dominant response to osmotic stress, and ultimately protein concentrations change only by carefully selected paths of protein-level regulation.

16.7 Characterizing the Link Between the Regulatory Processes

Next, we inspected the correlation between the regulatory patterns using the rate ratio profiles within the same molecules. Figure 16.3 shows the RNA and protein concentrations, and the rate ratios for four key proteins known to be up-regulated during osmotic stress response (Rep et al., 1999): glycerolphosphate dehydrogenase GPD1, cytosolic catalase T CTT1, and heat shock proteins HSP12 and HSP104. Time intervals where the rate ratios changed significantly were indicated by yellow rectangles – illustrating that the RNA level up-regulation was most active during the first 30 minutes and subsided afterward, with mRNA concentration recovering the stability within 60 minutes. By contrast, protein-level regulation was also the most active during the first time interval, but it counterbalanced the RNA-level regulation in the opposing direction (down) during the next time interval, resulting in stabilized protein-level concentrations. This pattern suggests that protein-level regulation buffered the abrupt change at the RNA level and contributed to the stable protein concentration levels.

The possible role of buffering by protein-level regulation was even more pronounced for down-regulated mRNAs, consistent with Lee et al.'s (2011) observation of less correlation between mRNA and protein concentrations for down-regulated RNAs. For example, PECA provides strong evidence of protein-level regulation that resulted in stable protein concentration for four members of the large subunit of ribosome (RPL9A, RPL9B, RPL16A, RPL19A) and several subunits of RNA polymerase I and III subunits (RPA43, RPA49, RPC19, RPC53, and RPC82). In these examples, mRNA concentration decreased significantly at 30 minutes and recovered to the pretreatment level at 60 minutes, whereas protein concentrations hardly changed. The rate ratio profiles reported by PECA showed that there was substantial protein-level up-regulation between 30 minutes and 60 minutes to fend off the effect of reduced mRNAs during the same time interval.

In sum, RNA-level and protein-level regulation were orchestrated together in the early response in this data, but the protein-level regulation clearly acted as the buffer to the vast transcriptome changes in this data set. Figure 16.4 shows the correlation patterns between RNA-level regulation and protein-level regulation across the time points. Because most regulation activities occurred in the

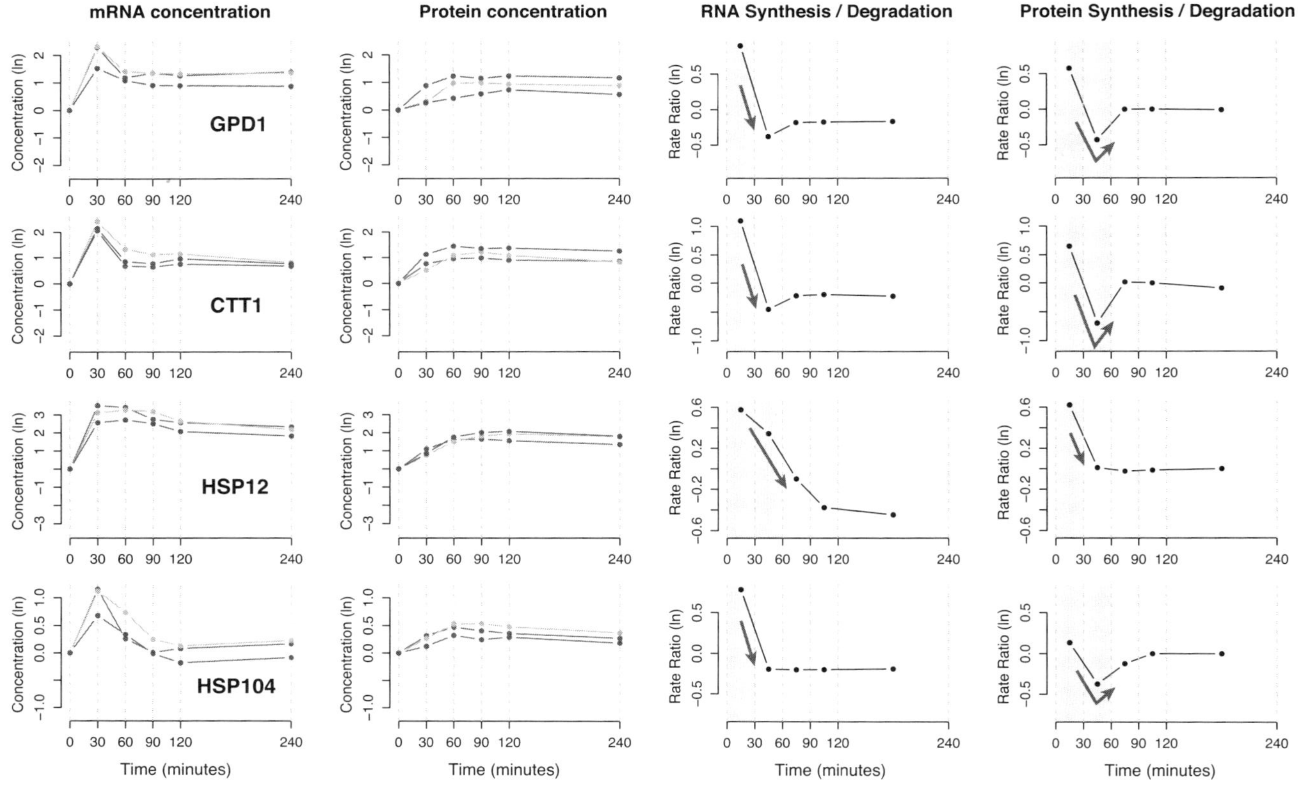

Figure 16.3   The mRNA and protein concentration data and estimated rate ratios at both levels of regulation for GPD1, CTT1, HSP12, and HSP104. These are four proteins with osmotic shock-induced expression. Blue, red, and green curves are time course data for each biological replicate. Yellow background indicates the time intervals during which the rate ratios deviated from the average range across the time course. Red arrows indicate significant regulation change at the RNA and protein level in each gene.

375

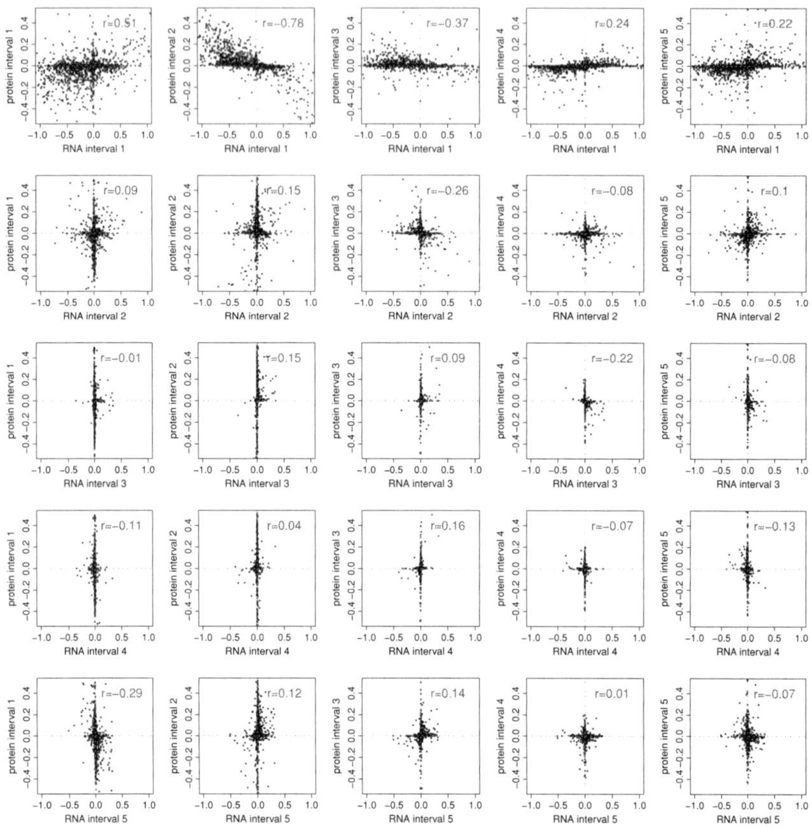

Figure 16.4 *S. cerevisiae* data with osmotic stress. The panels were arranged so that each row and column corresponds to each time point respectively. In each panel, the protein rate ratios were plotted against the RNA rate ratios (transformed by log base 2, then centered by median in each protein). The panels on diagonal positions show coupling at the same time point, whereas the panels on off-diagonal positions show buffering at different time points or time-delayed correlation.

first time interval in this data set, we focus on the first row of the figure, that is, on correlation between protein-level regulation with the RNA-level regulation during the first time interval. The top left panel shows that RNA and protein expression were consistently up- or down-regulated in many genes during the first time interval with positive correlation ($r = 0.51$). The negative correlations in the next two panels clearly illustrate that the RNA-level regulation of the first time interval was countered by protein-level regulation of the opposite direction of the second and third time intervals ($r = -0.78, -0.37$). In those intervals, the majority of the buffering effect was for the RNA-level down-regulation (countered by protein-level up-regulation), suggesting proteome-wide evidence of proteostasis through protein-level regulation. Interestingly, the positive correlations in the remaining two panels with large time lags (last two in the first row) suggest that the effect of RNA-level down-regulation takes a long time to come through at the protein concentration.

16.8 Conclusion

In this work, we proposed a statistical method to describe the patterns of gene expression regulation with respect to mass action kinetics of individual genes. Our method carries out probabilistic inference for the essential kinetic parameters of synthesis and degradation using time course data, extracting the proteins with statistically significant evidence of translational regulation. While the method has been demonstrated using paired gene and protein expression data, the kinetic relationship is also applicable to other types of paired data, such as DNA copy number and gene expression data, as illustrated in the analysis of osmotic shock, where one type of molecule is the precursor of the other by the central dogma. In this framework, we identify the signals of expression regulation in terms of synthesis/degradation rates instead of mean expression values, which provides biologically more interpretable results in temporal expression data.

We formulated the change point model to perform probabilistic inference and appropriate control of FDR. As illustrated in both simulation and yeast data analysis, the MCMC sampling procedure is straightforward and efficient with good mixing rates, and it showed swift convergence to the stationary distribution after starting from arbitrary initial points. In the simulation study, we showed that the method is able to detect protein-level regulation activities in scenarios with reasonably modest signal-to-noise ratios. We also validated this methodology using a yeast data set where cells were challenged to adapt to a sudden increase in osmolarity. Our method recovered a profile of translational regulation in a highly variable system where excessive gene expression changes, occurred yet not all of them led to protein expression changes, as expected.

A few components in the statistical model need further improvement. First, the model specification includes the constraint $k_t^d + k_t^s = 1$ for all t, which was introduced to address the identifiability problem. In the absence of this condition, both parameters must be estimated independently at each time point, which has no unique numerical solution, as explained earlier. The imposed condition mirrors the assumption that the total regulation activity (synthesis and degradation) adds up to a constant at all time points, which has to be modified if estimation of absolute rates of synthesis and degradation is of interest. Second, the prior for change points $\pi(C_i) \propto \varphi^{|C_i|}(1 - \varphi)^{T-1-|C_i|}$ with $\varphi = 0.5$ reflects the assumption that any change point arrangement with the same number of total change points has the same prior probability. This specification can deviate from biological reality in dynamic systems during perturbations, in which expression changes are induced in the early response more often than in the late response. Such prior information can be extracted from the data itself via an empirical Bayes approach, or careful elicitation on φ can also be an alternative remedy. We leave these aspects for future investigation.

References

Bar-Joseph, Z., Gitter, A., and Simon, I. 2012. Studying and modelling dynamic biological processes using time-series gene expression data. *Nat. Rev. Genet.*, **13**, 552–564.

Causton, H.C., Ren, B., Koh, S.S., Harbison, C.T., Kanin, E., Jennings, E.G., Lee, T.I., True, H.L., Lander, E.S., and Young, R.A. 2001. Remodeling of yeast genome expression in response to environmental changes. *Mol. Biol. Cell*, **271**(23), 323–337.

Conesa, A., Nueda, M. J., Ferrer, A., and Talon, M. 2006. maSigPro: a method to identify significantly differential expression profiles in time-course microarray experiments. *Bioinformatics*, **22**(**9**), 1096–1102.

Cox, B., Kislinger, T., and Emili, A. 2005. Integrating gene and protein expression data: pattern analysis and profile mining. *Methods (San Diego, Calif.)*, **35**(3), 303–314.

Crick, F. 1970. Central dogma of molecular biology. *Nature*, **227**(5258), 561–563.

Eisen, M.B., Spellman, P.T., Brown, P.O., and Botstein, D. 1998. Cluster analysis and display of genome-wide expression patterns. *Proc. Natl. Acad. Sci. USA*, **95**, 14863–14868.

Fournier, M.L., Paulson, A., Pavelka, N., Mosley, A.L., Gaudenz, K., Bradford, W.D., Glynn, E., et al. 2010. Delayed correlation of mRNA and protein expression in rapamycin-treated cells and a role for Ggc1 in cellular sensitivity to rapamycin. *Mol. Cell. Proteomics*, **9**, 271–284.

Garre, E., Romero-Santacreu, L., De Clercq, N., Blasco-Angulo, N., Sunnerhagen, P., and Alepuz, P. 2012. Yeast mRNA cap-binding protein Cbc1/Sto1 is necessary for the rapid reprogramming of translation after hyperosmotic shock. *Mol. Biol. Cell*, **23**(1), 137–150.

Genovese, C., and Wasserman, L. 2003. Bayesian and frequentist multiple testing. Pages 145–161 of: Bernardo, J.M., Berger, J.O., Bayarri, M., and Dawid, A.P. (eds.), *Bayesian Statistics 7*. Oxford: Oxford University Press.

Green, P.J. 1995. Reversible jump Markov chain Monte Carlo computation and Bayesian model determination. *Biometrika*, **82**, 711–732.

Lee, M., Topper, S., Hubler, S., Hose, J., Wenger, C., Coon, J., and Gasch, A. 2011. A dynamic model of proteome changes reveals new roles for transcript alteration in yeast. *Mol. Syst. Biol.*, **7**(1), 514.

Loven, J., Orlando, D.A., Sigova, A.A., Lin, C.Y., Rahl, P.B., Burge, C.B., Levens, D.L., Lee, T.I., and Young, R.A. 2012. Revisiting global gene expression analysis. *Cell*, **151**(3), 476–482.

Müller, P., Parmigiani, G., and Rice, K. 2006. FDR and Bayesian multiple comparison rules. *Johns Hopkins University Department of Biostatistics Working Paper*, 115.

Park, T., Yi, S.G., Lee, S., Lee, S.Y., Yoo, D.H., Ahn, J.I., and Lee, Y.S. 2003. Statistical tests for identifying differentially expressed genes in time-course microarray experiments. *Bioinformatics*, **19(6)**, 694–703.

Rep, M., Reiser, V., Gartner, U., Thevelein, J.M., Hohmann, S., Ammerer, G., and Ruis, H. 1999. Osmotic stress-induced gene expression in Saccharomyces cerevisiae requires Msn1p and the novel nuclear factor Hot1p. *Mole. Cell. Biol.*, **19**, 5474–5485.

Schwanhausser, B., Busse, D., Li, N., Dittmar, G., Schuchhardt, J., Wolf, J., Chen, W., and Selbach, M. 2011. Global quantification of mammalian gene expression control. *Nature*, **473**, 337–342.

Sonenberg, N., and Hinnebusch, A.G. 2009. Regulation of translation initiation in eukaryotes: mechanisms and biological targets. *Cell*, **136(4)**, 731–745.

Soufi, B., Kelstrup, C.D., Stoehr, G., Frohlich, F., Walther, T.C., and Olsen, J.V. 2009. Global analysis of the yeast osmotic stress response by quantitative proteomics. *Mole. Biosyst.*, **5**, 1337–1346.

Storey, J.D., Xiao, W., Leek, J.T., Tompkins, R.G., and Davis, R.W. 2005. Significance analysis of time course microarray experiments. *Proc. Natl. Acad. Sci. U.S.A.*, **102(36)**, 12837–12842.

Tai, Y.C., and Speed, T.P. 2006. Statistical tests for identifying differentially expressed genes in time-course microarray experiments. *Ann. Stat.*, **34(5)**, 2387–2412.

Vogel, C., Silva, G.M., and Marcotte, E.M. 2011. Protein expression regulation under oxidative stress. *Mole. Cell. Proteomics*, **10(12)**, M111.009217.

Warringer, J., Hult, M., Regot, S., Posas, F., and Sunnerhagen, P. 2010. The HOG pathway dictates the short-term translational response after hyperosmotic shock. *Mol. Biol. Cell*, **21**, 3080–3092.

17

From Transcription Factor Binding and Histone Modification to Gene Expression: Integrative Quantitative Models

CHAO CHENG

Abstract

Gene expression is precisely regulated by transcription factors and histone modifications. Most previous studies have focused on the regulation of specific genes by TFs or histone modifications. In recent years, genome-wide TF binding and histone modification data have been produced by high-throughput sequencing-based technologies such as ChIP-seq. With these data, we are able to construct statistical models to quantify the relationship between TF binding, histone modification, and gene expression. In this chapter, we first propose a statistical framework that integrates TF binding and/or histone modification signals to predict gene expression levels. We subsequently apply it to address several biological questions with the intention of providing new insight into the regulatory mechanisms of TFs and histone modifications. We show that more than 50% of variation in gene expression levels can be explained by TF binding or histone modification signals in the promoter region of genes. The TF binding and histone modification signals are highly redundant in terms of gene expression prediction. We also demonstrate that the TF model and histone model trained on protein-coding genes can accurately predict the expression levels of noncoding genes, suggesting a similar regulatory mechanism shared between the two gene classes. Moreover, the differential expression of genes under different conditions is largely reflected by their TF binding and histone modification changes. Altogether, the predictive models suggest that TF binding and histone modifications are highly interrelated, and both are involved in gene expression regulation in a cooperative manner.

17.1 Introduction

Gene expression is under precise regulation to ensure that genes are expressed at the right time and in the right tissues. Eukaryotic gene expression at the transcriptional level is mainly regulated by transcription factors (TFs) [1, 2] and histone modifications (HMs) [3, 4]. TFs are a family of proteins that bind to specific DNA sequences to activate or repress the transcriptional initiation of genes [5]. They can also affect gene expression by recruiting chromatin modifiers to induce chromatin structure changes [6, 7]. Eukaryotic chromosomes are

380

organized into chains of nucleosomes, which are composed of DNA wrapped around octamers of core histone proteins. Covalent modification of histone proteins can remodel local chromatin structure and thus affect the accessibility of TFs to DNA regions [1]. In addition, histone modifications can also participate in gene transcription by recruiting or interacting with transcriptional activators or repressors [3].

The majority of previous studies investigated the mechanisms of individual transcription factors or histone modifications in gene expression regulation under different biological themes. Owing to the wide application of next-generation sequencing techniques, a large number of TF binding, histone modification, and gene expression data have been produced in matched samples. For instance, the ENCODE project has published ChIP-seq data for $\sim$50 TFs, along with a handful of common histone marks and gene expression data for the human cell line K562 [8]. This provides us with an unprecedented opportunity to investigate the relationship between gene expression and TF binding/histone modification signals in a systematic and quantitative manner. In fact, quantitative models have been proposed to address this problem in a number of previous studies [9–16]. Specifically, these models aim to answer the following questions about gene transcription: (1) What percentage of variation of gene expression levels can be explained by TF binding and histone modification, respectively? (2) Do TF binding and histone modification provide complementary information for gene expression prediction? (3) At what genomic positions are TF binding and histone modification signals more informative for gene expression prediction? (4) How much of the differential expression of genes under different conditions or tissues can be determined by differential TF binding and/or histone modifications? (5) Can the expression of noncoding genes also be predicted by TF binding or histone modifications?

In this chapter, we first introduce the statistical framework and the computational methods used by quantitative models to integrate gene expression data with TF binding/histone modification data. Then, we describe several applications of the quantitative models and the resulting conclusions. Most of these analyses are based on the data sets from human K562 and GM12878 cell lines generated by the ENCODE project [8], worm early embryo development stage produced by the modENCODE project [17], and mouse embryonic stem cells (ESCs) collected from several pertinent publications.

17.2 Methods

We construct supervised learning models to integrate gene expression, TF binding, and histone modification data. In these models, the TF binding/histone

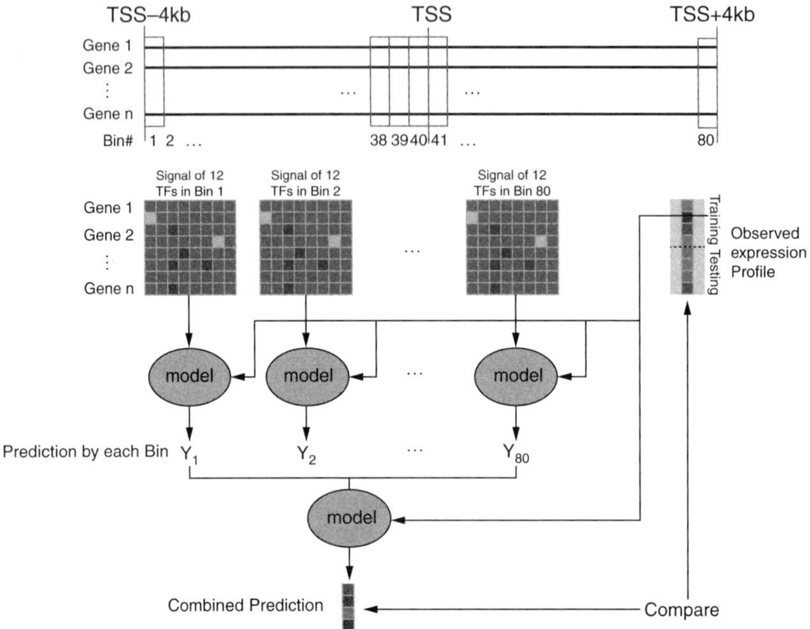

Figure 17.1 The framework for predicting gene expression levels from TF binding signals. DNA regions around transcription start sites of genes are divided into small bins of 100 bps in size. In each bin, a model is constructed to predict gene expression values based on the TF binding signals for a number of TFs. Then the predicted values for all bins are combined in a second-layer model to output the final expression prediction. The same framework can also be applied to histone modification data.

modification signals are taken as the predictor variables and gene expression level is taken as the response variable. The goodness of fit of the models indicates the predictive capacity of TF binding/histone modification signals in predicting gene expression levels.

17.2.1 A Framework for Integrating Transcription Factor Binding and Histone Modification Data with Gene Expression Data

To investigate the spatial effect of TF binding or histone modifications on gene expression, we proposed a statistical framework, as shown in Figure 17.1. First, we separated the DNA regions around the transcription start site (TSS) of genes (from −4kb upstream to 4kb downstream) into small bins, each of 100 bps in size. For each bin, we calculate the TF binding or histone modification signal as the mean coverage (the number of reads covering a position) of the 100 nucleotides. Second, in each bin we constructed a supervised learning model using the binding signals of a set of TFs (TF model) or the signals of a set of histone modifications (HM model) to predict gene expression values. Finally, the predicted values from all bins were combined with a second-layer model to output the final gene expression prediction [18].

Instead of dividing DNA regions into small bins, other strategies have also been utilized. Karlic et al. calculated the sum of tag counts for histone modifications in the 4001 bp regions surrounding the TSS and used their log-transformed values as the predictors in their linear regression models [10]. Ouyang et al. calculated the weighted sum of intensities of all of the binding peaks of a TF in a relatively broad DNA region (e.g., 5 MB) surrounding the TSS of a gene to represent the association strength of the TF to the gene [9]. Only TF signals within binding peaks (identified by a peak calling algorithm) were considered, and their contribution was weighted according to their distance to the TSS of a gene. The TF binding peak at the TSS was assigned to have the highest weight, and the weight decayed exponentially as the distance increased.

17.2.2 *Machine Learning Methods Used in the Predictive Models*

A number of supervised learning methods have been applied to model the quantitative relationship between gene expression and TF binding and/or histone modification signals. These methods include multivariate linear regression (MLR) [19], support vector regression (SVR) [20], random forest (RF) [21], and multivariate adaptive regression splines (MARS) [22]. In most models, TF binding or histone modification signals are directly used as the predictor variables. However, Ouyang et al. carried out principal component analysis (PCA) to extract principal components (PCs) from the TF-gene association matrix and then used these PCs as the predictive variables in a multiple linear regression model [9].

We have also applied supervised classification models such as support vector machine (SVM) to predict whether a gene is highly or lowly expressed according to their TF binding or histone modification signals [13, 18]. Particularly, in the landmark paper of the ENCODE, we proposed two-step models to predict the expression levels of human TSSs based on TF binding and histone modification signals at the TSS [8]. The TSS expression data were obtained from the Cap Analysis of Gene Expression (CAGE) experiments [23], in which a large fraction of TSSs (>40%) were not expressed – the TSSs associated with no CAGE tags. Thus, we first constructed a classification model to predict whether a TSS is expressed. For TSSs predicted to be expressed, we further predicted their expression values using a regression model, whereas the TSSs predicted to be nonexpressed were assigned an expression level of zero.

17.2.3 Performances Evaluation of Models

The performance of predictive models can be evaluated by cross-validation. We randomly selected a number of genes (2000 genes) as the training data and used the remaining as the testing data. A model was trained on the training data and subsequently applied to predict expression levels of genes in the testing data.

The predictive accuracy of the model can be measured by the Pearson correlation coefficient (r) between the predicted expression values ($\hat{y}_i$) and experimental measured levels (y_i) in the testing data. Predictive accuracy can also be measured by the coefficient of determination (R^2), the fraction of variance of gene expression explained by the model, which is defined as the following:

$$R^2 = 1 - \frac{\sum_i (y_i - \hat{y}_i)^2}{\sum_i (y_i - \bar{y}_i)^2},$$

where $\bar{y}$ is the mean gene expression level. We generated 10 groups of training data and testing data and then averaged the resulting r or R^2 to determine the predictive accuracy.

17.2.4 Data Sets

To show how we used the predictive models to relate gene expression with TF binding and histone modification signals, we applied the models to three sets of data: the human K562 and GM12878 data from the ENCODE project [8], the worm data from the modENCODE project [17], and mouse embryonic stem cells (ESCs) data.

The ENCODE K562 and GM12878 Data

The expression data contain expression levels (RPM, reads per million) of $\sim 130,000$ high-confidence TSSs defined by the GENCODE v7 annotation [24], which were measured by CAGE experiment. We choose the CAGE expression data for whole-cell Poly A+ RNA in K562 and GM12878 cell lines. The TF ChIP-seq data contain binding profiles for 40 and 35 sequence specific TFs in K562 and GM12878 cell lines, respectively [25]. The histone modification ChIP-seq data contain signals for 11 different types, including H3K4me1, H3K4me2, H3K4me3, H3K36me3, H3K9me1, H3K9me3,

H3K27me3, H4K20me1, H3K79me2, H3K9ac, and H3K27ac in the two cell lines.

The ModENCODE Early Embryo Data

The expression levels for all annotated worm transcripts were quantified using RNA-seq. Histone modification data for H3K4me2, H3K4me3, H3K9me2, H3K9me3, H3K27me3, H3K36me2, H3K36me3, H3K79me1, H3K7me2, and H3K7me3, and histone H3 occupation data, were obtained from ChIP-chip experiments. MicroRNA expression levels were measured by small RNA-seq experiments from Kato et al. [26].

The Mouse ESC Data

The data were collected from several different publications. Mouse gene expression levels in ESC were quantified using RNA-seq [27]. The microRNA expression data were quantified using small RNA-seq [28]. The TF data contain ChIP-seq binding profiles for 12 mouse TFs in ESC and were downloaded from the GEO database with the accession designation GSE11431 [29]. The TFs are E2f1, Esrrb, Klf4, Nanog, Oct4, Stat3, Smad1, Sox2, Tcfcp2l1, Zfx, c-Myc, and n-Myc. The ChIP-seq data for histone modifications in mouse ESC were downloaded from two data sources. The data for H3K4me1 and H3K4me2 were from Meissner et al. [30] and the data for H3K4me3, H3K4me9, H3K20me3, H3K27me3, and H3K36me3 were from Mikkelsen et al. [31].

Yeast and Fly Data

In yeast, the expression levels of genes were measured by microarrays and available from Wang et al. [32]; the histone modification data were performed by Pokholok et al. [33]. In fruit fly, the gene expression and chromatin data at 12 different developmental stages were obtained by using RNA-seq and ChIP-seq experiments, respectively, which are available from Gerstein et al. [17].

17.3 Applications

17.3.1 Predicting Gene Expression from TF Binding

The TF model aims to address the question that how much variation of gene expression can be explained by TF regulation? TFs are the key players in transcriptional regulation, which recognize and bind to specific DNA sequences called transcription factor binding motifs in promoters and enhancers. ChIP-seq experiments determine the binding events of a TF that occur throughout the whole genome and measure binding signals in both promoter and enhancer regions. However, it is still difficult to associate enhancers with genes regulated by them due to the following: (1) the enhancer confers distal regulation and (2) the enhancer-gene associations are very dynamic. Thus, the TF model described here focuses on binding signals of TFs in DNA regions that are proximal to genes.

The TF model has been used to quantify the relationship between TF binding and gene expression levels in K562 cells [34]. The model uses binding signals of 40 sequence specific TFs as predictor variables to predict the expression levels of $>100,000$ human TSSs (see Section 17.2.4 for details). TF binding was determined using ChIP-seq, and TSS expression was quantified using CAGE experiments by the ENCODE project. The TF binding signals were calculated as the average number of reads covering DNA regions of 100 bps surrounding a TSS. Figure 17.2A shows the scatter plot between the experimental expression levels and the predicted expression values of TSSs for protein-coding genes by a two-step model. The model first classifies the TSS into expressed and nonexpressed ones, assigns zero values to those classified to be nonexpressed, and then applies a regression model to predict the expression values of those classified as expressed. Random forest is used for both classification and regression steps. As shown, the predicted values are highly consistent with the actual expression levels if TSSs with a Pearson correlation coefficient $r = 0.81$. Both the classification and the regression steps achieve a fairly high accuracy with $AUC = 0.89$ for the classification step and $r = 0.62$ for the regression step. These results indicate that binding signals of a total of 40 TFs at the TSS are highly predictive to TSS expression levels, accounting for $>60\%$ of the variation in expression ($R^2 = 0.66$).

Binding signals of 40 TFs are used as predictors in the TF model. The contribution of each TF for gene expression prediction can be estimated by calculating its relative importance in the model. For the random forest method, the "mean decreased Gini" is often used to measure the relative importance of predictors. It measures the reduction in classification or regression accuracy (i.e., decrease in node purity) when the contribution of a specific predictor is removed from the model by permutation. A TF with larger "mean decreased Gini" in the model has higher importance for predicting the gene expression level. Figures 17.2B and 17.2C show the relative importance of TFs in the classification and the regression steps, respectively. As shown, some TFs such as Yy1 and Myc are more informative for predicting gene expression. These TFs generally bind to a large number of target genes and tend to regulate transcription through interacting with DNA elements in promoters. Many other TFs such as Znf274 exhibit limited contribution to gene expression prediction, presumably because they are only involved in the regulation of a small set

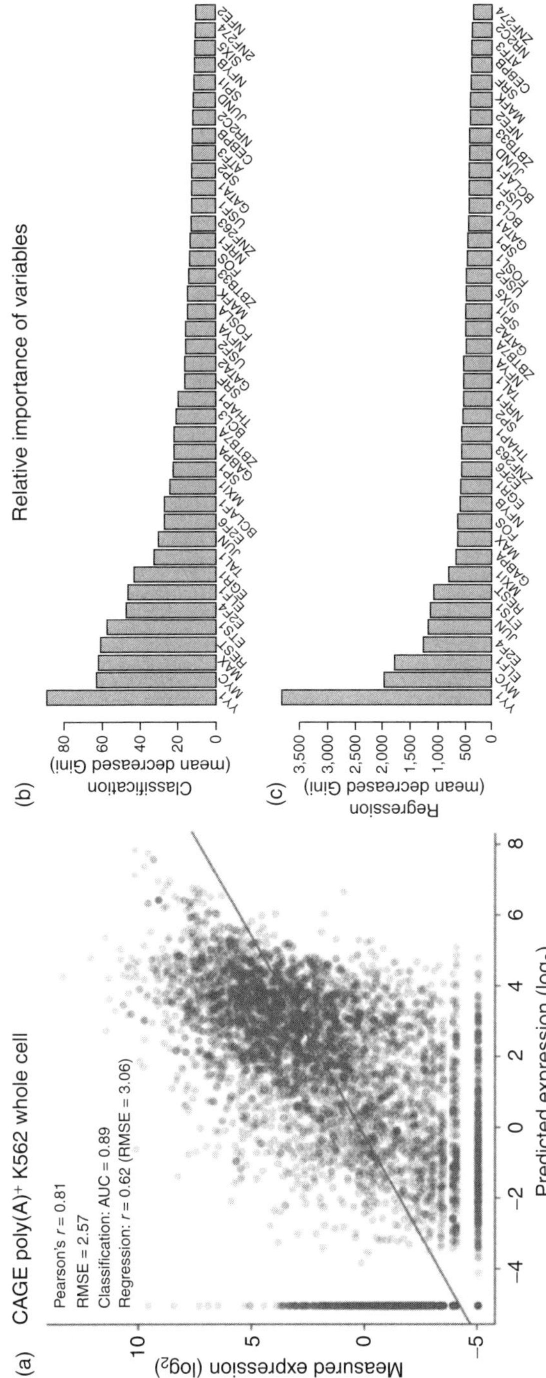

Figure 17.2 The TF model for predicting the expression levels of human promoters. (A) The consistency between predicted expression levels and experimental measurements. The expression levels of human promoters are measured by CAGE experiment as the tag density at TSSs in K562 cells. A two-step model is used to predict TSS expression values. First, a classification model is applied to classify TSSs into expressed or nonexpressed; then a regression model is applied to predict the expression values of TSSs classified as expressed. In both steps, the random forest method is used. (B) The relative importance of TFs in the classification model. (C) The relative importance of TFs in the regression model. AUC, area under curve; Gini, Gini coefficient; RMSE, root mean square error.

of genes. Alternatively, their regulation might be mediated via interacting with enhancers, which are distant from the TSS and cannot be captured by the binding signal at the TSS proximal region. Thus, the TF model provides insights into the mechanism of transcription regulation.

Interestingly, the TF model can achieve high predictive accuracy when only a small subset of TFs is used as predictors. A model with the top five TFs (Figure 17.2C) gives rise to almost the same accuracy. The individual predictive power of a TF can be examined using a degenerate model that takes this TF as the single predictor. It turns out that many of the TFs can individually predict TSS expression levels of genes with fairly high accuracy. For example, MYC alone can explain 55% of the variance in expression of all TSS, which is only 12% lower than the variance explained by the full model (66%). This is due to the fact that in the TSS-proximal region, the binding of many TFs is largely determined by the local chromatin structure suggesting high TF-TF correlation.

17.3.2 *Predicting Gene Expression from Histone Modifications*

The histone model aims to address the question, how much variation of gene expression can be explained by histone modification signals? Histone modifications can impact gene transcription directly by recruiting specific TFs or indirectly by modulating the local chromatin structure to change the accessibility to TFs. Again, we use the ENCODE data to show an example of predicting TSS expression levels (measured by CAGE experiment) based on histone modification signal. The model contains 14 predictors consisting of 11 histone marks, H3K4me1, H3K4me2, H3K4me3, H3K9me1, H3K9me3, H3K27me3, H3K36me3, H3K79me2, H4K20me1, H3K9ac and H3K27ac, one noncanonical histone type H2A.Z, DNase I hypersensitivity signal, and a DNA sequence – based feature (the normalized CpG content in all TSS regions) [16]. For each of the predictors, the signals at a 100 bps DNA region surrounding the TSS of genes are calculated based on ChIP-seq data in K562 cells. To be consistent with the TF model, a two-step procedure is used: applying a classification model for predicting whether a TSS is expressed, followed by applying a regression model to predict the expression values.

The histone model achieves high accuracy in predicting expression levels of human TSSs captured by CAGE experiments. As shown in Figure 17.3A (left), the predicted expression values for TSSs are highly correlated with their experimentally measured levels with a correlation coefficient $r = 0.9$. More specifically, both the classification and the regression steps result in very high accuracy with an AUC $= 0.95$ for classification and a correlation $r = 0.78$ for regression. These results indicate that the ability of a promoter to initiate

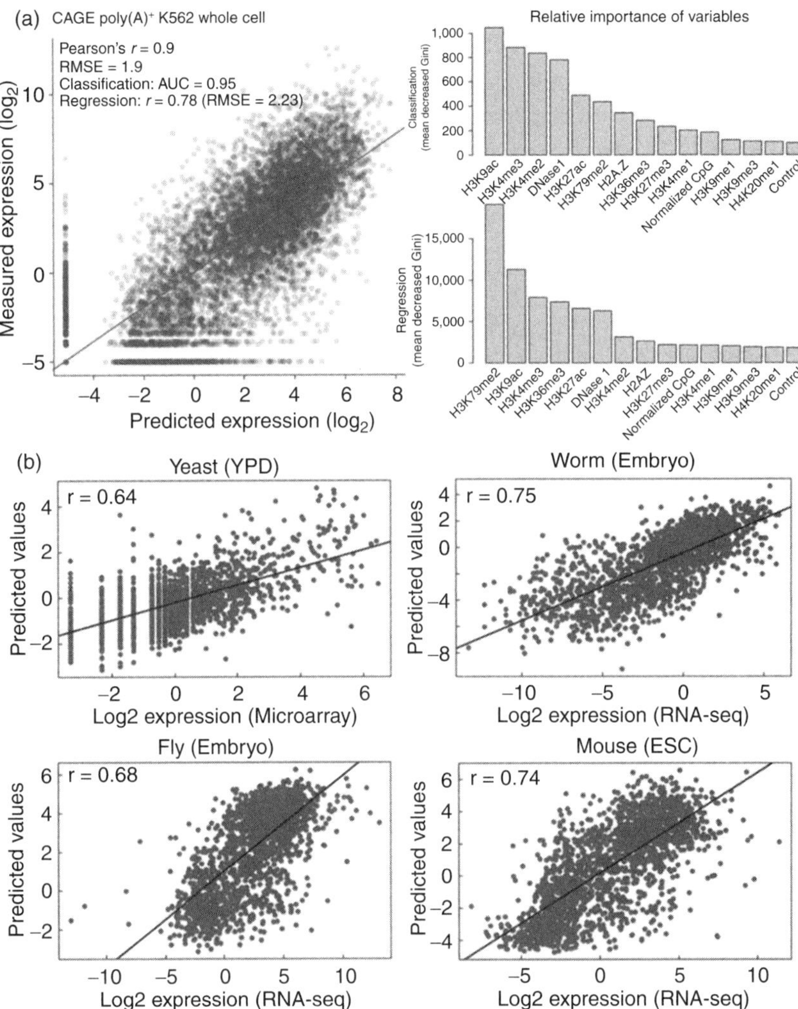

Figure 17.3 The histone model for predicting the expression levels of human promoters. (A) The consistency between predicted expression levels and experimental measurements (left); and the relative importance of different histone marks in the classification model and the regression model (right). (B) The predictive accuracy of the histone model in four different organisms: yeast cultured in YPD medium, worm at the early embryo stage, fly at the embryo stage, and mouse embryonic stem cells. Expression data are measured by microarray (yeast) or RNA-seq (worm, fly, and mouse) experiments. AUC, area under curve; Gini, Gini coefficient; RMSE, root mean square error.

transcription (reflecting by the TSS expression levels from CAGE data) is largely determined by the local chromatin status.

The relative importance of different chromatin features reflects the critical roles they play in regulating the transcription initiation of promoters. As shown in Figure 17.3A (right), the model indicates that activating acetylation marks (H3K27ac and H3K9ac) are roughly as informative as activating methylation marks (H3K4me3 and H3K4me2). In the classification step the acetylation mark H3K9ac contributes most, whereas in the regression step the methylation mark H3K79me2 contributes most to the prediction power. The DNase I signal also shows a high relative importance in the model. It should be noted that different histone marks play their roles in different DNA regions and have different density distributions along the gene. For example, H3K4me3 and H3K9ac are mainly associated with active promoters, whereas H3K4me1 and H3K27ac are mainly associated with enhancers. Both H3K79me2 and H3K36me3 mark gene bodies, but H3K79me2 occurs preferentially at the 5' ends of gene bodies, and H3K36me3 occurs more frequently at the 3' end. The model introduced in this section focuses on histone modification signals at the TSS sites. As a consequence, the relative importance of different histone marks might only be valid if they occur near the TSS. Different relative importance of histone marks would be expected for other DNA regions. It should also be noted that the TSS expression from CAGE is correlated but different from the transcript expression levels from RNA-seq experiment. The relative importance of histone marks is different between models for predicting TSS expression and transcript expression. For example, TSS expression quantified by CAGE are better predicted by promoter marks such as H3K4me3, whereas transcript expression measured with RNA-seq are better predicted by structural marks like H3K36me3.

The histone models have been constructed to predict gene expression levels based on histone modifications in several different organisms. As shown in Figure 17.3B, the models result in high accuracy in yeast, worm, fly, and human. In some organisms, gene expression data are available from both RNA-seq and microarray experiments. It has been shown that models using RNA-seq expression achieve significantly higher accuracy than those using microarray expression, suggesting higher accuracy of RNA-seq data. Overall, histone models in a variety of biological contexts in multiple different organisms indicate that overall, about 50% of variation of gene expression can be explained by histone modification signals in the promoter regions of genes. Models have also shown that the functions of different histone marks are fairly conserved in different organisms. First, the positive correlation between activating marks and the negative correlation between the repressive marks with gene expression levels are conserved. Second, the relative importance of different histone marks shows similar ranks. In this sense, the histone model provides a useful tool for comparing the function of histone marks between different species in a quantitative manner.

17.3.3 Predicting Gene Expression by Combining TF Binding and Histone Modification Signals

Having described the TF model and the histone model for predicting gene expression levels, we next ask the question, what is the relationship between TF binding and histone modifications in gene expression prediction? There are two possibilities. It is possible that in some situations, TF binding and histone modification may contribute independent signals to regulate different aspects of gene regulation, and consequently the predictive accuracy can be further improved by including features from both data types in a combined model. Alternatively, it is also possible they are redundant and do not provide additional information in terms of gene expression prediction.

This problem has been investigated using the data from mouse embryonic stem cells (ESCs) [18]. A number of genomic studies have been performed in mouse ESC cells that have generated ChIP-seq data for a number of TFs and different histone marks as well as RNA-seq expression data. We have collected the ChIP-seq data for 12 TFs and 7 histone marks to investigate the relationship between TF binding and histone marks using gene prediction models. The DNA regions around TSS and TTS (transcription terminal site) are divided into small bins of 100 bps, and for each bin the support vector regression (SVR) method is applied to predict gene expression based on the TF-binding signals (the TF model), histone modification signals (the HM model), or a combination of them both (the TF+HM model).

Both the TF model and the HM model can predict gene expression with high accuracy, but they display patterns that are quite different in the 160 bins. As shown in Figure 17.4A, the TF model achieves the highest predictive power ($r = 0.71$) at the TSS, but the predictive power decays quickly as the distance from TSS increases. TF-binding signals >2 kb away from the TSS provide very limited contribution to gene expression levels. In contrast, the HM model maintains high accuracy across whole transcribed regions, extending to upstream of the TSS and downstream of the TTS. The highest accuracy ($r = 0.72$) is achieved in the bins immediately downstream of the TSS. The substantial difference in the pattern of predictive powers between the TF models and the HM models results from the fact that most TFs mainly function at the TSS region, whereas distinct HMs function at different locations for gene expression regulation [35, 36]. A two-layer model is developed to integrate the

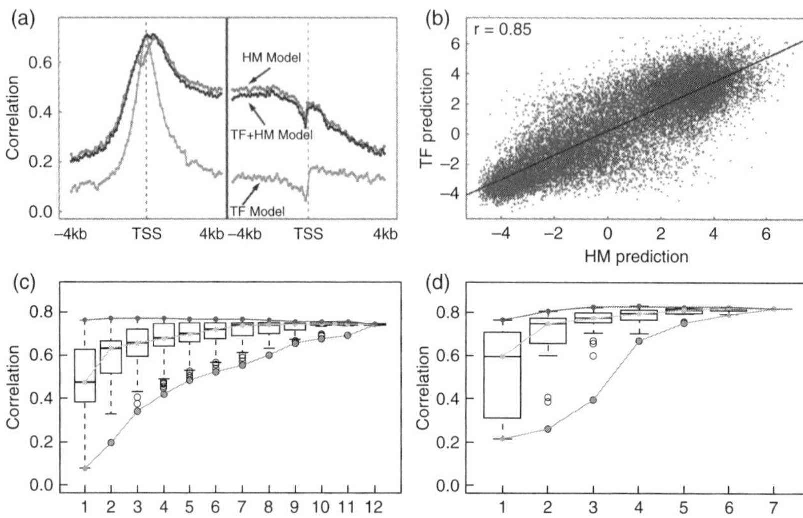

Figure 17.4 Redundancy between the TF binding and the histone modification signals for predicting gene expression in mouse ESC. (A) The prediction accuracy of three models: the TF model, the HM model, and a combined TF+HM model, in each of the 160 bins. (B) Consistency between TF model predictions and HM model predictions. The predicted expression values are based on the two-layer TF model (y-axis) and the two-layer model (x-axis). (C) Distribution of prediction accuracies of all m-TF models with m taken from 1 to 12. (D) Distribution of prediction accuracies of all m-HM models with m taken from 1 to 7. The maximum, the median, and the minimum prediction accuracy for m-TF (C) or m-HM (D) models overlap with the top, middle, and bottom curves, respectively.

binding signals of all TFs at different locations. In the first layer, signals of all the 12 TFs are combined to make predictions of expression separately at each bin. The predicted expression levels by distinct bins are then combined in the second layer to make the final prediction (see method 2.1 for details). SVR method is used in both layers. The two-layer TF model yields a correlation of $r = 0.77$ between the predicted and real expression levels. A similar two-layer HM model results in predictive accuracy of $r = 0.82$. These results indicate that both TF binding and histone modification signals account for $>50\%$ of the variation of gene expression individually.

The TF+HM model shows similar prediction accuracy as the HM model and the TF model across all bins. Consistently, the two-layer TF+HM model achieves similar prediction accuracy as the two-layer HM model with $r = 0.85$. Generally, these results suggest that the TF binding signals and HM signals are redundant for gene expression prediction. To further investigate their redundancy, we examine whether the HM model is able to predict expressions that have not been captured by the TF model, and vice versa. Specifically, the TF model is applied to predict gene expression, and then the difference between the predicted and real expression levels (expression residuals) is calculated for all genes. The expression residuals represent the expression levels that have not been explained by the TF model. Subsequently, the HM model is used to predict the expression residues. If the HM model provides additional predictive capability to the TF model, it would be expected to predict the expression residuals with fairly high accuracy. However, it turns out that the HM model is poorly predictive of the expression residuals resulting from the TF model. Similarly, the TF model is not able to predict the expression residuals from the HM model either, suggesting that the histone modification and the TF binding signals do not provide additional information to each other. Moreover, the prediction results by the TF model and the HM model are highly consistent with correlation coefficient $r = 0.85$ (Figure 17.4B). All these results suggest that the TF model and the HM model are statistically redundant for predicting gene expression. Their redundancy might be partially explained by the high correlation between the TF-binding signals and HM signals in the promoter regions.

The chromatin structure impacts the accessibility of all TFs, and therefore binding signals of different TFs are somehow correlated. To examine the redundancy of the 12 TFs, predictive models are tested for all possible combinations of TFs in the TF model. A total of 4,095 models are constructed by choosing m out of the 12 TFs ($m = 1, 2, \ldots, 12$). In each model, the maximum signals of TFs in the 160 bins are taken as the predictors. Figure 17.4C shows the distributions of the accuracies for models based on various numbers of TFs (denoted as m-TF model). As shown, although models with more factors are generally more predictive (the middle curve), there is no significant improvement for the maximum prediction accuracy of the m-TF models (the top curve). In fact, the one-TF model using E2f1 as the predictor resulted in a correlation of $r = 0.76$ between predicted and real expression levels, which is just slightly lower than the best prediction achieved by the four-TF model with predictors E2f1, Zfx, c-Myc, and n-Myc ($r = 0.77$). These results indicate high redundancy among these 12 TFs for expression prediction. Similarly, histone modifications in the HM model are also highly redundant (Figure 17.4D). The highest accuracy is achieved in a four-HM model with H3K4me2, H3K27me3, H3K36me3, and H3K4me3 as predictors ($r = 0.83$).

It should be noted that the redundancy only exists with regard to gene expression prediction. At the molecular level, distinct TFs or histone marks play very different roles in transcriptional regulation. For example, both H3K4me3 and H3K36me3 act as marks for active genes. While H3K4me3 acts in the promoter regions to facilitate the initiation of transcription, H3K36me3 functions mainly in the transcribed regions involved in transcriptional elongation. The predictive powers of individual histone marks are different in different genomic positions. H3K36me3 achieves the highest predictive accuracy in exonic regions, but in TSS proximal regions, H3K4me3 has the highest predictive accuracy. The redundancy between TFs and histone marks may suggest either a causal relationship (i.e., TFs function as the regulators for gene transcription, whereas histone modifications are simply the subsequent readout) or strong cooperativity between them with regard to transcriptional regulation.

17.3.4 *Predicting Differential Gene Expression*

Having shown that TF binding and histone modification signals are predictive to gene expression levels, we then ask, can differential gene expression be predicted by differential TF binding or differential histone modification signals between different conditions?

ChIP-seq data have shown that TF binding and histone modifications are dynamic: under different physiological conditions, their patterns of activity in the genome vary. Thus, we expect that the TF model and the histone model should be condition specific. That is, a model trained with data from one condition (e.g., tissue, cell line, development stage) will predict gene expression with higher accuracy in this condition than the others. It has been shown that the histone model is specific to developmental stage when using modENCODE worm data generated at different stages [13]. Specifically, an SVM-based (support vector machine) histone model is trained based on data at EEMB stage and applied to classify genes into highly and lowly expressed groups at EEMB, L1, L2, L3, L4 and Adult stages. As shown in Figure 17.5A, the model achieves the highest prediction accuracy in the matched stage, EEMB. Similarly, the random forest–based TF model trained in K562 or GM12878 data from ENCODE shows much higher accuracy for predicting TSS expression in the matched cell line (K Model→K, G Model→G) than the nonmatched one (K Model→G, G Model→K) (Figure 17.5B) [34].

Furthermore, the differential expression of TSSs in K562 versus GM12828 can be accurately predicted by differential TF binding between the two cell lines. The binding differences in K562 versus GM12878 arc calculated for 22 TFs for which the ChIP-seq data are available in both cell lines. As shown in Figure 17.5C, a model using those differences as predictors explains 53% of the variance in their TSS expression differences. These results suggest that the expression changes of genes under different conditions can largely be reflected by the TF binding changes in the promoter regions.

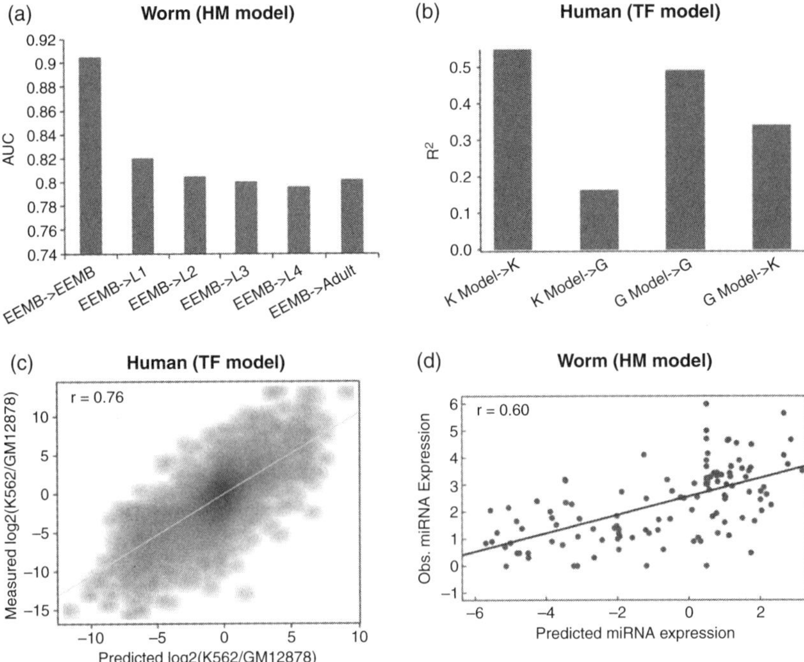

Figure 17.5 Model specificity, differential gene expression model, and microRNA expression model. (A) The histone model is developmental stage specific as shown in modENCODE worm data. Genes are classified into highly expressed or lowly expressed based on histone modification signals. Model trained by EEMB data achieves the highest classification accuracy in EEMB. (B) The TF model is cell line specific as shown in ENCODE K562 and GM12878 data. Models trained in K562 and GM12878 cell lines achieve significantly higher predictive accuracy in the matched than in the non-matched cell line. (C) Differential gene expression between K562 and GM12878 can be accurately predicted by the differential TF binding signals. (D) Histone model trained using protein-coding genes is also predictive to microRNA expression levels in worm.

17.3.5 Predicting the Expression of Noncoding Genes

We have shown that the TF binding and histone modification signals are predictive of expression levels of protein-coding genes. In this section, we apply the models to address the question, do noncoding RNAs share with protein-coding genes the same regulatory mechanism controlled by TFs and histone modifications? If they do share a similar regulatory mechanism, we would expect that the TF model and histone model trained on protein-coding genes are able to predict the expression levels of non-genes.

The CAGE data from the ENCODE projects provide the expression levels of >130,000 TSSs. These TSSs correspond to the promoters of protein-coding genes as well as noncoding RNAs. The TF model trained solely based on protein-coding TSSs predicts expression levels of noncoding TSSs with almost equal accuracy as it does with protein-coding TSSs. Similarly, the histone model trained on protein-coding TSSs also achieves similar accuracy when predicting protein-coding and noncoding TSSs. These results suggest that noncoding RNAs and protein-coding genes share similar regulatory mechanisms that are mediated by TFs and histone modifications.

The effectiveness of the histone model for predicting microRNA expression has been investigated using the modENCODE worm early embryo (EEMB) data. An SVR model is constructed using signals for 13 histone modifications or histone protein H3 occupation as the predictors. The model is trained on protein-coding genes and applied to predict the expression of 162 microRNAs, for which genomic locations are available from miRBase [37]. The predicted values are compared with the measurements in the small RNA-seq data set from Kato et al [26]. As shown in Figure 17.5D, the predictions are in good agreement ($r = 0.60$) with the experimental results. Some microRNAs locate within or near gene loci, which may confound the prediction of microRNA expression. To address this issue, the prediction accuracy is examined for microRNAs away from any known gene, and similar prediction accuracy is observed ($r = 0.62$). Similarly, the histone model trained solely on protein-coding data is predictive of microRNA expression in mouse ESC cells [18]. These results suggest that protein-coding and microRNA genes may share a similar mechanism of transcriptional regulation by histone modifications.

Conversely, the TF model failed to predict expression levels of microRNAs in mouse ESC cells. It might be the case that TF signals are predictive of microRNA expression only around the TSS region, as demonstrated in Figure 17.4A for coding gene expression. However, the annotation for worm and mouse microRNAs from the miRBase does not provide their actual transcriptional start sites but the start position of the corresponding pre-miRNAs ($\sim$100 nt). As the pre-miRNA DNA regions were in general distant (>1 kb) from the actual TSS of microRNAs (i.e., the TSS of pri-miRNAs), the TF-binding signals corresponding to the pre-microRNA regions contribute little to transcriptional regulation and are not able to predict gene expression levels. The promoter regions for most mouse microRNAs have been predicted using computational methods [28]. The signals of the earlier mentioned 12 TFs and 7 HMs in these predicted promoter regions are calculated and used to predict microRNA expression levels in mouse ESC cells. It turns out that, using models trained solely on data sets for protein-coding genes, both of the TF binding signals and the histone modification signals can distinguish highly and lowly expressed microRNAs [18].

17.3.6 Predicting Expression Levels for Genes with High and Low CpG Content

The predictive models can also be applied to model and compare the regulatory mechanisms of different gene classes. For instance, in many organisms, some promoters are associated with a nearby CpG island, while others are not. It has been shown that the normalized CpG content of human promoters follows a bimodal distribution, which classifies them into high CpG promoters (HCPs) and low CpG promoters (LCPs) [38].

In the mouse ESC data, the models have been constructed for the HCP genes and the LCP genes separately to predict their expression levels [18]. The results show that both the TF model and the histone models have higher performance in the HCP group than in the LCP group. When applied to all genes, the two-layer TF, HM, and TF+HM models achieve prediction accuracy of $r = 0.77$, $r = 0.82$, and $r = 0.85$, respectively. However, for the LCP group, the accuracy is $r = 0.63$, $r = 0.73$, and $r = 0.74$, and for the HCP group, the accuracy is $r = 0.70$, $r = 0.77$, and $r = 0.77$. Moreover, the relative importance of the TFs and histone modifications is different in models for the two gene groups. These results suggest that the expression of the LCP genes and the HCP genes might be regulated via different mechanisms.

17.4 Discussion

17.4.1 Interplay Between TF Binding, Histone Modification, and Other Chromatin Features for Regulating Gene Expression

TFs and histone modifications are two critical factors that regulate gene transcription in a cooperative manner. The interactions between them have been proposed as shown in Figure 17.6 based on their effects on gene expression revealed by the quantitative models. First, TFs and histone modifications can regulate the initiation of transcription directly by interacting with RNA polymerase and other general TFs and recruiting them to the TSS, or indirectly by controlling their accessibility to promoters via modulating chromatin structure [7, 39]. As a result, TF-binding data, histone modification data, and the data that capture local chromatin structure (e.g., DNase I hypersensitivity data) are all able to predict gene expression levels. Second, TFs and histone modifications are interrelated and cooperate in transcriptional regulation. For example, TFs can influence histone modifications by recruiting histone modifiers to a DNA region [40]; and conversely, histone modifications can affect TF binding by directly recruiting them or indirectly by changing their accessibility to DNA regions [7]. In line with this, TF-binding and histone-modification signals are

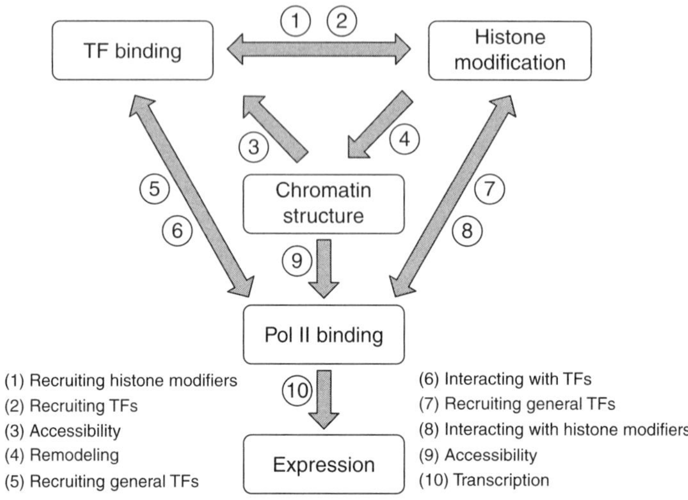

Figure 17.6 Regulatory mechanism of TF binding and histone modification on gene expression. TF binding and histone modification are interrelated and cooperative in gene expression regulation.

often highly correlated in promoter proximal regions. Owing to this high correlation, they share a similar amount of information and thus are redundant for "predicting" gene expression levels. Finally, the transcription status of genes can in turn affect TF-binding and histone modifications by interacting with TFs and histone marks [41], which further complicates the cause and effect relationship between TF binding, histone modifications, and gene expression. The prediction models introduced in this chapter suggest a highly coordinated system for transcriptional regulation that consists of TFs, histone modifications, RNA polymerase, and other chromatin-related proteins.

17.4.2 Regulatory Signals in Distal Regions

The models described here focus on TF binding or histone modification signals in DNA regions within or proximal to genes, however, the regulatory signals in distal regions have not been considered. As has been shown in multiple organisms ranging from yeast to human, TF binding and histone modification signals in promoter regions account for about 50% variation of gene expression levels. The explained variation might be substantially increased, as the signals in distal regulatory elements, for example, enhancers, can be included in the models. ChIP-seq experiments have demonstrated that for many TFs a considerably large proportion of their binding sites are located 2 kb away from any genes in human. It would be interesting to investigate how much of gene expression is determined by distal regulatory signals. The chromatin interaction data produced by the ChIA-PET or high-C experiments provide the connection between distal elements and their potential targets [42, 43]. It should be useful to incorporate this information into the models.

17.4.3 Cause or Consequence

The histone models show that histone modification signals are highly predictive to gene expression levels. However, they do not provide insight into whether histone modifications are the "cause" or "consequence" of transcription. In fact, both directions of causality have been previously reported. Studies have shown that some histone modifications are a direct consequence of previous active transcription, serving as a mark of past transcriptional events [44–46]. However, other studies have shown that chromatin modification changes precede changes in gene expression [47]. For example, it has been demonstrated that activating histone marks are already in place before induction of gene expression, and these marks can still be maintained even after the genes are silenced [48].

Similarly, the causal relationship between TF binding and gene expression is also not clear. Although the primary function of TFs is to regulate gene expression, not all of the TF binding events drive gene transcription. The binding signals of many TFs are highly correlated in promoter regions, and many TFs are highly predictive to gene expression individually. It might be the case that only a few "pioneering" TFs drive the initiation of gene transcription and concomitantly induce histone modification to modulate chromatin structure, which in turn increases the accessibility of active promoters to other TFs [49, 50]. To distinguish between causal and resulting TF binding events, additional information, for example, a time course of TF binding and gene expression data, would be required.

In summary, we introduced in this chapter a framework for predicting gene expression levels based on TF binding and histone modification signals. For implementation, different supervised learning methods such as multiple variable linear regression model, random forest, and support vector machine/ regression can be applied. We then described several applications of this framework for predicting expression levels of promoters (TSS expression measured by CAGE), protein-coding genes, and noncoding RNAs. We demonstrated how the models provide new insights into regulatory mechanisms mediated by TFs and histone modifications. With more genome-wide data from well-designed experiments (e.g., time course TF binding data or histone modification data) and new technologies (e.g., distal chromatin interaction data from ChIA-PET), these models can be modified and extended to address other relevant biological questions.

References

1 Farnham PJ: Insights from genomic profiling of transcription factors. *Nat Rev Genet* 2009, **10**:605–616.

2 Lobe CG: Transcription factors and mammalian development. *Curr Top Dev Biol* 1992, **27**:351–383.

3 Berger SL: The complex language of chromatin regulation during transcription. *Nature* 2007, **447**:407–412.

4 Kurdistani SK, Tavazoie S, Grunstein M: Mapping global histone acetylation patterns to gene expression. *Cell* 2004, **117**:721–733.

5 Latchman DS: Transcription factors: an overview. *Int J Biochem Cell Biol* 1997, **29**:1305–1312.

6 Kouzarides T: Chromatin modifications and their function. *Cell* 2007, **128**:693–705.

7 Li B, Carey M, Workman JL: The role of chromatin during transcription. *Cell* 2007, **128**:707–719.

8 Consortium EP, Dunham I, Kundaje A, Aldred SF, Collins PJ, Davis CA, Doyle F, et al: An integrated encyclopedia of DNA elements in the human genome. *Nature* 2012, **489**:57–74.

9 Ouyang Z, Zhou Q, Wong WH: ChIP-Seq of transcription factors predicts absolute and differential gene expression in embryonic stem cells. *Proc Natl Acad Sci U S A* 2009, **106**:21521–21526.

10 Karlic R, Chung HR, Lasserre J, Vlahovicek K, Vingron M: Histone modification levels are predictive for gene expression. *Proc Natl Acad Sci U S A* 2010, **107**:2926–2931.

11 Xu H, Lemischka IR, Ma'ayan A: SVM classifier to predict genes important for self-renewal and pluripotency of mouse embryonic stem cells. *BMC Syst Biol* 2010, **4**:173–182.

12 Xu X, Hoang S, Mayo MW, Bekiranov S: Application of machine learning methods to histone methylation ChIP-Seq data reveals H4R3me2 globally represses gene expression. *BMC Bioinformatics* 2010, **11**:396–415.

13 Cheng C, Yan KK, Yip KY, Rozowsky J, Alexander R, Shou C, Gerstein M: A statistical framework for modeling gene expression using chromatin features and application to modENCODE datasets. *Genome Biol* 2011, **12**:R15.

14 Park SJ, Nakai K: A regression analysis of gene expression in ES cells reveals two gene classes that are significantly different in epigenetic patterns. *BMC Bioinformatics* 2011, **12 Suppl 1**:S50.

15 Althammer S, Pages A, Eyras E: Predictive models of gene regulation from high-throughput epigenomics data. *Comp Funct Genomics* 2012, **2012**:284786.

16 Dong X, Greven MC, Kundaje A, Djebali S, Brown JB, Cheng C, Gingeras TR, Gerstein M, Guigo R, Birney E, Weng Z: Modeling gene expression using chromatin features in various cellular contexts. *Genome Biol* 2012, **13**:R53.

17 Gerstein MB, Lu ZJ, Van Nostrand EL, Cheng C, Arshinoff BI, Liu T, Yip KY, et al: Integrative analysis of the *Caenorhabditis elegans* genome by the modENCODE project. *Science* 2010, **330**:1775–1787.

18 Cheng C, Gerstein M: Modeling the relative relationship of transcription factor binding and histone modifications to gene expression levels in mouse embryonic stem cells. *Nucleic Acids Res* 2012, **40**:553–568.

19 Draper NR, Smith H: *Applied Regression Analysis* (3rd ed.). John Wiley; 1998.

20 Vapnik NV: *The Nature of Statistical Learning Theory.* Springer-Verlag; 1995.

21 Breiman L: Random forests. *Machine Learning* 2001, **45**:5–32.

22 Friedman JH: Multivariate adaptive regression splines. *Annals of Statistics* 1991, **19**:1–67.

23 Shiraki T, Kondo S, Katayama S, Waki K, Kasukawa T, Kawaji H, Kodzius R, et al: Cap analysis gene expression for high-throughput analysis of transcriptional starting point and identification of promoter usage. *Proc Natl Acad Sci U S A* 2003, **100**:15776–15781.

24 Harrow J, Frankish A, Gonzalez JM, Tapanari E, Diekhans M, Kokocinski F, Aken BL, et al: GENCODE: the reference human genome annotation for the ENCODE Project. *Genome Res* 2012, **22**:1760–1774.

25 Gerstein MB, Kundaje A, Hariharan M, Landt SG, Yan KK, Cheng C, Mu XJ, et al: Architecture of the human regulatory network derived from ENCODE data. *Nature* 2012, **489**:91–100.

26 Kato M, de Lencastre A, Pincus Z, Slack FJ: Dynamic expression of small non-coding RNAs, including novel microRNAs and piRNAs/21U-RNAs, during *Caenorhabditis elegans* development. *Genome Biol* 2009, **10**:R54.

27 Cloonan N, Forrest AR, Kolle G, Gardiner BB, Faulkner GJ, Brown MK, Taylor DF, et al: Stem cell transcriptome profiling via massive-scale mRNA sequencing. *Nat Methods* 2008, **5**:613–619.

28 Marson A, Levine SS, Cole MF, Frampton GM, Brambrink T, Johnstone S, Guenther MG, et al: Connecting microRNA genes to the core transcriptional regulatory circuitry of embryonic stem cells. *Cell* 2008, **134**:521–533.

29 Chen X, Xu H, Yuan P, Fang F, Huss M, Vega VB, Wong E, et al: Integration of external signaling pathways with the core transcriptional network in embryonic stem cells. *Cell* 2008, **133**:1106–1117.

30 Meissner A, Mikkelsen TS, Gu H, Wernig M, Hanna J, Sivachenko A, Zhang X, et al: Genome-scale DNA methylation maps of pluripotent and differentiated cells. *Nature* 2008, **454**:766–770.

31 Mikkelsen TS, Ku M, Jaffe DB, Issac B, Lieberman E, Giannoukos G, Alvarez P, et al: Genome-wide maps of chromatin state in pluripotent and lineage-committed cells. *Nature* 2007, **448**:553–560.

32 Wang Y, Liu CL, Storey JD, Tibshirani RJ, Herschlag D, Brown PO: Precision and functional specificity in mRNA decay. *Proc Natl Acad Sci U S A* 2002, **99**:5860–5865.

33 Pokholok DK, Harbison CT, Levine S, Cole M, Hannett NM, Lee TI, Bell GW, et al: Genome-wide map of nucleosome acetylation and methylation in yeast. *Cell* 2005, **122**:517–527.

34 Cheng C, Alexander R, Min R, Leng J, Yip KY, Rozowsky J, Yan KK, et al: Understanding transcriptional regulation by integrative analysis of transcription factor binding data. *Genome Res* 2012, **22**:1658–1667.

35 Guenther MG, Levine SS, Boyer LA, Jaenisch R, Young RA: A chromatin landmark and transcription initiation at most promoters in human cells. *Cell* 2007, **130**:77–88.

36 Kolasinska-Zwierz P, Down T, Latorre I, Liu T, Liu XS, Ahringer J: Differential chromatin marking of introns and expressed exons by H3K36me3. *Nat Genet* 2009, **41**:376–381.

37 Kozomara A, Griffiths-Jones S: miRBase: integrating microRNA annotation and deep-sequencing data. *Nucleic Acids Res* 2011, **39**:D152–157.

38 Saxonov S, Berg P, Brutlag DL: A genome-wide analysis of CpG dinucleotides in the human genome distinguishes two distinct classes of promoters. *Proc Natl Acad Sci U S A* 2006, **103**:1412–1417.

39 Mitchell PJ, Tjian R: Transcriptional regulation in mammalian cells by sequence-specific DNA binding proteins. *Science* 1989, **245**:371–378.

40 Yang WM, Yao YL, Sun JM, Davie JR, Seto E: Isolation and characterization of cDNAs corresponding to an additional member of the human histone deacetylase gene family. *J Biol Chem* 1997, **272**:28001–28007.

41 Okitsu CY, Hsieh JC, Hsieh CL: Transcriptional activity affects the H3K4me3 level and distribution in the coding region. *Mol Cell Biol* 2010, **30**:2933–2946.

42 Fullwood MJ, Ruan Y: ChIP-based methods for the identification of long-range chromatin interactions. *J Cell Biochem* 2009, **107**:30–39.

43 Dekker J: The three "C" s of chromosome conformation capture: controls, controls, controls. *Nat Methods* 2006, **3**:17–21.

44 Ng HH, Robert F, Young RA, Struhl K: Targeted recruitment of Set1 histone methylase by elongating Pol II provides a localized mark and memory of recent transcriptional activity. *Mol Cell* 2003, **11**:709–719.

45 Fischer JJ, Toedling J, Krueger T, Schueler M, Huber W, Sperling S: Combinatorial effects of four histone modifications in transcription and differentiation. *Genomics* 2008, **91**:41–51.

46 Li J, Moazed D, Gygi SP: Association of the histone methyltransferase Set2 with RNA polymerase II plays a role in transcription elongation. *J Biol Chem* 2002, **277**:49383–49388.

47 Chambeyron S, Bickmore WA: Chromatin decondensation and nuclear reorganization of the HoxB locus upon induction of transcription. *Genes Dev* 2004, **18**:1119–1130.

48 Barski A, Jothi R, Cuddapah S, Cui K, Roh TY, Schones DE, Zhao K: Chromatin poises miRNA- and protein-coding genes for expression. *Genome Res* 2009, **19**:1742–1751.

49 Zaret KS, Carroll JS: Pioneer transcription factors: establishing competence for gene expression. *Genes Dev* 2011, **25**:2227–2241.

50 Serandour AA, Avner S, Percevault F, Demay F, Bizot M, Lucchetti-Miganeh C, Barloy-Hubler F, et al: Epigenetic switch involved in activation of pioneer factor FOXA1-dependent enhancers. *Genome Res* 2011, **21**:555–565.

18

Data Integration on Noncoding RNA Studies

ZHOU DU, TENG FEI, MYLES BROWN, X. SHIRLEY LIU,
AND YIWEN CHEN

Abstract

Recent genome-wide studies revealed that the human genome encodes over 10,000 long
non-coding RNAs (lncRNAs) with little protein-coding capacity. Growing evidence
suggests that many lncRNAs may have important functions in complex diseases and are
potentially a new class of therapeutic targets for treating complex disease. In contrast
to the fast pace of cataloguing lncRNAs in the human genome, the function of the vast
majority of lncRNAs remain unknown. In this chapter, we described data integration
strategies for identifying lncRNA that are associated with cancer subtypes and clinical
prognosis, and predicted those that are potential drivers of cancer progression.

18.1 Introduction

The advancement in high-throughput technologies such as microarray, next-
generation sequencing (NGS) has greatly facilitated cost-effective large-scale
data generation. As a result, the amount of genomic data deposited into various
public data sources such as Gene Expression Omnibus (GEO) (http://www.ncbi.
nlm.nih.gov/geo/) and ArrayExpress (http://www.ebi.ac.uk/arrayexpress/) has
grown tremendously in the past several years. Taking NCBI short reads archive
database (http://www.ncbi.nlm.nih.gov/sra) as an example, the amount of data
in this database went from about 10 terabytes (TB) in 2008 to about 1000 TB in
2012, an around 100-fold increase in only four years. These public data sources
not only provide the raw data for the researchers to reproduce the discovery
that were reported in the original study but also provided opportunities for
using the same data for new discoveries. Moreover, integrating the data across
individual studies either horizontally or vertically offers unique opportunities
to make novel discoveries that would have been impossible based on the data
from a single study. The integration of genomic data from the same individual
under a specific disease condition is particularly powerful for disease-relevant

403

discoveries. In those genomics-based clinical studies, the orthogonal genomic data and corresponding clinical information were systematically collected from the same group of human subjects. These data can be integrated to discover genes that play important roles in the etiology of the disease and those that may serve as diagnostic, prognostic, and predictive biomarkers.

Recent transcriptome profiling in human cells from the ENCODE (http://encodeproject.org/ENCODE/) and GENCODE (http://www.gencodegenes.org/) projects showed that cumulatively ~70% of the human genome [1] can be transcribed, whereas only ~2% of the genome encodes proteins. In contrast to ~20,000 protein encoding genes (PCGs), there are ~35,000 (GENCODE) noncoding RNA genes in the human genome. The noncoding RNAs can be classified as either small noncoding RNAs (sncRNAs), which are shorter than or equal to 200 base-pair (bp), or long noncoding RNAs (lncRNAs), which are longer than 200 bp. Data integration has played a pivotal role in identifying the sncRNAs, especially microRNAs (miRNAs) in different species, and predicting the targets and biological function of miRNAs in physiology and disease [2–7]. Although significant knowledge has been accumulated on the sncRNA biology in the past decade with the joint effort of computational and experimental research, the identity and function of the lncRNAs in human genome are just beginning to be revealed. Data integration has played a critical role in identifying the lncRNA genes from a variety of genomic data in different biological contexts as well as in providing the evidence for lncRNA function [8–10]. Systematic efforts to catalog lncRNAs by traditional cDNA Sanger sequencing [11] and the integration of histone mark chromatin immunoprecipitation sequencing (ChIP-seq) [9, 12] and RNA sequencing (RNA-seq) [8, 13] data have revealed that the human genome encodes more than 10,000 lncRNAs. We refer the interested readers to other published reviews for data integration studies on both sncRNAs [2–4, 7] and lncRNAs [8–10]. This chapter is dedicated to describing the approaches to integrate the data from clinical studies for elucidating lncRNA function and uncovering its potential utility in diagnosis and prognosis in human diseases such as cancer [14].

Given their lower expression level compared with protein-coding genes (PCGs) [8], it has been debated whether the lncRNAs are simply the transcriptional noise in the cell or whether they may have biochemical function. Although we do not know how many of them are functional, growing evidence suggests that lncRNAs, similar to PCGs, may play important roles in both development [15] and human diseases such as cancer [16]. A growing list of lncRNAs has been shown to mediate oncogenic or tumor-suppressing effects in cancer, and they promise to be a new class of cancer therapeutic targets [17]. Although a handful of lncRNAs have been functionally characterized, little is known about the functions of most lncRNAs in normal physiology or disease [18]. LncRNAs may serve as cancer diagnostic or prognostic biomarkers that are independent of PCGs. A well-known example of a cancer diagnostic biomarker is PCA3 [19], a prostate-specific lncRNA gene that is significantly overexpressed in prostate cancer. Noninvasive monitoring of the ratio of urinary PCA3 and prostate-specific antigen (PCA) transcript level was recently approved by FDA as a diagnostic assay for prostate cancer [20].

In this chapter, we present a case study of data integration in a cancer-related lncRNA study [14], in which we identified lncRNA that are associated with cancer subtypes and clinical prognosis, and predicted those that are potential drivers of cancer progression in multiple cancers, including glioblastoma multiforme (GBM) [21], ovarian cancer (OvCa) [22], lung squamous cell carcinoma (lung SCC) [23], and prostate cancer [24]. We validated our predictions of two tumorgenic lncRNAs by experimentally confirming the prostate cancer cell growth dependence on these two lncRNAs. Our integrative analysis provided a resource of clinically relevant lncRNA for development of lncRNA biomarkers and identification of lncRNA therapeutic targets for human cancer.

18.2 Methods

18.2.1 Repurposing Microarray Data to Interrogate lncRNA Expression

As lncRNAs do not encode proteins, their functions are closely associated with their transcript abundance. Though RNA-seq is a comprehensive way to profile lncRNA expression, publicly available RNA-seq data sets of tumors are relatively limited compared to array-based expression profiles because of the high cost associated with the adoption of this technique. In addition, RNA-seq data sets with low sequencing coverage or small sample numbers have only limited statistical power to discover clinically relevant lncRNAs. In contrast, there are a large number of data sets that contain array-based gene expression profiles across hundreds of tumor samples. These array-based expression profiles are often accompanied by matched clinical annotation and/or genomic alteration profiles of tumors such as somatic copy number alteration (SCNA). Although lncRNAs are not the intended targets of measurement in the original array design, microarray probes can be reannotated for interrogating lncRNA expression [25–27]. Compared with RNA-seq data of low sequencing coverage, array-based expression data may have lower technical variation and better detection sensitivity for low-abundance transcripts [28, 29], which is a prominent feature of lncRNAs [8]. Moreover, array-based expression data contain strand information and allow for interrogating the expression of antisense single-exon lncRNAs, whereas most current RNA-seq data in clinical

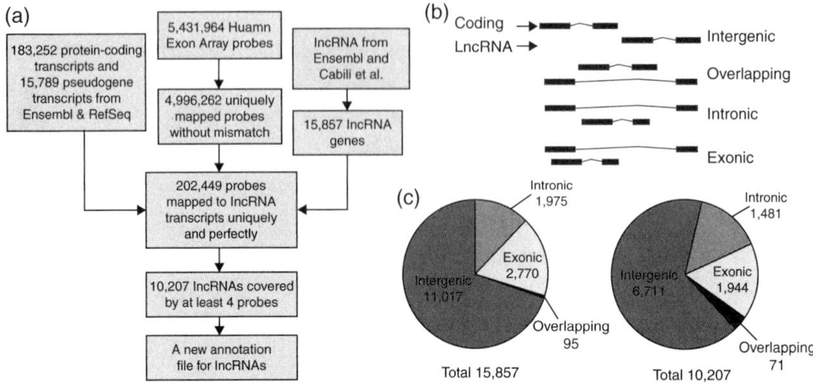

Figure 18.1 (A) Affymetrix Human Exon array probe reannotation pipeline for lncRNA. (B) Adopting the classification scheme from a previous study [34], lncRNA were classified into four categories, intergenic, overlapping, intronic, and exonic, on the basis of their relationship with protein-coding genes. (C) Pie charts showing the number of lncRNA in each category for all collected lncRNA and for those with at least four uniquely mapped exon array probes.

applications do not have strand information and thus are unable to accurately quantify the expression of this class of lncRNAs [30].

Among the different gene expression microarray platforms, we focused on reannotating the probes from the Affymetrix microarrays. These arrays not only have many more short probes that are likely to map to lncRNA genes but also have been the most widely used platforms for gene expression profiling of clinical studies. A computational pipeline was designed as follows to reannotate the probes from five major Affymetrix array types (Figure 18.1A) using the latest annotations of lncRNA and PCG. The lncRNA annotations were derived from two sources: the catalog of lncRNAs from the Ensembl database [31] (*Homo sapiens* GRCh37, release 67) and the catalog of lncRNAs generated on the basis of transcriptome assembly from RNA-seq data [8]. For those lncRNA transcripts with overlap on the same strand between these two sources, we only kept the Ensembl annotation to avoid redundancy. This resulted in a total of 15,857 lncRNA genes. We reannotated probe sets of the affymetrix microarrays for lncRNAs by mapping all probes to the human genome (hg19) by using SeqMap [32]. To avoid potential cross-hybridization of transcribed regions in the genome other than lncRNAs, we only kept those probes that mapped uniquely to the genome with no mismatch and removed all probes that mapped to protein-coding transcripts (183,252) or pseudogene transcripts (15,789) on the basis of the annotations from the Ensembl [31] and UCSC [33] databases.

Table 18.1 *Number of probes corresponding to lncRNAs and number of lncRNAs with at least four probes, coverage in five major Affymetrix array platforms*

	No. of probes corresponding to lncRNAs	No. of lncRNAs with at least four probes
Affymetrix Human Exon array	202,449	10,207
Affymetrix U95Av array	1865	76
Affymetrix U133 plus 2.0 array	43,752	2561
Affymetrix U133B array	21,880	1181
Affymetrix U133A array	2830	143

The preceding strategy was applied to generate the probes that corresponded to lncRNA transcripts for both Affymetrix exon array and the other 3′ IVT Affymetrix array platforms (Table 18.1). Among the five Affymetrix array types, the Affymetrix Human Exon 1.0 ST array has the most comprehensive coverage of the annotated human lncRNAs (Table 18.1), and we used the case of Affymetrix exon array for demonstration. By matching the selected probes to the lncRNA sequences, we obtained 202,449 probes from exon array and 10,207 corresponding lncRNA genes with at least four probes covering their annotated exons (Figure 18.1A), comprising approximately 64% of all 15,857 lncRNA genes (with over 60% coverage in each category [34] of the lncRNA genes) collected in this study (Figures 18.1B and 18.1C). The raw intensity of the exon array probes was corrected with a probe sequence–specific background model, and the expression level of a lncRNA gene was calculated by summarizing the background-corrected intensity of all probes corresponding to this gene [35]. The lncRNA expression was quantile normalized across different biological samples. The gene expression calculation was implemented with Jetta [36]. When batch information was available, Combat [37], an empirical Bayes method, was used to remove potential batch effects.

To gauge the reliability of our approach, we examined the correlation of both lncRNA and PCG expression between exon array and RNA-seq data on the same prostate cancer cell line LNCaP that were generated from two different laboratories [24, 38]. RNA-seq-based gene expression was calculated with Cufflinks 1.0.2 [39] (default parameters and the –G option), and the exon array–based gene expression was calculated by the same procedure as was described earlier. The Pearson correlation coefficient was used to quantify the strength of the associations between the exon array–based and RNA-seq-based expression levels. We found that both PCGs ($r = 0.70$, $P < 2.2 \times 10^{-16}$) and lncRNAs ($r = 0.29$, $P < 2.2 \times 10^{-16}$) showed significant concordance of expression between the exon array and RNA-seq data. This observation is consistent with the previous finding that the correlation between microarray and RNA-seq data is lower in genes with low expression [40], as lncRNAs are generally expressed at lower levels than PCGs [8]. As the level of probe coverage could also influence the accuracy of lncRNA expression derived from a microarray, we further investigated how the correlations of expression between the exon array and RNA-seq data change at different probe coverages by examining those PCGs with expression levels similar to those of lncRNAs. We found that the correlation between exon array- and RNA-seq-based expression showed a moderate increase when all probes (0.28) were used as compared with when only four probes (0.20) were used. The correlations were similar for PCGs (0.28) and lncRNAs (0.29) when we controlled for expression level. These results suggest that although probe coverage may influence the array-based lncRNA expression estimation, the dominant factor that governs the observed difference in correlation between array and RNA-seq data for PCGs and lncRNAs is their expression level. A recent study, in which a 60-mer custom oligonucleotide array was designed to investigate lncRNA expression, showed that the correlation of lncRNA expression between the custom array and RNA-seq data was between 0.24 and 0.31 [34]. Therefore, although the concordance between exon array and RNA-seq data is lower for lncRNA expression than for PCG expression, it may represent the typical performance in comparison of lncRNA expression between an array-based platform and RNA-seq. These examinations demonstrated the reliability of the usage of our reannotated exon array in measuring lncRNAs' expression and laid a foundation for our further study.

18.2.2 *Integrating lncRNA Expression, Somatic Copy Number Alteration Data, and Clinical Information*

One of the most important goals of disease research, especially in cancer research, is to identify driver genes that causally contribute to the disease initiation, progression, and maintenance, as these driver genes can potentially serve as targets for therapeutic interventions. Reliable identification of driver genes is challenging. The emergence of genomic technologies such as microarray and next-generation sequencing has greatly facilitated the identification of driver genes with the aid of computational methods. The expression data alone are insufficient for indentifying driver genes because the aberrant gene expression during the course of disease progression could be attributed to an indirect effect that is secondary to the major disease-causing events. Therefore it is important to integrate genomic data from different sources to enhance the specificity to indentify genes that may play a causal function in disease etiology. Aside from expression data, an important data source that is informative for identifying driver genes is genetic alteration data. For instance, in cancer, a disease with the hallmark of genomic instability [41, 42], many types of somatic genetic alterations are specific to the cancer genome but not to the genome of the normal tissue. These somatic genetic alterations include nucleotide substitution mutations and small insertion/deletions (indels), copy number gains and losses, and chromosomal rearrangements. The copy number gains and losses is a particularly interesting type of somatic genetic alteration because it can often be linked to aberrant gene expression, which makes it a powerful data source in combination with expression profile to identify concordant genetic and gene expression abnormality. The joint analysis of genome-wide somatic copy number alteration profile can lead to the discovery of driver genes by narrowing the vast number of genomic and expression changes in cancer to a small subset that may be more functionally relevant [43, 44]. It can also lead to improvements in cancer diagnosis by utilizing copy number alteration as additional biomarkers [43, 45].

The high-resolution characterization of the SCNA profile in the cancer genome has been made possible by the emergence of both array-based and NGS-based genomic technologies. Array comparative genomic hybridization (aCGH) is among the earliest techniques for characterizing genome-wide somatic copy number alternation in cancer genome. All aCGH arrays are two channel, and they work by first differentially labeling and hybridizing tumor genomic DNA and normal genomic DNA on a microarray that contains hundreds of thousands of probes [46–48]. The ratio between a tumor and the matched normal sample is then calculated for each probe. To quantify the change of copy number difference, the log of base 2 is usually used so that the log-ratio of 1 and -1 corresponds to double or half as many copies, respectively. The log-ratio of 0 corresponds to no change in the copy number in tumor sample compared to the normal sample at that genomic location. Using the ratio values from all the probes that correspond to different genomic locations, the copy number alteration profile along the chromosome can be inferred. There are two major types of aCGH. The first type of aCGH utilizes bacterial artificial chromosome (BAC) probes, which are typically several hundred bp in length [46]. The BAC aCGH has a median genomic resolution of several mega-bases [46]. The second type of aCGH is the oligonucleotide platform. Such oligonucleotide platforms as those from Agilent and Nimblegen have probes shorter than 100 bp, and each array has from hundreds of thousands to more than 1 million probes. Given the difference in design and manufacturing of the aCGHs (probe length, hybridization chemistry, etc.), BAC and oligonucletide aCGH have their own technical characteristics and may serve for different applications.

With the longer probe, the BAC aCGH in general has higher specificity in the hybridization signal of each probe, and each probe gives more accurate measurement, but it has lower resolution than oligonucleotide aCGH. However, for many applications, in which the aberration of interest is large, the resolution BAC aCGH is rather sufficient. In contrast, the oligonucleotide array has shorter probes and gives more noisy measurement on the individual probe level but provides higher genomic resolution.

In addition to the aCGH platforms, single nucleotide polymorphism (SNP) arrays can also be used to infer somatic copy number alterations in the cancer genome. The SNP arrays are mostly single-color arrays, in which only a tumor or a normal sample is hybridized on a microarray that contains oligonucleotide probes (25–50 bp). The two most popular SNP array platforms are the Affymetrix [49] and Illumina [50] SNP arrays. These arrays contained from hundreds of thousands to more than 1 million probes for inferring SNPs and/or copy number variations. The SNP arrays have the important advantage of measuring copy number alterations and loss of heterozygosity (LOH) simultaneously [51], but they have the disadvantage that the probe design and positioning are not optimal for the estimation of copy number. The advent of next-generation sequencing and the rapid increase in its throughput have made it possible to characterize copy number alteration with a much higher resolution (<10 kb) than aCGH or SNP arrays via whole-genome or whole-exome sequencing [52].

Characterizing somatic copy number amplifications and deletions in cancer genome with high resolution is only the first step in inferring genomic regions, the alteration of which are functionally important for the etiology of cancer. Once the genomic alterations have been detected, the next challenge is to distinguish between driver genomic alterations that confer a selective advantage for the tumor to initiate, grow, or persist and passenger genomic alterations that confer no selective advantages. To address this challenge, it is important to perform joint analysis of the somatic genomic alteration profiles across many tumors. Several algorithms [53] are designed for identifying those regions with aberrations that occur significantly more often than would be expected by chance, using permutation tests that are based on the overall pattern of aberrations seen across the genome. In the current study, we used two well-established algorithms, GISTIC [54, 55] and RAE [56], to identify the regions that harbor recurrent SCNAs by using SNP and aCGH data, respectively. As those regions with recurrent SCNAs often contain many lncRNA genes, we further integrate SCNA data and expression data to identify potential driver lncRNA genes based on the reasoning that functional SCNA should cause gene expression change and the driver lncRNAs should show higher or lower gene

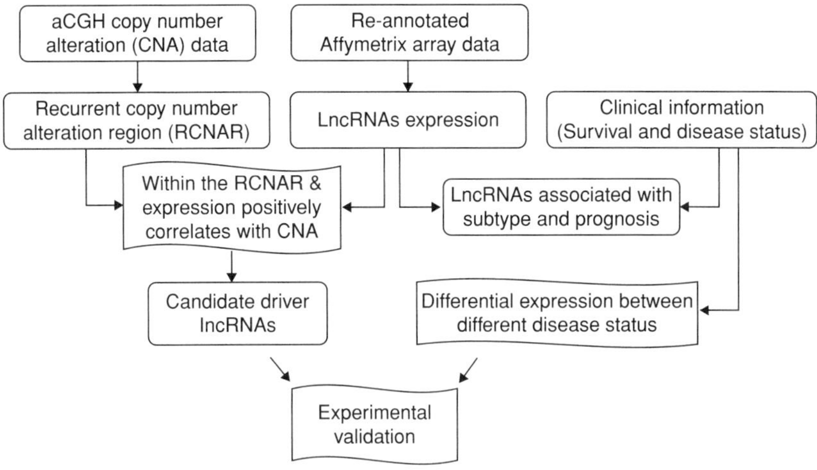

Figure 18.2 The workflow of integrating SCNA data, lncRNA expression data, and clinical information to identify the lncRNAs that are associated with cancer subtypes and clinical prognosis and/or those that are potential drivers of cancer progression.

expression in tumors with the corresponding genomic amplification or deletion compared with the rest tumors (Figure 18.2).

Besides the identification of potential driver lncRNAs, we integrated lncRNA expression with clinical information of individual patient samples including disease status (normal tissue vs. primary or metastatic tumor), subtype, and overall or progression-free survival information of the corresponding patient to predict those lncRNAs that showed different expression between disease status, subtype-specific expression, and/or associations with disease prognosis (Figure 18.2). The significance of differential expression between different statuses was assessed by Mann-Whitney U-test. To identify the lncRNAs that are associated with prognosis, the expression of which is associated with prognosis, we performed multivariate Cox proportional hazard (Cox regression) analyses to assess the associations between lncRNA expression with overall and progression-free survival while controlling for potentially cofounding clinical variables, including ethnicity, age, and gender.

18.3 Application

Using the earlier described approaches, we performed integrative analyses of lncRNA expression profiles, clinical information, and SCNA profiles of tumors in four different cancer types, including 150 tumor samples of prostate cancer from the Memorial Sloan-Kettering Cancer Center (MSKCC) Prostate

Oncogenome Project [24] and 451 tumor samples of glioblastoma multiforme (GBM) [21], 585 tumor samples of ovarian cancer (OvCa) [22], and 113 tumor samples of lung squamous cell carcinoma (lung SCC) [23] from the Cancer Genome Atlas Research Network (TCGA) project [21]. For prostate cancer, the data set includes exon array data, clinical annotation and SCNA data from the Gene expression Omnibus (GEO) (GSE21034). The SCNA regions were determined as the union of SCNA regions from two different studies [24, 57]. Recurrent SCNA regions across different tumors were identified by the algorithms, GISTIC [54] and RAE [56]. The magnitude of the SCNAs was estimated as the log2 ratios of segmented copy numbers between cancer and control DNAs. The exon array data, clinical annotations, and SCNA data of GBM, OvCa, and lung SCC were downloaded from TCGA (https://tcga-data.nci.nih.gov). We further obtained exon array data of 11 human normal tissues from Affymetrix (http://www.affymetrix.com/).

To validate the utility of exon array data in combination with clinical annotation to identify cancer-related lncRNAs, we examined the expression patterns of 13 literature-curated cancer-related lncRNAs [17] that have corresponding exon array probes in a prostate cancer data set [24]. This data set consists of 29 normal prostate samples, 131 primary prostate tumor samples, and 19 metastatic prostate tumor samples with exon array data [24] (Figure 18.3A). Notably, 9 out of these 13 known cancer-related lncRNAs showed significantly different expression between the tumor and normal prostate samples (Mann-Whitney U-test, $p < 0.05$). Three out of these nine lncRNAs were directly related to prostate cancer, including one known prostate cancer diagnostic biomarker, PCA3 [19, 20], and two lncRNAs, PCAT-1 [38] and PCGEM1 [58], that have been functionally implicated in prostate cancer progression. GAS5, a tumor-suppressive lncRNA known to be down-regulated in breast cancer [59], showed increased expression in prostate cancer (Table 18.2), a result suggesting complex and context-dependent functions of lncRNAs in different cancer types. Notably, several lncRNAs, such as NEAT1 [60], DANCR [61], HOTTIP [62], PRINS [63], and EGOT [64], that have established functions in forming nuclear speckles [60], in development [61] and in autoimmune disease [63], but were not previously known to be related to cancer, showed differential expression between tumor and normal prostate samples (Table 18.2), and this suggests their potential function in prostate cancer.

We next sought to identify lncRNAs that showed significant expression differences between tumors and normal prostate tissues and found 109 up-regulated and 104 down-regulated lncRNAs (Mann-Whitney U-test, false discovery rate < 0.05, fold change > 1.5) (Figure 18.3A). Notably, among the lncRNAs with sufficient exon array probe coverage, we rediscovered seven out

Table 18.2 *Known cancer-related lncRNAs or lncRNAs with established function in noncancer context and their regulation in cancer compared with normal prostate tissue*

Ensembl ID	Gene name	MW-U test p-value	Cancer vs. normal	Function annotation
ENSG00000225937	*PCA3*	9.50E-12	Up	Prostate cancer
ENSG00000234741	*GAS5*	1.77E-06	Up	Breast cancer
ENSG00000249859	*PVT1*	4.93E-11	Up	Multiple cancers
ENSG00000226950	*DANCR*	3.03E-08	Up	Development
ENSG00000253438	*PCAT1*	1.12E-05	Up	Prostate cancer
ENSG00000227418	*PCGEM1*	4.49E-04	Up	Prostate cancer
ENSG00000245532	*NEAT1*	0.00642	Up	Nuclear speckle
ENSG00000258492	*KCNQ1O T1*	0.0103	Up	Colon cancer
ENSG00000251164	*HULC*	0.0311	Up	Multiple cancers
ENSG00000251562	*MALAT1*	0.285	–	Multiple cancers
ENSG00000214548	*MEG3*	3.92E-08	Down	Multiple cancers
ENSG00000238115	*PRINS*	1.37E-07	Down	Autoimmune disease
ENSG00000243766	*HOTTIP*	1.95E-06	Down	Development
ENSG00000235947	*EGOT*	2.48E-05	Down	Development
ENSG00000214049	*UCA1*	2.11E-02	Down	Bladder cancer
ENSG00000228630	*HOTAIR*	0.0573	–	Multiple cancers
ENSG00000130600	*H19*	0.0842	–	Multiple cancers
ENSG00000240498	*ANRIL*	0.699	–	Prostate cancer

Note: The statistical significance of the expression difference between cancer and normal prostate tissue was evaluated by Mann-Whitney U-test (MW-U test)

of eight lncRNAs that were reported to show higher expression in prostate cancer from an independent study based on RNA-seq data [38]. Furthermore, we identified an additional 102 lncRNA genes that were up-regulated in prostate cancer but were missed by the other study [38], and this suggests that arrays and RNA-seq may be complementary methods to identify clinically relevant lncRNAs.

Cancer is a clinically heterogeneous disease, and individual cancer types can be further divided into molecular subtypes, each with specific biological and clinical behaviors. Previous studies established four subtypes of GBM (proneural, neural, classical, and mesenchymal) [21], four subtypes of OvCa (immunoreactive, proliferative, mesenchymal, and differentiated) [22], and four subtypes of lung SCC (basal, classical, primitive, and secretory) [23] on the basis of the expression profiles of PCGs, and six subtypes of prostate cancer on the basis of the SCNA profiles [24]. LncRNAs with subtype-specific expression may have an important function in individual molecular subtypes. We compared

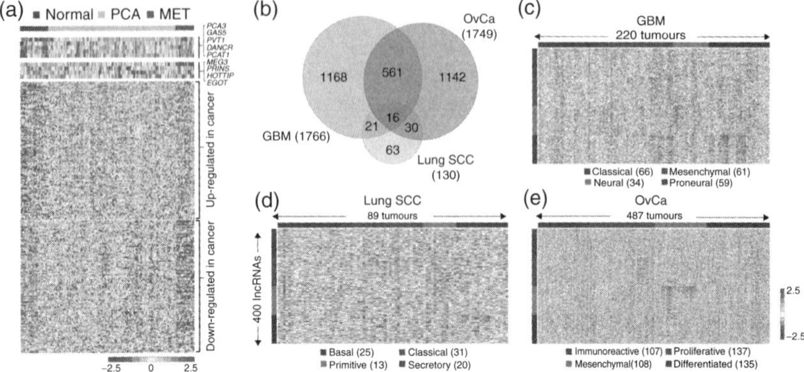

Figure 18.3 (A) The expression level of lncRNA that showed significantly differential expression between cancer and normal prostate tissues shown in heatmap across 29 normal prostate samples and 131 primary and 19 metastatic prostate tumor samples. Several known cancer-related lncRNA or lncRNA with established function in a noncancer context were highlighted. (B) Venn diagram representing the number of subtype-specific lncRNA in three cancers. The expression profile of the top 100 lncRNA that exhibited significantly higher expression in one subtype than the others for (C) GBM, (D) OvCa, and (E) Lung SCC shown in heatmap. (Note: the rank was based on the ascending order of the *p*-value.) Tumor samples were hierarchically clustered within each subtype.

lncRNA expression across different subtypes and identified hundreds of lncRNAs showing subtype-specific expression patterns in GBM, OvCa, and lung SCC (FDR < 0.05; Figures 18.3B–18.3E). The same approach did not yield any lncRNAs with significant subtype-specific expression in prostate cancer, which was reminiscent of the lack of a robust PCG expression–based subtype of prostate cancer [24]. In addition, 628 lncRNAs showed subtype-specific expression in more than one cancer type (Figure 18.3B), and some of these lncRNAs have been functionally implicated in other physiological or pathological processes. For example, MIAT, a lncRNA that showed specific expression in the mesenchymal subtype of OvCa and the proneural subtype of GBM, is known to confer risk of myocardial infarction [65] and regulate retinal cell fate specification [66]. In addition, RMST, a lncRNA known to be differentially expressed between rhabdomyosarcoma subtypes [67], also showed subtype-specific expression patterns in GBM, OvCa, and lung SCC. The lncRNAs that showed statistically higher expression (false discovery rate < 0.05) in only one subtype were considered to be subtype specific.

A previous study of HOTAIR [16, 68] showed that patients with higher HOTAIR expression had poorer prognosis in colorectal cancer [69]. To identify the lncRNAs that are associated with clinical outcome in prostate cancer,

GBM, OvCa, and lung SCC, we performed multivariate Cox regression analysis to evaluate the significance of the correlations between individual lncRNA expression and overall and progression-free survival in the presence of other confounding factors such as ethnicity, age, and gender. With these data, we are able to identify lncRNAs in prostate cancer, GBM, OvCa, and lung SCC whose expression was significantly correlated with overall or progression-free survival ($p < 0.01$). Notably, nine lncRNAs showed consistent positive or negative correlations between their expression and overall or progression-free survival in different cancer types, and this suggests their potential as more general prognostic biomarkers. The lncRNA gene with the Ensembl ID ENSG00000261582 is an example of a lncRNA that showed negative correlation between its expression and overall survival in both lung SCC and OvCa (Figure 18.4A). This lncRNA also showed subtype-specific expression in OvCa but not in lung SCC. Additionally, five lncRNAs showed marked and consistent positive or negative correlations between both overall and progression-free survival in OvCa (one such example, Ensembl ID ENSG00000225128, is shown in Figure 18.4B).

An important form of somatic genetic alteration in cancer is SCNAs, in which a genomic region is either amplified or deleted. Some of the genes within amplified (or deleted) regions show increased (or decreased) expression levels, leading to altered activity in cancer cells. Studies have suggested that the genes with causal roles in oncogenesis are often located in the SCNAs that are frequently altered across tumors [57, 69, 70]. To reveal the lncRNAs that may have tumor-promoting or -suppressing functions, we identified hundreds of lncRNAs that map to regions of recurrent SCNAs across tumors for prostate cancer, GBM, OvCa, and lung SCC (Figure 18.4C). Some of these lncRNAs also showed marked correlation between overall or progression-free survival [14]. In addition, we identified lncRNAs that were consistently located in regions of SCNAs across different cancers (Figure 18.4C) and found a significant overlap of the lncRNA genes that are located in SCNA gain or loss regions between some of the cancer types [14]. Among the many genes located within regions of SCNAs, probably only a fraction of them are drivers of cancer. To further distinguish driver from passenger lncRNAs in the regions of SCNAs, we integrated SCNA and expression profiles of lncRNAs in tumors. We reasoned that driver lncRNAs with SCNAs should result in corresponding gene expression changes [70, 71], as only those SCNAs that cause changes in transcript abundance could possibly alter lncRNA activity. Therefore, we selected lncRNAs whose SCNAs showed positive correlations with expression level changes as candidate drivers for prostate cancer, GBM, OvCa, and lung SCC. Among the lncRNAs in the SCNA regions, we selected those that showed significant and concordant expression changes (one-tailed Mann-Whitney U-test,

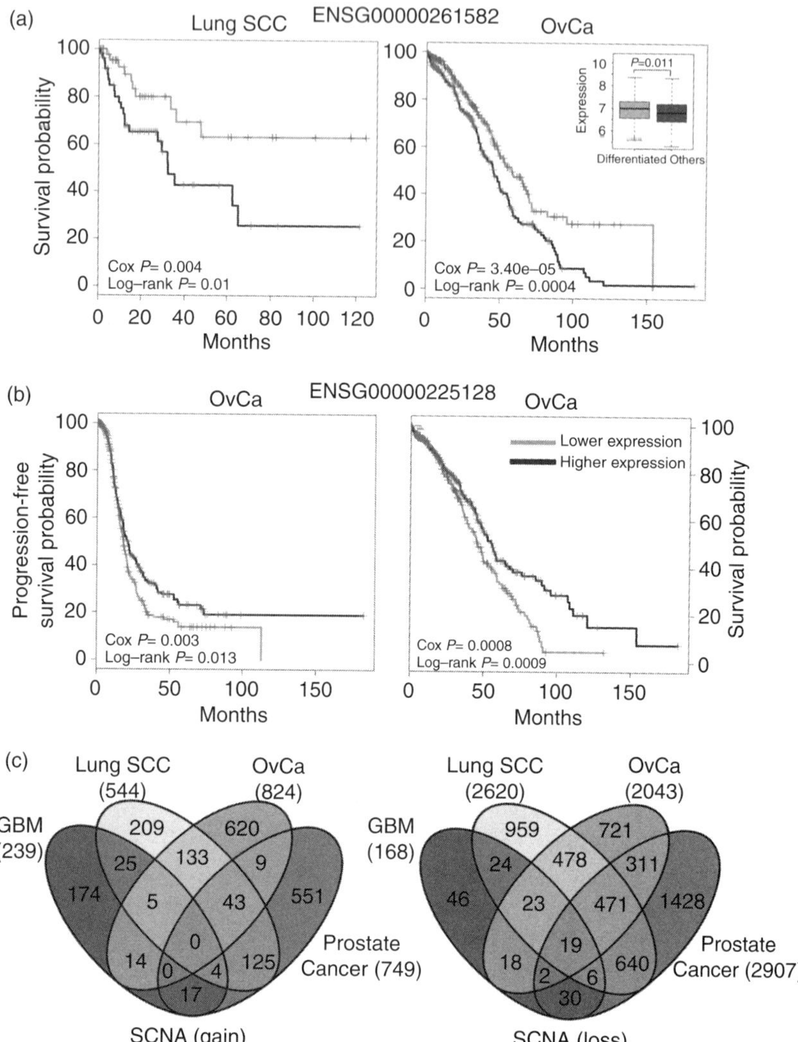

Figure 18.4 (A) Kaplan-Meier curve of two patient groups with higher (top 50%) and lower expression (bottom 50%) of ENSG00000261582 in Lung SCC and OvCa (red, higher expression; blue, lower expression). The box plot demonstrates that ENSG00000261582 was expressed higher in the "differentiated" subtype of OvCa than the other subtypes. Both the *p*-value of the multivariate Cox model for lncRNA expression and the *p*-value of the log-rank test were shown. (B) Kaplan-Meier curve for overall and progression-free survival of two patient groups with higher (top 50%) and lower expression (bottom 50%) of ENSG00000263041 in OvCa. (C) Number of lncRNA located in the SCNA (gain) and SCNA (loss) regions in different cancers shown as Venn diagrams.

$p < 0.05$) in tumor samples with a corresponding somatic copy number gain (log2 ratio > 0.2) or loss (log2 ratio < -0.2) compared to the other samples [14].

To further validate the reliability of the integrative studies, and as it is prohibitive to validate all candidate driver lncRNAs in the four cancer types, we focused our experimental validation and comprehensive annotation on candidate lncRNAs that may have tumor-promoting functions in prostate cancer (i.e., those in recurrent SCNA (gain) regions that showed positive correlations between their SCNAs and expression levels). Among all the candidate driver lncRNAs that showed increasing expression from normal to primary to metastatic prostate cancer, we chose the two that showed the most significant expression difference between tumor and normal prostate tissue (i.e., the two with the smallest p-values calculated by Mann-Whitney U-test) for experimental validation. The criterion of increasing expression from normal to primary to metastatic prostate cancer aimed to uncover lncRNAs that may be important therapeutic targets for both primary and metastatic cancers.

We named these two lncRNAs prostate cancer–associated noncoding RNAs 1 and 2, abbreviated as PCAN-R1 (Ensembl ID ENSG00000228288) and PCAN-R2 (Ensembl ID ENSG00000231806), respectively. Both lncRNAs showed positive correlations between gene expression and the advancement of the disease status and SCNAs (Figures 18.5A and 18.5B). To confirm that the two lncRNAs PCAN-R1 and PCAN-R2 are noncoding, we used two different methods, txCdsPredict from UCSC and phyloCSF [72], to calculate their coding potential. For coding-potential calculations with phyloCSF, we used the multiple sequence alignment of 29 mammalian genomes [73]. We chose the thresholds used previously (txCdsPredict $= 800$ [38] and phyloCSF $= 100$ [8]), below which the transcripts were considered to be noncoding. We found that the scores of all possible opening reading frames from the PCAN-R1 and PCAN-R2 transcripts were well below the thresholds (txCdsPredict scores: PCAN-R1, 470 and PCAN-R2, 359; phyloCSF scores: PCAN-R1, –123.1434 and PCAN-R2, –148.5448), supporting that these two lncRNA genes are noncoding.

We chose the prostate cancer cell line LNCaP, in which both lncRNAs have moderate or higher expression levels compared with their expression in other prostate cancer or non–prostate cancer cell lines, for experimental validation. Using 5′ and 3′ rapid amplification of cDNA ends (RACE), we found that for PCAN-R1, although one isoform (PCAN-R1-A) was almost identical to the Ensembl annotated transcript ENST00000425295 (Figure 18.5C), the other isoform (PCAN-R1-B) was a spliced variant of PCAN-R1-A with an intron

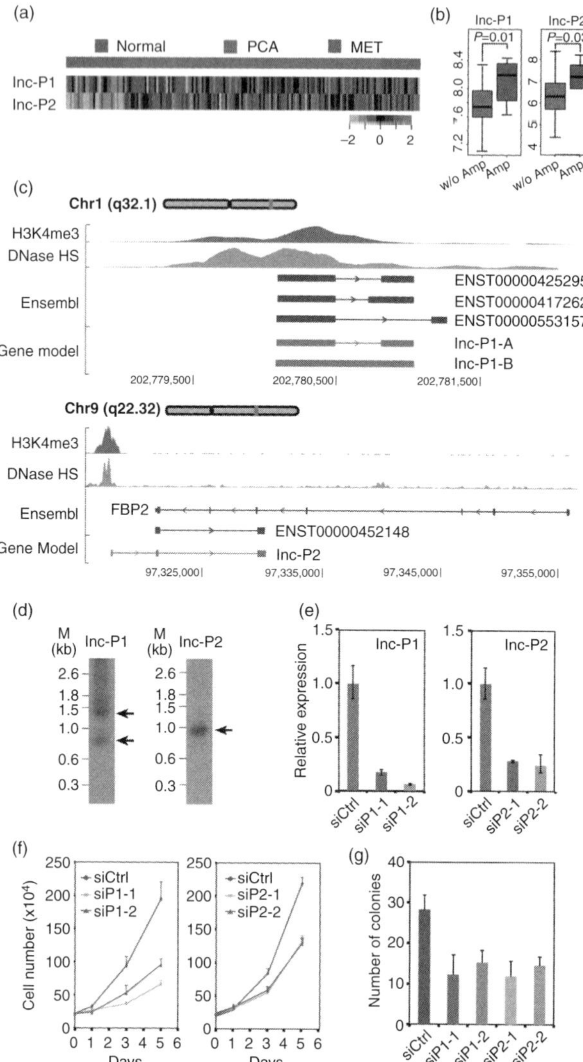

Figure 18.5 Experimental validation of lnc-P1 and lnc-P2 function. (A) Heatmap showing the expression of lnc-P1 and lnc-P2 in normal prostate tissue, primary and metastatic prostate cancer. (B) Box plot of lnc-P1 and lnc-P2 expression in tumors with genomic amplification and in the tumors without genomic amplification. (C) Transcript structure of lnc-P1 and lnc-P2 from Ensembl annotation and determined by 5′ and 3′ RACE experiments in LNCaP cell. In addition, the H3K4me3 and DNase I hypersensitive region profiles in the same cell line are shown. (D) The Northern blot of lnc-P1 and lnc-P2 transcripts. (E) Relative expression level of lnc-P1 and lnc-P2 upon knockdown by two different siRNA (purple and orange) and upon control siRNA treatment (green). (F) Growth curves of LNCaP cell with or without targeted siRNA-mediated knockdown of lnc-P1 or lnc-P2. The growth curves of control siRNA-treated cells and the growth curves of two targeted siRNA-treated cells plotted in purple, orange, and green, respectively. (G) Number of soft-agar colony formation of LNCaP cell with or without targeted siRNA-mediated knockdown of lnc-P1 or lnc-P2.

retention (Figure 18.5C). Notably, for PCAN-R2, the major isoform had an extra exon in the 5′ end, and the remaining two exons also had different lengths from the Ensembl annotation (Figure 18.5C). The new 5′ exon of PCAN-R2 was more consistent with the profile of histone H3 Lys4 trimethylation (H3K4me3), a histone mark of an active promoter and the profile of DNase I hypersensitive regions (i.e., the regions with an open chromatin state) in LNCaP cells.

We confirmed the transcript structures of PCAN-R1 and PCAN-R2 by northern blot and performed short interfering RNAs (siRNAs) knockdown experiments and observed the substantial decreases in cell growth. Additional experiments were further conducted and concordantly proved the influence on cancer cell growth caused by the expression of two lncRNAs. As a lncRNA may act in *cis* and influence the expression of its neighboring PCG, we investigated whether the expression of the neighboring PCG was regulated by PCAN-R1 or PCAN-R2. siRNA knockdown of PCAN-R1 or PCAN-R2 had no effect on the expression of their neighboring PCGs KDM5B and FBP2, respectively, and this suggests that the functional mechanisms of PCAN-R1 and PCAN-R2 are not directly through their neighboring PCGs. Notably, in normal tissues, PCAN-R1 and its neighboring PCG KDM5B showed the highest expression in testis. In contrast, although PCAN-R2 showed similar expression across different tissues, its neighboring PCG FBP2 showed a muscle-specific expression pattern, thus suggesting that the expressions of PCAN-R2 and FBP2 may be differently regulated.

18.4 Discussion

The case study presented in this chapter has demonstrated that integrating the orthogonal genomic data, such as lncRNA expression profiles, and somatic copy number alteration along with clinical information can greatly facilitate the discovery of lncRNA that may serve as therapeutic targets and diagnostic or prognostic biomarkers. Our analyses also indicate that repurposing microarray probes to construct a lncRNA expression profile in a patient sample is a cost-effective approach given the large number of such data sets available in public repositories. The constructed gene expression profiles of both lncRNAs and PCGs from our analyses are a valuable resource for understanding the similarities and differences of transcriptional (e.g., antisense RNA [74]) regulation of PCGs by lncRNAs across different cancer types. In the combination of matched SCNA profile and clinical information, these gene expression profiles also allow network models to be inferred [75, 76], which will help advance the understanding of lncRNA function in cancer etiology.

The experimental validation of two lncRNAs without previous implication in cancer suggests the effectiveness of our integrative analyses in finding functionally important lncRNAs in cancer. Our analyses predicted about 80–300 candidate driver lncRNAs that may have tumor-promoting functions in each of the four cancer types. An intersection of such a list of candidate driver lncRNAs with a list of lncRNAs generated from orthogonal functional genomic data sets, such as that generated by ribonucleoprotein immunoprecipitation followed by sequencing [77] (a genomic technique for identifying lncRNAs physically associated with the protein of interest), would greatly help prioritize their functional valuation in different biological contexts, including epigenetic regulation, and facilitate the discovery of lncRNA therapeutic targets.

In our current study, we only used SCNA and expression data in combination with clinical information for our integrative analysis. It is conceivable that other types of genomic data, such as SNP array [78] and genome sequencing data [52], can be further integrated to reveal the multifaceted relationship between the mutation spectrum and expression of lncRNAs, disease status, and clinical outcome.

In summary, we report a proof-of-principle study for identifying clinically relevant lncRNAs through integrative analyses of orthogonal genomic data sets and clinical information. Our study opens new avenues for leveraging publicly available genomic data to study the functions and mechanisms of lncRNAs in human disease.

References

1　Djebali, S., et al. Landscape of transcription in human cells. *Nature* **489**, 101–108 (2012).

2　Bartel, D. P. MicroRNAs: target recognition and regulatory functions. *Cell* **136**, 215–233 (2009).

3　Muniategui, A., Pey, J., Planes, F. J., & Rubio, A. Joint analysis of miRNA and mRNA expression data. *Brief Bioinform* **14**, 263–278 (2012).

4　Frampton, A. E., et al. Integrated analysis of miRNA and mRNA profiles enables target acquisition in human cancers. *Expert Rev Anticancer Ther* **12**, 323–330 (2012).

5　Berezikov, E. Evolution of microRNA diversity and regulation in animals. *Nat Rev Genet* **12**, 846–860 (2011).

6　Pritchard, C. C., Cheng, H. H., & Tewari, M. MicroRNA profiling: approaches and considerations. *Nat Rev Genet* **13**, 358–369 (2012).

7　Chen, K., & Rajewsky, N. The evolution of gene regulation by transcription factors and microRNAs. *Nat Rev Genet* **8**, 93–103 (2007).

8　Cabili, M. N., et al. Integrative annotation of human large intergenic noncoding RNAs reveals global properties and specific subclasses. *Genes Dev* **25**, 1915–1927 (2011).

9 Guttman, M., et al. Chromatin signature reveals over a thousand highly conserved large non-coding RNAs in mammals. *Nature* **458**, 223–227 (2009).

10 Guttman, M., & Rinn, J. L. Modular regulatory principles of large non-coding RNAs. *Nature* **482**, 339–346 (2012).

11 Ota, T., et al. Complete sequencing and characterization of 21,243 full-length human cDNAs. *Nat Genet* **36**, 40–45 (2004).

12 Khalil, A. M., et al. Many human large intergenic noncoding RNAs associate with chromatin-modifying complexes and affect gene expression. *Proc Natl Acad Sci U S A* **106**, 11667–11672 (2009).

13 Guttman, M., et al. Ab initio reconstruction of cell type–specific transcriptomes in mouse reveals the conserved multi-exonic structure of lincRNAs. *Nat Biotechnol* **28**, 503–510 (2010).

14 Du, Z., et al. Integrative genomic analyses reveal clinically relevant long noncoding RNAs in human cancer. *Nat Struct Mol Biol* **20**, 908–913 (2013).

15 Tian, D., Sun, S., & Lee, J. T. The long noncoding RNA, Jpx, is a molecular switch for X chromosome inactivation. *Cell* **143**, 390–403 (2010).

16 Gupta, R. A., et al. Long non-coding RNA HOTAIR reprograms chromatin state to promote cancer metastasis. *Nature* **464**, 1071–1076 (2010).

17 Prensner, J. R., & Chinnaiyan, A. M. The emergence of lncRNAs in cancer biology. *Cancer Discov* **1**, 391–407 (2011).

18 Wapinski, O., & Chang, H. Y. Long noncoding RNAs and human disease. *Trends Cell Biol* **21**, 354–361 (2011).

19 Lee, G. L., Dobi, A., & Srivastava, S. Prostate cancer: diagnostic performance of the PCA3 urine test. *Nat Rev Urol* **8**, 123–124 (2011).

20 Hessels, D., & Schalken, J. A. Urinary biomarkers for prostate cancer: a review. *Asian J Androl* **15**, 333–339 (2013).

21 Cancer Genome Atlas Research Network. Comprehensive genomic characterization defines human glioblastoma genes and core pathways. *Nature* **455**, 1061–1068 (2008).

22 Cancer Genome Atlas Research Network. Integrated genomic analyses of ovarian carcinoma. *Nature* **474**, 609–615 (2011).

23 Cancer Genome Atlas Research Network. Comprehensive genomic characterization of squamous cell lung cancers. *Nature* **489**, 519–525 (2012).

24 Taylor, B. S., et al. Integrative genomic profiling of human prostate cancer. *Cancer Cell* **18**, 11–22 (2010).

25 Liao, Q., et al. Large-scale prediction of long non-coding RNA functions in a coding-non-coding gene co-expression network. *Nucleic Acids Res* **39**, 3864–3878 (2011).

26 Mercer, T. R., Dinger, M. E., Sunkin, S. M., Mehler, M. F., & Mattick, J.S. Specific expression of long noncoding RNAs in the mouse brain. *Proc Natl Acad Sci U S A* **105**, 716–721 (2008).

27 Michelhaugh, S. K., et al. Mining Affymetrix microarray data for long non-coding RNAs: altered expression in the nucleus accumbens of heroin abusers. *J Neurochem* **116**, 459–466 (2010).

28 Raghavachari, N., et al. A systematic comparison and evaluation of high density exon arrays and RNA-seq technology used to unravel the peripheral blood transcriptome of sickle cell disease. *BMC Med Genomics* **5**, 28 (2012).

29 Xu, W., et al. Human transcriptome array for high-throughput clinical studies. *Proc Natl Acad Sci U S A* **108**, 3707–3712 (2011).

30 Levin, J. Z., et al. Comprehensive comparative analysis of strand-specific RNA sequencing methods. *Nat Methods* **7**, 709–715 (2010).

31 Flicek, P., et al. Ensembl 2012. *Nucleic Acids Res* **40**, D84–90 (2012).
32 Jiang, H., & Wong, W. H. SeqMap: mapping massive amount of oligonucleotides to the genome. *Bioinformatics* **24**, 2395–2396 (2008).
33 Kuhn, R. M., Haussler, D., & Kent, W. J. The UCSC genome browser and associated tools. *Brief Bioinform* **14**, 144–161 (2012)
34 Derrien, T., et al. The GENCODE v7 catalog of human long noncoding RNAs: analysis of their gene structure, evolution, and expression. *Genome Res* **22**, 1775–1789 (2012).
35 Kapur, K., Xing, Y., Ouyang, Z., & Wong, W. H. Exon arrays provide accurate assessments of gene expression. *Genome Biol* **8**, R82 (2007).
36 Seok, J., Xu, W., Gao, H., Davis, R. W., & Xiao, W. JETTA: junction and exon toolkits for transcriptome analysis. *Bioinformatics* **28**, 1274–1275 (2012).
37 Johnson, W. E., Li, C., & Rabinovic, A. Adjusting batch effects in microarray expression data using empirical Bayes methods. *Biostatistics* **8**, 118–127 (2007).
38 Prensner, J. R., et al. Transcriptome sequencing across a prostate cancer cohort identifies PCAT-1, an unannotated lincRNA implicated in disease progression. *Nat Biotechnol* **29**, 742–749 (2011).
39 Trapnell, C., et al. Transcript assembly and quantification by RNA-Seq reveals unannotated transcripts and isoform switching during cell differentiation. *Nat Biotechnol* **28**, 511–515 (2010).
40 Wang, Z., Gerstein, M., & Snyder, M. RNA-Seq: a revolutionary tool for transcriptomics. *Nat Rev Genet* **10**, 57–63 (2009).
41 Frohling, S., & Dohner, H. Chromosomal abnormalities in cancer. *N Engl J Med* **359**, 722–734 (2008).
42 Hanahan, D., & Weinberg, R. A. Hallmarks of cancer: the next generation. *Cell* **144**, 646–674 (2011).
43 Stratton, M. R. Exploring the genomes of cancer cells: progress and promise. *Science* **331**, 1553–1558 (2011).
44 Albertson, D. G., Collins, C., McCormick, F., & Gray, J. W. Chromosome aberrations in solid tumors. *Nat Genet* **34**, 369–376 (2003).
45 Hanash, S. Integrated global profiling of cancer. *Nat Rev Cancer* **4**, 638–644 (2004).
46 Pinkel, D., et al. High resolution analysis of DNA copy number variation using comparative genomic hybridization to microarrays. *Nat Genet* **20**, 207–211 (1998).
47 Pinkel, D. & Albertson, D. G. Array comparative genomic hybridization and its applications in cancer. *Nat Genet* **37 Suppl**, S11–S17 (2005).
48 Lee, C., Iafrate, A. J., & Brothman, A. R. Copy number variations and clinical cytogenetic diagnosis of constitutional disorders. *Nat Genet* **39**, S48–S54 (2007).
49 Matsuzaki, H., et al. Genotyping over 100,000 SNPs on a pair of oligonucleotide arrays. *Nat Methods* **1**, 109–111 (2004).
50 Shen, R., et al. High-throughput SNP genotyping on universal bead arrays. *Mutat Res* **573**, 70–82 (2005).
51 Beroukhim, R., et al. Inferring loss-of-heterozygosity from unpaired tumors using high-density oligonucleotide SNP arrays. *PLoS Comput Biol* **2**, e41 0323–0332 (2006).
52 Meyerson, M., Gabriel, S., & Getz, G. Advances in understanding cancer genomes through second-generation sequencing. *Nat Rev Genet* **11**, 685–696 (2010).
53 Yuan, X., Zhang, J., Zhang, S., Yu, G., & Wang, Y. Comparative analysis of methods for identifying recurrent copy number alterations in cancer. *PLoS ONE* **7**, e52516 (2013).

54 Beroukhim, R., et al. Assessing the significance of chromosomal aberrations in cancer: methodology and application to glioma. *Proc Natl Acad Sci U S A* **104**, 20007–20012 (2007).

55 Mermel, C. H., et al. GISTIC2.0 facilitates sensitive and confident localization of the targets of focal somatic copy-number alteration in human cancers. *Genome Biol* **12**, R41 (2011).

56 Taylor, B. S., et al. Functional copy-number alterations in cancer. *PLoS ONE* **3**, e3179 (2008).

57 Beroukhim, R., et al. The landscape of somatic copy-number alteration across human cancers. *Nature* **463**, 899–905 (2010).

58 Petrovics, G., et al. Elevated expression of PCGEM1, a prostate-specific gene with cell growth-promoting function, is associated with high-risk prostate cancer patients. *Oncogene* **23**, 605–611 (2004).

59 Mourtada-Maarabouni, M., Pickard, M. R., Hedge, V. L., Farzaneh, F., & Williams, G. T. GAS5, a non-protein-coding RNA, controls apoptosis and is downregulated in breast cancer. *Oncogene* **28**, 195–208 (2009).

60 Clemson, C. M., et al. An architectural role for a nuclear noncoding RNA: NEAT1 RNA is essential for the structure of paraspeckles. *Mol Cell* **33**, 717–726 (2009).

61 Kretz, M., et al. Suppression of progenitor differentiation requires the long noncoding RNA ANCR. *Genes Dev* **26**, 338–343 (2012).

62 Wang, K. C., et al. A long noncoding RNA maintains active chromatin to coordinate homeotic gene expression. *Nature* **472**, 120–124 (2011).

63 Szegedi, K., et al. The anti-apoptotic protein G1P3 is overexpressed in psoriasis and regulated by the non-coding RNA, PRINS. *Exp Dermatol* **19**, 269–278 (2010).

64 Wagner, L. A. et al. EGO, a novel, noncoding RNA gene, regulates eosinophil granule protein transcript expression. *Blood* **109**, 5191–5198 (2007).

65 Ishii, N., et al. Identification of a novel non-coding RNA, MIAT, that confers risk of myocardial infarction. *J. Human Genet.* **51**, 1087–1099 (2006).

66 Rapicavoli, N. A., Poth, E. M., & Blackshaw, S. The long noncoding RNA RNCR2 directs mouse retinal cell specification. *BMC Dev Biol* **10**, 49 (2010).

67 Chan, A. S., Thorner, P. S., Squire, J. A., & Zielenska, M. Identification of a novel gene NCRMS on chromosome 12q21 with differential expression between rhabdomyosarcoma subtypes. *Oncogene* **21**, 3029–3037 (2002).

68 Rinn, J. L., et al. Functional demarcation of active and silent chromatin domains in human HOX loci by noncoding RNAs. *Cell* **129**, 1311–1323 (2007).

69 Kogo, R., et al. Long noncoding RNA HOTAIR regulates polycomb-dependent chromatin modification and is associated with poor prognosis in colorectal cancers. *Cancer Res* **71**, 6320–6326 (2011).

70 Garraway, L. A., et al. Integrative genomic analyses identify MITF as a lineage survival oncogene amplified in malignant melanoma. *Nature* **436**, 117–122 (2005).

71 Akavia, U. D., et al. An integrated approach to uncover drivers of cancer. *Cell* **143**, 1005–1017 (2010).

72 Lin, M. F., Jungreis, I., & Kellis, M. PhyloCSF: a comparative genomics method to distinguish protein coding and non-coding regions. *Bioinformatics* **27**, i275–i282 (2011).

73 Lindblad-Toh, K., et al. A high-resolution map of human evolutionary constraint using 29 mammals. *Nature* **478**, 476–482 (2011).

74 Tran, V. G., et al. H19 antisense RNA can up-regulate Igf2 transcription by activation of a novel promoter in mouse myoblasts. *PLoS ONE* **7**, e37923 (2012).

75 Califano, A., Butte, A. J., Friend, S., Ideker, T., & Schadt, E. Leveraging models of cell regulation and GWAS data in integrative network-based association studies. *Nat Genet* **44**, 841–847 (2012).
76 Pe'er, D., & Hacohen, N. Principles and strategies for developing network models in cancer. *Cell* **144**, 864–873 (2011).
77 Zhao, J., et al. Genome-wide identification of polycomb-associated RNAs by RIP-seq. *Mol Cell* **40**, 939–953 (2010).
78 Syvanen, A. C. Accessing genetic variation: genotyping single nucleotide polymorphisms. *Nat Rev Genet* **2**, 930–942 (2001).

19

Drug-Pathway Association Analysis: Integration of High-Dimensional Transcriptional and Drug Sensitivity Profile

CONG LI, CAN YANG, GREG HATHER, RAY LIU, AND
HONGYU ZHAO

Abstract

Traditional drug discovery practices usually adopt the "one drug – one target" approach, which ignore the fact the disease occurrence is usually the result of an extremely complex combination of molecular events. Pathway-based approaches address this limitation by considering biological pathways as potential drug targets. A first step of pathway-based drug discovery is to identify associations between drug candidates and biological pathways. This has been made possible by the availability of high-dimensional transcriptional and drug sensitivity profile data. In this chapter, we describe two statistical methods, "iFad" and "iPad", which perform drug-pathway association analysis by integrating these two types high-dimensional data. We also demonstrate their utilities by applying them to the NCI-60 data set.

19.1 Introduction

Drug discovery is the process of identifying new candidate medications for diseases of interest. The common practice adopted by the pharmaceutical industry is to design maximally selective drug molecules to act on individual drug targets [11], which is usually referred to as the "one drug – one target" approach. This paradigm has indeed enjoyed some successes [27]. Yet, the last 15 years have witnessed a significant increase in the attrition rate of new candidate drugs due to their low efficacy and serious side effects [17, 29]. One fundamental reason for the decline in the productivity of the pharmaceutical industry may lie in the core philosophy of the "one drug – one target" approach [11]. Specifically, this philosophy ignores the fact that disease occurrence is usually the result of an extremely complex combination of molecular events [20] among certain sets of functionally related genes, usually referred to as "pathways". Targeting an individual drug target may not provide sufficient interference to the whole disease-related pathway and therefore usually results in unsatisfactory efficacy. Moreover, it fails to consider the mechanism of a candidate drug at a systems level, making it extremely difficult to evaluate drug safety and toxicity in the early developmental stages [14]. Due to these limitations of the "one drug – one target" approach, a new concept of drug discovery – polypharmacology [6] – is emerging as a promising alternative for drug developments. Instead of targeting individual drug targets, polypharmacology seeks to design or find candidate drugs that interfere multiple molecular targets. For example, pathway-based drug discovery, which pursues candidate drugs that interfere the activity of a whole biological pathway, has become increasingly appealing.

Recent advances in high-throughput technologies have enabled researchers to assay various aspects of many drug candidates and many drug targets simultaneously [9, 13, 26], which further expedited the paradigm shift from "one drug-one target" to polypharmacology or pathway-based drug discovery. However, data generated from these high-throughput technologies usually have very high dimensionality. In addition, data generated from different technologies are of distinct types. There is an increasingly urgent need for statistical methods that can effectively integrate these data types to delineate the complex relationships between large amount of candidate drugs and drug targets. A wide spectrum of approaches have been developed, depending on the types of data that are used. According to a recent review [23], these approaches can be grouped into three categories: 1) ligand-based approach, which predicts the binding affinity of candidate drugs to drug targets through its structural or topological similarities with other drug molecules with known target molecules [40, 38]; 2) target-based approach, which predicts drug-target interactions through the similarities between target molecules in protein structure, sequence, evolutionary and functional information [39, 24, 36, 10, 16]; 3) phenotype-based approach, which tries to identify the associations between drugs and targets by comparing biological phenotypes, e.g. gene expression levels and cell line responses to drugs. The major advantage of the first and second category of approaches is their ability to make use of the rich chemical and biological information about candidate drugs and targets, which are generally overlooked by the third category of approaches. However, these chemical and biological information is not available for every drug candidate or every drug target, which limits their utilities on a genome-wide scale. In contrast, the high-throughput technologies nowadays can easily assay various types of biological phenotypes at a large scale, enabling the third category of approaches to perform genome-wide analysis of drug-target associations in a high-throughput fashion. In this chapter, we focus on the third category of approaches, in particular, two recently developed pathway-based statistical methods that fall in this category. Before we present these two methods, we give a brief overview of the background of the data types that are used and the concept of biological pathways.

Two types of data are commonly used in drug target prediction - gene expression profiles and drug sensitivity data. Gene expression profiles consist of the expression levels of genome-wide transcripts. Although earlier studies predominantly used expression microarrays to gather transcript levels, RNA-seq data [37] are becoming more routinely collected recently due to the rapid development of the next-generation sequencing technology. Drug sensitivity analysis usually measures the cell's responses to drug treatments. A commonly used measurement of drug sensitivity is the "GI_{50}" value, which is the minimum concentration of the drug needed to inhibit the cell growth by 50%. Comparing the gene expression profiles and drug sensitivities of the same group of cells allows researchers to investigate the complex relationships between drugs and target genes. An early initiative that successfully integrated these two types of data to understand the drug mechanisms is the NCI-60 project [31], in which 60 human tumor cell lines were screened against more than 100,000 compounds to build a public repository of comprehensive gene expression profiles and drug sensitivity data.

According to the National Human Genome Research Institute (NHGRI), a biological pathway is defined as "a series of actions among molecules in a cell that leads to a certain product or a change in a cell" (http://www.genome.gov/27530687). Therefore genes/proteins and other molecules involved in the same pathway represent a tightly connected functional module and identifying the associations between drugs and the module as a whole is a critical task in pathway-based drug discovery. Among various biological pathways, three types of pathways are most commonly used: metabolic pathways, gene regulation pathways and signal transduction pathways. According to the definition of NHGRI, metabolic pathways refer to those that are involved in chemical reactions among various metabolites in human body. Gene regulation pathways control the "switches" that determine the expression levels of genes. Signal transduction pathways are responsible for transmitting signals from extracellular environment to a cell's interior in order to accomplish further cellular activities. With more and more pathway databases available, pathway-based analysis has become a common theme in many research areas. Accordingly, many bioinformatics tools have been developed to address various pathway-based analysis problems. A widely-known tool for pathway-based analysis is the "Gene Set Enrichment Analysis" (GSEA) [33]. The goal of GSEA is to identify biological pathways, or more generally speaking, any pre-defined gene sets that are enriched for genes significantly differentially expressed across different condition/treatment groups. It ranks all the genes according to their significance levels of differential expression and then walks down the list while recording a running sum statistic for each gene set which increases if a gene in the gene set is encountered and decreases otherwise. The maximum deviation from zero of this statistic is defined as the "enrichment score" for the corresponding gene set, whose significance is then assessed by permuting the group labels of the samples. Among the numerous pathway databases, popular examples are, the KEGG pathways (http://www.genome.jp/kegg/pathway.html), the BioCarta pathways (http://www.biocarta.com), the Reactome pathways (http://www.reactome.org) and many others.

In the following sections, we will introduce two recently developed and closely related methods for identifying associations between drugs and biological pathways. The first method is called "iFad" (integrative factor analysis model for drug-pathway association inference) [22]. The second method is called "iPad" (integrative penalized matrix decomposition for drug-pathway association analysis) [19]. Note that we adjusted the notations from the original articles for the purpose of unification.

19.2 The iFad Method

19.2.1 Model Description

The iFad method uses a Bayesian sparse factor analysis model to analyze paired gene expression data and drug sensitivity data generated from the same set of samples. It has been implemented as an R package and is publicly available on CRAN. The model considers two data matrices $Y^{(1)} \in \mathbb{R}^{N \times G^{(1)}}$ and $Y^{(2)} \in \mathbb{R}^{N \times G^{(2)}}$, representing the gene expression profiles and the drug sensitivity data (usually GI_{50} values), respectively, where N is the number of samples (usually cell lines), $G^{(1)}$ is the number of genes with expression levels available, and $G^{(2)}$ is the number drugs assayed. The key idea of the iFad method is to treat the activity levels of some biological pathways as a collection of latent factors underlying both gene expression data and drug sensitivity data. Mathematically, $Y^{(1)}$ and $Y^{(2)}$ are modeled as follows:

$$
\begin{aligned}
Y^{(1)} &= X B^{(1)} + E^{(1)}, \\
Y^{(2)} &= X B^{(2)} + E^{(2)},
\end{aligned}
\tag{19.1}
$$

where $X \in \mathbb{R}^{N \times K}$ represents the activity levels of K pathways in the N cell lines, and $B^{(1)} \in \mathbb{R}^{K \times G^{(1)}}$ and $B^{(2)} \in \mathbb{R}^{K \times G^{(2)}}$ are the factor loading matrices describing the effects of pathway activities on the gene expression levels and drug sensitivities, respectively. Note that the two feature spaces, namely gene expression levels and drug sensitivities, share the same collection of latent factors, i.e. the K pathways. The noise terms $E^{(1)} \in \mathbb{R}^{N \times G^{(1)}}$ and $E^{(2)} \in \mathbb{R}^{N \times G^{(2)}}$ are introduced to model any effects that are not captured by all the pathways here. The entries in both $E^{(1)}$ and $E^{(2)}$ are assumed to be i.i.d. (independent and identically distributed) normal with zero mean and variances of $\tau_e^{(1)-1}$ and $\tau_e^{(2)-1}$, respectively. Then the conjugate inverse-Gamma hyper priors are imposed on both $\tau_e^{(1)}$ and $\tau_e^{(2)}$,

$$
\begin{aligned}
\tau_e^{(1)-1} &\sim \text{Gamma}\,(\alpha_e^{(1)}, \beta_e^{(1)}), \\
\tau_e^{(2)-1} &\sim \text{Gamma}\,(\alpha_e^{(2)}, \beta_e^{(2)})
\end{aligned}
\tag{19.2}
$$

Usually a drug is primarily associated with only a few pathways, and vice versa. In the meanwhile, a typical pathway only involves several or tens of genes. Therefore, both $B^{(1)}$ and $B^{(2)}$ matrices should be sparse. With this recognition, the spike-and-slab mixture priors are used to infer the gene-pathway and drug-pathway associations.

$$
\begin{aligned}
P(B_{i,j}^{(1)}) &= (1 - \pi_{i,j}^{(1)})\delta_0(B_{i,j}^{(1)}) + \pi_{i,j}^{(1)}\mathcal{N}(B_{i,j}^{(1)}|0, \tau_b^{(1)-1}), \\
P(B_{i,j}^{(2)}) &= (1 - \pi_{i,j}^{(2)})\delta_0(B_{i,j}^{(2)}) + \pi_{i,j}^{(2)}\mathcal{N}(B_{i,j}^{(2)}|0, \tau_b^{(2)-1})
\end{aligned}
\tag{19.3}
$$

$$
\begin{aligned}
\tau_b^{(1)-1} &\sim \text{Gamma}\,(\alpha_b^{(1)}, \beta_b^{(1)}), \\
\tau_b^{(2)-1} &\sim \text{Gamma}\,(\alpha_b^{(2)}, \beta_b^{(2)})
\end{aligned}
\tag{19.4}
$$

where δ_0 is the unit point mass at zero. Instead of having the inverse-Gamma hyper priors on $\tau_b^{(1)-1}$ and $\tau_b^{(1)-1}$, the authors also provided an option of setting fixed values of $\tau_b^{(1)-1}$ and $\tau_b^{(2)-1}$ in their software package. A nice feature of the spike-and-slab prior is that it can naturally incorporate prior knowledge about gene-pathway and drug-pathway associations. To reflect such prior knowledge, two binary matrices $L^{(1)} \in \{0, 1\}^{K \times G^{(1)}}$ and $L^{(2)} \in \{0, 1\}^{K \times G^{(2)}}$ are introduced where $L_{i,j}^{(1)} = 1$ indicates that the j-th gene is involved in the i-th pathway and $L_{i,j}^{(2)} = 1$ indicates that the j-th drug is associated with the i-th pathway. Then the prior information is incorporated into the spike-and-slab prior as follows,

$$
P(Z_{i,j}^{(1)} = 1) = \pi_{i,j}^{(1)} =
\begin{cases}
\eta_0^{(1)}, & \text{if } L_{i,j}^{(1)} = 0 \\
1 - \eta_1^{(1)}, & \text{if } L_{i,j}^{(1)} = 1
\end{cases}
\tag{19.5}
$$

$$
P(Z_{i,j}^{(2)} = 1) = \pi_{i,j}^{(2)} =
\begin{cases}
\eta_0^{(2)}, & \text{if } L_{i,j}^{(2)} = 0 \\
1 - \eta_1^{(2)}, & \text{if } L_{i,j}^{(2)} = 1
\end{cases}
\tag{19.6}
$$

where $Z^{(1)} \in \{0, 1\}^{K \times G^{(1)}}$ and $Z^{(2)} \in \{0, 1\}^{K \times G^{(2)}}$ are auxiliary matrices representing the sparsity pattern of $B^{(1)}$ and $B^{(2)}$ respectively. η_0 is a prior parameter that controls the sparsity of the B matrices whereas η_1 controls the disbelief to the prior knowledge and is usually set to some small number or even 0 when the prior knowledge is mostly rigorously validated or reviewed by experts and fairly reliable. In practice, much more prior knowledge of gene-pathway associations is available than that of drug-pathway associations. And also because the ultimate goal of the iFad method is the identification drug pathway associations, the primary interest lies in the inference of $Z^{(2)}$. In fact, in their paper, $\eta_0^{(1)}$ and $\eta_1^{(1)}$ are both set to zero so that $Z^{(1)}$ is not the target of inference. For the purpose of identifiability, the pathway activity levels X are assumed to have standard normal prior distributions, i.e., $X_{i,j} \sim \mathcal{N}(0, 1)$.

19.2.2 Inference Algorithm

There are many parameters that need to be inferred in the iFad method, including $Z^{(1)}$, $Z^{(2)}$, $B^{(1)}$, $B^{(2)}$, $\tau_b^{(1)}$, $\tau_b^{(2)}$, $\tau_e^{(1)}$ and $\tau_e^{(2)}$. A commonly used approach in the community of Bayesian statistics to approximate their posterior distributions is the Markov Chain Monte Carlo (MCMC) method [8]. Recognizing the substantial dependence between Z and B, the authors employed a specific kind of MCMC, namely the collapsed Gibbs sampler [21], to sample from the posterior distribution of these parameters. The collapsed Gibbs sampling algorithm is detailed below.

Collapsed Gibbs Sampling Algorithm of iFad

Data Input: $Y^{(1)}$, $Y^{(2)}$, $\pi^{(1)}$, $\pi^{(2)}$

Parameters: $\alpha_e^{(1)}$, $\beta_e^{(1)}$, $\alpha_e^{(2)}$, $\beta_e^{(2)}$, $\tau_b^{(1)}$ (or $\alpha_b^{(1)}$, $\beta_b^{(1)}$), $\tau_b^{(2)}$ (or $\alpha_b^{(2)}$, $\beta_b^{(2)}$)

Initialization: randomly generate the following data

$Z^{(1)} \sim Bernoulli(\pi^{(1)})$, $Z^{(2)} \sim Bernoulli(\pi^{(2)})$

$X^{(1)}$, $X^{(2)} \sim \mathcal{N}(0, 1)$, $B^{(1)}$ and $B^{(2)}$ set to 0.

$\tau_e^{(1)} = \alpha_e^{(1)}/\beta_e^{(1)}$, $\tau_e^{(2)} = \alpha_e^{(2)}/\beta_e^{(2)}$

$\tau_b^{(1)} = \alpha_b^{(1)}/\beta_b^{(1)}$, $\tau_b^{(2)} = \alpha_b^{(2)}/\beta_b^{(2)}$ (if $\tau_b^{(1)}$ and $\tau_b^{(2)}$ are not set as fixed

values)

Sampling:

In each iteration,

(1) Sample $\tau_b^{(1)} \sim P(\tau_b^{(1)}|Z^{(1)}, B^{(1)}, \alpha_b^{(1)}, \beta_b^{(1)})$,

$\tau_b^{(2)} \sim P(\tau_b^{(2)}|Z^{(2)}, B^{(2)}, \alpha_b^{(2)}, \beta_b^{(2)})$

(2) Update matrices $Z^{(1)}$, $B^{(1)}$ and $Z^{(2)}$, $B^{(2)}$ separately as follows,

    For $g = 1$ to G

        For $k = 1$ to K, sample $Z_{k,g} \sim P(Z_{k,g}|Y, X, Z_{-k,g}, \tau_e, \pi_{i,j})$

        Sample $B_{:,g} \sim P(B_{:,g}|Y, X, Z_{:,g}, \tau_g, \tau_e)$

(3) Update matrix X,

    For $i = 1$ to N, sample $X_{i,:} \sim P(X_{i,:}|Y^{(1)}, B^{(1)}, \tau_e^{(1)}, Y^{(2)}, B^{(2)}, \tau_e^{(2)})$

(4) Sample $\tau_e^{(1)} \sim P(\tau_e^{(1)}|Y^{(1)}, B^{(1)}, X)$, $\tau_e^{(2)} \sim P(\tau_e^{(2)}|Y^{(2)}, B^{(2)}, X)$

(5) A permutation step to deal with label-switching of the latent factors

---

Note that at the end of each sampling iteration, a local permutation [32] is performed to address the label-switching problem.

19.3 The iPad Method

19.3.1 Model Description

Here we introduce another method for drug-pathway association analysis called "iPad" (integrative penalized matrix decomposition for drug-pathway association analysis). Similar to iFad, the iPad method also considers the biological pathways as a set of common latent factors underlying the gene expression and the drug sensitivity profiles. Therefore, the model for the gene expression profile matrix $Y^{(1)}$ and the drug sensitivity matrix $Y^{(2)}$ is still the same,

$$Y^{(1)} = XB^{(1)} + E^{(1)},$$
$$Y^{(2)} = XB^{(2)} + E^{(2)}, \tag{19.7}$$

However, when estimating X, $B^{(1)}$ and $B^{(2)}$, instead of adopting a full Bayesian approach, the authors cast the problem into a penalized matrix decomposition framework. Specifically, the iPad method seeks to minimize the sum of squared residuals,

$$||Y^{(1)} - XB^{(1)}||_F^2 + ||Y^{(2)} - XB^{(2)}||_F^2, \tag{19.8}$$

under certain constraints or penalties that will be detailed later. Here $|| \cdot ||_F$ stands for the Frobenius norm, i.e. $||A||_F = \sqrt{\sum_i \sum_j A_{ij}^2}$. As discussed earlier, prior knowledge about gene-pathway relationships are relatively rich and reliable, whereas the drug-pathway associations are of the primary interest. Moreover, the $B^{(2)}$ matrix is typically sparse since a drug is usually associated with only a few pathways and vice versa. Therefore, iPad explicitly incorporates known gene-pathway relationships and a sparse penalty is utilized to identify potential drug-pathway associations. Specifically, the iPad method seeks to solve the following optimization problem,

$$\operatorname*{minimize}_{X, B^{(1)}, B^{(2)}} \quad ||Y^{(1)} - XB^{(1)}||_F^2 + ||Y^{(2)} - XB^{(2)}||_F^2 + \lambda||B^{(2)}||_1$$

$$\text{subject to} \quad \sum_i X_{i,j}^2 \leq 1, \ \forall j = 1, \ldots, p; \ B_{i,j}^{(1)} = 0, \ \forall(i, j) : L_{i,j}^{(1)} = 0$$

$$(19.9)$$

Similar to the standard normal prior distribution on X in iFad, the ℓ_2 norm constraint on each column of X is for the purpose of identifiability. Note that we used the ℓ_1 norm ($||A||_1 = \sum_i \sum_j |A_{ij}|$) penalty to encourage the sparsity of $B^{(2)}$. Here $\lambda > 0$ is the penalty parameter. The great success of the ℓ_1 penalty goes back to the introduction of the "lasso" (least absolute shrinkage and selection operator) [34]. Since then, it has become a popular device in various high-dimensional statistical learning problems due to its attractive theoretical properties [3] and the availability of increasingly efficient algorithms [7, 35]. To better illustrate the iPad method, we first give a brief introduction of lasso.

Lasso refers to the following penalized regression problem,

$$\operatorname*{minimize}_{\beta} \quad ||Y - X\beta||^2 + \lambda||\beta||_1 \tag{19.10}$$

where $Y \in \mathbb{R}^{n \times 1}$ is a vector of responses, $X \in \mathbb{R}^{n \times p}$ is the design matrix, and $\beta \in \mathbb{R}^{p \times 1}$ is a vector of regression coefficients. Note that we assume Y has zero mean and therefore the intercept is omitted. When the penalty parameter $\lambda = 0$, this problem degenerates to the simple linear regression and every entry in the solution of β may be non-zero. However, a λ that is greater than zero will shrink the absolute values of regression coefficients and set those that are sufficiently small to exactly zero, achieving a sparse solution of β and allowing variable selection in a natural way. The magnitude of λ controls the sparsity of the lasso solution – a larger λ will result in fewer non-zero coefficients.

Note that in (19.9), only prior knowledge about gene-pathway associations but not the drug-pathway associations is incorporated. However, it can also be adapted to allow the incorporation of known drug-pathway associations by replacing the ℓ_1 norm penalty with a ℓ_2 norm penalty on their corresponding coefficients. We will discuss this in the following section.

19.3.2 Optimization Algorithm

The optimization problem (19.9) is a bi-convex problem. That is to say, when X is given, optimizing $B^{(1)}$ and $B^{(2)}$ is a convex optimization problem; when $B^{(1)}$

and $B^{(2)}$ are given, optimizing X is a convex optimization problem. This nice property naturally suggests the following alternating optimization algorithm:

Alternating Optimization Algorithm of iPad

Data Input: $Y^{(1)}$, $Y^{(2)}$, $L^{(1)}$

Parameter: λ

Initialization: Set $B^{(1)} = L^{(1)}$ and set $B^{(2)} = \mathbf{0}$

Optimization:

(1) Optimize X:

$$X = \underset{X}{\operatorname{argmin}} \quad ||Y^{(1)} - XB^{(1)}||_F^2 + ||Y^{(2)} - XB^{(2)}||_F^2$$

$$\text{subject to} \quad \sum_i X_{i,j}^2 \leq 1, \ \forall j = 1, \ldots, p \tag{19.11}$$

(2) Optimize $B^{(1)}$:

$$B^{(1)} = \underset{B^{(1)}}{\operatorname{argmin}} \quad ||Y^{(1)} - XB^{(1)}||_F^2$$

$$\text{subject to} \quad B_{i,j}^{(1)} = 0, \ \forall(i, j) : L_{i,j}^{(1)} = 0 \tag{19.12}$$

(3) Optimize $B^{(2)}$:

$$B^{(2)} = \underset{B^{(2)}}{\operatorname{argmin}} \quad ||Y^{(2)} - XB^{(2)}||_F^2 + \lambda||B^{(2)}||_1 \tag{19.13}$$

(4) Repeat steps (1)(2)(3) until convergence.

As mentioned earlier, problems (19.11 $\sim$ 19.13) are three convex optimization problems. We go through them one by one.

Problem (19.11) can be rewritten in the following way,

$$\underset{X}{\operatorname{minimize}} \quad ||Y - XB||_F^2$$

$$\text{subject to} \quad \sum_i X_{i,j}^2 \leq 1, \ \forall j = 1, \ldots, p \tag{19.14}$$

where $Y = [Y^{(1)} \ Y^{(2)}]$ and $B = [B^{(1)} \ B^{(2)}]$. This problem is solved using an iterative projected gradient descent algorithm. The gradient of the objective function is easily obtained as $(XB - Y)B'$. At each descent step, X takes a step along the negative direction of the gradient and is then projected to the feasible region $\{X : \sum_i X_{i,j}^2 \leq 1, \forall j = 1, \ldots, p\}$. The Nesterov's method was used to accelerate the convergence [25].

Problem (19.12) is relatively trivial because it essentially amounts to $G^{(1)}$ separate ordinary least squares (OLS) problems. To see this, we observe that each column of $B^{(1)}$ can be optimized separately,

For $g \in \{1, 2, \ldots, G^{(1)}\}$,

$$\underset{B^{(1)}_{:,g}}{\text{minimize}} \ ||Y^{(1)}_{:,g} - X_{:,L^{(1)}_{:,g}} B^{(1)}_{L^{(1)}_{:,g},g}||_2^2 \tag{19.15}$$

where $Y^{(1)}_{:,g}$ refers to the g-th column of $Y^{(1)}$, $X_{:,L^{(1)}_{:,g}}$ refers to the columns of X corresponding to the non-zero entries of $L^{(1)}_{:,g}$ and $B^{(1)}_{L^{(1)}_{:,g},g}$ refers to the g-th columns and the rows of $B^{(1)}$ corresponding to the non-zero entries of $L^{(1)}_{:,g}$.

Similar to (19.12), problem (19.13) can also be separated into $G^{(2)}$ small problems because each column of $B^{(2)}$ can be optimized separately. In fact, optimizing each column of $B^{(2)}$ is a lasso problem,

For $g \in \{1, 2, \ldots, G^{(2)}\}$,

$$\underset{B^{(2)}_{:,g}}{\text{minimize}} \ ||Y^{(2)}_{:,g} - X B^{(2)}_{:,g}||_2^2 + \lambda ||B^{(2)}_{:,g}||_1 \tag{19.16}$$

Efficient coordinate descent algorithm is readily available for solving the lasso problem [7, 35]. As mentioned earlier, this setup does not allow incorporating known drug-pathway associations. When such prior knowledge is available and needs to be incorporated, problem (19.16) can be modified as follows,

For $g \in \{1, 2, \ldots, G^{(2)}\}$,

$$\underset{B^{(2)}_{:,g}}{\text{minimize}} \ ||Y^{(2)}_{:,g} - X B^{(2)}_{:,g}||_2^2 + \lambda(||B^{(2)}_{(1-L^{(2)}_{:,g}),g}||_1 + ||B^{(2)}_{L^{(2)}_{:,g},g}||_2) \tag{19.17}$$

Thus the ℓ_2 norm penalty can avoid the coefficients in $B^{(2)}$ corresponding to the known drug-pathway associations to be shrunk to zero.

Compared with the iFad method, it is relatively non-trivial to deal the missing values in $Y^{(1)}$ and $Y^{(2)}$ for the iPad method, especially when optimizing X. When all the values for a whole row or column are missing in $Y^{(1)}$ and $Y^{(2)}$, that row/column shall be completely removed. When a row or a column is partially missing, optimizing $B^{(1)}$ and $B^{(2)}$, i.e. solving problem (19.15) and (19.16) is still relatively easy because each column can be optimized separately. One can simply omit the missing elements in the column being optimized when solving the OLS or lasso problem. However, optimizing X is less straightforward. Suppose the observed values in Y are indexed by $\Omega \in \{0, 1\}^{N \times (G^{(1)} + G^{(2)})}$.

We define an operator $\mathcal{P}_\Omega$ that projects matrix X onto the linear space supported by Ω,

$$\mathcal{P}_\Omega(X)_{i,j} = \begin{cases} X_{i,j}, & \text{if } (i,j) \in \Omega. \\ 0, & \text{if } (i,j) \notin \Omega. \end{cases}$$

Then problem (19.14) in the presence of missing values can be written as:

$$\begin{aligned} & \underset{X}{\text{minimize}} \quad ||\mathcal{P}_\Omega(Y) - \mathcal{P}_\Omega(XB)||_F^2 \\ & \text{subject to} \quad \sum_i X_{i,j}^2 \le 1, \ \forall j = 1, 2, \ldots, p \end{aligned} \tag{19.18}$$

Define $\Omega_\perp = 1 - \Omega$ as the index for the missing values in Y and re-write the objective function in (19.18):

$$\begin{aligned} & ||\mathcal{P}_\Omega(Y) - \mathcal{P}_\Omega(XB)||_F^2 \\ =& ||\mathcal{P}_\Omega(Y) - (XB - \mathcal{P}_{\Omega_\perp}(XB))||_F^2 \\ =& ||(\mathcal{P}_\Omega(Y) + \mathcal{P}_{\Omega_\perp}(XB)) - XB||_F^2 \end{aligned} \tag{19.19}$$

This observation naturally suggests the following iterative algorithm to solve problem (19.18), in which the estimate of X at each iteration is plugged into $\mathcal{P}_{\Omega_\perp}(XB)$ for optimizing (19.19) in the next iteration.

Algorithm for optimizing X with missing values in Y

Data Input: Y, B, Ω

Initialization: Initialize X

Optimization:

    (1) Set $Y = \mathcal{P}_\Omega(Y) + \mathcal{P}_{\Omega_\perp}(XB)$
    (2) Optimize X:

$$\begin{aligned} & X = \underset{X}{\text{argmin}} \quad ||Y - XB||_F^2 \\ & \text{subject to} \quad \sum_i X_{i,j}^2 \le 1, \ \forall j = 1, 2, \ldots, p \end{aligned}$$

    (3) Repeat steps (1)(2) until convergence.

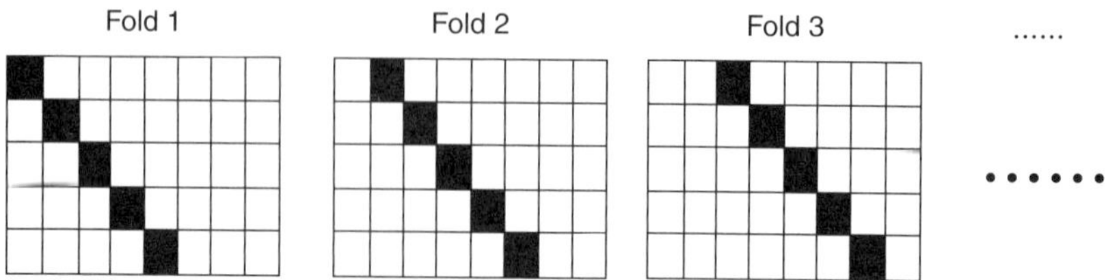

Figure 19.1 An illustration of data partitioning in the cross validation procedure. The black entries in each figure represent one fold of the data. In each round of the cross validation, the black entries are masked as missing values whereas the white entries are treated as observed data.

19.3.3 Parameter Tuning and Significance Test

In the iPad method, λ is a critical parameter because it controls the number of non-zero coefficients in the drug-pathway association matrix, $B^{(2)}$. Similar to the original lasso regression, one way to assess the relative importance of the coefficients in $B^{(2)}$ is to solve the problem for a decreasing sequence of λ values and record the order of the coefficients in which they became non-zero. This order provides an indication of the relative importance of the coefficients – the more important coefficients are supposed to become non-zero earlier than the less important ones. However, this procedure does not allow evaluation of the significance of the coefficients. Alternatively, one can first find an appropriate value of λ and then perform permutation test to assess the significance of the coefficients.

In iPad, cross-validation is used to find an appropriate λ value. Specifically, the $Y^{(2)}$ matrix is evenly partitioned into n folds as illustrated in Figure 19.1. In each round of cross-validation, each of the ten folds is masked as missing values while the other entries are treated as observed values for iPad, in which a sequence of λ values are solved. Then the mean squared errors (MSE) of the masked values for the ten folds are evaluated for each λ. The λ with the smallest MSE is chosen. Given the λ value, the significance of the coefficients in $B^{(2)}$ can be assessed via permutation test. In each permutation, the rows in $Y^{(2)}$ are shuffled whereas $Y^{(1)}$ is kept unchanged. After the estimates of the $B^{(2)}$ in the permuted data sets are obtained, the p-value of each coefficient in the $B^{(2)}$ matrix is calculated as follows,

$$p_{i,j} = \frac{\sum_{t=1}^{T} \mathbb{1}(|\tilde{B}_{i,j}^{(2)(t)}| \geq |\hat{B}_{i,j}^{(2)}|)}{T} \tag{19.20}$$

where $\tilde{B}^{(2)(t)}$ is the $B^{(2)}$ estimate in the t-th permutation and $\hat{B}^{(2)}$ is the $B^{(2)}$ estimate in the original data.

19.4 Applications to the NCI-60 Data Sets

19.4.1 Data Description

The NCI-60 project provides a comprehensive resource of both gene expression and drug sensitivity profiles of 60 human cancer cell lines from nine different types of tissues. Ma et al [22] analyzed the gene expression and drug sensitivity data from CellMiner database [30] consisting gene expression levels and drug sensitivity data for 57 cell lines, 6958 genes and 101 molecules. One of the 60 cell lines was removed because of the unavailability of its gene expression profile. Two other cell lines were excluded because they were the only two prostate cancer cell lines. The 6958 genes were included because they were either in the 776 cancer-related genes from Chen et al [5] or the 8919 genes from the Integrated Druggable Genome Database Project [12, 28]. The 101 molecules were chosen because 1) they have known 2D structures; 2) they have been tested at least twice, and 3) they are annotated in the CancerResource database [1].

The prior knowledge of gene-pathway and drug-pathway association were obtained from the KEGG MEDICUS database [15], which consists of 58 pathways that are either known to be cancer-related or have drug targets. The 58 pathways overlap with 1863 of the 6958 genes described previously. Therefore, only these 1863 were kept in further analysis. In the end, they compiled the following data set: a gene expression data matrix $Y^{(1)} \in \mathbb{R}^{57 \times 1863}$, a drug sensitivity data matrix $Y^{(2)} \in \mathbb{R}^{57 \times 101}$, a gene-pathway association matrix $L^{(1)} \in \mathbb{R}^{58 \times 1863}$, and a drug-pathway association matrix $L^{(2)} \in \mathbb{R}^{58 \times 101}$. The goal is to identify unknown drug-pathway associations, i.e., the drug-pathway associations that are not present in $L^{(2)}$.

19.4.2 Results from iFad

The compiled NCI-60 data described above consist of 57 human cancer cell lines from eight types of tissues, namely, breast cancer (BR), central neural system (CNS), colon cancer (CO), lung cancer (LC), leukemia (LE), melanoma (ME), ovarian cancer (OV), and renal cancer (RE). To avoid the complication of heterogeneity between different cancer types, Ma et al analyzed each tumor type separately. The $Y^{(1)}$ and $Y^{(2)}$ matrices were both partitioned into eight sub-matrices corresponding to the eight types of cancer. Both $\eta_0^{(1)}$ and $\eta_1^{(1)}$ were set to 0 so that the gene-pathway associations were fixed during the inference. $\eta_1^{(2)}$ was set to 0 and $\eta_0^{(2)}$ was varied from 0.05 to 0.25. A total of 100,000 MCMC iterations were performed to obtain the posterior distributions of the parameters to be inferred. The posterior probability of $Z_{i,j}^{(2)} = 1$ was used to measure

Table 19.1 *The iFad-identified drug-pathway associations that are validated by the CancerResource Database*

KEGG pathways	Drug	Posterior probability	Cancer type
Glutathione metabolism	Vincristine	0.9987	LC
ErbB signaling pathway	Mitoxantrone	0.9937	LC
Thyroid cancer	Doxorubicin	0.991	RE
Glutathione metabolism	6-Mercaptopurine	0.9867	RE
Bladder cancer	Tamoxifen	0.982	LC
VEGF signaling pathway	Carmustine	0.9803	RE
Thyroid cancer	Doxorubicin	0.9713	OV
ErbB signaling pathway	Camptothecin	0.9583	LC
Bladder cancer	Edelfosine	0.958	LC
Bladder cancer	Chlorambucil	0.9473	RE
Melanoma	Chlorambucil	0.947	ME
VEGF signaling pathway	6-Mercaptopurine	0.932	ME
Bladder cancer	Geldanamycin	0.9273	CO
Thyroid cancer	Dactinomycin	0.9263	OV
Apoptosis	Thymidine	0.923	CO
Cell cycle	Tiazofurin	0.919	LC
Drug metabolismother enzymes	Daunorubicin	0.9157	LC
VEGF signaling pathway	Lomustine	0.9103	RE
Focal adhesion	Geldanamycin	0.91	BR
Endometrial cancer	Doxorubicin	0.909	CO
VEGF signaling pathway	Quinacrine	0.9023	RE
Base excision repair	Decitabine	0.9017	LC

Note: Adapted from Ma et al. [22].

the probability that the j-th molecule is associated with the i-th pathway. Associations with posterior probabilities greater than 0.9 were checked against the known drug-pathway association in the CancerResource Database [1]. At $\eta_0^{(2)} = 0.05$, 76 associations have posterior probabilities greater than 0.9 across the 8 cancer types, among which 22 can be validated by the CancerResource Database. Table 19.1 shows the 22 validated associations. A complete list of these identified associations can be found in Table S4 of Ma et al [22].

19.4.3 Results from iPad

Although separating the cell lines from different cancer types could avoid the complication of across cancer type heterogeneity, the sample size is extremely limited for each cancer type, with usually about 6 or 7 samples per cancer type. Therefore previous analysis by iFad may be of limited power. One advantage of the iPad method over iFad is the computational efficiency. The computation time of iPad is much less than iFad since it avoids the time-consuming MCMC simulations. In fact, if the 57 cell lines from the eight cancer types were pooled together in the analysis, iFad could not finish 10,000 MCMC iterations in two weeks, whereas iPad was able to finish 10,000 permutations within one day. If the p-values are not needed, then iPad can simply obtain an order of relative importance of the drug-pathway pairs by solving a sequence of 100 lambda values within 20 minutes.

To have a larger sample size, we used iPad to analyze the pooled data set consisting of 57 cell lines. We first obtained a λ value through a ten-fold cross validation. Then we performed 1000 permutations with this λ value to obtain the p-value for each drug-pathway pair. The most significant 50 drug-pathway pairs are shown in Table 19.2, among which 18 pairs can be validated by the CancerResource Database. The results are shown in Table 19.2. The false discovery rates (FDR) were obtained using the Benjamini-Hochberg procedure [2]. We note that there were not much overlap between the drug-pathway pairs that were identified by iPad and those identified by iFad. This is probably because the results from iFad consist of primarily the cancer type – specific associations, whereas iPad identified associations that are common for multiple cancer types. Or probably the sample size in the iFad analysis was too small to support reliable statistical inference.

19.5 Discussion

Drug target identification is an important task in the early stage of drug discovery and development. In recent years, more and more high-throughput technologies have become available to massively measure various molecular phenotypes at the genome level. However, there exist enormous challenges in effectively integrating the large data sets generated by these technologies to facilitate drug discovery. In addition, the traditional "one drug – one target" paradigm has been gradually supplanted by the concept of polypharmacology, which seeks to find or design candidate drug molecules that interfere multiple drug targets. An example that carries this philosophy is pathway-based drug discovery, which pursues candidate drugs that interfere the activity of a whole biological pathway.

In this chapter, we introduced two statistical methods, namely, iFad and iPad, for pathway-based drug discovery. Both of them try to identify associations between drugs and biological pathways by integrating gene expression and drug sensitivity profiles of the same set of cell lines. Both of them assume the gene expression and drug sensitivity profiles are controlled by the same set of latent factors, namely, the activity levels of some biological pathways.

Table 19.2 *The drug-pathway associations that are identified by iPad*

KEGG pathways	Drug	p-value	FDR	Validated
Focal adhesion	Mechlorethamine	0	0	True
Tight junction	Lucanthone	0	0	True
Acute myeloid leukemia	Diallyl Disulfide	0	0	False
One carbon pool by folate	Azacitidine	0	0	False
Gap junction	Geldanamycin	0	0	False
Focal adhesion	Bleomycin	0	0	False
Cell cycle	Mycophenolic Acid	0	0	False
One carbon pool by folate	Aclacinomycins	0	0	False
Cell cycle	Tiazofurin	0	0	True
Primary immunodeficiency	Aminoglutethimide	0	0	False
Cell cycle	Selenazofurin	0	0	True
T cell receptor signaling pathway	Rebeccamycin	0	0	False
Tight junction	Chloroquine Phosphate	0.001	0.234	True
Glioma	Mitotane	0.001	0.234	False
Tight junction	Lomustine	0.001	0.234	True
Small cell lung cancer	Doxorubicin	0.001	0.234	True
Gap junction	Tegafur	0.001	0.234	False
Natural killer cell mediated cytotoxicity	Primaquine	0.001	0.234	False
Regulation of actin cytoskeleton	Pimozide	0.001	0.234	False
Gap junction	Plumbagin	0.001	0.234	False
Wnt signaling pathway	Avarol	0.001	0.234	False
Vibrio cholerae infection	Fludarabine phosphate (USAN)	0.001	0.234	False
One carbon pool by folate	Cyclopentenyl Cytosine	0.001	0.234	False
Cell cycle	Carmustine	0.001	0.234	False
Endometrial cancer	Combretastatin A4	0.001	0.234	True
Metabolism of xenobiotics by cytochrome P450	Vitamin K 3	0.002	0.317	False
T cell receptor signaling pathway	Melphalan	0.002	0.317	True
Glutathione metabolism	Curcumin	0.002	0.317	True
Small cell lung cancer	Daunorubicin Hydrochloride	0.002	0.317	True
Focal adhesion	Geldanamycin	0.002	0.317	True
T cell receptor signaling pathway	Teniposide	0.002	0.317	True
Jak-STAT signaling pathway	Tegafur	0.002	0.317	False
Prostate cancer	Coralyne	0.002	0.317	False
T cell receptor signaling pathway	Dexrazoxane	0.002	0.317	True
Tight junction	Nocodazole	0.002	0.317	False
Small cell lung cancer	Piroxantrone	0.002	0.317	False

KEGG pathways	Drug	p-value	FDR	Validated
Acute myeloid leukemia	Cholecalciferol	0.002	0.317	False
Gap junction	Vitamin K 3	0.003	0.366	False
Small cell lung cancer	Mitomycin	0.003	0.366	False
Tight junction	Vinblastine	0.003	0.366	True
Small cell lung cancer	Daunorubicin	0.003	0.366	True
One carbon pool by folate	Decitabine	0.003	0.366	False
Renal cell carcinoma	Razoxane	0.003	0.366	False
Gap junction	Pimozide	0.003	0.366	False
Pyrimidine metabolism	Tamoxifen	0.003	0.366	False
Regulation of actin cytoskeleton	Edelfosine	0.003	0.366	True
Base excision repair	Tanespimycin	0.003	0.366	False
Insulin signaling pathway	Cholecalciferol	0.003	0.366	False
T cell receptor signaling pathway	Pipobroman	0.004	0.418	True
Tight junction	Diallyl Disulfide	0.004	0.418	False

The major difference between the two models is that iFad adopts a full Bayesian framework and hence requires intensive MCMC simulations whereas iPad solves penalized matrix decomposition problem to identify drug-pathway associations. The latter requires substantially less computational resource and can easily handle relatively large data sets. However, statistical inference is much easier and more natural for iFad because the posterior distributions obtained from the MCMC simulations naturally serve as a measure of the statistical evidences for the associations. Whereas statistical inference or hypothesis testing for penalized methods is very challenging and relatively under-developed. Hence iPad uses permutation tests to assess the significance of drug-pathway associations.

Besides the two methods described in this chapter, there are other methods that jointly model gene expression profiles and drug-related data, although they may not incorporate knowledge about biological pathways. For example, Kutalik et al [18] developed a bi-clustering method called iterative signature algorithm (ISA), to search for "co-modules" that represent gene-drug associations. Chang et al [4] proposed a Bayesian network-based method to infer gene-drug dependencies. In this chapter, only two types of data are integrated. In the future, with the development of technologies, more and more types of high-throughput data will be generated. Hence novel statistical methods are greatly needed to tackle the new challenge of effectively integrating multiple types of high-throughput data in drug discovery.

References

1 Ahmed, Jessica, Meinel, Thomas, Dunkel, Mathias, Murgueitio, Manuela S, Adams, Robert, Blasse, Corinna, Eckert, Andreas, Preissner, Saskia, and Robert, Preissner. 2011. CancerResource: a comprehensive database of cancer-relevant proteins and compound interactions supported by experimental knowledge. *Nucleic acids research*, **39**(suppl 1), D960–D967.

2 Benjamini, Yoav, and Hochberg, Yosef. 1995. Controlling the false discovery rate: a practical and powerful approach to multiple testing. *Journal of the Royal Statistical Society. Series B (Methodological)*, **57**(1), 289–300.

3 Bühlmann, Peter Lukas, van de Geer, Sara A, and Van de Geer, Sara. 2011. *Chapter 6 and 7 of: Statistics for high-dimensional data*. New York: Springer.

4 Chang, Jeong-Ho, Hwang, Kyu-Baek, June Oh, S, and Zhang, Byoung-Tak. 2005. Bayesian network learning with feature abstraction for gene-drug dependency analysis. *Journal of bioinformatics and computational biology*, **3**(01), 61–77.

5 Chen, Bo-Juen, Causton, Helen C, Mancenido, Denesy, Goddard, Noel L, Perlstein, Ethan O, and Pe'er, Dana. 2009. Harnessing gene expression to identify the genetic basis of drug resistance. *Molecular systems biology*, **5**(310).

6 Csermely, Péter, Agoston, Vilmos, and Pongor, Sandor. 2005. The efficiency of multi-target drugs: the network approach might help drug design. *Trends in Pharmacological Sciences*, **26**(4), 178–182.

7 Friedman, Jerome, Hastie, Trevor, Höfling, Holger, and Tibshirani, Robert. 2007. Pathwise coordinate optimization. *The Annals of Applied Statistics*, **1**(2), 302–332.

8 Gelman, Andrew, Carlin, John B, Stern, Hal S, and Rubin, Donald B. 2003. *Bayesian data analysis*. CRC press.

9 Giuliano, Kenneth A, Haskins, Jeffrey R, and Taylor, D Lansing. 2003. Advances in high content screening for drug discovery. *Assay and drug development technologies*, **1**(4), 565–577.

10 He, Zhisong, Zhang, Jian, Shi, Xiao-He, Hu, Le-Le, Kong, Xiangyin, Cai, Yu-Dong, and Chou, Kuo-Chen. 2010. Predicting drug-target interaction networks based on functional groups and biological features. *PloS one*, **5**(3), e9603.

11 Hopkins, Andrew L. 2008. Network pharmacology: the next paradigm in drug discovery. *Nature chemical biology*, **4**(11), 682–690.

12 Hopkins, Andrew L, and Groom, Colin R. 2002. The druggable genome. *Nature reviews Drug discovery*, **1**(9), 727–730.

13 Hughes, Joanne E. 1999. Genomic technologies in drug discovery and development. *Drug discovery today*, **4**(1), 6.

14 Iskar, Murat, Zeller, Georg, Zhao, Xing-Ming, van Noort, Vera, and Bork, Peer. 2012. Drug discovery in the age of systems biology: the rise of computational approaches for data integration. *Current opinion in biotechnology*, **23**(4), 609–616.

15 Kanehisa, Minoru, Goto, Susumu, Furumichi, Miho, Tanabe, Mao, and Hirakawa, Mika. 2010. KEGG for representation and analysis of molecular networks involving diseases and drugs. *Nucleic acids research*, **38**(suppl 1), D355–D360.

16 Kitchen, Douglas B, Decornez, Hélène, Furr, John R, and Bajorath, Jürgen. 2004. Docking and scoring in virtual screening for drug discovery: methods and applications. *Nature reviews Drug discovery*, **3**(11), 935–949.

17 Kola, Ismail, and Landis, John. 2004. Can the pharmaceutical industry reduce attrition rates? *Nature reviews Drug discovery*, **3**(8), 711–716.

18 Kutalik, Zoltán, Beckmann, Jacques S, and Bergmann, Sven. 2008. A modular approach for integrative analysis of large-scale gene-expression and drug-response data. *Nature biotechnology*, **26**(5), 531–539.

19 Li, Cong, Yang, Can, Hather, Greg, Liu, Ray, and Zhao, Hongyu. Integrative penalized matrix decomposition for drug-pathway association analysis. *in preparation.*

20 Lindsay, Mark A. 2005. Finding new drug targets in the 21st century. *Drug discovery today*, **10**(23), 1683–1687.

21 Liu, Jun S. 1994. The collapsed Gibbs sampler in Bayesian computations with applications to a gene regulation problem. *Journal of the American Statistical Association*, **89**(427), 958–966.

22 Ma, Haisu, and Zhao, Hongyu. 2012. iFad: an integrative factor analysis model for drug-pathway association inference. *Bioinformatics*, **28**(14), 1911–1918.

23 Ma, Haisu, and Zhao, Hongyu. 2013. Drug target inference through pathway analysis of genomics data. *Advanced drug delivery reviews.*

24 Nagamine, Nobuyoshi, Shirakawa, Takayuki, Minato, Yusuke, Torii, Kentaro, Kobayashi, Hiroki, Imoto, Masaya, and Sakakibara, Yasubumi. 2009. Integrating statistical predictions and experimental verifications for enhancing protein-chemical interaction predictions in virtual screening. *PLoS computational biology*, **5**(6), e1000397.

25 Nesterov, Yurii. 1983. A method of solving a convex programming problem with convergence rate O (1/k2). *Soviet Mathematics Doklady*, **27**(2), 372–376.

26 Petriz, Bernardo A, Gomes, Clarissa P, Rocha, Luiz AO, Rezende, Taia, and Franco, Octávio L. 2012. Proteomics applied to exercise physiology: A cutting-edge technology. *Journal of cellular physiology*, **227**(3), 885–898.

27 Pujol, Albert, Mosca, Roberto, Farrés, Judith, and Aloy, Patrick. 2010. Unveiling the role of network and systems biology in drug discovery. *Trends in pharmacological sciences*, **31**(3), 115–123.

28 Russ, Andreas P, and Lampel, Stefan. 2005. The druggable genome: an update. *Drug discovery today*, **10**(23), 1607–1610.

29 Schadt, Eric E, Friend, Stephen H, and Shaywitz, David A. 2009. A network view of disease and compound screening. *Nature Reviews Drug Discovery*, **8**(4), 286–295.

30 Shankavaram, Uma, Varma, Sudhir, Kane, David, Sunshine, Margot, Chary, Krishna, Reinhold, William, Pommier, Yves, and Weinstein, John. 2009. CellMiner: a relational database and query tool for the NCI-60 cancer cell lines. *BMC genomics*, **10**(1), 277.

31 Shoemaker, Robert H. 2006. The NCI60 human tumour cell line anticancer drug screen. *Nature Reviews Cancer*, **6**(10), 813–823.

32 Stegle, Oliver, Sharp, Kevin, and Winn, John. 2000. A Comparison of Inference in Sparse Factor Analysis. *Journal of Machine Learning Research*, **1**, 1–48.

33 Subramanian, Aravind, Tamayo, Pablo, Mootha, Vamsi K, Mukherjee, Sayan, Ebert, Benjamin L, Gillette, Michael A, Paulovich, Amanda, Pomeroy, Scott L, Golub, Todd R, Lander, Eric S, et al. 2005. Gene set enrichment analysis: a knowledge-based approach for interpreting genome-wide expression profiles. *Proceedings of the National Academy of Sciences of the United States of America*, **102**(43), 15545–15550.

34 Tibshirani, Robert. 1996. Regression shrinkage and selection via the lasso. *Journal of the Royal Statistical Society. Series B (Methodological)*, 267–288.

35 Tibshirani, Robert, Bien, Jacob, Friedman, Jerome, Hastie, Trevor, Simon, Noah, Taylor, Jonathan, and Tibshirani, Ryan J. 2012. Strong rules for discarding predictors in lasso-type problems. *Journal of the Royal Statistical Society: Series B (Statistical Methodology)*, **74**(2), 245–266.

36 Vina, Dolores, Uriarte, Eugenio, Orallo, Francisco, and Gonzlez-Daz, Humberto. 2009. Alignment-Free Prediction of a Drug- Target Complex Network Based on

Parameters of Drug Connectivity and Protein Sequence of Receptors. *Molecular pharmaceutics*, **6**(3), 825–835.

37 Wang, Zhong, Gerstein, Mark, and Snyder, Michael. 2009. RNA-Seq: a revolutionary tool for transcriptomics. *Nature Reviews Genetics*, **10**(1), 57–63.

38 Xie, Lei, Xie, Li, Kinnings, Sarah L, and Bourne, Philip E. 2012. Novel computational approaches to polypharmacology as a means to define responses to individual drugs. *Annual review of pharmacology and toxicology*, **52**, 361–379.

39 Yamanishi, Yoshihiro, Araki, Michihiro, Gutteridge, Alex, Honda, Wataru, and Kanehisa, Minoru. 2008. Prediction of drug–target interaction networks from the integration of chemical and genomic spaces. *Bioinformatics*, **24**(13), i232–i240.

40 Yamanishi, Yoshihiro, Kotera, Masaaki, Kanehisa, Minoru, and Goto, Susumu. 2010. Drug-target interaction prediction from chemical, genomic and pharmacological data in an integrated framework. *Bioinformatics*, **26**(12), i246–i254.

Index

Note: Page numbers followed by 'f' and 't' indicate figures and tables.

accelerated failure time (AFT) model, 227
accuracy quality control (AQCg; AQCp),
    MetaQC package, 41
aCGH (array comparative genomic
    hybridization), 333, 403
AD (Alzheimer's disease), 102–103
adaptively weighted Fisher's (AW-Fisher's)
    method, 43
adjusted Rand index, 242
Affymetrix arrays, 400–401
AFT (accelerated failure time) model, 227
allele frequencies, comparison of, 14f
allele-specific binding (ASB), in ChIP-seq,
    109, 123–128
Alzheimer's disease (AD), 102–103
Amandine, E., 202
AMD GWAS, 19–20
AQC (accuracy quality control), MetaQC
    package, 41
ARACNe algorithm, 297
Argonaute protein, 285–286, 292
array comparative genomic hybridization
    (aCGH), 333, 403
ArrayExpress, 397
ASB (allele-specific binding), in ChIP-seq,
    109, 123–128
autologistic regression, 206, 208–209
AW-Fisher's (adaptively weighted Fisher's)
    method, 43

background window, ChIP-chip peak calling,
    119
bacterial artificial chromosome (BAC) probes,
    aCGH, 403–404
Bayesian Consensus Clustering (BCC)
    method, 251–260
  application to TCGA data, 257–260

defined, 238
Dirichlet mixture model, 252–253
estimation, 254–256
illustrative example, 256–257
multisource model, 253–254
overview, 251–252
Bayesian false discovery rate (BFDR), 362
Bayesian graphical models
  BayesGraph for TCGA integration,
      201–214
    graphical models, 203
    markov random fields, 204–205
    MCMC Simulations, 207–208
    overview, 201–203
    posterior inference using false discovery
        rates, 208
    probability model, 205–207
    simulation study, 208–211
    TCGA integrative analysis, 211–214
  data description, 226–227
  iBAG models, 220–226
    linear, 221–224
    non-linear extensions, 224–226
    overview, 220–221
  illustrations, 226
  overview, 4–5, 217–220
  results, 227–233
Bayesian information criterion (BIC)
    correlation motif model, 117–118
    directed acyclic graph, 273
BBID database, 335
BCC method. *See* Bayesian Consensus
    Clustering method
Benjamini-Hochberg (BH) procedure, 51
BFDR (Bayesian false discovery rate), 362
BH (Benjamini-Hochberg) procedure, 51
BIC. *See* Bayesian information criterion

binding, miRNAs
ComiR targeting method, 292–294
effect of single and multiple targets, 290
Fermi-Dirac combination of targets, 290–292
thermodynamics of, 289–290
binding, TF
cause or consequence relationship between gene expression, 393–394
ENCODE K562 and GM12878 data, 378–379
framework for integrating with gene expression data, 376–377
interplay between histone modification and other chromatin features, 391–392
machine learning methods used in predictive models, 377–378
ModENCODE Early Embryo data, 379
Mouse ESC data, 379
overview, 374–375
performance evaluation of models, 378
predicting differential gene expression, 388–389
predicting expression levels for genes with HCP and LCP content, 391
predicting expression of noncoding genes, 389–390
predicting gene expression by combining with histone modifications, 385–388
predicting gene expression from, 379–382
regulatory signals in distal regions, 392–393
Yeast and Fly data, 379
BioCarta database, 335, 336f, 422
biological pathways
defined, 421
iFad method, drug-pathway association analysis, 422
iPad method, drug-pathway association analysis, 425
BioPAX pathway format, 306
burden tests, meta-analysis of GWAS
that assume distribution of variant effect sizes, 25–26
that assume variants have similar effect sizes for a simple burden test in study k, the impact of multiple rare variants, 24

cancer (sub)type analysis, 350–352, 352f
cancer genomics. *See also* latent variable approach, integrative clustering analysis; somatic mutations in cancer genomes
active subnetwork search and discovery, 309–310
joint NMF, 134–139
network-regularized joint NMF Method, 139–143
overview, 131–134, 304–305
PARADIGM pathway method, 310–319
applications of, 317, 319
interaction parameters, 316–317
interactions and probabilistic factors, 314–316
matrix of, 318f
overview, 310–312
variables, 312–314
PARADIGM-SHIFT pathway method, 319–322
applications of, 320–322
overview, 319–320
pathway databases, 305–306
pathway methods, 307–308
pathway-based mutation assessment, 308–309
sparse Multiple Block PLS method, 143–147
TieDIE pathway method, 322–328
cancer-related lncRNAs, 406, 407f, 408f
CancerResource Database, 432
canonical correlation analysis (CCA), 241
causal inference, eQTLs, 270–271
CCA (canonical correlation analysis), 241
CD4:CD3 ratio, 172
analysis results by applying MCP to each outcome separately, 191t–192t
analysis results of gMCP, 185t, 193t–194t
analysis results of gMCP with Laplacian penalty, 195t–198t
analysis results of sparse gMCP, 186t–187t
overlaps of different analysis methods, 184t
CD4/CD8 ratio (T-Lymphocyte Helper/Suppressor Profile), 171–172
analysis results by applying MCP to each outcome separately, 191t–192t
analysis results of gMCP, 185t, 193t–194t
analysis results of gMCP with Laplacian penalty, 195t–198t
analysis results of sparse gMCP, 186t–187t
overlaps of different analysis methods, 184t
centered parametrization, 206
change-point model, gene expression regulation, 358–360
ChIP (chromatin immunoprecipitation), 108. *See also* ChIP-X data
ChIP-chip analysis, 118–123. *See also* ChIP-X data
ChIP-seq (chromatin immunoprecipitation sequencing), 108, 378–380, 385, 388, 398. *See also* ChIP-X data
ChIP-X data
allele-specific binding in ChIP-seq, 123–128

ChIP-chip analysis, 118–123
ChIP-seq, 108, 378–380, 385, 388, 398
correlation motif approach, 112–118
general problem setting and motivations, 110–112
overview, 108–110
chromatin immunoprecipitation (ChIP), 108. *See also* ChIP-X data
chromatin immunoprecipitation sequencing (ChIP-seq), 108, 378–380, 385, 388, 398. *See also* ChIP-X data
chromosome instability (CIN), 156, 164
CIFA (common and individual feature analysis), 242
CIMP (CpG island methylator phenotype), 164
CIN (chromosome instability), 156, 164
cis-eQTLs, 87, 94, 268–271, 269f
clinical iBAG model, 222–224
clipper method, analysis of gene expression, 307–308
clustering. *See also* latent variable approach, integrative clustering analysis
Bayesian Consensus Clustering method, 251–260
differential clustering algorithm, 101
exploratory methods for multisource data, 242–243
cMCP (composite MCP), 176–177, 179–180
CNAs (copy number aberrations), 333–334
CNVs (copy number variations), 132, 143–147, 201–203, 207, 214
Cochran-Mantel-Haenszel method, single-variant association test statistics, 24
coexpression clusters, 76
coexpression network, 58, 59f, 67, 68, 69f, 70, 100–102
coherent FFLs, 295f
collapsed Gibbs sampling algorithm, iFad method, 424–425
colon cancer, 156
colorectal carcinoma (CRC) study, 163–165, 164f, 165f
combining effect sizes analysis, microarrays, 44
combining *p*-values analysis, microarrays
evidence aggregation methods, 42–44
order statistics methods, 44
combining ranks analysis, microarrays, 44–45
ComiR targeting prediction algorithm, 287t, 288, 292–294, 293f
common and individual feature analysis (CIFA), 242
complete conditionals, 234–235
complex correlation structures, 219

complex diseases
disease subtype discovery, 56–57
MetaNetwork for differential network detection, 58
MetaPath for pathway analysis, 52–55
network integration of genetically regulated gene expression
diabetes genes, 93–100
differential connectivity in coexpression network, 100–102
late-onset Alzheimer's disease brain study, 102–103
LINKER approach, 92
modeling genetic information flow, 88–91
overview, 86–88
PRINCE approach, 91–92
prize collecting Steiner tree problem, 92–93
random walk approach, 91
network integration of genetically regulated gene expression to study, 86–104
composite MCP (cMCP), 176–177, 179–180
composite penalization, 180
computational burden, exploratory methods for multisource data, 261
computational cancer genomics. *See* cancer genomics
computational methods. *See* integrative analysis; integrative quantitative models; latent variable approach, integrative clustering analysis
conditional analyses
meta-analysis of GWAS, 26–28
results of conditional association analysis for LDL and variants in LDLR, 31t
consensus clustering (ensemble clustering), 242–243, 260f
consensus PCA, 241
consistency quality control (CQCg; CQCp), MetaQC package, 41
context score, TargetScan targeting prediction algorithm, 287, 294
Conway, A. R. A., 156
cooperative functional effects, mdmodules, 136–138
copy number aberrations (CNAs), 333–334
copy number data, *IGF1R* gene, 225f
copy number variations (CNVs), 132, 143–147, 201–203, 207, 214
core modules, 347, 350f
correlation motif model
data generative process, 113f
integrative analysis of ChIP-X data, 112–118

448 *Index*

coupled transcription-splicing modules
  mechanisms of, 81
  methods and materials, 77–80
CpG island methylator phenotype (CIMP), 164
CQC (consistency quality control), MetaQC
    package, 41
CRC (colorectal carcinoma) study, 163–165,
    164f, 165f
cross validation
  data partitioning, 430f
  generalized cross validation, 225
  V-fold, 178
Cytoscape biological network viewer, 306

DAG. *See* directed acyclic graph
DANCR lncRNA, 406
data generative process, correlation motif
    model, 113f
data partitioning, cross validation procedure,
    430f
data set membership vector, FCC, 79
databases
  BBID, 335
  BioCarta, 335, 336f, 422
  CancerResource Database, 432
  hmChIP, 109
  JASPAR, 81
  KEGG, 335, 336f
  KEGG MEDICUS, 431
  pathway, 305–306
  Pathway Commons, 306
  Pathway Interaction, 306
DAVID tool
  pathways with lowest FDR, 336f
  somatic mutations in cancer genomes,
    335
DCA (differential clustering algorithm), 101
DCA (differential coexpression analysis),
    281–282
DCGs (directed cyclic graphs), 266, 276–277
De Novo Driver Exclusivity
  Dendrix algorithm, 343–346, 347–349
  Multi-Dendrix algorithm, 343–344,
    346–347, 349–351
Dendrix algorithm, 309, 343–346, 345f,
    347–349
dependent clustering, 257, 258f
diabetes genes, 93–100
DiffCoEx method, differential coexpression,
    102
differential clustering algorithm (DCA), 101
differential coexpression analysis (DCA),
    281–282
differential connectivity
  in coexpression network, 100–102
  modular differential connectivity, 103, 103f
differential gene expression, 389f

differential principal component analysis
    (dPCA), 128
dimension reduction. *See also* joint and
    individual variation explained method
  iCluster method, 3, 162f, 220, 243
  MetaPCA, 3, 39t, 55–56
  nonnegative matrix factorization, 3, 55,
    133f, 134–139
  overview, 3
  partial least squares, 3, 241–242
  sparse multi-block partial least squares
    regression, 3, 143, 144, 148–149
directed acyclic graph (DAG)
  Bayesian framework for inference, 278–279
  hybrid methods, 274–276
  Markov equivalence classes, 272f
  method and software, 281t
  overview, 271–273
  PC algorithm, 274
  search-and-score methods, 273–274
  structure equation models, 279–280
directed cyclic graphs (DCGs), 266, 276–277
directed graphical models
  Bayesian framework for DAG inference,
    278–279
  directed acyclic graph
    hybrid methods, 274–276
    method and software, 281t
    overview, 271–273
    PC algorithm, 274
    search-and-score methods, 273–274
  directed cyclic graphs, 266, 276–277
  overview, 265–266, 276–277
  QTL directed dependency graph, 277–278
Dirichlet mixture model, BCC method,
    252–253
distant eQTL, 267
DM. *See* DNA methylation
DNA. *See also* Encyclopedia of DNA
    Elements
  CNVs, 132, 143–147, 201–203, 207, 214
  GWAS
    AMD GWAS, 19–20
    GWAS-tailored software, 29–32
    imputation, 8–9
    methods for single marker test, 9–19
    number of publications by year, 8f
    overview, 7–8
    plasma lipid levels, 28–29
    rare variant associations, 20–28
    workflow of, 10t
  targeted cancer treatment and, 218
DNA methylation (DM), 201–202
  mdmodules, 136
  TCGA project, 132
dPCA (differential principal component
    analysis), 128

driver mutations, 308, 331, 400–401
Drosophila targets, 286, 293
drug discovery, 419. *See also* drug-pathway
    association analysis
drug sensitivity, 421–422. *See also*
    drug-pathway association analysis
drug-pathway association analysis
    iFad method, 422–425
    iPad method, 425–430, 434t–435t
    iterative signature algorithm, 435
    NCI-60 project, 431–433
    overview, 419–422

EGOT lncRNA, 406
EM (expectation-maximization) algorithm,
    115–117, 311
embryonic stem cells (ESCs), 385
EMT (epithelial-to-mesenchymal transition),
    296
Encyclopedia of DNA Elements (ENCODE),
    71, 81, 87, 109, 375, 382
    CAGE data from, 389–390
    K562 and GM12878 data, 378–379, 389f
    transcriptome profiling in human cells
        from, 398
ensemble clustering (consensus clustering),
    242–243, 260f
epigenomic analysis, 75–76
epithelial-to-mesenchymal transition (EMT),
    296
EQC (external quality control), MetaQC
    package, 40–41
eQTL mapping, 267–270, 281t
eQTL meta-analysis, 43
eQTLs. *See* expression quantitative trait loci
ESCs (embryonic stem cells), 385
E-step, EM algorithm, 116
estimates
    analysis results by applying MCP to each
        outcome separately, 191t–192t
    analysis results of gMCP, 185t, 193t–194t
    analysis results of gMCP with Laplacian
        penalty, 195t–198t
    analysis results of sparse gMCP, 186t–187t
    BCC method, 254–256
    estimated graph, simulated data sets, 210f
    JIVE method, 245–246, 247f
    mass-action-based model for gene
        expression regulation, 360–363
eukaryotic gene expression, 374–375
evidence aggregation methods, combining
    p-values analysis, 42–44
exon membership vector, FCC, 79
exonic lncRNAs, 400f
exons
    co-splicing mechanisms and, 76
    co-splicing networks, 78f

expectation-maximization (EM) algorithm,
    115–117, 311
exploratory methods for multisource data. *See*
    multi-source data, exploratory methods
    for
expression quantitative trait loci (eQTLs)
    causal inference and, 270–271
    cis-, 87, 94, 268–271, 269f
    diabetes genes, 94–97
    differential coexpression analysis, 281–282
    directed acyclic graph, 271–276, 281t
        hybrid methods, 274–276
        overview, 271–273
        PC algorithm, 274
        search-and-score methods, 273–274
    directed graphical model estimation using,
        276–280
        Bayesian framework for DAG inference,
            278–279
        overview, 276–277
        QTL directed dependency graph,
            277–278
        structure equation models, 279–280
    eQTL mapping, 267–270, 281t
    gene transcripts, 95f
    identifying regulatory SNPs, 132
    local eQTL versus distant eQTL, 267
    modeling genetic information flow in
        network, 88–91
    overview, 265–266
    protein QTL data, 281
    trans-, 87, 94, 268–271
external quality control (EQC), MetaQC
    package, 40–41

factorization methods, 241–242
false discovery rate (FDR), 202, 336, 336f,
    337f
    LOAD brain study, 103, 103f
    posterior inference using, 208
FCC (frequent coupled cluster), 77–80, 78f
FDR. *See* false discovery rate
feed-forward loop (FFL), 294–296, 295f
FEM. *See* fixed effects model
Fermi-Dirac combination of targets, 290–292,
    291f
FFL (feed-forward loop), 294–296, 295f
Fisher's method, microarray meta-analysis,
    42–43
    meta-analysis of GWAS, 16
    MetaDE package, 42–44, 46t, 47–48, 49t,
        51
fixed effects model (FEM)
    meta-analysis of GWAS, 16
    MetaDE package, 44
formal framework, exploratory methods for
    multisource data, 240–241

FOS regulatory factor, 75–76
fragments per kilobase of exon per million fragments mapped (FPKM), 74
frequent coupled cluster (FCC), 77–80, 78f

GA (genetic algorithm), 280
GABP regulatory factor, 75
gain-of-function (GOF), PARADIGM-SHIFT pathway method, 319
GAM (generalized additive models), 224–225, 225f
GAS5 lncRNA, 406
Gaussian graphical models (GGMs), 203
Gaussian mixture model (GMM), 153
GBM (Glioblastoma Multiforme), 219, 226, 406, 407, 409
GCV (generalized cross validation), 225
GE. *See* gene expression
GENCODE, 398
gene expression (GE). *See also* histone modifications; mass-action-based model for gene expression regulation; transcription factor binding
    allele-specific gene expression, 269–270
    eQTLs and, 268f
    JIVE method and, 248–251
    mdmodules, 136
    pathway methods for analysis of, 307–308
    TCGA project, 132
Gene Expression Omnibus (GEO), 37, 109, 397, 406
gene expression profiles
    drug-pathway association analysis, 421
    iFad method, drug-pathway association analysis, 422
    iPad method, drug-pathway association analysis, 425
gene expression regulation
    analysis of osmotic shock in yeast, 366–367
    change-point model, 358–360
    characterizing link between regulatory processes, 368–371
    data integration to study, 5–6
    estimation and inference, 360–363
    overview, 356–358
    scoring protein-level regulation changes, 367–368
    simulation study, 363–366
gene membership vector, FCC, 79
gene regulation pathways, 421
Gene Set Enrichment Analysis (GSEA), 307, 323
    gene sets with lowest FDR, 337f
    overview, 421–422
    somatic mutations in cancer genomes, 335–336
gene set methods, 307

gene sets, 337f
generalized additive models (GAM), 224–225, 225f
generalized cross validation (GCV), 225
genetic algorithm (GA), 280
genetic interaction, 201–203
GenMiR++ algorithm, 297–298
genome-wide association studies (GWAS)
    meta-analysis of, 7–33
        AMD GWAS, 19–20
        GWAS-tailored software, 29–32
        imputation, 8–9
        methods for single marker test, 9–19
        overview, 7–8
        plasma lipid levels, 28–29
        rare variant associations, 20–28
        workflow of, 10t
    number of publications by year, 8f
genomics. *See also* cancer genomics; latent variable approach, integrative clustering analysis
    epigenomic analysis, 75–76
    iBAG models
        linear, 221–224
        non-linear extensions, 224–226
        overview, 220–221
    Roadmap Epigenomics project, 109
Genotype of Tissue Expression (GTEx) project, 87
GEO (Gene Expression Omnibus), 37, 109, 397, 406
germline variants, somatic mutations in cancer genomes, 333
GES (greedy equivalence search) algorithm, DAG, 273
GGMs (Gaussian graphical models), 203
GI$_{50}$ value, drug sensitivity, 421, 422
Gibbs sampling procedure, 254–255
GISTIC2 algorithm, 334
Glioblastoma Multiforme (GBM), 219, 226, 406, 407, 409
Glymour, Clark, 274
GM12878 data, 378–379
gMCP. *See* group MCP
GMM (Gaussian mixture model), 153
GOF (gain-of-function), PARADIGM-SHIFT pathway method, 319
graphical (regression) networks, miRNAs, 297
graphical models. *See also* Bayesian graphical models; directed graphical models
    graphical model and network analysis, 4
    probabilistic, 310
greedy equivalence search (GES) algorithm, DAG, 273
group MCP (gMCP), 102, 180
    analysis results of, 185t, 193t–194t

analysis results of, with Laplacian penalty, 195t–198t
defined, 175
marker selection under heterogeneity model, 175–176
GSEA. *See* Gene Set Enrichment Analysis
GTEx (Genotype of Tissue Expression) project, 87
GWAS. *See* genome-wide association studies

H3K27me3 transcription factor, 146
H3K4me3 mark, 387–388
Hammersley Clifford theorem, 204
HCPs (high CpG promoters), 391
heat kernel model, cancer genomics, 310
HER2 tumor subtype, 161–162
heterogeneity model
    meta-analysis of GWAS, 16–17
    MetaDE package and, 49
    penalized integrative analysis of high-dimensional omics data, 174–180
heterogeneous stock mice, 171, 183–188
high CpG promoters (HCPs), 391
high-dimensional transcriptional and drug sensitivity profile. *See* drug-pathway association analysis
high-dimensionality, 219
high-order cooperativity, in transcription regulatory networks, 71–73
histone modifications (HMs), 109
    cause or consequence relationship between gene expression, 393–394
    ENCODE K562 and GM12878 data, 378–379
    framework for integrating with gene expression data, 376–377
    interplay between TF binding and other chromatin features, 391–392
    machine learning methods used in predictive models, 377–378
    ModENCODE Early Embryo data, 379
    Mouse ESC data, 379
    overview, 374–375
    performance evaluation of models, 378
    predicting differential gene expression, 388–389
    predicting expression levels for genes with HCP and LCP content, 391
    predicting expression levels of human promoters, 383f
    predicting expression of noncoding genes, 389–390
    predicting gene expression by combining with TF binding, 385–388
    predicting gene expression from, 382–385
    regulatory mechanism of, 392f

regulatory signals in distal regions, 392–393
Yeast and Fly data, 379
hmChIP database, 109
HMs. *See* histone modifications
homogeneity model
    Cochran's homogeneity test, 49
    penalized integrative analysis of high-dimensional omics data, 174–175
horizontal meta-analysis, 1, 2f
hot spots, eQTL, 267
HOTAIR lncRNA, 408–409
HotNet algorithm, 310, 323, 326
    applying to mutation data, 340–341
    config file used for running on TCGA GBM data, 353
    diffusion time used for PPI networks, 343f
    overview, 336–340
    parameter selection, 341–343
HOTTIP lncRNA, 406
HOX family genes, 145
hybrid methods, DAG, 274–276
hypothesis settings, MetaDE package, 45, 46t, 47–48

iASeq model, ASB, 123–125, 127f
iBAG. *See* integrative Bayesian analysis of genomics data models
iCluster method, 220
    exploratory methods for multisource data, 243
    joint analysis with lasso iCluster method, 162f
iFad method, drug-pathway association analysis, 422–425
ILP (integer linear program), 347
imputation, meta-analysis of GWAS, 8–9
incoherent FFLs, 295f
individual structures
    JIVE method, 244–248, 247f
    miRNA and gene expression, 248–251, 250f
inference
    Bayesian framework for DAG inference, 278–279
    causal inference, eQTLs, 270–271
    iFad method, 424–425
    mass-action-based model for gene expression regulation, 360–363
    network inference algorithms, 298
    posterior inference using false discovery rates, 208
information flow, modeling, 88–91
Ingenuity PathwayAnalysis (IPA) system, 142–143, 146, 146f
inner lasso penalty, SNP, 176
inner MCP penalty, 176

integer linear program (ILP), 347
Integrated Druggable Genome Database
    Project, 431
integrated pathway level (IPL), 316
integrated subtypes of colorectal cancer,
    164–165
integration with biological pathway
    information, 3
integrative analysis. *See also* Bayesian
    graphical models; latent variable
    approach, integrative clustering
    analysis; penalized integrative analysis
    of high-dimensional omics data
  of ChIP-X data, 108–128
    allele-specific binding in ChIP-seq,
      123–128
    ChIP-chip peak calling, 118–123
    correlation motif approach, 112–118
    general problem setting and motivations,
      110–112
    overview, 108–110
  of gene regulation
    coupled transcription-splicing modules,
      77–81
    overview, 66–68
    splicing modules, 73–77
    transcriptional modules, 68–73
integrative Bayesian analysis of genomics data
    models (iBAG models)
  linear, 221–224
  non-linear extensions, 224–226
  overview, 220–221
integrative quantitative models. *See also*
    histone modifications; transcription
    factor binding
  cause or consequence relationship between
    gene expression, 393–394
  ENCODE K562 and GM12878 data,
    378–379
  framework for integrating with gene
    expression data, 376–377
  interplay between histone modification and
    other chromatin features, 391–392
  machine learning methods used in
    predictive models, 377–378
  ModENCODE Early Embryo data, 379
  Mouse ESC data, 379
  overview, 374–375
  performance evaluation of models, 378
  predicting differential gene expression,
    388–389
  predicting expression levels for genes with
    high and low CpG content, 391
  predicting expression of noncoding genes,
    389–390
  predicting gene expression by combining
    with histone modifications, 385–388
  predicting gene expression from, 379–382

regulatory signals in distal regions,
    392–393
Yeast and Fly data, 379
interaction potential (IP), TIE score, 97
intergenic lncRNAs, 400f
internal quality control (IQC), MetaQC
    package, 40
interpretation, 219
  of meta-analysis results, 19
  patient-specific, 304–328
interventional Markov equivalence classes,
    273–274
intronic lncRNAs, 400f
inverse Wishart (IW) prior, 204
IP (interaction potential), TIE score, 97
IPA (Ingenuity PathwayAnalysis) system,
    142–143, 146, 146f
iPad method, drug-pathway association
    analysis, 425–430, 434t–435t
IPL (integrated pathway level), 316
IQC (internal quality control), MetaQC
    package, 40
iterative signature algorithm (ISA), 435
IW (inverse Wishart) prior, 204

jActiveModules plugin, 309
JAMIE method
  correlation motif model, 118–123
  joint peak calling by, 121f
JASPAR database, 81
Ji, H., 122
JIVE method. *See* joint and individual
    variation explained method
joint analysis with lasso iCluster method, 162f
joint and individual variation explained (JIVE)
    method, 244–251
  application to TCGA data, 248–251
  defined, 238
  estimation, 245–246, 247f
  gene expression (GE) and, 250f
  illustrative example, 246–248
  MetaPCA package, 55, 56f
  microRNA (miRNA) and, 250f
  model, 244–245
joint clustering, 256, 258f
joint NMF, 133f, 134–139
joint structure
  JIVE method, 244–248, 247f
  miRNA and gene expression, 248–251,
    250f

Kaplan-Meier curve
  lung squamous cell carcinoma (lung SCC),
    410f
  ovarian cancer (OvCa), 410f
Kaplan-Meier survival analysis, 138f
KEAP1 mutation, 321–322, 321f
KEGG database, 335, 336f

KEGG MEDICUS database, 431
KEGG pathways, 422
  drug-pathway associations identified by iPad, 434t–435t
  network-regularized joint NMF method and, 142
K-means clustering method, 153–154

Laplace prior, 223
Laplacian penalty, 182–183, 195t–198t
"large d, small n" data, 170
lasso (least absolute shrinkage and selection operator)
  inner lasso penalty, SNP, 176
  joint analysis with lasso iCluster method, 162f
  lasso iCluster method, 162f
  Lasso prior, 223
  overview, 426
lasso iCluster method, 162f
Lasso prior, 223
latent variable approach, integrative clustering analysis
  example, 161–162
  Gaussian mixture model, 153
  integrated subtypes of colorectal cancer, 164–165
  K-means clustering method, 153–154
  latent variable models, 156–159
  model selection, 160–161, 163–164
  overview, 151–153
  principal component analysis, 154–155
  subtype analysis and, 156
  TCGA colorectal cancer data set, 163
late-onset Alzheimer's disease (LOAD) brain study, 102–103, 103f
LCPs (low CpG promoters), 391
LD (linkage disequilibrium), 10t, 171
least absolute shrinkage and selection operator. *See* lasso
ligand-based approach, drug-pathway association analysis, 420
linear iBAG model
  clinical model, 222–224
  mechanistic model, 222
  overview, 221–224
  posterior probabilities, 228f–230f
  prognostic markers, 231–232
linkage disequilibrium (LD), 10t, 171
LINKER approach, network integration of genetically regulated gene expression, 92
linker genes, 322–323, 324f
liver tissue
  gene transcripts in tissue containing eQTLs that overlap with insulin QTLs, 95f
  top five genes ranked by TIE scores, 99t
lncRNAs. *See* long noncoding RNAs

LOAD (late-onset Alzheimer's disease) brain study, 102–103, 103f
local eQTL, 267
Lock, E. F., 152, 245, 248
LOF (loss-of-function), PARADIGM-SHIFT pathway method, 319
log-ratios of copy number, lung cancer sample, 159f
Logsdon, Benjamin, 279
long noncoding RNAs (lncRNAs)
  identifying, 405f
  integrating lncRNA expression, 402–405
  integrative analyses of in four cancer types, 406–413
  overview, 398–399
  repurposing microarray data to interrogate lncRNA expression, 399–402
long tail phenomenon, 332f
loss-of-function (LOF), PARADIGM-SHIFT pathway method, 319
low CpG promoters (LCPs), 391
lung squamous cell carcinoma (lung SCC), 406, 407, 409

macular degeneration, 20f
MAPE (meta-analysis pathway enrichment) methods, 52–55, 53f
markers, integrative analysis, 170
Markov chain Monte Carlo (MCMC), 204, 207–208, 255, 345, 424
Markov equivalence classes (MECs), 272f
Markov property, MRF, 204–205
Markov random fields (MRF), BayesGraph for TCGA integration, 204–205
MASS software package, 32
mass spectrometry (MS), 357, 366–367
mass-action-based model for gene expression regulation
  analysis of osmotic shock in yeast, 366–367
  change-point model, 358–360
  characterizing link between regulatory processes, 368–371
  estimation and inference, 360–363
  overview, 356–358
  scoring protein-level regulation changes, 367–368
  simulation study, 363–366
MAT peak calling method, 122
matrix decomposition, 422, 425, 435
Matrix eQTL software, 270
maximum p-value (maxP) statistic method, MetaDE package, 44, 46t, 47–48, 49t, 51
max-min hill-climbing (MMHC) algorithm, 274–275
MCC (multiclass correlation) method, 51
mCCA (multiple canonical correlation analysis), 242

MCMC (Markov chain Monte Carlo), 204, 207–208, 255, 345, 424
MCP. *See* minimax concave penalty
MDC (modular differential connectivity), LOAD brain study, 103, 103f
MDI (multiple data set integration), 243
MDL (minimum description length), DAG, 273
mdmodules (multi-dimensional modules)
   biological relevance of, 136–138
   clinical associations of, 138–139
MDRM. *See* multidimensional regulatory module
MDS (multidimension scaling), 55
ME (microRNA expression), 132, 136, 379, 389f, 390
mean decreased Gini, TF binding, 380
mean squared errors (MSE), 430
mechanistic iBAG model, 222
MECs (Markov equivalence classes), 272f
MEMo method, 309
Memorial Sloan-Kettering Cancer Center (MSKCC) Prostate Oncogenome Project, 405–406
Mendelian randomization, 271
Menezes, R. X., 202
messenger RNA (mRNA)
   concentration, 369f
   gene expression, 201–203
   iBAG models, 220–226
   targeted cancer treatment and, 218
meta-analysis methods, 3–4. *See also* Bayesian Consensus Clustering method; meta-analysis of GWAS; principal components analysis
meta-analysis of GWAS, 7–33
   age-related macular degeneration, 20f
   AMD GWAS, 19–20
   GWAS-tailored software, 29–32
   imputation, 8–9
   methods for single marker test, 9–19
   overview, 7–8
   plasma lipid levels, 28–29
   rare variant associations, 20–28
     approaches, 23
     burden tests that assume a distribution of variant effect sizes, 25–26
     burden tests that assume variants have similar effect sizes for a simple burden test in study k, the impact of multiple rare variants, 24
     conditional analyses, 26–28
     meta-analysis of single-variant association test statistics, 24
     Monte Carlo method for empirical assessment of significance, 26
     overview, 20, 22–23

sharing summary statistics, 23–24
summary of loci, 21t
variable threshold tests with an adaptive frequency threshold, 25
workflow of, 10t
meta-analysis pathway enrichment (MAPE) methods, 52–55, 53f
meta-analytic framework for the liquid association (MetaLA) method, 58
metabolic pathways, 421
MetaClust package, 39t, 56–58
MetaDE package, 39t, 42–52, 59–61
MetaDiffNet network, 58
MetaGeneModule approaches, MetaClust package, 57
MetaLA (meta-analytic framework for the liquid association) method, 58
MetaNetwork package, 39t, 58
MetaOmics software
   MetaClust package, 39t, 56–58
   MetaDE package, 42–52, 59–61
   MetaNetwork package, 39t, 58
   MetaPath package, 39t, 52–55, 59–61
   MetaPCA package, 3, 39t, 55–56
   MetaPredict package, 39t, 58–59
   MetaQC package, 38–42, 59–61
   overview, 37–38
MetaPath package, 39t, 52–55, 59–61
MetaPCA package, 3, 39t, 55–56
MetaPredict package, 39t, 58–59
MetaQC package, 38–42, 39t, 59–61
MetaSKAT software package, 32
MetaSparseKmeans method, 56–57, 57f
methods and materials. *See also names of specific methods*
   BayesGraph for TCGA integration, 205–208
   coupled transcription-splicing modules, 77–80
   splicing modules, 73–74
   transcriptional modules, 68–70
methylation data
   DNA methylation, 132, 136, 201–202
   IGF1R gene, 225f
Metropolis-Hastings ratio, 279, 360–362
MIAT lncRNA, 408
microarrays. *See also* clustering; latent variable approach, integrative clustering analysis
   combining effect sizes analysis, 44
   combining *p*-values analysis, 42–44
     evidence aggregation methods, 42–44
     order statistics methods, 44
   combining ranks analysis, 44–45
   conventions, 2
   for detecting differentially expressed genes, 38, 40–42

modeling data sets, 69f
repurposing microarray data to interrogate
lncRNA expression, 399–402
sequencing cancer genomes, 333–334
microRNA expression (ME), 132, 379, 389f,
390
microRNAs (miRNAs)
binding, 289–292
ComiR targeting method, 292–294
effect of single and multiple targets, 290
Fermi-Dirac combination of targets,
290–292
thermodynamics of, 289–290
cooperation between genes and, 148
JIVE method and, 248–251
network inference algorithms, 298
as network regulators, 294–298
network-regularized joint NMF method
and, 141–143
overexpressing genes, 298–299
overview, 285–286
sponge effect, 298
target prediction algorithms, 286–289,
294
microsatellite instability (MIN), 156, 164
minimax concave penalty (MCP)
analysis results by applying MCP to each
outcome separately, 191t–192t
defined, 174
marker selection under heterogeneity
model, 175–176
mismatched penalties, 180
sparse group MCP (gMCP), 175–176
minimum description length (MDL), DAG,
273
minimum p-value (minP) statistic method,
MetaDE package, 44, 46t, 47–48,
49t, 51
miRanda targeting prediction algorithm, 287,
287t, 291f, 293f
mirConnX algorithm, 297
miRNA expression (ME), mdmodules, 136
miRNAs. *See* microRNAs
mirSVR targeting prediction algorithm, 287t,
289, 293f
mirWIP targeting prediction algorithm, 287t,
289
MMHC (max-min hill-climbing) algorithm,
274–275
Mo, Qianxing, 152, 157
model-based approach, integrative clustering.
See also latent variable approach,
integrative clustering analysis
integrative clustering analysis, 160–161,
163–164
modeling genetic information flow, 88–91
modENCODE project, 109, 379, 389f

modular differential connectivity (MDC),
LOAD brain study, 103, 103f
molecular interaction network, 146f
Monte Carlo method, meta-analysis of GWAS,
26
Mouse ESC data, 379
MRF (Markov random fields), BayesGraph for
TCGA integration, 204–205
mRNA. *See* messenger RNA
MS (mass spectrometry), 357, 366–367
MSE (mean squared errors), 430
MSigDB gene set collection, 335
MSKCC (Memorial Sloan-Kettering Cancer
Center) Prostate Oncogenome Project,
405–406
M-step, EM algorithm, 116–117
multi-cancer markers, 171
multiclass correlation (MCC) method, 51
Multi-Dendrix algorithm, 343–344, 346–347,
346f, 349–351
multidimension scaling (MDS), 55
multi-dimensional modules (mdmodules)
biological relevance of, 136–138
clinical associations of, 138–139
multidimensional regulatory module (MDRM)
regulatory analysis and, 146–147
synergistic functions across multiple
dimensions, 145–146
multimodality, TCGA, 201
multi-platform datasets, schematic
representation of, 218f
multiple canonical correlation analysis
(mCCA), 242
multiple data set integration (MDI), 243
multiple data sets, 172, 189–190. *See also*
integrative analysis
multi-source data, exploratory methods for
BCC method, 251–260
application to TCGA data, 257–260
Dirichlet mixture model, 252–253
estimation, 254–256
illustrative example, 256–257
multisource model, 253–254
overview, 251–252
clustering methods, 242–243
computational burden, 261
factorization methods, 241–242
formal framework, 240–241
JIVE method, 244–251
application to TCGA data, 248–251
estimation, 245–246
illustrative example, 246–248
model, 244–245
overview, 238–240
MutationAssessor, 308
mutations. *See also* somatic mutations in
cancer genomes

MutSig method, 308, 327f
MutSigCV algorithm, 334

National Human Genome Research Institute
    (NHGRI), 421
NCI-60 project, 132, 131–133
NEAT1 lncRNA, 406
negative markers, iBAG models, 227–232,
    232t
Neto, Elias Chaibub, 277, 278
network analysis, SNP data, 182–183
network inference algorithms, miRNAs,
    298
network integration of genetically regulated
        gene expression
    diabetes genes, 93–100
    differential connectivity in coexpression
        network, 100–102
    LINKER approach, 92
    LOAD brain study, 102–103
    modeling genetic information flow, 88–91
    overview, 86–88
    PCST problem, 92–93
    PRINCE approach, 91–92
    random walk approach, 91
network regulators, miRNAs as, 294–298
network-regularized joint NMF method, 133f
    IPA system, 142–143
    KEGG pathways, 142
    miRNAs, 141–143
    overview, 139–141
    sparse network-regularized NMF
        algorithm, 140–141
network-regularized multiple NMF (NRNMF)
        framework, 140
Newton, Michael A., 208
next-generation sequencing (NGS), 37, 397,
    402, 404
NFE2L2 (Nrf2) oncogene, 320–322, 321f
NFYB regulatory factor, 75–76
NGS (next-generation sequencing), 37, 397,
    402, 404
NHGRI (National Human Genome Research
    Institute), 421
NMF. *See* nonnegative matrix factorization
noncoding RNA studies, data integration on
    application, 405–413
    clinical information, 405
    integrating lncRNA expression, 402–405
    overview, 397–399
    repurposing microarray data to interrogate
        lncRNA expression, 399–402
    somatic copy number alteration data,
        403–405
non-linear iBAG model
    posterior probabilities, 228f–230f
    prognostic markers, 231–232

nonnegative matrix factorization (NMF), 55
    defined, 3
    joint NMF, 133f, 134–139
Normal-Exponential prior, 223
Normal-Gamma prior, 223, 225–226
not allele specific (NS) state, ASB, 124
Nrf2 (*NFE2L2*) oncogene, 320–322, 321t
NRNMF (network-regularized multiple NMF)
        framework, 140
NS (not allele specific) state, ASB, 124
nucleosome positioning, 109

observed occurrence index (OOI)
    analysis results by applying MCP to each
        outcome separately, 191t–192t
    analysis results of gMCP, 185t, 193t–194t
    analysis results of gMCP with Laplacian
        penalty, 195t–198t
    analysis results of sparse gMCP, 186t–187t
    defined, 188
oligonucletide aCGH, 403–404
Oncodrive FM method, 308–309
"one drug – one target" approach, 419–420
OOI. *See* observed occurrence index
optimization algorithm, iPad method, 426–430
order statistics methods, combining *p*-values
        analysis, 44
outer MCP penalty, 176
ovarian cancer (OvCa), 406, 407, 409
overexpressing genes, 298–299
overlapping lncRNAs, 400f
overlapping subjects, 17
OWL (Web Ontology Language), 306

p53 protein, 313–314
PageRank teleporting random walk, 92
pairwise correlation matrices (PCMs), 101
pancreatic islets, 95f
PARADIGM pathway method, 298
    applications of, 317, 319
    interaction parameters, 316–317
    interactions and probabilistic factors,
        314–316
    matrix of activities, 318f
    modeling components, 312f
    overview, 310–312
    variables, 312–314
PARADIGM-SHIFT pathway method
    analysis of *NFE2L2* and *KEAP1* mutations,
        321f
    applications of, 320–322
    calculating shift scores, 319f
    overview, 319–320
parameter tuning, iPad method, 430
partial correlation, 211
partial least squares (PLS), 3, 241–242
partitioning explained variation, 235

passenger mutations, 308–309, 331
Pathifier method, 307
PathOlogist method, 307
PathScan approach, 334
pathway analysis, 181–182, 307–308, 336f.
    See also drug-pathway association
    analysis
Pathway Commons database, 306
pathway databases, cancer genomics, 305–306
Pathway Interaction Database, 306
pathway-based drug discovery
    (polypharmacology), 420. *See also*
    drug-pathway association analysis
pathway-based mutation assessment, 308–309
PBR (potential binding regions), ChIP-chip
    peak calling, 120, 121f
PC algorithm, DAG, 274–275, 281t
PCA. *See* principal components analysis
PCAN-R1 lncRNA, 411, 412f, 413
PCAN-R2 lncRNA, 411, 412f, 413
PCGs (protein encoding genes), 398
PCMs (pairwise correlation matrices), 101
PCST (prize-collecting Steiner tree) method,
    92–93, 309
PDIs (protein-DNA interactions) genome, 108
peak calling, ChIP-chip, 118–123
PECA (Protein Expression Control Analysis),
    357–358
penalization
    composite, 180
    methods, 5
penalized integrative analysis of
        high-dimensional omics data
    data quality control and processing,
        189–190
    examples, 170–173
    heterogeneity model
        marker selection, 175–180
        overview, 174
    heterogeneous stock mice, WTCCC,
        183–188
    homogeneity model
        marker selection, 175
        overview, 174
    interplay among SNPs, 181–183
        network analysis, 182–183
        pathway analysis, 181–182
    overview, 170
PenPC algorithm, 275
phenotype-based approach, drug-pathway
        association analysis
    defined, 420
    iFad method, 422–425
    iPad method, 425–430
phyloCSF method, 411
PicTar targeting prediction algorithm, 287t,
    288

Ping-Pong algorithm, 132
PITA targeting prediction algorithm, 287t,
    288, 291f, 293f
plasma lipid levels, meta-analysis of GWAS,
    28–29
PLS (partial least squares), 3, 241–242
polypharmacology (pathway-based drug
        discovery), 420. *See also* drug-pathway
        association analysis
positive markers, iBAG models, 227–232,
    233t
posterior inference
    for genes, 212t
    using false discovery rates, 208
posterior probabilities
    linear iBAG model, 228f–230f
    non-linear iBAG model, 228f–230f
potential binding regions (PBR), ChIP-chip
        peak calling, 120, 121f
PPI (protein-protein interaction) networks, 81,
    87, 265–266, 336–338, 338f, 343f
pQTL (protein QTL) data, 281
PR (Product of ranks) method, MetaDE
    package, 45
predicting gene expression
    by combining with TF binding and histone
        modifications, 385–388
    differential gene expression, 388–389
    with high and low CpG content, 391
    from histone modifications, 382–385
    of noncoding genes, 389–390
    from TF binding, 379–382
PRINCE approach, 91–92
principal components analysis (PCA), 55
    consensus clustering, 260f
    exploratory methods for multisource data,
        241, 247f
    integrative clustering analysis, 154–155
    mechanistic iBAG model, 222
    MetaQC package, 41, 60f
PRINS lncRNA, 406
prize-collecting Steiner tree (PCST) method,
    92–93, 309
probabilistic factors, PARADIGM pathway
    method, 314–316
probabilistic graphical models, PARADIGM
    pathway method, 310
probability model, BayesGraph for TCGA
    integration, 205–207
Product of ranks (PR) method, MetaDE
    package, 45
prognostic markers, iBAG models, 227–232
protein concentration, 369f
protein encoding genes (PCGs), 398
Protein Expression Control Analysis (PECA),
    357–358
protein QTL (pQTL) data, 281

protein synthesis (translation), 356–357, 364f, 369f
protein-DNA interactions (PDIs) genome, 108
protein-level regulation changes, scoring, 367–368
protein-protein interaction (PPI) networks, 81, 87, 265–266, 336–338, 338f, 343f
proteomics, 357, 359. *See also specific protein entries*

QC measures. *See* quality control measures
QDG (QTL directed dependency graph), 277–278
QTL directed dependency graph (QDG), 277–278
QTLnet method, 278
quality control (QC) measures
    MetaQC package, 40–41
    Single Nucleotide Polymorphisms, 15t

RACE (rapid amplification of cDNA ends), 411, 412f, 413
random effects model (REM)
    meta-analysis of GWAS, 16–17
    MetaDE package, 44
random forest (RF) method, 377, 380, 381f, 388, 393
random walk approach, 89f, 91
RankProd (RP) method, MetaDE package, 44
rapid amplification of cDNA ends (RACE), 411, 412f, 413
rare variant associations, meta-analysis of GWAS
    approaches, 23
    burden tests that assume distribution of variant effect sizes, 25–26
    burden tests that assume variants have similar effect sizes for a simple burden test in study k, the impact of multiple rare variants, 24
    conditional analyses, 26–28
    meta-analysis of single-variant association test statistics, 24
    Monte Carlo method for empirical assessment of significance, 26
    overview, 20, 22–23
    results for meta-analysis of gene-level rare variant association test, 30t
    sharing summary statistics, 23–24
    summary of loci, 21t
    variable threshold tests with an adaptive frequency threshold, 25
RAREMETAL software package, 32
RDF (Resource Description Framework), 306
Reactome pathways, 422
recurrent heavy subgraphs (RHSs), 68, 69–70, 69f

regression (graphical) networks, miRNAs, 297
regularization methods, 5
regularization parameters, 177
regulatory processes, gene expression, 368–371
relevance networks, miRNAs, 296
REM. *See* random effects model
Resource Description Framework (RDF), 306
reversible edge (REV) proposal, 279
reversible-jump Markov chain Monte Carlo (MCMC), 360, 367
RF (random forest) method, 377, 380, 381f, 388, 393
RHSs (recurrent heavy subgraphs), 68, 69–70, 69f
RMST lncRNA, 408
RNA. *See also* microRNAs
    lncRNAs
        identifying, 405f
        integrating lncRNA expression, 402–405
        integrative analyses of in four cancer types, 406–413
        overview, 398–399
        repurposing microarray data to interrogate lncRNA expression, 399–402
    mRNA
        concentration, 369f
        gene expression, 201–203
        iBAG models, 220–226
        targeted cancer treatment and, 218
    siRNAs, 412f, 413
    sncRNAs, 398
RNA sequencing (RNA-seq), 87, 398–402, 406
RNA synthesis (transcription), 356–357, 369f
rna22 targeting prediction algorithm, 287t, 288
RNA-seq (RNA sequencing), 87, 398–402, 406
Roadmap Epigenomics project, 109
RP (RankProd) method, MetaDE package, 44
rth ordered p-value (rOP) statistic method, MetaDE package, 44, 46t, 47–48, 49t, 51

S. cerevisiae data with osmotic stress, 370f
sample statistical analysis plan, 12t
SCNA (somatic copy number alteration) data, 399, 403–406, 409
search-and-score methods, DAG, 273–274
SEMs (structure equation models), 277, 279–280
separate clustering, 256, 258f
sequence kernel association tests (SKAT), 22
Sequence Read Archive (SRA), 37
sequencing cancer genome/exome, 333–334
short interfering RNAs (siRNAs), 412f, 413

SIF (Simple Interchange Format), 306
SIFT, 308
signal transduction pathways, 421
Signaling Pathway Impact Analysis (SPIA),
    307
significance test, iPad method, 430
significantly mutated subnetworks, 336–343
Simple Interchange Format (SIF), 306
simulated data sets, 210f
simulation study
    BayesGraph for TCGA integration,
        208–211
    mass-action-based model for gene
        expression regulation, 363–366, 365f
single marker test, meta-analysis of GWAS,
    9–19
single nucleotide polymorphisms (SNPs), 9,
    333
    analysis results by applying MCP to each
        outcome separately, 191t–192t
    analysis results of gMCP, 185t, 193t–194t
    analysis results of gMCP with Laplacian
        penalty, 195t–198t
    analysis results of sparse gMCP, 186t–187t
    arrays, 404
    eQTLs and, 267, 268f
    meta-analysis of GWAS, 18t
    MetaDE for marker gene detection, 42
    penalized integrative analysis of
        high-dimensional omics data, 181–183
        network analysis, 182–183
        pathway analysis, 181–182
    quality control, 15t
    TCGA project, 132
single-data-set analysis, 171
single-nucleotide variants (SNVs), 333–334
single-variant association test statistics
    Cochran-Mantel-Haenszel method, 24
    meta-analysis of GWAS, 24
singular value decomposition (SVD), 246
siRNAs (short interfering RNAs), 412f, 413
SKAT (sequence kernel association tests), 22
skeleton, DAG, 272
skewed to the nonreference allele (SN) state,
    ASB, 124
small noncoding RNAs (sncRNAs), 398
sMBPLS. *See* sparse multi-block partial least
    squares regression
SMR (standardized mean ranks), MetaQC
    package, 41
SN (skewed to the nonreference allele) state,
    ASB, 124
sncRNAs (small noncoding RNAs), 398
SNMRMF (sparse network-regularized NMF)
    algorithm, 140–141
SNPs. *See* single nucleotide polymorphisms
SNVs (single-nucleotide variants), 333–334

software, 29–32. *See also* MetaOmics software
somatic copy number alteration (SCNA) data,
    399, 403–406, 409
somatic mutations in cancer genomes
    cancer (sub)type analysis, 350–352
    DAVID tool, 335
    Dendrix algorithm, 343–346, 347–349
    GSEA algorithm, 335–336
    Multi-Dendrix algorithm, 343–344,
        346–347, 349–351
    overview, 331–333
    sequencing, 333–334
    significantly mutated subnetworks,
        336–343
sparse group MCP (gMCP), 178–180,
    186t–187t
sparse multi-block partial least squares
    regression (sMBPLS), 3, 133f, 143,
    144, 148–149
    multidimensional regulatory module,
        145–147
    overview, 143–144
sparse network-regularized NMF (SNMRMF)
    algorithm, 140–141
SPIA (Signaling Pathway Impact Analysis),
    307
splicing modules
    exons, 76–77
    identifying novel functions associated with
        co-splicing but not coexpression, 76
    methods and materials, 73–74
    transcriptional and epigenomic analysis,
        75–76
sponge effect, miRNAs, 298
squared Euclidean error function, NMF,
    134–135
SR (Sum of ranks) method, MetaDE package,
    45
SRA (Sequence Read Archive), 37
SRF transcription factor, 145
SSC (sum of squared cosines), MetaPCA
    package, 55
standardized mean ranks (SMR), MetaQC
    package, 41
STAT1 transcription factor, 145
statistical methods. *See* integrative analysis;
    latent variable approach, integrative
    clustering analysis; MetaOmics
    software
Stouffer's method
    MetaDE package, 43, 46t, 47–48, 49t, 51
    microarray meta-analysis, 43
structural variants (SVs), 333–334
structure equation models (SEMs), 277,
    279–280
structured pathway methods, analysis of gene
    expression, 307

subtype analysis, integrative clustering analysis, 156
Sum of ranks (SR) method, MetaDE package, 45
sum of squared cosines (SSC), MetaPCA package, 55
sum of variance (SV), MetaPCA package, 55
support vector machine (SVM), 293, 377, 388, 393
SV (sum of variance), MetaPCA package, 55
SVD (singular value decomposition), 246
SVM (support vector machine), 293, 377, 388, 393
SVs (structural variants), 333–334

TAF8 regulatory factor, 75
target exclusivity, miRNAs, 289
target prediction algorithms, miRNAs, 294
  ComiR, 287t, 288
  miRanda, 287, 287t
  mirSVR, 287t, 289
  mirWIP, 287t, 289
  overview, 286–287
  PicTar, 287t, 288
  PITA, 287t, 288
  rna22, 287t, 288
  TargetScan, 287–288, 287t
target-based approach
  differential coexpression, 101
  drug-pathway association analysis, 420
targets, miRNAs, 290
TargetScan targeting prediction algorithm, 287–288, 287t
TCGA project. *See* The Cancer Genome Atlas project
TF binding. *See* transcription factor binding
TF+HM model, 386–387, 386f
TFBSs (transcription factor binding sites), 108
TFs. *See* transcription factors
The Cancer Genome Atlas (TCGA) project, 132, 152, 163, 165f. *See also* Bayesian graphical models; somatic mutations in cancer genomes
  application of BCC method to data, 257–260
  application of JIVE method to data, 248–251
  BayesGraph for TCGA integration, 211–214
  cancer types and samples for integrative analysis, 212t
  integrative clustering analysis, 163
  TCGA GBM study, 334
third-order tensor, transcriptional regulatory modules, 69f
TIE score, 97–98, 99t
TieDIE pathway method, 322–328, 327f
TileMap peak calling method, 122

time course experiments
  analysis of osmotic shock in yeast, 366–367
  change-point model, 358–360
  characterizing link between regulatory processes, 368–371
  estimation and inference, 360–363
  overview, 356–358
  scoring protein-level regulation changes, 367–368
  simulation study, 363–366
TIPC (trait-IP correlation), 97–98
T-Lymphocyte Helper/Suppressor Profile. *See* CD4/CD8 ratio
top scoring pair (TSP) algorithm, prediction analysis, 58–59
Tpi1 gene, 99t
trait-IP correlation (TIPC), 97–98
transcription (RNA synthesis), 356–357, 369f
transcription, defined, 201
transcription factor binding sites (TFBSs), 108
transcription factor (TF) binding
  cause or consequence relationship between gene expression, 393–394
  ENCODE K562 and GM12878 data, 378–379
  framework for integrating with gene expression data, 376–377
  interplay between histone modification and other chromatin features, 391–392
  machine learning methods used in predictive models, 377–378
  ModENCODE Early Embryo data, 379
  Mouse ESC data, 379
  overview, 374–375
  performance evaluation of models, 378
  predicting differential gene expression, 388–389
  predicting expression levels for genes with HCP and LCP content, 391
  predicting expression of noncoding genes, 389–390
  predicting gene expression by combining with histone modifications, 385–388
  predicting gene expression from, 379–382
  regulatory signals in distal regions, 392–393
  Yeast and Fly data, 379
transcription factors (TFs), 71–72, 72f
  gene expression, 87, 89f
  GLI1, 110, 111f, 112
  GLI3, 110, 111f, 112
  H3K27me3, 146
  predicting expression levels of human promoters, 381f
  regulatory mechanism of, 392f
  SRF, 145
  STAT1, 145

transcriptional analysis, splicing modules,
75–76
transcriptional modules
high-order cooperativity and regulation in
transcription regulatory networks,
71–73
methods and materials, 68–70
transcriptional regulation, 72f, 379, 387–388,
391–392
transcriptomics meta-analysis. *See also*
mass-action-based model for gene
expression regulation
for differential network detection, 58
MetaClust package, 56–58
MetaDE package, 42–52, 59–61
MetaNetwork, 58
MetaPath package, 52–55, 59–61
MetaPCA, 55–56
MetaPredict, 58–59
MetaQC package, 38–42, 59–61
overview, 37–38
trans-eQTLs, 87, 268–271, 269f
translation (protein synthesis), 356–357, 364f,
369f
trans-regulated gene expression, 87
TRe-CASE model, 270
TSP (top scoring pair) algorithm, prediction
analysis, 58–59
txCdsPredict method, 411

undirected networks, 190
uniform design (UD), sampling method, 160
unsupervised analysis
Bayesian consensus clustering (BCC), 3
cluster analysis, 3
iCluster method, 3
MetaSparseKmeans method, 3
overview, 3
untargeted approach, differential coexpression,
101

variable selection, 170, 174
variables
PARADIGM pathway method, 312–314
variable threshold tests with an adaptive
frequency threshold, 25
Venn diagram
enriched pathways identified by MAPE, 60f
lncRNA located in SCNA regions of
cancer, 410f
subtype-specific lncRNA in cancers, 408f
vertical multi-omics analysis, 1, 2f
V-fold cross-validation, 178
v-structures, DAG, 272–273

walking in gene network, 89f
Web Ontology Language (OWL), 306
Wellcome Trust Case Control Consortium
(WTCCC), 171, 183–188
WGCNA package, 102
white adipose tissue
gene transcripts in tissue containing eQTLs
that overlap with insulin QTLs, 95f
top five genes ranked by TIE scores, 99t
whole-cell pathway model
active subnetwork search and discovery,
309–310
PARADIGM pathway method, 310–319
PARADIGM-SHIFT pathway method,
319–322
pathway databases, 305–306
pathway methods, 307–308
pathway-based mutation assessment,
308–309
TieDIE pathway method, 322–328
whole-exome sequencing, 333
whole-genome sequencing, 333
WTCCC (Wellcome Trust Case Control
Consortium), 171, 183–188

Yeast and Fly data, 379